The Foundations of Quantum Mechanics –
Historical Analysis and Open Questions

Fundamental Theories of Physics

An International Book Series on The Fundamental Theories of Physics:
Their Clarification, Development and Application

Editor: ALWYN VAN DER MERWE
University of Denver, U.S.A.

The Foundations of Quantum Mechanics – Historical Analysis and Open Questions

Lecce, 1993

edited by

Claudio Garola and Arcangelo Rossi

Department of Physics,
University of Lecce,
Lecce, Italy

SPRINGER SCIENCE+BUSINESS, MEDIA, B.V.

A C.I.P. Catalogue record for this book is available from the Library of Congress.

ISBN 978-94-010-4017-4   ISBN 978-94-011-0029-8 (eBook)
DOI 10.1007/978-94-011-0029-8

Printed on acid-free paper

TABLE OF CONTENTS

INTRODUCTION

The Conference entitled *The Foundations of Quantum Mechanics. Historical Analysis and Open Questions* (Lecce, 5-8 October 1993) was conceived as an interdisciplinary meeting among Italian researchers (physicists, logicians, mathematicians, historians and epistemologists) concerned with the subject indicated by the name of the Conference.

The idea of collecting together scholars looking at this subject from different cultural perspectives had already proved successful in 1988, when a Conference with the same title was organized at the University of Camerino, the proceedings of which (published in Italian by EditEl, Commenda di Rende, 1991) clearly witness the widespread need of debating the theme of structure and history of Quantum Mechanics (QM), which is a central topic in the scientific culture of this century, with students of different disciplines. Of course, the liveliness of the discussion was enhanced by the well known fact that QM cannot be simply classified as a stable piece of exact knowledge, to be only applied and refined in some minor details, since its interpretation and epistemological conception are still controversial despite its seventy years' spectacular success, and a number of problems occur in each of the alternative interpretations proposed in the literature.

The proceedings that we are presenting here collect most talks given at the Conference, that have been widely refined and improved by the authors for publication in the proceedings themselves. Of course, the responsibility for the content of the papers belongs to the authors. But the editors retain that the average quality of the articles published here is decidedly good, and hope that they will be a stimulating lecture for the reader. This does not mean that all purposes of the conference have been attained. Indeed, the languages of historians, physicists, logicians and epistemologists are still very different, and laborious translation procedures must often be undertaken in order to start the communication process. Furthermore, there are different ways of interpreting the words *foundations of QM*, since some researchers retain that they mean solving basic problems in QM (as stability

ix

of matter, and so on), others interpret them as denoting the effort of going deeper in understanding the epistemological, logical and mathematical roots of the theory (hence criticizing and changing it, if needed). Finally, the links between the technical aspects and some philosophical conclusions or physical interpretations are sometimes rather loose, which proves that a long way must still be covered before we can say that some foundational problems have been solved. But, notwithstanding these critical remarks, we believe that the Lecce Conference promoted an advancement in the right direction.

Because of the interdisciplinary character of the Conference, we have collected the articles in alphabetic order, avoiding to divide them in groups according to thematic or methodological criteria. However, we would like to present them briefly here, and this will be done by grouping and ordering the papers according to our perception of similarities and relationships, that we recognize in advance to be subjective.

Let us begin with a first set of papers that can be classified as falling within the so-called *logico-algebraic approach to QM*. G. **Cattaneo** and F. **Laudisa** consider the question whether it is possible to single out the logico-algebraic structures that underly the unsharp generalization of QM, and assert that an affirmative answer to this question is provided by the family of *state-effect-probability* structures; then, they present two possible axiomatic formulations for the structure of the effects in QM, and briefly investigate the problem of their equivalence, proving some propositions that establish a link between the two formulations and legitimate the conjecture that they are actually equivalent. R. **Giuntini** considers the mathematical problem of extending in a natural way the structures called *unsharp ortho-algebras*, that are defined by suitably formalizing and generalizing the algebraic properties of the set of all effects of any Hilbert space; then, the author shows that natural extensions exist, called quantum multi-valued (QMV) algebras (that generalize the multi-valued algebras introduced by other authors), and proves that every quasi-linear QMV algebra corresponds to an unsharp orthoalgebra: this allows him to associate a QMV algebra to the class of all effects of any given Hilbert space, which makes QMV algebras physically interesting. M. **Dalla Chiara** and R. **Giuntini** introduce *Lukasiewicz quantum logic* (LQL) as the logic that is semantically characterized by the class of all QMV algebras and show that there are interesting physical interpretations for the different conjunctions that arise in the standard models of LQL based on effects in a Hilbert space. Finally, M. **Piazza** discusses the reasons that suggest the introduction of the partially ordered algebraic structures called *quantales*, which were conceived as a possible common setting for noncommutative C*-algebras, constructive foundations of QM and noncommutative logics; the author considers a possible link between the noncommutativity of observations in QM and the noncommutative logics in which the formulas of the language are strictly sequential, focusing on *noncommutative linear logic*, for which quantales provide an appropriate algebraic semantics.

Within the same cultural framework of the above papers, but more concerned with epistemological and interpretational problems, we find the articles by Garola and Cattaneo - Nisticò. **C. Garola** shows that Bell's theorem, which states that QM is not compatible with the simultaneous assumptions of realism and locality (thus implying, according to some authors, that QM is necessarily a nonlocal theory), depends on assuming implicitly a classical viewpoint regarding the truth mode of quantum laws (quantum laws are valid in every physical context), and proves that, if a new principle is adopted which is more respectful of the operational philosophy of QM (quantum laws cannot be asserted to be valid in non-observable physical contexts), Bell's theorem can be invalidated; this result opens the way to the construction of a language for QM endowed with a Tarskian theory of truth, which is compatible with some forms of realism, and to the avoidance of the verificationist truth theory underlying the standard interpretation of QM, that can be seen as the deep source of many quantum paradoxes. **G. Cattaneo** and **G. Nisticò** discuss the principle (which is a theorem in von Neumann's approach to QM) according to which two observables are simultaneously measurable iff their corresponding operators commute, and show that this principle does not prohibit that in some cases the outcomes of two noncommuting observables be simultaneously known; this result is obtained by considering an experimental situation similar to that envisaged by Einstein, Podolski and Rosen (EPR) in their 1935 article and introducing the concept of *mirror* observable, and the authors also study how this knowledge must be interpreted according to QM and when it is actually attainable.

The EPR thought experiment and Bell's theorem also play an important role in the paper by **G.C. Ghirardi** and **R. Grassi**. Indeed, these authors, following a widespread interpretation of Bell's theorem, agree that QM implies nonlocality, show that counterfactual statements naturally enter the problem of attributing elements of physical reality to individual physical systems and reconsider, on the basis of these premises, the EPR argument aiming to show the incompleteness of QM; then they deduce that nonlocality prohibits to attribute objective properties by means of an EPR-type argument in a relativistic context, which, surprisingly, invalidates the claim that there are in nature *spooky actions at a distance* and shows that a peaceful coexistence of standard QM with relativity holds in some nonstandard and deeper sense if nonlocality of QM is accepted.

Another group of authors is concerned with the aim of devising experimental tests that allow physicists to settle the controversy between QM and some kinds of realism. Thus, some papers consider the possibility of testing the validity of QM by means of macroscopic measurements on superconducting devices that has been recently pointed out by several authors. In particular, **T. Calarco** and **R. Onofrio** analyze the concept of non-invasive measurement introduced by Leggett and Garg, and discuss in some detail a quantum measurement model that reconciles this kind of measurements with Heisenberg's original interpretation of QM, which forbids to imagine a quantum measurement as one that does not influence the subsequent dynamics. On the other side, **L. Chiatti, M. Cini** and **M. Serva** investigate the possibility of testing the validity of QM by means of a macroscopic system such that the coherence effects of linear superpositions of (pure) states can be detected (SQUID), and show that there is a threshold above which QM cannot be discriminated from a suitable macro-realistic model that assumes Macroscopic Realism but drops the requisite of non-invasive measurability. Other papers refer to a more conventional experimental approach. Thus, **A. Garuccio** and **L. de Caro** generalize the Clauser, Horne, Shimony and Holt (CHSH) approach to the case of three-valued observables, aiming to obtain a general Bell-type inequality for the joint detection probabilities that does not depend on further additional assumptions (as it occurs, on the contrary, in the case of the CHSH inequality); then, they add additional hypotheses in order to obtain inequalities that can be experimentally tested for comparing the predictions of local realism with the predictions of QM, and also discuss the role and meaning of the additional hypotheses that have been introduced.

The problem of measurement in QM is also the main topic in the paper by **A. Afriat** and **F. Selleri**, who deal with the conservation of angular momentum in various quantum measurement theories and argue that Bohr's theory entails instantaneous signalling over any distance; this problem does not arise, according to the authors, if the apparatus is described in quantum mechanical terms following von Neumann, but total angular momentum should not be conserved in this case, which suggests to consider a third theory, according to which the measurement may disturb the state of the system even in the case of eigenstates: in this theory, angular momentum is conserved and instantaneous signalling does not occur, but eigenstates lose their usual meaning.

Some further papers deal with quantum probabilities and their relationships with classical probabilities. Thus, **L. Accardi** criticizes the existing axiomatic approaches to QM, since they introduce *ad hoc* axioms in order to recover the successful model of the theory, and attempts to construct an axiomatization of QM based uniquely on physically meaningful axioms, taking the indeterminacy principle as a basic starting-point and axiomatizing measurements rather than events; Accardi's axioms lead him to introduce a structure called *algebra of measurements*, which is commutative in classical physics, noncommutative in QM, and to deduce in the quantum case, from a qualitative formulation of Heisenberg's principle only, the intrinsic stochasticity of the model together with the explicit form of the superposition principle. **E.G. Beltrametti, C. del Noce** and **M.J. Maçzinski** discuss the problem of defining "the boundary that marks the possibility of representing classically the observed probabilities", and state a theorem that provides sufficient and necessary conditions for a Bell-type inequality to be a condition of classical representability of a set of observed joint probabilities; furthermore, they also provide an algoritm for generating Bell-type inequalities for given empirical situations. **D. Costantini** and **U. Garibaldi** show that classical probabilities can be used in order to derive some quantum correlations if stochastic dependence is taken in account, and prove that some canonical quantum results usually obtained by assuming indistinguishability of particles can be obtained by means of probabilistic assumptions only.

The suggestion that some interesting hints for the solution of problems in the foundations of QM can be found in classical theories is put forward by other authors. In particular, **A. Carati, L. Galgani** and **J. Sassarini** refer on some recent studies on classical electrodynamics of point particles in order to show that, in the model corresponding to the Abraham-Lorenz-Dirac equation, one obtains qualitative similarities with the quantum tunnel effect, with the acceleration playing the role of hidden parameters; according to the authors, this result provides some unexpected and probably useful suggestions for the discussion of the problem of hidden parameters in QM.

As we have anticipated above, some authors choose to cope with fundamental problems in QM rather than with foundational problems. Thus, **G.F. De Angelis** and **M. Serva** deal with the classical problem of stability for a system of point charges, and supply a path integral representation for the semigroups generated by the quantum Hamiltonian of a relativistic spinless particle in an external electromagnetic field that allows them to compare one-body relativistic Schrödinger operators with one-particle Hamiltonians in the external field approximation of QED. **R. Fedele** deepens the comparision between QM and wave optics and shows that, in the non-relativistic limit, a correspondence between these two theories can be established whenever the *paraxial approximation* is introduced; this correspondence can be implemented by considering the conventional paraxial description of particle beams and introducing the recently proposed *Thermal Wave Model* for charged - particle beam transport, which allows one to describe, at least in paraxial approximation, different optical and dynamical phenomena in a formally unified framework (some physical uses of this unified framework are also outlined).

Let us come now to some papers that deal with theories that can be considered alternatives to standard QM. **F. Guerra** presents Nelson's stochastic mechanics (NSM) as a generalization of classical mechanics based on the theory of stochastic processes and stochastic variational principles, shows that it can be connected with QM by means of a simple physical interpretation scheme (hence, it can be seen as a quantization procedure for mechanical systems), and shows that the Brownian motion that generates quantum fluctuations can be derived as a consequence of stochastic variational principles; the author also proves that different representations of NSM are connected through stochastic measure-preserving transformations and considers the basic aspects of the measurement problem from the viewpoint of NSM, showing that the collapse of the wave function is not instantaneous, being ruled by a well defined dynamical scheme with a time asymptotic relaxation behaviour (the relaxation being slower if the result is more uncertain). **G. Peruzzi** discusses Bohm's theory according to some recent approaches, and recalls that a basic feature of Bohm's mechanics is the recovery of determinism, since the wave function is interpreted as a real physical object, guiding particles in their deterministic trajectories and only seeming to collapse because of our ignorance of its real evolution as a kind of force field; our ignorance also accounts for the probabilistic interpretation of the wave function as an instrument for dealing with quantum effective measurements, without necessarily assuming, as Bohm himself initially did, a subquantum essential randomness or a quantum potential, but rather assuming the quantum probability flux as a purely phenomenological representation of the trajectories of particles: these actually fluctuate inside the global force field described by the wave function, which is interpreted as a sort of Cartesian aether that also explains quantum nonlocality. **M. Roncadelli** presents a new approach to QM, called *random path quantization* (RPQ), based on structural similarities (quantum amplitudes satisfy a calculus similar to the calculus of probabilities in the theory of classical stochastic diffusion processes), and proves that, starting from classical mechanics according to the Hamilton-Jacobi formulation and introducing quantum fluctuations as a Fresnel white noise in the first order equation that yields the configuration space trajectories, one obtains a Langevin equation that provides *quantum random paths*; then, the author shows that the quantum mechanical propagator can be expressed as a Fresnel noise average involving quantum random paths, and also offers an intuitive picture of QM, suggested by RPQ, based on the assumption that any environment contains background quantum fluctuations.

Coming to more philosophical questions, the paper by **G. Tagliaferri** on the mysteries of quantum theory, which reports the talk that introduced the Conference, synthetizes the reasons that have led to organize the Conference itself by recording the conflict that still occurs in QM between the heuristic fruitfulness and empirical successfulness of the theory, from one side, and the deep interpretative problems which do not allow scientists to accept fully QM as the ultimate theory of microcosm, from the other side; then, Tagliaferri recalls the first formulation of some basic problems, as the questions of determinism and physical reality, that were proposed at the birth of the theory, and considers the up-to-date status of QM, maintaining that the seeming experimental validation of it against Bell's criticism does not surmount all interpretative and philosophical difficulties: on the contrary, it makes them even deeper, urging a stronger research effort in order to overcome the mysteries of QM.

An interesting attempt of coping with ontological problems taking into account QM is presented in the paper by **S. Bergia** and **V. Fano**. These authors defend the idea that the aim of building a satisfactory realistic interpretation of QM is scientifically fruitful, though it may imply radical changes in the ordinary way of thinking, and suggest that Husserl's phenomenology may allow one to overcome Kant's prohibition of a scientific ontology and provide criteria (as space-time coherence and causal uniformity) in order to select among different ontologies; then, they examine some views of quantum reality in the literature, dealing in particular with the approaches by De Broglie-Bohm, Nelson and Roncadelli, and conclude that none of the existing interpretations of QM can lead to a quantum ontology that entirely satisfies the criteria quoted above (though the authors retain that Roncadelli's theory is "the one that gives expression to most aspects of quantum reality"). The issue of quantum ontology recurs in the paper by **V. Fano**, who compares Kant's constructive *a priori* view of physical theories with QM, where he also finds *a priori* schemes, yet lacking in the full space-time continuity and visualizability that Kant retains to be a necessary attribute of physics; then, the author suggests that a quantum reality which does not comply with Kant's trascendental schematism can be conceived, circumventing Kant's prohibition of any scientific ontology.

Also **A. Rebaglia** looks for a conceptual ontological model of QM, starting from the fathers of the theory's avowed philosophical biases: Heisenberg's Aristotelian notion of potentiality and effectivity, and Bohr's Kierkegaardian overcoming plain empiricism by introducing individual free choices among different open possibilities that are involved in quantum jumps (the two viewpoints together being considered a completion and overcoming of Berkeley's immaterialism and Hume's skepticism); she then proposes a hermeneutical ontological model derived from Heidegger as a further, more precise (though still analogical) step towards a quantum ontology, deepening the referential nature of Being and the dynamical circularity between the observer and the observed event that occur both in Heidegger's existentialism and in quantum reality.

A problem that is strictly connected with the purpose of constructing a quantum ontology is the problem of defining physical objects at the microscopic level. **E. Castellani** agrees in her paper with the idea that the notion of physical object, conceived as a *carrier of properties*, depends on the context (for instance classical mechanics or QM) and briefly illustrates the approach to the problem based on the concepts of invariance and group of transformations (*group theoretical approach*), also pointing out the connection of this approach with the problem of objectivity of scientific knowledge; in this context, the use of the notion of imprimitivity system for selecting observable quantities that define quantum particles is also briefly commented on.

Some proposals on quantum ontology are also contained in the paper by **M. Dorato**, who compares QM and special relativity with respect to the issue of ontological determinateness of events, considering the experiments showing that Bell's inequalities are violated as a proof of non-separability of spatially distant events; the author asserts that these experiments are compatible with special relativity only if one interprets all events as co-determinate, without any ontological distinction between past and future, thus essentially accepting Costa De Beauregard's interpretation according to which EPR correlations imply space-time perfect reversibility.

The suggestion of adopting a humbler philosophical attitude is promoted in the paper by **E. Giannetto**. This author traces the loss of individuality in QM back to quantum indeterminacy, interpreted as a breakdown of any deterministic or reductionistic hypothesis regarding individual objects and based on their separability; moreover, according to Giannetto, separability has been shown to be impossible by QM, but it is already proved impossible by a deeper understanding of classical and relativistic physics, so that the supposed incompleteness of QM can be seen as a characteristic of any physical theory and as a recall to overcome man's will of dominating the world through the acceptance of his own being intimately connected with the physical world itself.

An attempt of constructing a purely operational approach to QM is supplied by **A. Drago**, who proposes to solve the QM paradoxes through a reformulation of the theory based on constructive finite mathematics and nonclassical logic; in this reformulation all physical quantities have a strictly operational meaning and doubly negated propositions cannot be identified with positive assertions, which should overcome, according to the author, Von Neumann's and Schrodinger's paradoxes regarding the wave packet reduction. Then, **A. Drago** and **A. Pirolo** explicitly identify Drago's approach to QM with Jordan's, interpreting the commutation relations on which matrix mechanics is based in terms of mere mathematical symmetries, which generate in particular Heisenberg's uncertainty relations, excluding waves differential equations and accepting nonclassical logic.

Finally, **G. Tarozzi** discusses quantum acausality in order to show that it must not be considered an intrinsic feature of the theory but rather a consequence of the standard interpretation of QM, and maintains that one can get over paradoxes and contradictions by adopting a more realistic and causal viewpoint; this goes back to the classical definitions of causality provided by Laplace, Kant, Mill and Hume (but one must give up with the mechanistic aspects of Laplace's definition), which are contradicted by the basic principles of the Copenhagen interpretation, in order to incorporate Einstein's realistic version of the wave-particle dualism and to avoid the retroaction in time implied by Wheeler's delayed choice experiments.

The papers that we have presented so far were mainly dedicated to foundational or philosophical problems in QP. We would like to introduce now a number of articles that are more closely devoted to the other main topic in our Conference, that is, the historical analysis of QM.

An overall view on the issue of individuality in QM is the content of the paper by **A. Rossi**, who treats this issue as the last development of a problem started with classical physics and regarding the existence in this theory of a distinguishability, in principle even if not in fact, among individual objects, as shown by Maxwell-Boltzmann statistics; coming to modern physics, the author observes that, after attributing a relativistic variability of properties as mass and dimensions to individual objects, Einstein introduced with Bose a quantum statistics that apparently violated Leibniz's principle of the identity of indiscernibles, but interpreted it as the expression of a new positive attractive correlation among particles that are still distinguishable in principle; instead, the negative correlation (repulsion) inherent in Fermi-Dirac statistics does not allow one to make any distinction between two particles in different states, each of them being able to occupy exclusively either state: this loss of identity is however only an interpretative model of a specific statistical correlation, even if Fermi identified it with empirical reality, conforming to the naive realistic attitude typical of the philosophically disengaged Italian scientific milieu of his time.

The Fermi-Dirac statistics recurs in the paper by **N. Guicciardini** and **G. Introzzi**, who stress the difference that occurs between Fermi's and Dirac's contributions to the discovery of their quantum statistics, analyzing the conceptual framework and mathematical methods used by these authors; they show that Fermi focused his work on an extension of Pauli's exclusion principle, on combinatorial calculus and on the use of the correspondence principle, while Dirac started from his generalization of Heisenberg's matrix mechanics in order to find out the antisymmetric eigenfunctions of the new statistics without considering intuitive correspondences among classical and quantum concepts: the authors conclude that the essential equivalence of Fermi's and Dirac's approaches can be seen as an indirect proof of the validity of their statistics notwithstanding their mathematical and methodological differences.

Planck's work is the central topic in the papers by Campogalliani and Cerreta. According to **P. Campogalliani**, Planck focused on the question of thermodynamical irreversibility in terms of his fundamental natural radiation hypothesis, which is an extension of Boltzmann's molecular chaos hypothesis, interpreted as the deep source of quantum indeterminacy, i. e., a microscopical indeterminacy that is only added and not opposed to the macroscopic deterministic thermodynamical evolution. According to **P. Cerreta**, on the contrary, Planck's position essentially reduces to Boltzmann's, as asserted in Kuhn's work on black body theory; indeed this author retains, even more explicitly than Kuhn, that Planck and Boltzmann shared the same physical and even mathematical paradigm, that is, the paradigm of mathematical finitism and physical discreteness.

Coming to Einstein, **S. D'Agostino** traces back Einstein's conviction of the problematical character of QM to its early sources, before the EPR paper and the Bohr-Einstein debate, that is to Einstein's long methodological reflection on classical physics; this culminated in Einstein's discussion on general relativity and unified field theory, according to which the attempt of belabouring fully overdetermined theories whose correspondence with observables is overall and independent of single concepts and propositions is limited by the introduction of *ad hoc* postulates defining single terms and propositions, among which operational and instrumental terms giving a metrical meaning to the theory are included; if, then, according to Einstein, the problem of the meaning of the concepts is still open, not only in QM but also in classical physics and relativity, Einstein's criticism regarding the incompleteness of QM is connected with his search for maximum empirical overdetermination of theories, following from a realistic or, better, neo-Kantian inspired philosophical attitude.

Finally, **B. Carazza**, **G.P. Guidetti** and **N. Robotti** stress the crucial role of the dispersion formula in classical and quantum spectroscopy quoting the discovery of X rays (1895) and the transition from old quantum theory to QM as examples; indeed, it led to recognize the wave nature of X rays, from one side, and to reduce quantum phenomena to strictly observable quantities according to matrix mechanics, from the other side.

We have thus ended our presentation of the papers collected in this book. Before closing this introduction we would like to mention all the interesting talks that have been given at the Conference and that do not appear here for reasons that do not depend on ourselves, as the talks by

R. Ascoli, C. Dalla Pozza, L. Lanz, M. Papini and A. Rimini. We would also like to remind with pleasure the Conference's cooperative and tolerant climate, which has favoured a fruitful intellectual exchange and lively and sincere discussions; we only regret that the texts of the debates following the lectures are not available for publication.

Finally, we wish to thank the National History of Physics Group of CNR, the Department of Mathematics of the University of Cosenza and the Department of Information Sciences of the University of Milan for their financial contributions, the Department of Physics of the University of Lecce for its financial and organizative support. We are also grateful to Dr V. Elia for helping in revising the manuscripts, and to the secretary M. C. Gerardi, whose competence and concern have been one of the reasons of the success of the Conference.

Claudio Garola and Arcangelo Rossi

ON THE AXIOMS OF PROBABILITY

L. ACCARDI
Università Tor Vergata
Centro Vito Volterra

Abstract. A set of physically meaningful axioms is introduced, which allows to deduce the mathematical structure of quantum theory, the superposition principle and the Schrödinger equation included.

1. Introduction

The main conceptual difference between classical and quantum probability is the Heisenberg principle which we formulate in a way that abstracts its specific quantum features (in particular $\hbar$), and lifts it to a universal principle of any experimental science:

There exist pairs of observables which cannot be simultaneously measured with arbitrary precision on the same system.

The first basic implication of the Heisenberg principle is that, while in classical probability the identification *event* $\equiv$ *measured event* is always implicitly assumed, quantum probability must distinguish between these two notions. Equivalently one could say that classical probability deals with *platonic entities*, while quantum probability must distinguish between the statements:

- *The observable A has the value a.*
- *The result of an experiment says that the observable A has the value a.*

In other words: in classical probability the experiment is **neutral** and there is no distinction between platonic statements and experimental statements. In quantum probability the experiment is **active**.

While all this has been well understood in the literature on the foundations of quantum theory, the fact that from these statements one can

1

C. Garola and A. Rossi (eds.), The Foundations of Quantum Mechanics, 1-18.
© *1995 Kluwer Academic Publishers.*

deduce some very detailed information on the mathematical structure of quantum theory seems to be much less appreciated.

The goal of the present paper is to show that *these apparently qualitative statements include in them a strong rigidity* in the sense that *practically the whole mathematical structure of quantum theory is uniquely defined by them.*

In the past thirty years several attempts have been made to deduce the quantum mechanical formalism from physically meaningful requirements (axioms): quantum logics, the von Neumann–Segal algebraic approach, Landé axiomatization of amplitudes, ... (cf. [Ac'77] for a survey of the main conceptual features of these approaches). Especially the quantum logic approach has given rise to several generations of variants (the various *operational approaches* by Ludwig, Foulis and Randall and further subsequent formal extensions ... are only a few examples of these variants).

The main conceptual difficulty with these theories is, in the author's opinion, their difficulty in singling out some *physically meaningful* condition to temperate the amorphous character coming from a too weak set of axioms. In these approaches the jump from a too wide, hence too structurally poor, class of models to the specific models used in quantum theory is always accomplished by the introduction of *ad hoc* mathematical axioms which are justified *not a priori*, by their intrinsic physical meaning, but *a posteriori*, from the fact that by their use one can construct the empirically successful model. This circumstance seriously weakens the interest of these axiomatic approaches because the most important goal of an axiomatic theory is to throw light on the physical roots (axioms) of a mathematical formalism, while in the above mentioned cases it is the mathematical formalism that justifies a posteriori some of the axioms.

The approach to be discussed here, even if it absorbs some terminology previously developed (e.g. the notion of *filter*), is based on quite different ideas. It arose from the author's attempt to construct an axiomatization of quantum theory based uniquely on physically meaningful axioms. The success of any such attempt has to be measured by its ability to *deduce* the mathematical formalism of the theory.

Clearly also in the present approach there will be idealizations and even technical assumptions. However, while the use of idealizations is intrinsic to every mathematical axiomatization (in some sense a model is *by definition* an idealization), the technical assumptions will enter here only as *genericity assumptions* (cf. the conditions (6.2) and (6.3) below) which simplify the proof of the main structure theorem without radically restricting the class of models considered. There is the sound hope that a refinement of the proof of the main classification theorem shall eventually dispense of these technical assumptions.

On the other hand the fact that such a qualitative statement, as the above introduced formulation of the indeterminacy principle, includes in itself such specific features as the use of amplitudes rather than probabilities and even a generalization of both the Schrödinger equation and the theory of group representations, is surely to be interpreted as a strong indication in favour of the present approach, at least in a relative sense, i.e. when compared with approaches in which these features are not deduced from physically meaningful postulates, but introduced *by hands*.

2. Conditioning

The notion of *conditioning* is a crucial one both in classical and in quantum probability. The term *conditioning* in mathematics means *acquisition of new information*. In Physics it means *preparation of the experiment*.

The latter is a particular case of the former because to *prepare* an experiment, means to be sure that for some , experimentally produced, systems certain observables shall have certain predetermined values. Thus to *prepare* an experiment means to *acquire experimental information* on these observables.

In Classical Probability the acquisition of information is a cumulative process: new information do not destroy old ones.

In Quantum Probability, acquisition of new information may destroy previously acquired information.

This difference suggests that *quantum conditioning should be different from classical conditioning.*

In connection with this difference, I would like to mention the following question: *Is quantum physics the only situation in which acquisition of new information may destroy previous information?*

My feeling in that the answer to this question should be *no*. Consider for example an experiment in which one wants to correlate the vision capacity of the eye with both its electrical and chemical activity. So in this case too a simultaneous measurement with arbitrary precision should be excluded.

It is clear that the measurement of one of these quantities might affect the other one. However a thorough, scientific analysis of the validity of the indeterminacy principle (in statements such as the above one) is still absent from the literature.

In any case, if the answer to the above question is "*no*", then we should expect that non Kolmogorovian probabilistic models shall play a role also outside quantum physics.

It is a fact that, when a single conditioning event (preparation) is fixed, then all the models reduce to the usual Kolmogorovian model. The same is true when several, *different but compatible conditioning events* are considered. Thus the problems only arise in presence of several different, and in fact incompatible, conditionings.

Notice that a preparation, in the sense described above, is a particular kind of measurement. Moreover, the measurement of the observable A giving the result a is also an event, precisely the event $[A = a]$. Conversely, any event in a physical theory can be described by the fact that a certain number of observables, at a certain time have some values. These obsevables shall be said to be *associated* to the given event. *Platonic events* are distinguished by *experimental events* because in the former case the values of the associated observables are *asserted*, in the latter – *measured*.

An event is called *classical* if there exist some measurements of the associated observables which do not alter the values of any other observable of the theory in consideration. For example, the (macroscopic) trajectory of a billiard ball is indifferent to the facts that we turn on or off the light in the room or that we photograph it. This means that the theory under consideration only considers a restricted class of observables: in our example some electrons of the ball surely shall be affected by the interaction with the beam of photons generated by turning on the light, but these microscopic effects are neglected as non observable ones. For classical events there is no difference between *platonic* and *experimental* statements, but quantum theory teaches us that there exist nonclassical events.

This suggest the following program:

Do not axiomatize events, axiomatize measurements.

The Axiomatization of events is accomplished by classical logic through the one-to-one the correspondence that could be schematized as follows:

Event $A \Longleftrightarrow$ Proposition: *the event A happens.*

The one-to-one correspondence:

Events $\Longleftrightarrow$ Propositions

is then lifted to a correspondence of structures:

Events and operations on them $\Longleftrightarrow$ Boolean algebras.

This allows us to use the representation theorem for Boolean algebras as algebras of subsets of a given set (Stone's theorem) and to associate, at least in a *platonic* way, a state space (sample space) to any classical family of events (the term *classical* means here that the usual boolean operations can be applied without restrictions, giving rise to other events of the same family).

This is the precise meaning of the statement that *classical logic is the logic of classical events.* But in our enlarged context we have seen that an event should be identified with *results of acts of measurement.* This suggests to axiomatize the measures rather than the events themselves. Moreover any event A can be associated to a a 2–valued observable namely the observable which takes the value 0 if the event A does not happen, 1 if the event A happens.

In this way the set of events can be identified with a subset of the set of observables.

The basic new feature, introduced by the present approach in order to enlarge the classical context to an extent that allows the introduction of the new quantum–theoretical features, can be summarized in the following:

PROPOSAL: Replace the *logic of events* by the *logic of measurements*.

Notice that also to a measurement $\mathcal{M}$ one can associate the proposition: *The measurement $\mathcal{M}$ has produced the result m*. However now there is no a priori reason to associate the elementary logical connectives to the operations that one can perform on measurements.

Our axioms shall deal precisely with these operations and with their counterparts in an abstract model. Our goal shall be the developement of analogues, for measurements, of the structure theory of Boolean algebras as well as analogues of their classification theorem. This can be considered as a joint program for *Logic and Probability*.

3. Axioms of probability

Let $\mathcal{M}$ be a set whose elements shall be denoted $X, Y, Z, \ldots \in \mathcal{M}$. An element of $\mathcal{M}$ is called indifferently a *measurement, measurement apparatus, apparatus* or an *instrument*.

The measurements we have in mind are the so called 1–st kind measurements: the measured system is not destroyed but emerges from the apparatus and can again be subjected to measurements. This allows to introduce, as a first set of axioms, the *composition* of measurements and its main properties.

– (**A1.**) There exists a binary composition law

$$(X, Y) \in \mathcal{M} \times \mathcal{M} \Longrightarrow X \cdot Y \in \mathcal{M}$$

called *multiplication*.

Interpretation: The multiplication of two instruments X, Y corresponds to the consecutive performance of each of them (in series):

$$input \implies X \implies Y \implies output$$

For non 1–st kind measurements multiplication is meaningless.

– (**A2.**) Associativity of the multiplication

$$(X \cdot Y) \cdot Z = X \cdot (X \cdot Z)$$

Interpretation: The definition of *instruments* is largely arbitrary: one can always consider two consecutive measurements as a single one according to the following scheme:

$$\Longrightarrow (X \Longrightarrow Y) \Longrightarrow Z \Longrightarrow \; = \; \Longrightarrow X \Longrightarrow (Y \Longrightarrow Z) \Longrightarrow$$

– **(A3.)** There exist a measurement, denoted 1, characterized by the property:
$$X \cdot 1 = 1 \cdot X = X \quad ; \quad \forall X \in \mathcal{M}$$

Interpretation: 1 is the trivial measurement in which there is no interaction between the system and the apparatus therefore every system *passes through* the apparatus.

– **(A4.)** There exists a measurement, denoted 0, caracterized by the property:
$$X \cdot 0 = 0 \cdot X = 0 \quad ; \quad \forall X \in \mathcal{M}$$

Interpretation: 0 is the trivial measurement which destroys every system, therefore no system *passes through* the apparatus.

– **(A5.)** There exist an operation

$$* : \mathcal{M} \Longrightarrow \mathcal{M}$$

called *time reversal*, such that

$$(X \cdot Y)^* = Y^* X^*$$

$$(X^*)^* = X$$

Moreover
$$1^* = 1; \qquad 0^* = 0$$

Interpretation: The time reversal of X corresponds to do in reverse order all the physical operations corresponding to the measurement X.

$$(\Longrightarrow X \Longrightarrow)^* \quad = \quad \Longleftarrow X^* \Longleftarrow$$

A *symmetric instrument* is one in which the order of the sequential operations is irrelevant.

$$X^* = X$$

- **(A6.)** **RANDOMIZATION AXIOM** For each instrument X and each number $p \in [0, 1]$ there exists an instrument, denoted

$$pX$$

with the following properties:

$$(pX) \cdot Y = p(X \cdot Y) = Y \cdot (pX)$$

$$(1 \cdot X) = X; \qquad (0 \cdot X) = 0$$

Interpretation: For any input, pX either produces the same output as X or no output. In many trials of pX with the same preparation X the ratio

$$\frac{\#\text{outputs of } p\,X}{\#\text{outputs of } X}$$

is approximatively p.

DEFINITION Two instruments X, Y are called _compatible_ if:

$$X \cdot Y = Y \cdot X$$

Interpretation:

$$\Rightarrow X \Rightarrow Y \Rightarrow \quad = \quad \Rightarrow Y \Rightarrow X \Rightarrow$$

The previous axioms concerned measurements _in series._ Now we discuss measurements _in parallel._ Because of the indeterminacy principle not all measurements can be performed in parallel. This means that the corresponding composition law cannot be everywhere defined.

- **(A7.)** (Sum Axiom) There exist a binary composition law among compatible measurements

$$X \cdot Y = Y \cdot X \Rightarrow X + Y \in \mathcal{M}$$

satisfying the following conditions:

- **(A7.1)** COMMUTATIVITY:

$$X + Y = Y + X$$

- **(A7.2)** ASSOCIATIVITY:

$$(X + Y) + Z = X + (Y + Z)$$

– (**A7**.3) NEUTRALITY OF ZERO

$$X + 0 = 0 + X = X$$

– (**A7**.4) CANCELLATION LAW:

$$X + Y = X + Z \Rightarrow Y = Z$$

The following Lemma shows that the distributive property is meaningful:

LEMMA If X_o, is compatible with $X_1, \ldots, X_n$ then it is also compatible with $X_1 \cdot X_2 \cdots \cdots X_n$.

Proof: Simple computation.

– (**A7**.5) Distributivity: if X, Y, Z are pairwise compatible, then

$$(X + Y) \cdot Z = X \cdot Z + Y \cdot Z$$

DEFINITION An algebra of measurements is a quintuple:

$$\{\mathcal{M}, \cdot, +, *, \text{multiplication} \ \ \text{by} \ p \in [0, 1]\}$$

where $\mathcal{M}$ is a set and the operations $\cdot, +, *$ and the multiplication by $p \in [0, 1]$ satisfy the Axioms (**A1**.), $\ldots$, (**A7**.).

The interpretation of the additivity axioms is straightforward.

Now we begin to explore the mathematical consequences of the axioms. The first main remark is that the physical operations can be uniquely extended to all measurements and give rise to a $*$–algebra.

THEOREM Given an algebra of measurements $\mathcal{M}$, there exists a, unique up to isomorphism, $*$–algebra $\mathcal{A}$ over the reals and an injective map $j : \mathcal{M} \Rightarrow \mathcal{A}$ which preserves the algebraic structure, i.e. such that

$$j(X)^* = j(X)^*$$

$$j(XY) = j(X) \cdot j(Y)$$

$$j(pX) = pj(X), \qquad \forall p \in [0, 1]$$

$$j(1) = 1$$

$$j(0) = 0$$

and if $XY = YX$ then

$$j(X + Y) = j(X) + j(Y)$$

and such that if B is a $*$–algebra and $k : M \Rightarrow B$ is a map satisfying the above identities, then there exists a $*$–homomorphism $\alpha : A \Rightarrow B$ such that:

$$k = \alpha \circ j$$

The pair $\{A, j\}$ is called the $*$–algebra generated by M. Since A is determined by M up to isomorphism, in the following we shall also call A a *measurement algebra*.

DEFINITION An algebra of measurements M is called *classical* if the multiplication in M is commutative.

THEOREM If M is a commutative measurement algebra, then there exist a topological space S and an $*$–homorphism

$$j : M \Rightarrow C(S)$$

where $C(S)$ denotes the $*$–algebra of continuous complex valued functions on S with the pointwise operations. If the states on M separate the points, then the homomorphism can be taken to be injective.

DEFINITION Let A be a $*$–algebra of measurements. A subalgebra A_o of A is called *maximal Abelian* if it is Abelian and not properly contained in any Abelian subalgebra of A.

REMARK(1.) If $A_o \subseteq A$ is maximal Abelian then necessarily A_o contains the center of A.

REMARK(2.) Any Abelian subalgebra $A_o \subseteq A$ is contained in a maximal Abelian subalgebra (by Zorn's lemma).

DEFINITION $A \in M$ is called a *projection* if

$$A = A^*; \qquad A^2 = A$$

In the following we shall use the notation: $Proj(M) =$ family of all projections on M.

LEMMA Let M be a classical measurement algebra generated (algebraically) by its projections. Then $Proj(M)$ is a Boolean algebra with the operations:

$$p \wedge q = pq$$

$$p \vee q = p + q - pq$$

THEOREM The two categories
i) Boolean algebras,
ii) Classical algebras of measurements generated by projections,

are isomorphic.

From now on we shall limit ourselves to algebras which are algebraically generated by the projectors.

Not all measurement algebras have this property (not all even posses projections). In the following we shall formulate axioms for this class of algebras.

REMARK No probability has been introduced up to now! Only Boolean algebras. This is natural since a classical theory needs not be a priori a statistical one. We shall see that the situation with a quantum theory is quite different.

4. Non classical measurement algebras

Let M be a measurement algebra. The following requirement shall be called, for evident reasons, _local Kolmogorovianity_:

Any $X \in M$ is contained in a maximal classical subalgebra of measurements $M_o \subseteq M$.

All classical subalgebras of measurements contain the center (in particular the multiples of 1 are in the common intersection).

DEFINITION A _partition of the identity_ in a $*$-algebra A is a family (A_α), of elements of A, such that each (A_α) is a projector and

$$A_\alpha A_\beta = \delta_{\alpha,\beta} A_\alpha \tag{4.1}$$

$$\sum_{\alpha=1}^{n} A_\alpha = 1 \tag{4.2}$$

The interpretation of the above properties is simple: the property of being an elementary filter means that each A_α is a a device which filters the value a_α, schematically:

$$A = ? \Rightarrow A_\alpha \Rightarrow [A = a_\alpha]$$

The property of being a 1-st kind measurement is reflected in the condition that repetition of the same operation gives the same result, schematically:

$$\Rightarrow A_\alpha \Rightarrow A_\alpha \Rightarrow \quad \equiv \quad \Rightarrow A_\alpha \Rightarrow$$

The self-adiointness denotes reversibility of the apparatus:

$$\Rightarrow A_\alpha \Rightarrow \quad = \quad \Leftarrow A_\alpha \Leftarrow$$

Condition (4.1) means that in any moment an observable assumes only one value.

$$A_\alpha A_\beta = 0, \qquad \text{if } \alpha \neq \beta \tag{4.3}$$

Condition (4.2) means that in any moment an observable assumes at least one value.

DEFINITION A partition of the identity (A_α) is called _maximal_ if the $*$–algebra generated by it is maximal, i.e. if any measurement X commuting with it:

$$X A_\alpha = A_\alpha X; \qquad \forall \alpha$$

is essentially either a function of it or an essentially compatible measurement. More precisely such an X must have the form:

$$X = \sum_\alpha \lambda_\alpha A_\alpha$$

with

$$\lambda_\alpha \in \kappa := \text{Center of} \mathcal{A} := \{\lambda \in \mathcal{A} : \lambda \kappa = \kappa \lambda, \forall \kappa \in \mathcal{A}\}$$

Summing up, in the general correspondence:
    physical quantities $\Rightarrow$ mathematical objects
one has the more precise correspondences:
    Filter $\equiv$ precise value of an observable,
    elementary filter $\Rightarrow A_\alpha \Rightarrow \equiv$ projection $A_\alpha = A_\alpha^2 = A_\alpha^*$,
    an observable assumes one value at the time $\Longleftrightarrow \alpha \neq \beta \Rightarrow A_\alpha A_\beta = 0$,
    all values are present $\Longleftrightarrow \sum_\alpha A_\alpha = 1$,
    concatenation of filters $\Longleftrightarrow$ multiplication,
    simultaneous action of compatible filters $\Longleftrightarrow$ addition,
    observable $\Longleftrightarrow$ maximal partition of the identity.

5. The quantum model

As stated in the introduction, the fundamental new qualitative feature of quantum physics is the Heisenberg principle whose qualitative essence has been individuated by us in the mere statement of existence of incompatible observables. The formulation of this principle in terms of measurements is the following:

  – (A8.) **HEISENBERG PRINCIPLE (weak form)** There exist two
     partitions of the identity $(A_\alpha), (B_\beta)$ such that

     i) $A_\alpha B_\beta \neq 0 \qquad \forall \alpha, \beta = 1, \cdots, n$

     ii) Both (A_α) and (B_β) are maximal.

     (condition (ii) is a genericity requirement and is not essential, but
     makes several computations much more transparent).

Now let T be a set, labeling the observables of the theory and let

$$S = \{1, \cdots, n \ (\leq \infty)\}$$

be a set labeling the values of each observable in the sense that to every $x \in T$ one associates an observable, i.e. a partition of the identity:

$$\forall x \in T \ , \ \forall \alpha \in S \ , \ \exists A_\alpha(x) \in \mathcal{A}_o$$

$$A_\alpha(x)A_{\alpha'}(x) = \delta_{\alpha\alpha'}A_\alpha(x)$$

$$\sum_\alpha A_\alpha(x) = 1; \qquad A_\alpha(x) = A_\alpha(x)^*$$

6. Deduction of the superposition principle

<u>DEFINITION</u> A **Schwinger algebra** of rank n over a set T is a triple

$$\{\mathcal{A}, T, (A(x))_{x \in T}\} \tag{6.1}$$

where $\mathcal{A}$ is a real associative $*$-algebra , T is a set and , for every x in T , $A(x) = \{A_1(x), ..., A_n(x)\}$ is a maximal partition of the identity in $\mathcal{A}$ such that for any x, y in T , for any $j, k = 1, ..., n$ and for any $\gamma \in \kappa = $ *center of* $\mathcal{A}$, the following conditions hold :

$$\gamma A_j(x) A_k(y) = 0 \Longleftrightarrow \gamma = 0 \tag{6.2}$$

$$\gamma A_j(x) \geq 0 \Longleftrightarrow \gamma \geq 0 \tag{6.3}$$

where the order is the usual one in a $*$–algebra. Writing $A_j(x)$ as $A_j(x) \cdot A_j(x)$ (cf. (4.1)), it follows that, because of (6.2) equality holds in the right hand side of (6.3) if and only if $\gamma = 0$.

First we show that the very definition of Schwinger algebra implies a well defined form of *stochasticity* intrinsic in the definition itself.

<u>DEFINITION</u> Let κ be a real $*$-algebra and let n be an integer or $+\infty$. A n-dimensional κ-valued **stochastic matrix** is a matrix $P = (p_{ij})$ $(i, j = 1, ..., n)$ such that

$$p_{ij} \in \kappa; \qquad p_{ij} \geq 0; \qquad \sum_{j=1}^n p_{ij} = 1 \tag{6.4}$$

where 1 denotes the identity in κ. If also the condition

$$\sum_{i=1}^n p_{ij} = 1 \tag{6.5}$$

is satisfied , then we say that P is a κ-valued **bi-stochastic matrix.**

THEOREM Let $\{\mathcal{A}, T, (A(x))_{x \in T}\}$ be a Schwinger algebra. Then for any pair of elements x, y of T there exists a κ-valued n -dimensional bi-stochastic matrix $P = (p_{ij})$ $(i, j = 1, ..., n)$ such that for any $i, j = 1, ..., n$ one has

$$A_i(x) A_j(y) A_i(x) = p_{ij}(x, y) A_i(x) \tag{6.6}$$

$$p_{ij}(x, y) = p_{ji}(y, x) \tag{6.7}$$

Proof: Let $x, y \in T$ and $i, j = 1, ..., n$. Then, because of the maximality of $A(x)$, $A_i(x) A_j(y) A_i(x)$ must be a linear combination of the $A_i(x)$ with coefficients in the center and, because of the orthogonality of the $A_i(x)$ one easily checks that it must have the form

$$A_i(x) A_j(y) A_i(x) = p_{ij}(x, y) A_i(x) \tag{6.8}$$

for some elements $p_{ij}(x, y) \in \kappa$. Multiplying both sides of (6.7) on the right by $A_i(x)$ $(i = 1, ..., n)$ and using (4) one finds that

$$p_{ij}(x, y) A_i(x) = A_i(x) A_j(y) A_i(x) = \big(A_j(y) A_i(x)\big)^* \big(A_j(y) A_i(x)\big) \geq 0$$

and because of (6.3) this implies that $p_{ij}(x, y)$ is positive. Summing (6.8) over j and using (4.1) , (4.2) leads to

$$A_i(x) = \Big(\sum_{j=1}^{n} p_{ij}(x, y)\Big) A_i(x) \tag{6.9}$$

and this implies (6.4) because of (6.3). The symmetry relation (6.7) follows from associativity, in fact :

$$p_{ij}(x, y) A_i(x) A_j(y) = A_i(x) \Big(A_j(y) A_i(x) A_j(y)\Big) = p_{ji}(y, x) A_i(y) A_j(x)$$

which implies (6.3) because of (6.2)

We can now formulate the structure axiom which defines the natural mathematical model for the first kind measurements.

SAQ Let T be a set and let $\{A(x) : x \in T\}$ be a family of n-valued maximal observables and denote $a_j(x)$ $(j = 1, ..., n)$ the values of $A(x)$. Then there exists a Schwinger algebra $\{\mathcal{A}, T, (A(x))_{x \in T}\}$ with the property that for any pair of elements x, y in T and for any $i, j = 1, ..., n$ one has

$$Prob\bigg\{A(y) = a_j(y) | A(x) = a_i(x)\bigg\} = p_{ij}(x, y) \tag{6.10}$$

where $P(x, y) = (p_{ij}(x, y))$ is the transition matrix associated to the pair of partitions of the identity $A(x)$, $A(y)$ according to the above Theorem.

In view of the structure axioms we shall identify, in the following, the maximal discrete observables with their mathematical models in the Schwinger algebra (i.e., the partitions of the identity) and for two such observables $A(x)$, $A(y)$ the associated bistochastic matrix $P(x, y) = (p_{ij}(x, y))$ will be called the **transition probability** **matrix** between $A(x)$ and $A(y)$.

Let us start from observable quantities $A, B, C, \ldots$, their values: $a_\alpha, b_\beta, c_\gamma, \ldots$ and their transition probabilities defined by the above theorem

$$P(A|B), P(B|C), \ldots$$

It can be proved that the Heisenberg principle is not contained in the transition probabilities: it is a genuine physical principle that cannot be read off from the only knowledge of the transition probabilities. Therefore it is natural to ask oneself: *Beyond the Heisenberg indeterminacy principle (in weak form), what else is needed to deduce the structure of complex Hilbert space and the statistical interpretation of its vectors?*

The answer ([Ac'82]) is: *Essentially Nothing!*

To prove this statement, we introduce, motivated by the following lemma, a smaller class of Schwinger algebras.

LEMMA Let $\{\mathcal{A}, T, (A(x))_{x \in T}\}$ be a Schwinger algebra. For any pair of elements x, y of T the set

$$\{A_i(x) A_j(y) : i, j = 1, ..., n\}$$

is linearly independent over κ.

Proof: If, for given $x, y \in T$, there exist elements $\gamma_{ij} \in \kappa$ (i,j = 1,..., n) such that

$$\sum_{i,j=1}^{n} \gamma_{ij} A_i(x) A_j(y) = 0 \qquad (6.11)$$

then fixing i and j in $\{1, ..., n\}$ and multiplying (6.11) on the left by $A_i(x)$ and on the right by $A_j(y)$ one finds

$$\gamma_{ij} A_i(x) A_j(y) = 0 \qquad \forall i, j = 1, ..., n$$

Hence all the γ_{ij} are zero by condition (6.2).

DEFINITION Let n be a natural integer or $+\infty$; let κ be a commutative $*$-algebra and let $P = (p_{ab})$ be a κ-valued $n \times n$ bistochastc matrix.

An **Heisenberg algebra** with center κ and transition probability P is an associative $*$-algebra $\mathcal{A}$ such that:

i) The center of $\mathcal{A}$ is isomorphic to κ.

ii) There exist two maximal partitions of the identity in $\mathcal{A}$

$$A = (A_a); \qquad B = (B_b); \qquad a, b = 1, \dots, n \qquad (6.12)$$

such that:

$$A_a B_b A_a = p_{ab} A_a; \qquad B_b A_a B_b = p_{ba} B_b \qquad (6.13)$$

iii) Any element $x \in \mathcal{A}$ can be written in the form

$$x = \sum_{ab} \lambda_{ab} A_a B_b \qquad (6.14)$$

for some $\lambda_{ab} \in \kappa$.

Let $\mathcal{A}$ be an Heisenberg algebra. Then, because of (6.14), there exist elements $\gamma_{ab}^{cd} \in \kappa$ $(a, b, c, d = 1, \dots, n)$ such that

$$B_b \cdot A_a = \sum_{c,d=1}^{n} \gamma_{ab}^{cd} A_c \cdot B_d \qquad (6.15)$$

these elements will be called the **structure constants** of $\mathcal{A}$ in the $(A_a B_b)$-basis.

THEOREM Let $\mathcal{A}$ be an associative algebra with identity and let κ denote its center. Let $(A_a), (B_b)$ $(a, b = 1, \dots, n)$ be partitions of the identity in $\mathcal{A}$ such that the set $\{A_a \cdot B_b : a, b = 1, \dots, n\}$ is a κ-basis of $\mathcal{A}$, and let γ_{ab}^{cd} $(a, b = 1, \dots, n)$ be elements of κ such that the identity (6.15) holds. Then

$$\sum_{a=1}^{n} \gamma_{ab}^{a'b'} = \delta_{bb'} \qquad (6.16)$$

$$\sum_{b=1}^{n} \gamma_{ab}^{a'b'} = \delta_{aa'} \qquad (6.17)$$

$$\gamma_{a'b}^{ab'} \gamma_{a''b'}^{ab''} = \gamma_{a'b}^{ab''} \gamma_{a''b'}^{a'b''} \qquad (6.18)$$

If moreover $\mathcal{A}$ is a $*$-algebra then

$$\sum_{c,d=1}^{n} (\gamma_{ab}^{cd})^* \cdot \gamma_{cd}^{c'd'} = \delta_{ac'} \cdot \delta_{bd'} \qquad (6.19)$$

$$(\gamma_{a'b}^{ab'})^* \cdot \gamma_{ab'}^{a''b''} = \sum_{e,d=1}^{n} \gamma_{a'b'}^{a''d} \cdot \gamma_{ed}^{a''b''} \cdot \gamma_{ab}^{eb''} \qquad (6.20)$$

Conversely , if κ is a commutative associative real *-algebra with identity and γ_{ab}^{cd} $(a, b = 1, ..., n)$ are elements of κ satisfying (6.16), (6.17), (6.18), then there exist an associative algebra $\mathcal{A}$ with center κ and two partitions of the identity in $\mathcal{A}$, $A = (A_a)$, $B = (B_b)$, such that $A_a \cdot B_b$ is a basis of $\mathcal{A}$ over κ and (6.19) holds. If moreover (6.20) and (6.15) hold then $\mathcal{A}$ has a unique structure of *-algebra whose involution is characterized by the property that its restriction on κ coincides with the original involution on κ and for all a, b

$$A_a = A_a^* \quad ; \quad B_b = B_b^* \tag{6.21}$$

Proof: cf. [Ac'82]. The equations (6.16),..., (6.20), are called the *structure equations* of the given Heisenberg algebra. The classification of the Heisenberg algebras is achieved by the solution of the structure equations, provided by the following theorem.

<u>THEOREM</u> Let $\mathcal{A}$ be an associative real algebra generated by the maximal partitions of the identity $A = (A_a)$; $B = (B_b)$. Assume that the transition probability matrix $P = (p_{ab})$ between A and B is strictly positive in the sense of (6.11) and denote γ_{ab}^{cd} the structure constants of $\mathcal{A}$ in the $(A_a B_b)$-basis. Then there exists a κ-valued matrix $U = (u_{ab})$ such that

$$\sum_{b=1}^{n} u_{a'b}\left(\frac{p_{ab}}{u_{ab}}\right) = \delta_{a,a'} \quad ; \quad a, a' = 1, ..., n \tag{6.22}$$

$$\sum_{a=1}^{n} \left(\frac{p_{ab}}{u_{ab}}\right) u_{ab'} = \delta_{bb'} \quad ; \quad b, b' = 1, ..., n \tag{6.23}$$

$$\gamma_{ab}^{a'b'} = \frac{u_{ab'} u_{a'b}}{u_{ab} u_{a'b'}} p_{ab} \quad ; \quad a, b, a', b' = 1, ..., n \tag{6.24}$$

Conversely if κ is a real commutative *-algebra then , given a κ-valued strictly positive bi-stochastic matrix $P = (p_{ab})$ and a κ-valued matrix $U = (u_{ab})$ satisfying (6.22), (6.23), (6.24) then there exist :
- an associative real algebra $\mathcal{A}$ with center κ
- two maximal partitions of the identity $A = (A_a)$, $B = (B_b)$ in $\mathcal{A}$ with transition matrix P such that the γ_{ab}^{cd} , defined by the right hand side of (6.24), are the structure constants of $\mathcal{A}$ in the $(A_a B_b)$-basis.

Proof: cf. [Ac'82].

The results above allow to solve the following problem: let $n = 1, 2, ..., +\infty$, given a a commutative associative real *-algebra with identity κ_o , a set T and a family $\{P(x, y) : x, y \epsilon T\}$ of κ_o-valued $n \times n$ transition probability matrices, find :

— i) an Heisenberg algebra $\mathcal{A}$ of dimension n^2 over its center κ and such that κ contains a subalgebra isomorphic to κ_o.

– ii) for each $x \in T$ a maximal Abelian partition of the identity $\{A_a(x) : = 1, ..., n\}$ in $\mathcal{A}$ such that for each $x, y \in T$ the transition probability matrix associated to the pair $(A_a(x))$, $(A_b(y))$ is $P(x, y)$

Notice that, since the symmetry condition

$$p_{ab}(x, y) = p_{ba}(y, x) \qquad (6.25)$$

has been shown to be a necessary condition for the solution of the problem, we can assume that it is satisfied. Moreover we will assume that:

$$P(x, x) = 1; \qquad \forall x \in T \qquad (6.26)$$

$$p_{ab}(x, y) > 0; \qquad \forall x \in T, \qquad \forall a, b = 1, ..., n \qquad (6.27)$$

and we shall look only for generic solutions (i.e. such that the structure constants of $\mathcal{A}$ in all the $(A_a(x)B_b(y))$-bases are invertible. Under these assumptions we have.

THEOREM The following assertions are equivalent:

i1) There exist an Heisenberg algebra with centre κ satisfying conditions (i), (ii) above.

i2) For each $x, y \in T$ there exists a κ-valued transition amplitude matrix $U(x, y)$ for $P(x, y)$ such that:

$$U(x, x) = 1; \qquad \forall x \in T \qquad (6.28)$$

$$U(x, y) \cdot U(y, z) = U(x, z); \qquad x, y, z \in T \qquad (6.29)$$

i3) There exists a κ-module H and for each $x \in T$- a κ-basis $(a_j(x))$ $(j = 1, ..., n)$ of H such that the operators $A_j(x)$ defined by

$$A_j(x)a_k(x) = \delta_{jk}a_j(x) \qquad (6.30)$$

satisfy

$$A_j(x)A_k(y)A_j(x) = p_{jk}(x, y)A_j(x) \qquad (6.31)$$

Proof: cf. [Ac'82].

REMARK The Hilbert space model of quantum theory is recovered when $\kappa = \mathbf{C}$ and

$$p_{ij}(x, y) = |U_{ij}(x, y)|^2 \qquad (6.32)$$

in this case one immediately recognizes in (9), (10) the conditions of unitarity of the matrix $U(x, y)$.

REMARK Equation (6.29) is a generalization of Schrödinger's evolution. In our theory it appears as a compatibility condition for a set of transition probability matrices $\{P(x, y)\}$ to admit an Heisenberg algebra model. If the index set T is acted upon by a group G so that probabilities are preserved $(P(x, y) = P(gx, gy))$, one might study the corresponding generalized unitary representation of G on H. Thus equation (6.29) is also a generalization of the notion of *unitary representation*. Examples are easily constructed.

REMARK The reversibility of the *generalized evolution* $U(x, y)$, implicit in equation (6.29), has a purely statistical origin, stemming from the symmetric role that two maximal observables $A(x)$ and $A(y)$ play in their mutual conditioning.

References

[Ac'77] Accardi, L. (1977). "L'edificio matematico della meccanica quantistica: situazione attuale", in *Matematica e Fisica*, Donini, E. et al. (eds.), De Donato, pp. 175–210.
[Ac'82] Accardi, L. (1984). "Some trends and problems in quantum probability", in *Quantum probability and applications to the quantum theory of irreversible processes*, Accardi, L. et al. (eds.), Springer LNM n.1055, pp. 1–19.

THREE QUANTUM THEORIES OF MEASUREMENT

A. AFRIAT and F. SELLERI
Dip. di Fisica, Università di Bari
Via Amendola 143 - 70126 - Bari, Italy

Abstract: We consider the conservation of angular momentum in the measurement theories of Bohr (apparatus described classically), von Neumann (standard quantum theory) and Wigner, Araki and Yanase.

1. Introduction

We first show that the average angular momentum of a pair of spin - 1/2 particles in the quantum mechanical singlet state is changed by interaction with a classically described spin meter; this, together with the requirement of the conservation of angular momentum, could give rise to the possibility of instantaneous signalling over any distance. In the second part we establish that the total angular momentum of a spin - 1/2 particle and a quantum - mechanically described measuring apparatus is not conserved, if it is assumed – as in standard quantum theory – that the interaction between apparatus and particle does not disturb the state of the particle (theorem of Wigner, Araki and Yanase). Therefore instantaneous signals do not arise in this approach, but other problems do. We then establish, in the third part, that this total angular momentum can be conserved if the state of the particle can be altered by the measurement process, even when the particle is in an eigenstate of the measured quantity. The possibility of instantaneous signalling is not a consequence of this approach, but eigenstates lose their usual meaning.

2. Apparatus described in classical terms

Our concepts, according to Bohr [1], are formed by our experience, the range of which, however, is circumscribed by the limitations of our senses. As these only

C. Garola and A. Rossi (eds.), The Foundations of Quantum Mechanics, 19-34.

give us direct access to objects of certain kinds, our ordinary experience will be limited to such objects, and our concepts will be limited correspondingly.

The objects of microphysics, for instance, are individually inaccessible to our unassisted senses, and hence there is no reason why, argues Bohr, the concepts we have formed on an entirely different level should be applicable to those objects. The devices, however, that give us access to those objects – measuring apparata – are constructed so as to be directly accessible. Hence, although there is no point in trying to describe the objects of microphysics using concepts formed from the experience of objects of a very different scale, we can use those concepts to describe measuring apparata. We can *and must*, according to Bohr, because not only our concepts but our language and therefore our means of description are formed by our ordinary experience. Fock [2] asserts this unambiguously: "Measuring instruments must be described classically". We can also quote Jammer [3]:

> For Bohr the absence of a formal theory of measurement did not indicate any imperfection or incompleteness of his epistemological analysis of quantum mechanics but was rather required for reasons of consistency. In his view, classical concepts, representing the ultimately immediate data of common experience, are in the last resort not formalizable, for any formal elaboration becomes physically meaningful only if it is interpreted in terms of classical concepts. Bohr's insistence on the logical (though not physical) necessity of drawing a sharp distinction between object and measuring instrument can therefore never be replaced by any formal treatment.
> For this very reason Bohr never showed real interest in an axiomatic formulation of quantum mechanics. For such an axiomatization cannot dispense with undefined primitive concepts and relations whose concrete meaning can be conveyed only in terms of the language of ordinary experience.

In this first part, then, we shall consider the total angular momentum of a pair of particles in singlet state and apparati described in 'classical' terms, as wished by Bohr [4].

For the spin operators $S_{\alpha 3}$ and $S_{\beta 3}$ of particles α and β, corresponding to measurements along the z - direction, we have the eigenvalue equations

$$S_{\alpha 3} \otimes I_\beta \left| u_0^\pm \right\rangle \otimes \left| v \right\rangle = \pm \frac{\hbar}{2} \left| u_0^\pm \right\rangle \otimes \left| v \right\rangle \qquad (1)$$

and

$$I_\alpha \otimes S_{\beta 3} \left| u \right\rangle \otimes \left| v_0^\pm \right\rangle = \pm \frac{\hbar}{2} \left| u \right\rangle \otimes \left| v_0^\pm \right\rangle, \qquad (2)$$

where I_α and I_β are the identity operators for the two spaces, and $|u\rangle$ and $|v\rangle$ are arbitrary descriptions of α and β, respectively. With respect to these z-direction bases the singlet state can be represented thus:

$$|\Psi_s\rangle = \frac{1}{\sqrt{2}}\left\{|u_0^+\rangle \otimes |v_0^-\rangle - |u_0^-\rangle \otimes |v_0^+\rangle\right\}. \tag{3}$$

If $\vec{S}_\alpha$ and $\vec{S}_\beta$ are the vectors $(S_{\alpha 1}, S_{\alpha 2}, S_{\alpha 3})$ and $(S_{\beta 1}, S_{\beta 2}, S_{\beta 3})$, we can form the vector

$$\vec{S} = \vec{S}_\alpha \otimes I_\beta + I_\alpha \otimes \vec{S}_\beta \tag{4}$$

and the operator

$$\vec{S}^2 = \sum_{i=1}^{3}\left[\left(S_{\alpha i} \otimes I_\beta\right) + \left(I_\alpha \otimes S_{\beta i}\right)\right]^2. \tag{5}$$

We can also form the operator

$$S_3 = \left(S_{\alpha 3} \otimes I_\beta\right) + \left(I_\alpha \otimes S_{\beta 3}\right). \tag{6}$$

The singlet state is an eigenstate of both $\vec{S}^2$ and S_3, belonging to the eigenvalue zero in either case, so zero is also the average value of the observables represented by these operators, for $|\Psi_s\rangle$. Hence

$$\vec{S}^2|\Psi_s\rangle = 0 = S_3|\Psi_s\rangle, \tag{7}$$

and

$$\langle\Psi_s|\vec{S}^2|\Psi_s\rangle = 0 = \langle\Psi_s|S_3|\Psi_s\rangle. \tag{8}$$

It will be convenient to introduce the triplet state

$$|\Psi_t\rangle = \frac{1}{\sqrt{2}}\left\{|u_0^+\rangle \otimes |v_0^-\rangle + |u_0^-\rangle \otimes |v_0^+\rangle\right\} \tag{9}$$

so that we can express the vectors

$$|u_0^+\rangle \otimes |v_0^-\rangle \quad \text{and} \quad |u_0^-\rangle \otimes |v_0^+\rangle, \tag{10}$$

of which the singlet state is a superposition, as superpositions of $|\Psi_s\rangle$ and $|\Psi_t\rangle$:

$$|u_0^+\rangle \otimes |v_0^-\rangle = \frac{1}{\sqrt{2}}\left\{|\Psi_t\rangle + |\Psi_s\rangle\right\},$$

$$|u_0^-\rangle \otimes |v_0^+\rangle = \frac{1}{\sqrt{2}}\left\{|\Psi_t\rangle - |\Psi_s\rangle\right\}. \tag{11}$$

The vectors $|\Psi_t\rangle, |\Psi_s\rangle, |u_0^+\rangle \otimes |v_0^-\rangle$ and $|u_0^-\rangle \otimes |v_0^+\rangle$, then, are all in the eigensubspace belonging to the eigenvalue zero of the operator S_3. So much for preliminaries.

If we make a spin measurement corresponding to the operator $S_{\alpha 3} \otimes I_\beta$ and discover the eigenvalue $\hbar/2$, we will collapse the superposition $|\Psi_s\rangle$ onto the vector $|u_0^+\rangle \otimes |v_0^-\rangle$, *and change the average value of the operator* $\vec{S}^2$, which was zero before the measurement,

$$\left\langle u_0^+ \otimes v_0^- | \vec{S}^2 | u_0^+ \otimes v_0^- \right\rangle = \left[\frac{\langle \Psi_t | + \langle \Psi_s |}{\sqrt{2}} \right] \vec{S}^2 \left[\frac{|\Psi_t\rangle + |\Psi_s\rangle}{\sqrt{2}} \right]$$

$$= \frac{1}{2} \left(\langle \Psi_t | \vec{S}^2 | \Psi_t \rangle + \langle \Psi_t | \vec{S}^2 | \Psi_s \rangle + \langle \Psi_s | \vec{S}^2 | \Psi_t \rangle + \langle \Psi_s | \vec{S}^2 | \Psi_s \rangle \right). \tag{12}$$

As $\vec{S}^2 | \Psi_s \rangle = 0$, we can eliminate all terms involving Ψ_s, which leaves us with

$$\frac{1}{2} \langle \Psi_t | \vec{S}^2 | \Psi_t \rangle, \tag{13}$$

which, as the triplet state is an eigenstate of $\vec{S}^2$ belonging to the eigenvalue $2\hbar^2$, is equal to $\hbar^2$. Hence

$$\left\langle u_0^+ \otimes v_0^- | \vec{S}^2 | u_0^+ \otimes v_0^- \right\rangle = \hbar^2. \tag{14}$$

Similarly

$$\left\langle u_0^- \otimes v_0^+ | \vec{S}^2 | u_0^- \otimes v_0^+ \right\rangle = \hbar^2. \tag{15}$$

The first measurement (it could just as well have been on β), therefore, changes the angular momentum of the state.

In this first case we are considering, as the interaction between apparatus and system collapses the superposition, the angular momentum acquired by the system - if angular momentum is to be conserved, as it ought to be - must be drawn from the apparatus. If this is indeed the case, the apparatus must react somehow, perhaps by rotating.

A later measurement of

$$S_{\alpha 3} \otimes I_\beta \quad \text{or} \quad I_\alpha \otimes S_{\beta 3} \tag{16}$$

will change nothing according to quantum mechanics, as both $\left|u_0^+\right\rangle \otimes \left|v_0^-\right\rangle$ and

$\left|u_0^-\right\rangle \otimes \left|v_0^+\right\rangle$ are eigenstates of the operators in question, and hence cannot be affected by measurements represented by them.

Therefore the apparatus making the first measurement collapses the superposition and presumably provides the angular momentum hence gained by the system, whereas the apparatus making the later measurement is not affected. This means that an experimenter can induce an instantaneous reaction of a remote apparatus, say A, with his apparatus, say B. If he collapses the superposition with B before A has a chance to, A will be unaffected; otherwise A will react.

3. Standard quantum mechanical treatment of apparatus

We could, however (following von Neumann [5]), describe the apparatus in quantum mechanical rather than classical terms, and assume that the measurement interaction throws it into a superposition. Let us see what happens if we make this assumption.

Consider an apparatus aligned to measure spin along the z-direction, which can be in three macroscopically distinguishable states: $\left|A^0\right\rangle, \left|A^+\right\rangle, \left|A^-\right\rangle$. Before measurement the apparatus is in its ground state $\left|A^0\right\rangle$ and the particle whose spin is to be measured is in some arbitrary state $\left|u^+\right\rangle$. These circumstances can be represented by the vector

$$\left|A^0\right\rangle \otimes \left|u^+\right\rangle. \tag{17}$$

If $\left|u^+\right\rangle$ is an eigenstate of the operator $\vec{S} \cdot \hat{a}$ (where $\hat{a}$ is a unit vector and $\vec{S}$ is now the vector of spin operators for the only particle in question here), we can take the sum of $\vec{S} \cdot \hat{a}$ and the corresponding operator $\vec{M} \cdot \hat{a}$ for the apparatus to form

$$\vec{J} \cdot \hat{a} = \vec{M} \cdot \hat{a} \otimes \mathbf{I}_S + \mathbf{I}_M \otimes \vec{S} \cdot \hat{a}, \tag{18}$$

representing the total angular momentum of system and apparatus, where $\mathbf{I}_S$ and $\mathbf{I}_M$ are the identity operators in the two spaces. It will be easier to write $M(a)$ instead of $\vec{M} \cdot \hat{a} \otimes \mathbf{I}_S$, and $S(a)$ instead of $\mathbf{I}_M \otimes \vec{S} \cdot \hat{a}$.

We can compare $\left\langle \vec{J} \cdot \hat{a} \right\rangle_b$ and $\left\langle \vec{J} \cdot \hat{a} \right\rangle_a$, the average values of $\vec{J} \cdot \hat{a}$ before and after measurement. We have

$$\left\langle \vec{J}\cdot\hat{a}\right\rangle_b=\left\langle A^0\otimes u^+\left|\vec{J}\cdot\hat{a}\right|A^0\otimes u^+\right\rangle=\left\langle A^0\left|M(a)\right|A^0\right\rangle+\left\langle u^+\left|S(a)\right|u^+\right\rangle. \quad (19)$$

But $S(a)$ was chosen so that

$$S(a)|A\rangle\otimes|u^+\rangle=\frac{\hbar}{2}|A\rangle\otimes|u^+\rangle, \quad (20)$$

where $|A\rangle$ is an arbitrary description of the apparatus, and hence

$$\left\langle \vec{J}\cdot\hat{a}\right\rangle_b=\left\langle A^0\otimes u^+\left|\vec{J}\cdot\hat{a}\right|A^0\otimes u^+\right\rangle=\left\langle A^0\left|M(a)\right|A^0\right\rangle+\frac{\hbar}{2}. \quad (21)$$

As the apparatus is aligned along the z-axis, the particle will find itself in an eigenstate of S_3 (again, as there is only one particle here, we can drop the α) *by the end of the measurement process*[1], so we should express $|u^+\rangle$ with respect to the basis $|u_0^\pm\rangle$:

$$|A^0\rangle\otimes\left(|u_0^+\rangle\langle u_0^+|+|u_0^-\rangle\langle u_0^-|\right)|u^+\rangle=|A^0\rangle\otimes\left(c_1|u_0^+\rangle+c_2|u_0^-\rangle\right). \quad (22)$$

The apparatus is arranged so that measurement will give rise to the evolution

$$|A^0\rangle\otimes\left(c_1|u_0^+\rangle+c_2|u_0^-\rangle\right)\to c_1|A^+\otimes u_0^+\rangle+c_2|A^-\otimes u_0^-\rangle, \quad (23)$$

thus enabling us to distinguish between $|u_0^+\rangle$ and $|u_0^-\rangle$.

Let us see, then, if the above evolution leaves the average value of $\vec{J}\cdot\hat{a}$ unchanged.

$$\left\langle \vec{J}\cdot\hat{a}\right\rangle_a$$

$$=\left\langle c_1\left(A^+\otimes u_0^+\right)+c_2\left(A^-\otimes u_0^-\right)\left|M(a)+S(a)\right|c_1\left(A^+\otimes u_0^+\right)+c_2\left(A^-\otimes u_0^-\right)\right\rangle$$

$$=\left\langle c_2\left(A^-\otimes u_0^-\right)\left|S(a)\right|c_2\left(A^-\otimes u_0^-\right)\right\rangle+\left\langle c_2\left(A^-\otimes u_0^-\right)\left|S(a)\right|c_1\left(A^+\otimes u_0^+\right)\right\rangle$$

$$+\left\langle c_2\left(A^-\otimes u_0^-\right)\left|M(a)\right|c_2\left(A^-\otimes u_0^-\right)\right\rangle+\left\langle c_2\left(A^-\otimes u_0^-\right)\left|M(a)\right|c_1\left(A^+\otimes u_0^+\right)\right\rangle$$

$$+\left\langle c_1\left(A^+\otimes u_0^+\right)\left|S(a)\right|c_2\left(A^-\otimes u_0^-\right)\right\rangle+\left\langle c_1\left(A^+\otimes u_0^+\right)\left|S(a)\right|c_1\left(A^+\otimes u_0^+\right)\right\rangle$$

$$+\left\langle c_1\left(A^+\otimes u_0^+\right)\left|M(a)\right|c_2\left(A^-\otimes u_0^-\right)\right\rangle+\left\langle c_1\left(A^+\otimes u_0^+\right)\left|M(a)\right|c_1\left(A^+\otimes u_0^+\right)\right\rangle$$

$$=|c_2|^2\left\langle u_0^-\left|S(a)\right|u_0^-\right\rangle+|c_2|^2\left\langle A^-\left|M(a)\right|A^-\right\rangle+|c_1|^2\left\langle u_0^+\left|S(a)\right|u_0^+\right\rangle+|c_1|^2\left\langle A^+\left|M(a)\right|A^+\right\rangle. \quad (24)$$

---

[1] We shall not address the issue of where exactly that process ends.

We can represent $|u_0^+\rangle$ and $|u_0^-\rangle$ as $\begin{pmatrix} 1 \\ 0 \end{pmatrix}$ and $\begin{pmatrix} 0 \\ 1 \end{pmatrix}$, respectively, $S(a)$ as

$\begin{pmatrix} \cos\vartheta & \sin\vartheta e^{-i\varphi} \\ \sin\vartheta e^{i\varphi} & -\cos\vartheta \end{pmatrix}$, $|u^+\rangle$ and $|u^-\rangle$ as $e^{i\varphi_1}\begin{pmatrix} \sqrt{\dfrac{1+\cos\vartheta}{2}} \\ e^{i\varphi}\sqrt{\dfrac{1-\cos\vartheta}{2}} \end{pmatrix}$ and $e^{i\varphi_2}\begin{pmatrix} -\sqrt{\dfrac{1-\cos\vartheta}{2}} \\ e^{i\varphi}\sqrt{\dfrac{1+\cos\vartheta}{2}} \end{pmatrix}$,

respectively, where $\hat{a} = (\sin\vartheta\cos\phi,\ \sin\vartheta\sin\phi,\ \cos\vartheta)$. Hence we have

$$c_1 = \langle u_0^+ | u^+\rangle = (1,0)e^{i\varphi_1}\begin{pmatrix} \sqrt{\dfrac{1+\cos\vartheta}{2}} \\ e^{i\varphi}\sqrt{\dfrac{1-\cos\vartheta}{2}} \end{pmatrix} = e^{i\varphi_1}\sqrt{\dfrac{1+\cos\vartheta}{2}} \qquad (25)$$

and

$$c_2 = \langle u_0^- | u^+\rangle = (0,1)e^{i\varphi_1}\begin{pmatrix} \sqrt{\dfrac{1+\cos\vartheta}{2}} \\ e^{i\varphi}\sqrt{\dfrac{1-\cos\vartheta}{2}} \end{pmatrix} = e^{i(\varphi_1+\varphi)}\sqrt{\dfrac{1-\cos\vartheta}{2}}. \qquad (26)$$

Also,

$$\langle u_0^- | S(a) | u_0^-\rangle = (0,1)\frac{\hbar}{2}\begin{pmatrix} \cos\vartheta & \sin\vartheta e^{-i\varphi} \\ \sin\vartheta e^{i\varphi} & -\cos\vartheta \end{pmatrix}\begin{pmatrix} 0 \\ 1 \end{pmatrix} = -\frac{\hbar}{2}\cos\vartheta \qquad (27)$$

and

$$\langle u_0^+ | S(a) | u_0^+\rangle = (1,0)\frac{\hbar}{2}\begin{pmatrix} \cos\vartheta & \sin\vartheta e^{-i\varphi} \\ \sin\vartheta e^{i\varphi} & -\cos\vartheta \end{pmatrix}\begin{pmatrix} 1 \\ 0 \end{pmatrix} = \frac{\hbar}{2}\cos\vartheta. \qquad (28)$$

Hence, from equation (24),

$$
\begin{aligned}
\langle \bar{J}\cdot\hat{a}\rangle_a &= -\frac{\hbar}{2}\frac{1-\cos\vartheta}{2}\cos\vartheta + \frac{1-\cos\vartheta}{2}\langle A^- | M(a) | A^-\rangle \\
&\quad + \frac{\hbar}{2}\frac{1+\cos\vartheta}{2}\cos\vartheta + \frac{1+\cos\vartheta}{2}\langle A^+ | M(a) | A^+\rangle \\
&= \frac{\hbar}{2}\cos^2\vartheta + \frac{1-\cos\vartheta}{2}\langle A^- | M(a) | A^-\rangle + \frac{1+\cos\vartheta}{2}\langle A^+ | M(a) | A^+\rangle.
\end{aligned}
\qquad (29)
$$

By virtue of

$$M(a) = M_+ \sin\vartheta e^{-i\varphi} + M_- \sin\vartheta e^{i\varphi} + M_3 \cos\vartheta \qquad (30)$$

$\left(\text{where } M_\pm = \frac{1}{2}(M_1 \pm iM_2)\right)$ however,

$$\langle \vec{J} \cdot \hat{a} \rangle_a = \frac{\hbar}{2}\cos^2\vartheta + \sin\vartheta e^{-i\varphi}\left(\frac{1-\cos\vartheta}{2}\langle A^-|M_+|A^-\rangle + \frac{1+\cos\vartheta}{2}\langle A^+|M_+|A^+\rangle\right)$$

$$+ \sin\vartheta e^{i\varphi}\left(\frac{1-\cos\vartheta}{2}\langle A^-|M_-|A^-\rangle + \frac{1+\cos\vartheta}{2}\langle A^+|M_-|A^+\rangle\right) \tag{31}$$

$$+ \cos\vartheta\left(\frac{1-\cos\vartheta}{2}\langle A^-|M_3|A^-\rangle + \frac{1+\cos\vartheta}{2}\langle A^+|M_3|A^+\rangle\right).$$

This will not generally be equal to the value before measurement, (21), which, using (30), can be expressed as

$$\frac{\hbar}{2} + \sin\vartheta e^{-i\varphi}\langle A^0|M_+|A^0\rangle + \sin\vartheta e^{i\varphi}\langle A^0|M_-|A^0\rangle + \cos\vartheta\langle A^0|M_3|A^0\rangle. \tag{32}$$

For $\langle \vec{J} \cdot \hat{a} \rangle_b$ to be equal to $\langle \vec{J} \cdot \hat{a} \rangle_a$, we can require that corresponding terms in the two expressions be equal. While there is no problem in identifying the terms proportional to $\sin\vartheta e^{-i\varphi}$, $\sin\vartheta e^{i\varphi}$ and $\cos\vartheta$, and to impose that the term proportional to $\cos^2\vartheta$ gives a vanishing contribution to (31), we see that there is no ϑ independent term in (31) that can be set equal to the $\hbar/2$ of (32). We can conclude that angular momentum is generally not conserved in this case.

4. Araki-Yanase treatment of the measurement process

Let us now consider the possibility that measurement disturbs the system, in the sense of Wigner, Araki and Yanase [6]. Until now we have assumed that if, say, the system is in the state $|u_0^+\rangle$, measurement will result in the evolution (23), with $c_1 = 1$, $c_2 = 0$. It is unlikely, however, that measurement should produce no change in the system at all; here it could introduce a component of $|u_0^-\rangle$ into the state of the system, in which case the evolution would be better represented thus:

$$|A^0\rangle \otimes |u_0^+\rangle \rightarrow |A^+\rangle \otimes |u_0^+\rangle + |\varepsilon^+\rangle \otimes |u_0^-\rangle, \tag{33}$$

where $|\varepsilon^+\rangle$ is the state of the apparatus associated with $|u_0^-\rangle$, when the system began in the state $|u_0^+\rangle$ (these considerations are clearly symmetrical in + and -).

Presumably $|\varepsilon^+\rangle$ is one of three things: or measurement always produces the connections $|A^+\rangle \otimes |u_0^+\rangle, |A^-\rangle \otimes |u_0^-\rangle$, and hence $|\varepsilon^+\rangle$ is a multiple of $|A^-\rangle$, which means that even if measurement affects the system, we are aware it has been disturbed; or it is a multiple of $|A^+\rangle$, in which case we have no way of knowing that the state has been changed; or $|\varepsilon^+\rangle$ corresponds to yet another state of the apparatus. We can express the first two possibilities as follows:

$$|A^0\rangle \otimes |u_0^+\rangle \rightarrow c_1'|A^+\rangle \otimes |u_0^+\rangle + c_2'|A^-\rangle \otimes |u_0^-\rangle$$
$$|A^0\rangle \otimes |u_0^+\rangle \rightarrow c_1''|A^+\rangle \otimes |u_0^+\rangle + c_2''|A^+\rangle \otimes |u_0^-\rangle .$$

$$(34)$$

The formulation we have adopted is general enough to cover all of the above possibilities.

We can assume that $|u_0^+\rangle$ and $|u_0^-\rangle$ are normalized; were we to assume that $|A^+\rangle$ and $|\varepsilon^+\rangle$ are as well, we would have to associate coefficients c_A and c_ε (where $|c_A|^2 + |c_\varepsilon|^2 = 1$) with the vectors $|A^+\rangle \otimes |u_0^+\rangle$ and $|\varepsilon^+\rangle \otimes |u_0^-\rangle$ to normalize their superposition. Instead we shall shorten the vectors $|A^+\rangle$ and $|\varepsilon^+\rangle$, and hence require that the sum of their squared lengths be equal to unity:

$$\langle A^+|A^+\rangle + \langle \varepsilon^+|\varepsilon^+\rangle = 1, \qquad (35)$$

and of course

$$\langle A^-|A^-\rangle + \langle \varepsilon^-|\varepsilon^-\rangle = 1. \qquad (36)$$

We assume that:

$$\langle \varepsilon^+|\varepsilon^+\rangle, \langle \varepsilon^-|\varepsilon^-\rangle << 1,$$
$$\langle A^+|A^+\rangle, \langle A^-|A^-\rangle \approx 1. \qquad (37)$$

Angular momentum, however, will only be conserved if $|\varepsilon^+\rangle$ and $|\varepsilon^-\rangle$ are different from zero, as we shall presently see.

If we take an arbitrary state $\left|u^{+}\right\rangle$, eigenstate of $S(a)$, before measurement we have, once again:

$$\left|A^{0}\right\rangle\otimes\left|u^{+}\right\rangle = e^{i\varphi_{1}}\left(r^{+}\left|A^{0}\right\rangle\otimes\left|u_{0}^{+}\right\rangle + r^{-}e^{i\varphi}\left|A^{0}\right\rangle\otimes\left|u_{0}^{-}\right\rangle\right), \tag{38}$$

where

$$r^{\pm} = \sqrt{\frac{1\pm\cos\vartheta}{2}}. \tag{39}$$

After measurement this state becomes

$$r^{+}e^{i\varphi_{1}}\left(\left|A^{+}\right\rangle\otimes\left|u_{0}^{+}\right\rangle + \left|\varepsilon^{+}\right\rangle\otimes\left|u_{0}^{-}\right\rangle\right) + r^{-}e^{i(\varphi_{1}+\varphi_{2})}\left(\left|A^{-}\right\rangle\otimes\left|u_{0}^{-}\right\rangle + \left|\varepsilon^{-}\right\rangle\otimes\left|u_{0}^{+}\right\rangle\right)$$
$$= e^{i\varphi_{1}}\left(r^{+}\left|A^{+}\right\rangle + r^{-}e^{i\varphi}\left|\varepsilon^{-}\right\rangle\right)\otimes\left|u_{0}^{+}\right\rangle + e^{i\varphi_{1}}\left(r^{-}e^{i\varphi}\left|A^{-}\right\rangle + r^{+}\left|\varepsilon^{+}\right\rangle\right)\otimes\left|u_{0}^{-}\right\rangle. \tag{40}$$

We can also take $\left|u^{-}\right\rangle$, the other eigenstate of $S(a)$:

$$\left|A^{0}\right\rangle\otimes\left|u^{-}\right\rangle = e^{i\varphi_{2}}\left(-r^{-}\left|A^{0}\right\rangle\otimes\left|u_{0}^{+}\right\rangle + r^{+}e^{i\varphi}\left|A^{0}\right\rangle\otimes\left|u_{0}^{-}\right\rangle\right)$$
$$= -r^{-}e^{i\varphi_{2}}\left(\left|A^{+}\right\rangle\otimes\left|u_{0}^{+}\right\rangle + \left|\varepsilon^{+}\right\rangle\otimes\left|u_{0}^{-}\right\rangle\right) + r^{+}e^{i(\varphi+\varphi_{2})}\left(\left|A^{-}\right\rangle\otimes\left|u_{0}^{-}\right\rangle + \left|\varepsilon^{-}\right\rangle\otimes\left|u_{0}^{+}\right\rangle\right) \tag{41}$$
$$= e^{i\varphi_{2}}\left(-r^{-}\left|A^{+}\right\rangle + r^{+}e^{i\varphi}\left|\varepsilon^{-}\right\rangle\right)\otimes\left|u_{0}^{+}\right\rangle + e^{i\varphi_{2}}\left(r^{+}e^{i\varphi}\left|A^{-}\right\rangle - r^{-}\left|\varepsilon^{+}\right\rangle\right)\otimes\left|u_{0}^{-}\right\rangle.$$

We can write

$$\left|A^{0}\right\rangle\otimes\left|u^{+}\right\rangle \rightarrow \left|\phi_{1}\right\rangle\otimes\left|u_{0}^{+}\right\rangle + \left|\phi_{2}\right\rangle\otimes\left|u_{0}^{-}\right\rangle$$
$$\left|A^{0}\right\rangle\otimes\left|u^{-}\right\rangle \rightarrow \left|\phi_{2}'\right\rangle\otimes\left|u_{0}^{+}\right\rangle + \left|\phi_{1}'\right\rangle\otimes\left|u_{0}^{-}\right\rangle, \tag{42}$$

where

$$\left|\phi_{1}\right\rangle = e^{i\varphi_{1}}\left(r^{+}\left|A^{+}\right\rangle + r^{-}e^{i\varphi}\left|\varepsilon^{-}\right\rangle\right),$$
$$\left|\phi_{2}\right\rangle = e^{i\varphi_{1}}\left(r^{-}e^{i\varphi_{1}}\left|A^{-}\right\rangle + r^{+}\left|\varepsilon^{+}\right\rangle\right),$$
$$\left|\phi_{1}'\right\rangle = e^{i\varphi_{2}}\left(r^{+}e^{i\varphi}\left|A^{-}\right\rangle - r^{-}\left|\varepsilon^{+}\right\rangle\right),$$
$$\left|\phi_{2}'\right\rangle = e^{i\varphi_{2}}\left(-r^{-}\left|A^{+}\right\rangle + r^{+}e^{i\varphi}\left|\varepsilon^{-}\right\rangle\right). \tag{43}$$

It will be useful in the sequel to know the matrix elements between the states $\left|\phi_{1}\right\rangle, \left|\phi_{2}\right\rangle, \left|\phi_{1}'\right\rangle, \left|\phi_{2}'\right\rangle$ for a general operator Γ, which will be later specified to be either the unity operator or an angular momentum operator of the apparatus.

$$\langle\phi_1|\Gamma|\phi_1\rangle=(r^+)^2\langle A^+|\Gamma|A^+\rangle+(r^-)^2\langle\varepsilon^-|\Gamma|\varepsilon^-\rangle+r^+r^-e^{i\varphi}\langle A^+|\Gamma|\varepsilon^-\rangle+c.c.$$

$$\langle\phi_2|\Gamma|\phi_2\rangle=(r^-)^2\langle A^-|\Gamma|A^-\rangle+(r^+)^2\langle\varepsilon^+|\Gamma|\varepsilon^+\rangle+r^+r^-e^{-i\varphi}\langle A^-|\Gamma|\varepsilon^+\rangle+c.c.$$

$$\langle\phi_1'|\Gamma|\phi_1'\rangle=(r^+)^2\langle A^-|\Gamma|A^-\rangle+(r^-)^2\langle\varepsilon^+|\Gamma|\varepsilon^+\rangle-r^+r^-e^{-i\varphi}\langle A^-|\Gamma|\varepsilon^+\rangle+c.c.$$

$$\langle\phi_2'|\Gamma|\phi_2'\rangle=(r^-)^2\langle A^+|\Gamma|A^+\rangle+(r^+)^2\langle\varepsilon^-|\Gamma|\varepsilon^-\rangle-r^+r^-e^{i\varphi}\langle A^+|\Gamma|\varepsilon^-\rangle+c.c.$$

(44)

(where c.c. refers to the previous term). It will also be useful to know that

$$\langle\phi_1|\Gamma|\phi_1\rangle+\langle\phi_2'|\Gamma|\phi_2'\rangle=\langle A^+|\Gamma|A^+\rangle+\langle\varepsilon^-|\Gamma|\varepsilon^-\rangle$$

$$\langle\phi_1'|\Gamma|\phi_1'\rangle+\langle\phi_2|\Gamma|\phi_2\rangle=\langle A^-|\Gamma|A^-\rangle+\langle\varepsilon^+|\Gamma|\varepsilon^+\rangle$$

(45)

and

$$\langle\phi_1|\Gamma|\phi_1\rangle-\langle\phi_2'|\Gamma|\phi_2'\rangle=\cos\vartheta\big(\langle A^+|\Gamma|A^+\rangle-\langle\varepsilon^-|\Gamma|\varepsilon^-\rangle\big)+\sin\vartheta e^{i\varphi}\langle A^+|\Gamma|\varepsilon^-\rangle+c.c.$$

$$\langle\phi_1'|\Gamma|\phi_1'\rangle-\langle\phi_2|\Gamma|\phi_2\rangle=\cos\vartheta\big(\langle A^-|\Gamma|A^-\rangle-\langle\varepsilon^+|\Gamma|\varepsilon^+\rangle\big)-\sin\vartheta e^{-i\varphi}\langle A^-|\Gamma|\varepsilon^+\rangle+c.c.$$

(46)

and

$$\langle\phi_1|\Gamma|\phi_2\rangle=$$
$$r^+r^-\big(e^{i\varphi}\langle A^+|\Gamma|A^-\rangle+e^{-i\varphi}\langle\varepsilon^-|\Gamma|\varepsilon^+\rangle\big)+(r^+)^2\langle A^+|\Gamma|\varepsilon^+\rangle+(r^-)^2\langle\varepsilon^-|\Gamma|A^-\rangle,$$

$$\langle\phi_2'|\Gamma|\phi_1'\rangle=$$
$$-r^+r^-\big(e^{i\varphi}\langle A^+|\Gamma|A^-\rangle+e^{-i\varphi}\langle\varepsilon^-|\Gamma|\varepsilon^+\rangle\big)+(r^-)^2\langle A^+|\Gamma|\varepsilon^+\rangle+(r^+)^2\langle\varepsilon^-|\Gamma|A^-\rangle.$$

(47)

From these last equations we can deduce that

$$\langle\phi_1|\Gamma|\phi_2\rangle+\langle\phi_2'|\Gamma|\phi_1'\rangle=\langle A^+|\Gamma|\varepsilon^+\rangle+\langle\varepsilon^-|\Gamma|A^-\rangle,$$

(48)

$$\langle\phi_1|\Gamma|\phi_2\rangle-\langle\phi_2'|\Gamma|\phi_1'\rangle=$$
$$\cos\vartheta\big(\langle A^+|\Gamma|\varepsilon^+\rangle-\langle\varepsilon^-|\Gamma|A^-\rangle\big)+\sin\vartheta\big(e^{i\varphi}\langle A^+|\Gamma|A^-\rangle+e^{-i\varphi}\langle\varepsilon^-|\Gamma|\varepsilon^+\rangle\big).$$

(49)

Let us now, as before, see if angular momentum is conserved. We already know that

$$\langle A^0\otimes u^+|\vec{J}\cdot\hat{a}|A^0\otimes u^+\rangle=\langle A^0|M(a)|A^0\rangle+\frac{\hbar}{2}.$$

(50)

If we had taken $|u^-\rangle$ instead of $|u^+\rangle$, we would have obtained

$$\langle A^0 \otimes u^- | \vec{J} \cdot \hat{a} | A^0 \otimes u^- \rangle = \langle A^0 | M(a) | A^0 \rangle - \frac{\hbar}{2}. \tag{51}$$

Here $\langle \vec{J} \cdot \hat{a} \rangle_a$, for $|u^+\rangle$, will be

$$\langle \phi_1 \otimes u_0^+ + \phi_2 \otimes u_0^- | \vec{J} \cdot \hat{a} | \phi_1 \otimes u_0^+ + \phi_2 \otimes u_0^- \rangle, \tag{52}$$

which is equal to

$$\frac{\hbar}{2}\cos\vartheta(\langle\phi_1|\phi_1\rangle - \langle\phi_2|\phi_2\rangle) + \langle\phi_1|\phi_2\rangle\frac{\hbar}{2}\sin\vartheta e^{-i\varphi} + c.c. + \langle\phi_1|M(a)|\phi_1\rangle + \langle\phi_2|M(a)|\phi_2\rangle. \tag{53}$$

The average value for $|u^-\rangle$ after measurement is

$$\begin{aligned}
&\frac{\hbar}{2}\cos\vartheta(\langle\phi_2'|\phi_2'\rangle - \langle\phi_1'|\phi_1'\rangle) + \langle\phi_2'|\phi_1'\rangle\frac{\hbar}{2}\sin\vartheta e^{-i\varphi} + c.c. \\
&+ \langle\phi_1'|M(a)|\phi_1'\rangle + \langle\phi_2'|M(a)|\phi_2'\rangle.
\end{aligned} \tag{54}$$

If angular momentum is to be conserved, we can set (53) and (54), respectively, equal to

$$\langle A^0 | M(a) | A^0 \rangle + \frac{\hbar}{2} \tag{55}$$

and

$$\langle A^0 | M(a) | A^0 \rangle - \frac{\hbar}{2} \tag{56}$$

(the average values before measurement) and add and subtract the equations thus formed, giving:

$$\begin{aligned}
2\langle A^0 | M(a) | A^0 \rangle &= \frac{\hbar}{2}\cos\vartheta(\langle\phi_1|\phi_1\rangle + \langle\phi_2'|\phi_2'\rangle - \langle\phi_2|\phi_2\rangle - \langle\phi_1'|\phi_1'\rangle) \\
&+ \frac{\hbar}{2}\sin\vartheta e^{-i\varphi}(\langle\phi_1|\phi_2\rangle + \langle\phi_2'|\phi_1'\rangle) + c.c. + \langle\phi_1|M(a)|\phi_1\rangle + \langle\phi_2|M(a)|\phi_2\rangle \\
&+ \langle\phi_1'|M(a)|\phi_1'\rangle + \langle\phi_2'|M(a)|\phi_2'\rangle
\end{aligned} \tag{57}$$

and

$$\hbar = \frac{\hbar}{2}\cos\vartheta\big(\langle\phi_1|\phi_1\rangle - \langle\phi_2'|\phi_2'\rangle + \langle\phi_1'|\phi_1'\rangle - \langle\phi_2|\phi_2\rangle\big)$$

$$+\frac{\hbar}{2}\sin\vartheta e^{-i\varphi}\big(\langle\phi_1|\phi_2\rangle + \langle\phi_2'|\phi_1'\rangle\big) + c.c. + \langle\phi_1|M(a)|\phi_1\rangle - \langle\phi_2'|M(a)|\phi_2'\rangle \quad (58)$$

$$-\langle\phi_1'|M(a)|\phi_1'\rangle + \langle\phi_2|M(a)|\phi_2\rangle .$$

Let us consider equation (57) first. The apparatus matrix elements figuring in it can all be deduced from equations (45) and (48) by taking either $\Gamma = I$ or $\Gamma = M(a)$. So we have

$$2\langle A^0|M(a)|A^0\rangle = \hbar\cos\vartheta\big(\langle\varepsilon^-|\varepsilon^-\rangle - \langle\varepsilon^+|\varepsilon^+\rangle\big) + \frac{\hbar}{2}\sin\vartheta e^{-i\varphi}\big(\langle A^+|\varepsilon^+\rangle - \langle\varepsilon^-|A^-\rangle\big) + c.c.$$

$$+\langle A^+|M(a)|A^+\rangle + \langle\varepsilon^+|M(a)|\varepsilon^+\rangle + \langle A^-|M(a)|A^-\rangle + \langle\varepsilon^-|M(a)|\varepsilon^-\rangle . \tag{59}$$

This condition is easily satisfied, even in the limit $\big(|\varepsilon^+\rangle = 0 = |\varepsilon^-\rangle\big)$ of ordinary quantum mechanics, where it becomes

$$2\langle A^0|M(a)|A^0\rangle = \langle A^+|M(a)|A^+\rangle + \langle A^-|M(a)|A^-\rangle . \tag{60}$$

This last requirement can be met, for instance, by making the rather natural assumption that the expectation value of the angular momentum over the states $|A^0\rangle, |A^+\rangle, |A^-\rangle$ is always the same.

The situation is different for equation (58). The apparatus matrix elements figuring in it can be calculated from equations (46) and (49), with either $\Gamma = 1$ or $\Gamma = M(a)$. The result, to lowest order in $|\varepsilon^\pm\rangle$, and using

$$\langle A^+|A^-\rangle = 0 = \langle\varepsilon^+|\varepsilon^-\rangle \tag{61}$$

(different apparatus states being macroscopically distinguishable and hence orthogonal) and relation (30), is

$$\hbar = \hbar\cos^2\vartheta\big(1 + \langle A^+|M_3|A^+\rangle - \langle A^-|M_3|A^-\rangle\big)$$

$$+\sin^2\vartheta\big(\langle A^+|M_+|\varepsilon^-\rangle - \langle\varepsilon^+|M_+|A^-\rangle + \langle\varepsilon^-|M_-|A^+\rangle - \langle A^-|M_-|\varepsilon^+\rangle\big)$$

$$+\sin\vartheta\cos\vartheta e^{i\varphi}\frac{\hbar}{2}\big(\langle A^+|\varepsilon^-\rangle - \langle\varepsilon^+|A^-\rangle + \langle\varepsilon^+|A^+\rangle - \langle A^-|\varepsilon^-\rangle\big) \tag{62}$$

$$+\sin\vartheta\cos\vartheta e^{i\varphi}\big(\langle A^+|M_-|A^+\rangle - \langle A^-|M_-|A^-\rangle + \langle A^+|M_3|\varepsilon^-\rangle - \langle\varepsilon^+|M_3|A^-\rangle\big)$$

$$+c.c. + \sin^2\vartheta e^{2i\varphi}\big(\langle A^+|M_-|\varepsilon^-\rangle - \langle\varepsilon^+|M_-|A^-\rangle\big) + c.c.$$

This equality can be satisfied only if different powers of $e^{i\varphi}$ on the two sides are equal. This gives

$$\hbar = \hbar\cos^2\vartheta\left(1+\langle A^+|M_3|A^+\rangle - \langle A^-|M_3|A^-\rangle\right)$$
$$+\sin^2\vartheta\left(\langle A^+|M_+|\varepsilon^-\rangle - \langle\varepsilon^+|M_+|A^-\rangle + \langle\varepsilon^-|M_-|A^+\rangle - \langle A^-|M_-|\varepsilon^+\rangle\right),\qquad(63)$$

$$0 = \frac{\hbar}{2}\left(\langle A^+|\varepsilon^-\rangle - \langle\varepsilon^+|A^-\rangle + \langle\varepsilon^+|A^+\rangle - \langle A^-|\varepsilon^-\rangle\right)$$
$$+\langle A^+|M_-|A^+\rangle - \langle A^-|M_-|A^-\rangle + \langle A^+|M_3|\varepsilon^-\rangle - \langle\varepsilon^+|M_3|A^-\rangle,\qquad(64)$$

$$0 = \langle A^+|M_-|\varepsilon^-\rangle - \langle\varepsilon^+|M_-|A^-\rangle.\qquad(65)$$

Notice that of these three equalities only the second and third ones can be satisfied in the limit $|\varepsilon^+\rangle = 0 = |\varepsilon^-\rangle$. Relation (63) cannot instead be satisfied in this limit. This implies, as first noticed by Wigner [5], that angular momentum is not conserved in the usual quantum theory of measurement. Wigner's terms $|\varepsilon^\pm\rangle$ are however such as to restore angular momentum conservation provided one has

$$1+\langle A^+|M_3|A^+\rangle - \langle A^-|M_3|A^-\rangle = 1\qquad(66)$$

$$\langle A^+|M_+|\varepsilon^-\rangle - \langle\varepsilon^+|M_+|A^-\rangle + \langle\varepsilon^-|M_-|A^+\rangle - \langle A^-|M_-|\varepsilon^+\rangle = \hbar.\qquad(67)$$

Equation (66) is easily satisfied if $\langle A^+|M_3|A^+\rangle = \langle A^-|M_3|A^-\rangle$. Equation (67) can be written

$$\mathrm{Re}\langle A^+|M_+|\varepsilon^-\rangle - \mathrm{Re}\langle\varepsilon^+|M_+|A^-\rangle = \frac{\hbar}{2}.\qquad(68)$$

Now equation (65) and its complex conjugate summed together give

$$\mathrm{Re}\langle A^+|M_-|\varepsilon^-\rangle - \mathrm{Re}\langle\varepsilon^+|M_-|A^-\rangle = 0.\qquad(69)$$

If we add and subtract (68) and (69), we obtain

$$\mathrm{Re}\langle A^+|M_1|\varepsilon^-\rangle - \mathrm{Re}\langle\varepsilon^+|M_1|A^-\rangle = \frac{\hbar}{4},\qquad(70)$$

and

$$-\mathrm{Im}\langle A^+ |M_2| \varepsilon^- \rangle + \mathrm{Im}\langle \varepsilon^+ |M_2| A^- \rangle = \frac{\hbar}{4}, \tag{71}$$

respectively, since

$$M_1 = M_+ + M_-, \tag{72}$$

and

$$iM_2 = M_+ - M_-. \tag{73}$$

Conditions (68), (69), (70), and (71) must be met for angular momentum to be conserved in measurement processes.

Equation (68) gives Planck's constant a new and completely unforeseen role in quantum mechanics: it becomes the magnitude of the error which is necessarily present in every process of spin measurement. This equation can be interpreted, if taken by itself, as follows: (33) tells us that the third component of angular momentum can be expected to be the same for the states $|A^0\rangle, |A^+\rangle, |A^-\rangle$. This is not so, however, for the spin-flip amplitudes $|\varepsilon^\pm\rangle$: $|\varepsilon^+\rangle$ can be expected to have one unit of M_3 *more* than $|A^0\rangle, |A^+\rangle, |A^-\rangle$, while $|\varepsilon^-\rangle$ should have one unit of M_3 *less*. Remembering the physical interpretation of $M_\pm$ as third component of angular momentum raising and lowering operators one sees therefore that $M_+|\varepsilon^-\rangle$ is a state with the same amount of M_3 as $|A^+\rangle$. In the same way $M_-|\varepsilon^+\rangle$ is a ket with the same amount of M_3 as $|A^-\rangle$.

5. Concluding remarks

We have shown that if the measurement apparatus is described classically and angular momentum is assumed to be conserved, the validity of quantum mechanics implies the possibility of instantaneous signalling over arbitrary distances. If, on the other hand, the apparatus is described quantum-mechanically, the total angular momentum of system and apparatus can only be conserved if it is assumed that measurement may disturb the state of the system, even in the case of eigenstates. We will show in a forthcoming paper that the Wigner-Araki-Yanase theory applied to correlated pairs of particles avoids the action at a distance present in the Bohr approach, but that it cannot prevent the violation of Bell-type inequalities.

References

[1] Bohr, N. (1934), *Atomic Theory and the Description of Nature*, Cambridge University Press, Cambridge, London; (1928) *Atti del Congresso Internazionale dei Fisici*, Como, 11-20 Settembre 1927, Zanichelli, Bologna, Vol. 2, pp. 565-588; (1928) "Das Quantenpostulat und die neuere Entwicklung der Atomistik," *Die Naturwissenschaften*, **16** pp. 245--257.

[2] Fock, V.A. quoted by E. P. Wigner, in "The subject of our discussions," *Rendiconti della Scuola Internazionale di Fisica "Enrico Fermi"* IL, 7, 1970.

[3] Jammer, M. (1974). The Philosophy of Quantum Mechanics, Wiley, New York.

[4] Cufaro-Petroni, N., Garuccio, A., Selleri, F., Vigier, J.P. (1980). "Sur la contradiction entre la théorie quantique classique (idéalisée) de la mesure et la conservation du carré du moment angulaire total dans le paradoxe d'Einstein, Podolski et Rosen," Comptes Rendus Acad. Sci. Paris, **290B** (111).

[5] Von Neumann, J. (1932). *Mathematische Grundlagen der Quantenmechanik*, Springer, Berlin.

[6] see Wigner, E.P. (1952), "Die Messung quantenmechanischer Operatoren," *Zeitschrift für Physik*, **133** pp. 101--108; Araki, H.A. and Yanase, M.M. (1960), "Measurement of quantum mechanical operators," *Physical Review*, **120** pp. 622--626; Yanase, M.M. (1961), "Optimal measuring apparatus," *Physical Review*, **123** pp. 666--668.

CHARACTERIZATION AND DEDUCTION
OF BELL-TYPE INEQUALITIES

E.G. BELTRAMETTI and C. DEL NOCE
Dipartimento di Fisica Università and INFN - Genova
Genova, Italy

and

M.J. MĄCZYŃSKI
Institute of Mathematics, Technical University of Warsaw
Warsaw, Poland

Abstract. The problem of the intrinsic characterization of classical probabilities is considered. A theorem is given which specifies when an inequality of the form $0 \leq L \leq 1$, where L is a linear combination of the observed probabilities, is a condition for classical representability. An algorithm, based on Boole's theory, that generates Bell-type inequalities for given empirical situations is illustrated.

1. Introduction

The difference between classical and quantum behaviour can be traced back to certain constraints on the statistics of the observed phenomena, that mark the boundary between the classical and the quantum case. Bell inequalities are the paradigmatic example of such constraints, when the statistics under discussion is the one arising from the so-called EPR correlation.

The problem of defining the boundary that marks the possibility, or impossibility, of representing classically the observed probabilities has been widely studied, but still has open questions. It can be already recognized in Boole's study (1862) of the "conditions of possible experience", and has received great attention after Bell's work on hidden variable interpretations of the EPR correlation. A geometric characterization of the conditions of classical representability has been provided by the remarkable polytope approach by Pitowsky (1989), suggesting that these conditions can in ge-

35

C. Garola and A. Rossi (eds.), The Foundations of Quantum Mechanics, 35-41.
© *1995 Kluwer Academic Publishers.*

neral be written as inequalities of the form $0 \leq L \leq 1$ where L is a linear combination with real coefficients of the probabilities p_i of the n events under examination and the joint probabilities $p_{ij}, p_{ijk}, \ldots, p_{1\ldots n}$. We shall call them Bell-type inequalities, noticing that the usual EPR correlation corresponds to $n = 4$. Pitowsky has also shown that the problem of making explicit these inequalities has a complexity that increases exponentially with n. Partial overcomings of this difficulty have however been obtained (Beltrametti and Mączyński, 1991, 1993).

In section 2 we shall report on the characterization of Bell-type inequalities; more explicitly on a theorem (Beltrametti and Mączyński, 1994) that provides sufficient and necessary conditions for the inequality $0 \leq L \leq 1$ to be a condition of classical representability of the probabilities $p_{ij}, p_{ijk}, \ldots$, $p_{1\ldots n}$.

In section 3 we shall briefly report on an algorithm, developed by one of the present authors (Del Noce, 1994), based on Boole's method and able to deduce complete sets of Bell-type inequalities for any given empirical situations. Some examples will be discussed.

2. A theorem on the characterization of Bell-type inequalities

Let $\mathbf{A}$ be a Boolean algebra and p a probability measure on $\mathbf{A}$. Let $N = \{1, \ldots, n\}$ and let ε be a function from the set of subsets of N into the real numbers:

$$\varepsilon : 2^N \longrightarrow \mathbf{R}$$

with the convention that $\varepsilon(\emptyset) = 0$. For each ε we define a correlation function p_ε from the cartesian product $\mathbf{A}^n$ into the real numbers:

$$p_\varepsilon(A_1, \ldots, A_n) := \sum_{S \subseteq N} \varepsilon(S) p \left(\bigcap_{i \in S} A_i \right) \tag{1}$$

for all $A_1, \ldots, A_n \in \mathbf{A}$.

Observe that the right-hand side of (1) can be interpreted as a linear combination of probabilities and correlations with coefficients -1, 0, 1 if we define the correlations by

$$p_{ij} := p(A_i \cap A_j)$$
$$p_{ijk} := p(A_i \cap A_j \cap A_k)$$
$$\vdots$$
$$p_{1\ldots n} := p(A_1 \cap A_2 \cap \ldots \cap A_n).$$

Now the following theorem can be proved:

Theorem 1 . *The following conditions are equivalent:*

(i) *for every Boolean algebra* **A** *and every probability measure* p *defined on* **A** *the inequality*

$$0 \le p_\varepsilon(A_1, \ldots, A_n) \le 1$$

holds for all $A_1, \ldots, A_n \in$ **A**.

(ii) $\forall S \subseteq N \quad 0 \le \sum_{T \subseteq S} \varepsilon(T) \le 1.$

The inequality in (i) is called a Bell-type inequality.

The idea of the proof is as follows. First we show that (ii) $\Rightarrow$ (i). For each non-empty $S \subseteq N$ we define the following element of **A**:

$$X_S := \bigcap_{i \in S} A_i \cap \bigcap_{j \in N \setminus S} A_j';$$

If $S = \emptyset$ we define $X_\emptyset = 0$. The elements $X_S, S \subseteq N$ are pairwise disjoint and each element $\bigcap_{i \in S} A_i$ can be expressed as a disjoint union of elements X_T with $S \subseteq T \subseteq N$. Hence, by the additivity of p we obtain

$$p\left(\bigcap_{i \in S} A_i\right) = \sum_{S \subseteq T \subseteq N} p(X_T) \tag{2}$$

for each $S \subseteq N$. This implies

$$p_\varepsilon(A_1, \ldots, A_n) = \sum_{T \subseteq N} \left(\sum_{S \subseteq T} \varepsilon(S)\right) p(X_T)$$

Since by assumption (ii) $0 \le \sum_{S \subseteq T} \varepsilon(S) \le 1$, we obtain again by the additivity of p:

$$0 \le p_\varepsilon(A_1 \ldots, A_n) \le \sum_{T \subseteq N} p(X_T) =$$

$$= p\left(\bigcup_{T \subseteq N} X_T\right) \le p(1) = 1$$

which mean that the inequalities in (i) holds.

To prove the implication (i) $\Rightarrow$ (ii) we prove by contradiction that non (ii) $\Rightarrow$ non (i). Hence assume that (ii) does not hold. This means that $\sum_{T \subseteq S_0} \varepsilon(T) \notin [0, 1]$ for some $S_0 \subseteq N$. We now take as **A** the Boolean algebra of all subsets of N, i. e. we take **A** $= 2^N$ and we define a probability measure on **A** by setting

$$p(S) = \begin{cases} 0 \text{ for } S \ne S_0 \\ 1 \text{ for } S = S_0 \end{cases}$$

If we define

$$A_i = \bigcup_{\{i\} \subseteq S \subseteq N} S$$

then

$$p_\varepsilon(A_1, \ldots, A_n) = \sum_{T \subseteq N} \left(\sum_{S \subseteq T} \varepsilon(S) \right) p(X_T)$$

This implies that $p_\varepsilon(A_1, \ldots, A_n) = \sum_{T \subseteq S_0} \varepsilon(T)$ and (i) cannot hold. Hence the implication (i) $\Rightarrow$ (ii) has been proved by contradiction.

The criterion provided by Theorem 1 becomes especially simple when the coefficients $\varepsilon(S)$ take only the values -1, 0, 1, as it is the case of the Bell-Clauser-Horne inequalities. In such a situation the condition (ii) reads $\forall S \subseteq N, \sum_{T \subseteq S} \varepsilon(S) = 0, 1$ and its verification goes as follows: for each $S \subseteq N$ we count the number P of positive coefficients and the number M of negative coefficients at all terms whose subscripts belong to S: if $P - M = 0$ or 1 for all $S \subseteq N$, this means that the condition (ii) of Theorem 1 holds and consequently by the theorem (i) also holds, i. e. the considered Bell-type inequality holds in classical probability theory. For example, the inequality

$$0 \leq p_1 + p_3 + p_5 - p_{13} - p_{14} - p_{15} - p_{23} - p_{35} + p_{24} + p_{135} \leq 1$$

satisfies condition (ii) and it is a Bell-type inequality valid in classical probability theory. Of course all standard Bell and Clauser-Horne inequalities can be easily verified by this method.

3. A method for the deduction of Bell-type inequalities

Let $\mathbf{A}$ be a Boolean algebra, and $A_1, A_2, \ldots, A_n \in \mathbf{A}$. Let p be a probability measure defined on $\mathbf{A}$. As in the equation (2) of section 2, we have

$$p_S := p \left(\bigcap_{i \in S} A_i \right) = \sum_{S \subseteq T \subseteq N} p(X_T) \quad \forall S \subseteq N.$$

If $\forall T \subseteq N$ we set $\lambda_T := p(X_T)$ and if we let S take the values $\{1\}, \{2\},$ $\ldots, \{n\}, \{i, j\}, \ldots,$ corresponding to those sets of indexes of probabilities and joint probabilities that interest us, we may consider the system of v equations

$$\begin{cases} \sum_{S \subseteq T \subseteq N} \lambda_T = p_S, & \forall S \text{ considered} \\ \sum_{T \subseteq N} \lambda_T = 1. \end{cases} \quad (3)$$

Let us write the above system as $Y\lambda = p$. It has more unknowns than equations, in most cases considered in common applications. If rank $Y = v$, then v variables λ may be expressed as linear combinations of the remaining $2^n - v$ variables λ and of the p's. The case rank $Y < v$ may be discussed just with a slight generalization (Del Noce, 1994). Let us subdivide the matrices Y and λ into submatrices so that $Y = (Y_1|Y_2)$ where Y_1 is a non-singular submatrix of dimensions $v \times v$ and $\lambda = \begin{pmatrix} \sigma \\ \tau \end{pmatrix}$ where σ is a column vector with v components. The system $Y_1\sigma + Y_2\tau = p$ gives $\sigma = Y_1^{-1}p - Y_1^{-1}Y_2\tau$. A theorem proved by Boole (1862) states that the conditions

$$\lambda_i \geq 0, \quad i = 1, 2, \ldots, 2^n \tag{4}$$

i. e.

$$\begin{cases} (Y_1^{-1}p - Y_1^{-1}Y_2\tau)_i \geq 0, & i = 1, 2, \ldots, v \\ \tau_j \geq 0, & j = 1, 2, \ldots, 2^n - v \end{cases} \tag{5}$$

are necessary and sufficient for the classical representability of the probabilities p_S.

Now, the problem is that the above conditions involve the p's but also some variables λ (the vector τ), and it is necessary to get rid of these variables in order to obtain conditions on the p's only. Boole (1862) states that the elimination can be performed, but his directions how to proceed are not complete, and nor the way is evident by itself. One of the present authors (Del Noce, 1994) has proposed and discussed a general method for solving the problem. Here we shall limit ourselves to showing the method applied to the physical example of the EPR correlation. We shall thus come to the Bell-Clauser-Horne inequalities.

Let us consider a particle of spin 0 that decays into a pair of spin-1/2 particles moving in opposite directions. For one of the two particles, say a, we can measure the spin component along the direction $\hat{n}_1$ or along $\hat{n}_2$. For the other particle, say b, along $\hat{m}_1$ or $\hat{m}_2$. We define the events

A_1: the spin component of a along $\hat{n}_1$ is measured to be up;
A_2: the spin component of a along $\hat{n}_2$ is measured to be up;
A_3: the spin component of b along $\hat{m}_1$ is measured to be up;
A_4: the spin component of b along $\hat{m}_2$ is measured to be up.

We regard the events A_1, A_2, A_3, A_4 as elements of a Boolean algebra A.

The system (3) may be written down explicitly as follows

$$\begin{cases}
\lambda_{\{1\}} + \lambda_{\{1,2\}} + \lambda_{\{1,3\}} + \lambda_{\{1,4\}} + \lambda_{\{1,2,3\}} + \lambda_{\{1,2,4\}} + \lambda_{\{1,3,4\}} + \lambda_{\{1,2,3,4\}} = p_1 \\
\lambda_{\{2\}} + \lambda_{\{1,2\}} + \lambda_{\{2,3\}} + \lambda_{\{2,4\}} + \lambda_{\{1,2,3\}} + \lambda_{\{1,2,4\}} + \lambda_{\{2,3,4\}} + \lambda_{\{1,2,3,4\}} = p_2 \\
\lambda_{\{3\}} + \lambda_{\{1,3\}} + \lambda_{\{2,3\}} + \lambda_{\{3,4\}} + \lambda_{\{1,2,3\}} + \lambda_{\{1,3,4\}} + \lambda_{\{2,3,4\}} + \lambda_{\{1,2,3,4\}} = p_3 \\
\lambda_{\{4\}} + \lambda_{\{1,4\}} + \lambda_{\{2,4\}} + \lambda_{\{3,4\}} + \lambda_{\{1,2,4\}} + \lambda_{\{1,3,4\}} + \lambda_{\{2,3,4\}} + \lambda_{\{1,2,3,4\}} = p_4 \\
\lambda_{\{1,3\}} + \lambda_{\{1,2,3\}} + \lambda_{\{1,3,4\}} + \lambda_{\{1,2,3,4\}} = p_{13} \\
\lambda_{\{1,4\}} + \lambda_{\{1,2,4\}} + \lambda_{\{1,3,4\}} + \lambda_{\{1,2,3,4\}} = p_{14} \\
\lambda_{\{2,3\}} + \lambda_{\{1,2,3\}} + \lambda_{\{2,3,4\}} + \lambda_{\{1,2,3,4\}} = p_{23} \\
\lambda_{\{2,4\}} + \lambda_{\{1,2,4\}} + \lambda_{\{2,3,4\}} + \lambda_{\{1,2,3,4\}} = p_{24} \\
\lambda_{\emptyset} + \lambda_{\{1\}} + \lambda_{\{2\}} + \lambda_{\{3\}} + \lambda_{\{4\}} + \lambda_{\{1,2\}} + \lambda_{\{1,3\}} + \lambda_{\{1,4\}} + \lambda_{\{2,3\}} + \\
+ \lambda_{\{2,4\}} + \lambda_{\{3,4\}} + \lambda_{\{1,2,3\}} + \lambda_{\{1,2,4\}} + \lambda_{\{1,3,4\}} + \lambda_{\{2,3,4\}} + \lambda_{\{1,2,3,4\}} = 1
\end{cases}$$

If we solve it for the 9 unknowns $\lambda_{\emptyset}$, $\lambda_{\{1\}}$, $\lambda_{\{2\}}$, $\lambda_{\{3\}}$, $\lambda_{\{4\}}$, $\lambda_{\{1,3\}}$, $\lambda_{\{1,4\}}$, $\lambda_{\{2,3\}}$, $\lambda_{\{2,4\}}$ and impose the condition (4), we get the following inequalities:

$$p_1 + p_2 + p_3 + p_4 - p_{13} - p_{14} - p_{23} - p_{24} - 1 \leq -\lambda_{\{1,2,3,4\}} + \lambda_{\{1,2\}} + \lambda_{\{3,4\}} \quad (6)$$

$$\begin{aligned}
-p_1 + p_{13} + p_{14} &\leq \lambda_{\{1,2,3,4\}} - \lambda_{\{1,2\}} + \lambda_{\{1,3,4\}} & (7) \\
-p_2 + p_{23} + p_{24} &\leq \lambda_{\{1,2,3,4\}} - \lambda_{\{1,2\}} + \lambda_{\{2,3,4\}} & (8) \\
-p_3 + p_{23} + p_{13} &\leq \lambda_{\{1,2,3,4\}} + \lambda_{\{1,2,3\}} - \lambda_{\{3,4\}} & (9) \\
-p_4 + p_{14} + p_{24} &\leq \lambda_{\{1,2,3,4\}} + \lambda_{\{1,2,4\}} - \lambda_{\{3,4\}} & (10) \\
-p_{13} &\leq -\lambda_{\{1,2,3,4\}} - \lambda_{\{1,2,3\}} - \lambda_{\{1,3,4\}} & (11) \\
-p_{14} &\leq -\lambda_{\{1,2,3,4\}} - \lambda_{\{1,2,4\}} - \lambda_{\{1,3,4\}} & (12) \\
-p_{23} &\leq -\lambda_{\{1,2,3,4\}} - \lambda_{\{1,2,3\}} - \lambda_{\{2,3,4\}} & (13) \\
-p_{24} &\leq -\lambda_{\{1,2,3,4\}} - \lambda_{\{1,2,4\}} - \lambda_{\{2,3,4\}} & (14)
\end{aligned}$$

$$0 \leq \lambda_{\{1,2\}}, \quad 0 \leq \lambda_{\{3,4\}}, \quad 0 \leq \lambda_{\{1,2,3\}}, \quad 0 \leq \lambda_{\{1,2,4\}},$$
$$0 \leq \lambda_{\{1,3,4\}}, \quad 0 \leq \lambda_{\{2,3,4\}}, \quad 0 \leq \lambda_{\{1,2,3,4\}}$$

We are now left with the problem of eliminating the variables λ from this system. Observe that single variables λ as well as linear combinations of them appear in the above inequalities. The combinations are not simple sums nor involve always the same variables. One can think of eliminating a variable λ just by writing it at one side of the inequalities in which it appears and carrying all other variables at the other side, and then comparing the inequalities just found, because if $u \leq \lambda_S$ and $\lambda_S \leq v$, then $u \leq v$. But the inequalities built up by this procedure will still contain some variables λ. A similar procedure carried out with suitable combinations of the λ's rather than with single variables λ is equally useless. Therefore the direct elimination of the variables λ is not an easy task. The solution of this problem can be done in the following way (Del Noce, 1994). Consider a choice of some inequalities among the (6)–(14), sum the right-hand sides of the chosen inequalities. If the coefficient of each λ in the sum is 0 or -1, we write that the sum of the left-hand sides of the chosen inequalities is less or equal 0, thus obtaining a valid inequality among the variables p_S; else we discard the choice as not useful. Then we repeat the procedure for all possible choices. In terms of the system (5), we consider all possible row vectors s (with v components, each one equal only to 0 or 1), and search those such that the vector $sY_1^{-1}Y_2$ has components equal only to 0 or 1.

The inequalities

$$sY_1^{-1}p \geq 0$$

built by the vectors s so found are valid inequalities for the probabilities p_S. They are all the necessary and sufficient conditions for the classical representability of these probabilities. In particular, they contain a minimal subset of necessary and sufficient inequalities. In our example all the vectors s considered are 512, and the inequalities found are 88. Among them there are the well-known inequalities (24 in number) of Bell-Clauser-Horne:

$$0 \leq p_{ij} \leq p_i, \, p_j \quad i = 1, 2; \quad j = 3, 4$$
$$p_i + p_j - p_{ij} \leq 1 \quad i = 1, 2; \quad j = 3, 4$$
$$0 \leq p_1 + p_3 - p_{13} - p_{14} - p_{23} + p_{24} \leq 1$$
$$0 \leq p_1 + p_4 - p_{13} - p_{14} - p_{24} + p_{23} \leq 1$$
$$0 \leq p_2 + p_3 - p_{13} - p_{23} - p_{24} + p_{14} \leq 1$$
$$0 \leq p_2 + p_4 - p_{14} - p_{23} - p_{24} + p_{13} \leq 1.$$

The application of this algorithm to physically meaningful cases has led to interesting results, like e. g., the characterization of probabilities in an extension of the Clauser-Horne problem (the so-called 4×4 Clauser-Horne problem) and in an experiment of decay into three spin 1/2 particles (Del Noce, 1994).

References

1. Beltrametti E.G. - Mączyński M.J. (1991). On a characterization of classical and nonclassical probabilities, *J. Math. Phys.*, **32** (5), 1280-1286.
2. Beltrametti E.G. - Mączyński M.J. (1993). On the characterization of probabilities: A generalization of Bell's inequalities, *J. Math. Phys.*, **34** (11), 4919-4929.
3. Beltrametti E.G. - Mączyński M.J. (1994). On Bell-type inequalities, *to be published in the August 1994 issue of Foundations of Physics*.
4. Boole G. (1862). On the Theory of Probabilities, *Phil. Trans. R. Soc. London*, **152** 225.
5. Del Noce C. (1994). An algorithm for the deduction of Bell-type inequalities based upon Boole's method, *submitted for publication*.
6. Pitowsky I. (1989). *Quantum Probability-Quantum Logic*, Lecture Notes in Physics Vol. 321, Springer-Verlag, Berlin.

THE SEARCH FOR A QUANTUM REALITY

SILVIO BERGIA

Dip. di Fisica Università and INFN - Bologna
Via Irnerio, 46 - 40126 - Bologna, Italy

and

VINCENZO FANO

Istituto di Filosofia - Università di Urbino
61100 - Urbino, Italy

Abstract. We argue that pursuing the aim of building satisfactory "views of quantum reality", or "quantum ontologies", is epistemologically legitimate and not subject to no-go theorems. The question as to what extent existing interpretations may be considered satisfactory from this point of view is dealt with on the basis of criteria elaborated within the post-Kantian philosophical tradition. The aim may also prove worth the effort from a purely scientific point of view, although it may imply such radical changes of the ordinary ways of thinking as to require very long times.

1. INTRODUCTION

We are aware of various recent studies which cast serious doubts on the opinion, widespread in the scientific community, that experiments have conclusively refuted local realism [1]. It is of course perfectly possible that their authors are right, and that our mourning over local realism is premature. We tend however to share the attitude that has been recently expressed by Cushing: "Although one can, in principle, attempt to undermine the empiric [*sic*] leg of the triad upon which the argument rests, each successive experiment forecloses more such loopholes and makes such a line of attack ever less plausible. So the arrow of *modus tollens* appears more reasonably directed at the assumptions of locality and/or determinism." [2] We will thus adopt here the majority's viewpoint, that is we will assume the conclusivness of the experiments.

C. Garola and A. Rossi (eds.), The Foundations of Quantum Mechanics, 43-58.
© *1995 Kluwer Academic Publishers.*

The world of quantum phenomena has always presented great difficulties for those who feel that observations do not exhaust the realm of objectivity. Violation of local realism [3] makes the difficulty even greater [4]. Of course, one could always renounce the idea of interpreting quantum mechanics (QM) in terms of some underlying reality. Although no epistemology can *enforce* on us a realistic attitude, physicists often share the view R.Penrose has recently expressed as follows:

"I have taken for granted that any 'serious' philosophical viewpoint should contain at least a good measure of realism. It always surprises me when I learn of apparently serious-minded thinkers, often physicists concerned with implications of QM, who take the strongly subjective view that there is, in actuality, *no* real world 'out there' at all! The fact that I take a realistic line wherever possible is not meant to imply that I am unaware that such subjective views are often seriously maintained - only that I am unable to make sense of them." (Penrose 1989, p.299)

Due to psychological reasons such as those expressed by Penrose, the search for a quantum reality lying beneath or accompanying the world of quantum phenomena is to some extent unavoidable. We further argue that it is legitimate from an epistemological point of view and not subject to no-go theorems.

As far as the first issue is concerned, we will limit ourselves to quoting A.Shimony, who has answered the question as to "how good is the case for instrumentalism" in these terms: "The *a priori* arguments of Berkeley and his successors against physical realism rest upon unjustifiably restrictive and arbitrary semantical, epistemological and methodological theses, as a vast body of critical literature [5] has attempted to demonstrate. If the *a priori* arguments fail, then instrumentalism must be judged *a posteriori* in competition with various versions of physical realism, preference being given to the point of view with the greatest explanatory power" (Shimony, 1989, p.32). As far as the second issue is concerned, one only needs recalling that realistic interpretations of QM, like the de Broglie-Bohm non-local interpretation, have been formulated. This is sufficient to show that the old prohibition, "Thou shalt not dare to describe the quantum world", fails as well. There remains open the question as to how satisfactory existing realistic interpretations may be considered.

We will here further argue that pursuing the aim of formulating a satisfactory realistic interpretation of QM

i) may prove worth the effort from a purely scientific point of view;

ii) may imply such radical changes of the ordinary ways of thinking as to require very long times.

These will be the main issues to be debated in this paper.

Before entering into any detail, we should, however, express in a sufficiently clear way what exactly our purpose is. This implies, as we will argue, intruding into the field of philosophy. Section 2 of the paper is devoted to this excursus, where we will argue that the purpose can only be that of building "views of quantum reality", or "quantum ontologies", essentially meaning that the world of quantum phenomena can be *described* as if it really existed independently of observations. Evidently, this implies getting involved in principle with the measurement problem. Dealing with it in any detail would lead us too far, and we will limit ourselves to a very preliminary analysis, conforted by the existence of interpretations of QM according to which the problem is less central than it used to be [6]. We argue that, if quantum ontologies are at all possible, there will be a plurality of them, among which, by definition, no experimental criterion can discriminate. We further argue that criteria for a selection can be derived from requirements indicated by philosophers like Husserl.

A very preliminary list of possible quantum ontologies will be given in the next section (Section 3), with a few comments on the way they come to terms with non-locality. A first attempt at classifying them according to general epistemological criteria leads us to argue that most of them are either incomplete or unsatisfactory.

In section 4 we argue, on epistemological and historical grounds, that the search for an ontology lying beneath or accompanying a physical theory may prove scientifically fruitful, and, on the basis of examples taken from the history of physics, that the search for a satisfactory ontology may take a very long time indeed, and that we should not expect to find one for QM just round the corner.

2. REALISM VERSUS ONTOLOGY

We have already freely spoken of "realistic" interpretations of QM. A closer look at the meaning that can be attributed to this expression must lead to the conclusion that they are interpretations which allow for a description of a world of *interphenomena* side by side with that of the world of *phenomena* (Reichenbach 1944). Formulating hypotheses about interphenomena does not necessarily mean getting involved in the philosophical issue as to the existence of a "real" world lying behind the phenomena. The issue seems rather to be whether the world of quantum (inter)phenomena may be described as if it really existed independently of observations. The emphasis is on the expression "described as if", whereby we put aside the basic realistic option, and stress the circumstance that reality enters into our considerations only insofar as it is described.

The term "quantum reality" seems to have become popular in a sense that seems to be very close to the one just outlined. It appears in the subtitle of a book by Peat (Peat 1990) and it forms the title of a fine book by N.Herbert (1985) [7]. The expression quantum ontology has recently become very popular in this context as well [8].

What has been said thus far has its counterpart in the Kantian and post-Kantian philosophical tradition. To begin with, we may recall the distinction between "appearances" and "things in themselves". As Norman Kemp Smith has shown, this juxtaposition must not be considered in ontological terms (Kemp Smith 1962, p. 204 ff.). Indeed, things in themselves are not the *causes* of the appearances; nor are they the only realm of objectivity, which is to be found in the realm of appearances as well. Thus, the problem of the Being is not anymore an *a priori* analysis of the possible structures of things, like in the Wolffian tradition, but an *a priori* analysis of the categories the intellect applies in order to interpret the appearances. As a consequence, Kant denies ontology the status of a science [9]. In order to deal with ontology circumventing the Kantian prohibition, the phenomenological school of Brentano, Meinong and Husserl drew a distinction between what they called the *Dasein* and the *Sosein*, the "being there" and the "being thus"; in other terms between "existence in principle" and "existence in some specific way." [10] The latter, however, is not to be intended as the mere collection of our experiences: although our knowledge begins with experience, when a fact presents itself to our conscience, we grasp, together with it, its essence. The *Dasein* may thus be identified with the essential structure of experience; as such, it is not the product of the categories of the intellect, but an *intrinsic* peculiarity of our experience. As a consequence, the analysis of the *Dasein* is not any more an *a priori* task, but rather an empirical one.

We suggest that the search for a quantum reality should be dealt with having in mind a concept of ontology close to the one that emerges from what has just been said. In our short discussion of Section 3 we will have recourse to instruments elaborated by the above mentioned authors whenever they will appear suited to the purpose.

The legitimacy of the purpose appears to be questioned by the celebrated distinction, drawn by Carnap, between *internal* and *external* ontological questions. For instance, the answer to the question whether "Olympus" is a real object, can only be: *within the scope of* mythology, Olympus is a mythological object; *within the scope of* geography, it is a geographical object (internal questions). But it makes no sense to ask (external question) whether Olympus is an object independent of our consciousness *outside* any linguistic framework (Carnap 1950, pp. 20-22). The indiscriminate search for a quantum reality, that is a search carried on independently of any definite linguistic framework, would be therefore meaningless from Carnap's viewpoint. We are aware that we are here violating Carnap's prohibition. However, we show, in Section 4, that this violation may be scientifically fruitful.

Before entering into any detail, we must however touch upon some further points.

In the first place, in the light of what has been said, we must be prepared to the idea that, if quantum ontologies are at all possible, there will be a plurality of them. Actually, as we shall argue, there will be nearly as many views of quantum reality as there are interpretations of QM. By definition, no experimental criterion can discriminate between them [11]. Philosophy of science is not wanting as regards the problem of choosing between more theories compatible with a given experimental picture and mathematical setup: economy of hypotheses, greater predictivity, greater logical simplicity and formal elegance are all elements to which one may appeal, on epistemological grounds, in order to choose among theories which happen to be equivalent in the sense just clarified. Our task here, however, is of selecting among *ontologies* rather than between *theories*. The above mentioned requirements, although in principle useful to establishing a hierarchy among different ontologies, will in general be superseded by other criteria. In the first place, a good quantum ontology must describe an *objective* world. However, how is one to evaluate objectiveness? One way is to associate it with the permanent, rather than with the elusive, characters of our experiences. This question would deserve a very accurate discussion. We will limit ourselves to enunciating the essential characteristics of objective experience according to Husserlian phenomenology (Husserl 1952, §§ 15-16): they are space-time coherence (unitary space-time scheme) and causal uniformity (the same circumstances produce the same results). We will later on briefly discuss how these considerations apply to the case of QM. We may also draw a distinction between weak (the general notion) and strong ontologies, the latter being characterized by the circumstance that the interphenomena are described as taking place in the space-time continuum. Strong ontologies are the more so insofar as they make reference to geometrical structures defined on the continuum and/or to the extent that the description of physical quantities is given in terms close to those used to describe sensible experience. Preference given to stronger ontologies cannot be justified on general epistemological grounds, but, possibly, on the basis of a historical survey, showing the fruitfulness of speculations aiming at obtaining a description of interphenomena in the space-time continuum. This aspect will be dealt with in Section 4. In the second place, as we have anticipated in the Introduction, the aim pursued implies confronting oneself with the measurement problem. Indeed, a quantum ontology may arise only within a framework, if it exists at all, in which measurements can be described as operations recording a preexistent reality, that is something which can be *described as* preexistent. As is well known, current views of the measurement problem do not conform to this requirement. Such is, for instance, Bohr's conception of QM, which implies an inseparable link between the systems being observed and the measuring devices. Neither does conform to it the view of quantum measurement originated with von Neumann [12] and further elaborated upon by London and Bauer (1939; exp. p. 251 of English transl. reprinted in Wheeler and Zurek 1983) and by Wigner (1961; reprinted in Wheeler and Zurek 1983; exp. pp. 172, 173, 175, 177), implying the idea that the human mind must play a specific role in the collapse of the wave function [13]. Reality seems here to be to some extent created by consciousness, and thus not to pre-exist observation. A way out seems to be suggested by the alleged circumstance that one may identify a preexisting reality as consisting of the wave function itself [14]. However, measurements do not simply record reality as described by the wave function, since they produce its collapse. Therefore, even disregarding its being only associable with the weakest conceivable ontology [15], this view of the measurement process cannot be further considered in this paper. Everett's theory [16], which claims not to require the collapse of the wave function, seems to avoid the difficulty encountered by the NLBW theory. We will however maintain (Section 3) that the ontology foreshadowed by this theory is not acceptable on general grounds. All the difficulties seem to be avoided by the Ghirardi, Rimini, Weber theory (Ghirardi, Rimini, Weber, 1986; hereafter to be referred to as GRW theory), which we consider a self-consistent quantum mechanical description of macroscopic bodies and of the measurement process. The GRW theory, however, is a modified form of QM. We will therefore not discuss it here since we are only interested in quantum ontologies referring to QM in its codified form. In our short discussion of existing quantum ontologies in the next section, we will briefly comment on their way to deal with the measurement problem.

Dealing with the problem in general would lead us too far, and, as we have anticipated, we will limit ourselves to discussing in some detail possible quantum ontologies with respect to the sole problem on "non-locality." [17]

3. HOW MANY QUANTUM REALITIES?

As we have anticipated, we will only dwell on views of quantum reality formalized in terms either coinciding with, or included in, QM.

Within this framework, Herbert (1985) recognizes 8 distinct views of quantum reality: 1) The Copenhagen interpretation, Part 1 (there is no deep reality); 2) The Copenhagen interpretation, Part 2 (reality is created by observation); 3) Reality is an undivided wholeness (the de Broglie-Bohm causal interpretation); 4) The many worlds interpretation (in De Witt's version: reality consists of a steadily increasing number of parallel universes); 5) Quantum logic (the world obeys a non-human kind of reasoning); 6) Neorealism (the world is made of ordinary objects); 7) Consciousness creates reality; 8) The duplex world of Werner Heisenberg (the world is twofold, consisting of potentialities and actualities).

We find it expedient to make use of this classification, but with some suppression, addition and modification. To begin with, we can immediately dismiss 1), which, more than an ontology, sounds like the refusal of any possible ontology. 2) and 7) can be, and to to some extent have been, dealt with together, namely they are only compatible with the degenerate wave function ontology that we have already discarded. We have dealt to some extent with 4) in our preliminary discussion of the measurement problem. It should be added that De Witt's version of the MWI has ontological implications. We wish here to point out that a MWI ontology, beside being based on an *ad hoc* hypothesis, appears to directly violate the two basic Husserlian requirements of space-time coherence and causal uniformity. For this reason we will not consider it here any further.

Let us briefly comment on some of the remaining views with specific reference to the "non-locality" problem. According to Herbert, "Heisenberg's description [8] is no full-fledged model of reality, but just a one man's attempt to convey the flavour of the deep reality symbolized by a ψ wave" (Herbert 1985, p. 194). In particular, it seems to us that it conveys in fact the "flavour" of a non-separate ψ; however, in no way does it provide an "explanation" of non locality. In fact, it does not even provide a description of it. It introduces a category which might characterize a full-fledged ontology still to come, but it does not in itself provide one. It may be added that Heisenberg's conception does not respect causality. Also in the Aristotelic system the actualization of potentialities requires the intervention of some actual entity: what will this entity be? Once again, the observer's consciousness? Furthermore, what is the empirical basis of such an ontology?

Let us now turn to 5). Most of the work on quantum logic has not been concerned with the nature of reality (Herbert 1985, p. 181), so that no ontology seems to be involved; or, as an ontology, quantum logic is either incomplete (Herbert 1985, p. 243) or not typical. [18] In particular, it seems to have very little to say about non-locality (Herbert 1985, p. 243) (and vice versa) [19].

The basic statement of 6), "the world is made of ordinary objects", needs some specification. The way it is intended by Herbert is explained by the following passage: "Suppose reality consists of ordinary objects which possess their attributes innately. Bell's theorem requires for such a world that its objects be connected by non-local influences...Without faster-than-light connections, an ordinary object model of reality simply cannot explain the facts. Suppose reality consists of contextual entities which do not possess attributes of their own but acquire them in the act of measurement...Bell's theorem requires for such entities that the context which determines their attributes must include regions beyond light-speed range of the actual measurement site. In other words, only contextual realities which are non-local can explain the facts." (Herbert 1985, p. 51) As we have anticipated, in agreement with D.Howard (1989), we tend to single out, as the basic assumptions at stake in the recent experiments, those of separability and locality; we also agree with Howard on giving preference to locality as the requirement that we would prefer to keep. We must therefore exclude view 5) as a possible ontology.

Let us now briefly consider the only view of Herbert's list that has been left out: the one that is considered by many people as the most viable, perhaps the only viable one, view of reality. It is the one based on Bohm's version of the pilot wave idea. Schrödinger's equation is obtained adding the quantum potential, a term in $\hbar^2$, to the classical Hamilton-Jacobi equation. The underlying ontology is definitely realistic [20], although non mechanistic, if "the essence of mechanism is to say that basic reality consists of the parts of a system which are in a preassigned interaction." (Bohm and Hiley 1993, p. 58), but the proposed view of the quantum world is in any case revolutionary, its essential element being non-separability, accounted for by the quantum potential .

Before briefly commenting on this interpretation, we want to add two more items to Herbert's list, namely two views of quantum reality which emerge in connection with specific versions of QM, namely those respectively due to Nelson (1966) and Roncadelli (1991, 1992; for a detailed account, see also Roncadelli and Defendi 1992). Nelson's formulation is in terms of a stochastic differential equation suggesting that particles undergo a kind of universal Brownian motion. Schrödinger's equation emerges as a consequence. Stochasticity and classical probabilities provide elements for a "realistic" interpretation. The description of the "Brownian motion", however, is in terms of the Einstein-Smoluchowski approximation, a feature which defies attempts at a really dynamical interpretation; and, in particular, the osmotic part of the drift term has no natural interpretation in term of classical physics, and is in fact of a quantum nature in so far as it contains $\hbar$. From our point of view, Nelson's formulation seems also to have little to say with respect to "non locality".

Roncadelli's formulation suggests a view of quantum world, a quantum ontology, close to that implied by Nelson's formulation, namely that the quantum world is the effect of a universal noise. However, it does not coincide with Nelson's in one crucial aspect: the quantum behaviour arises here as an additive correction (the noise) to the classical Hamilton-Jacobi formulation of mechanics, specifically as an additive correction to the equation determining the trajectories. Thus, on the one hand, the classical limit is built in and trivially achievable; on the other hand, the quantum correction is a real background noise; on the contrary, the quantum correction in Nelson's formulation implies, beside the noise, the osmotic contribution to the drift term, which must be considered as a reaction of the quantum system to the noise (Roncadelli 1993). Roncadelli's formulation allows an almost immediate comparison with view 3) (the de Broglie-Bohm causal interpretation) as well. The two approaches are actually very close from a purely formal point of view, namely they are both based on a Hamilton-Jacobi formulation of classical dynamics, the quantum correction being obtained in Bohm's version in terms of a correction to the Hamilton-Jacobi equation, the equation determining the trajectories remaining purely classical, whereas in Roncadelli's version the surfaces on which Hamilton's principal function is constant are completely determined by classical mechanics, and the quantum correction manifests itself only by introducing an element of randomness in the trajectories.

As far as non locality, or, better said, non separability, is concerned, Bohm's version seems preferable (see, for instance, Bohm and Hiley 1988), in so far as Roncadelli's formulation does not *directly* bear on the subject. However, the latter has the advantage of clearly exhibiting the purely quantum correction to classical physics. Moreover, the quantum potential is *ad hoc*. The universal noise may seem *ad hoc* as well, but one should not forget that quantum physics is characterized by an element of indeterminacy that is here clearly isolated and exhibited. It is in fact often suggested that the indeterminism of QM arises only in the case of systems subject to observation, the evolution being otherwise deterministic. In this approach, indeterminism is inherent. The debate on determinism is particularly relevant in the framework of evolutionistic cosmology, where the deterministic views of Laplace and Einstein seem untenable. A universal quantum noise seems relevant in this context.

Roncadelli's formulation suggests the possibility of interpreting in somewhat physical terms the particles quantum paths, and thus seems to add a further realistic feature to the causal interpretation. From the foregoing discussion, it could then be concluded that, among the existing formulations, it is the one that gives expression to most aspects of quantum reality.

However, *neither Roncadelli's universal noise nor his basic stochastic equation allow a mechanistic interpretation*. Firstly, as in Nelson's formulation, the nature of the quantum fluctuations is unknown: one simply must assume the existence of a universal quantum noise. On the other hand, *the universal noise has an amplitude, and not a probability, distribution,* with a significant factor of i in the exponent. It seems very unlikely that a mechanism whatsoever producing such an effect could ever be conceived. Due to these reasons, even this formulation can hardly be said to meet Husserl's first criterion requiring a coherent space-time description. Therefore, in conclusion, even though Roncadelli's formulation goes some way along the direction of providing a "quantum ontology", in the sense alluded to in the Introduction, it does not lead to a strong ontology. If we keep feeling that a satisfactory ontology should make reference to geometrical structures defined on the continuum and/or to the extent that the description of physical quantities is given in terms close to those used to describe sensible experience, we must conclude that such an ontological vision does not appear to have been reached as yet.

4. CAN THE SEARCH FOR A SATISFACTORY ONTOLOGY PROVE FRUITFUL FROM A PURELY SCIENTIFIC POINT OF VIEW?

This question can be tackled both from a historical and an epistemological point of view. We will limit ourselves to deal with the problem from the first point of view, mentioning a few cases in which ontological "explanations" have led to important scientific progress. This suggests that it may be fruitful to escape the Carnapian prohibition to deal with external questions (in this way the historical point of view acquires an epistemological value as well).

There is a certain similarity between the situation concerning quantum physics after the Aspect experiments and classical electromagnetism after the ether experiments. Think, for instance, of the latter as looked at in 1905. One then had what we would like to call a *codified formalism* or *a formalized theory* (Maxwell-Lorentz's electrodynamics), the ether (first and second order) experiments, and various possible interpretations of the experiments, which we will refer to as *theories*, or simply *interpretations*, in agreement with the codified formalism. Among them, various ether theories, such as Stokes's theory, or the Lorentz-Poincaré refined version of Fresnel's theory, and, of course, Einstein's relativity. To each such interpretation there corresponded different views of the reality of nature (if any).

In the first two *views of reality*, there was an all-pervading ether, dragged by celestial bodies in the first instance; the ether concept was instead disposed of in the relativistic view of reality. The speed of light had a definite value only in the ether in the Lorentz-Poincaré view of reality, and the null result of the Michelson-Morley experiment was explained in terms of a real contraction of bodies in motion through the ether. It was instead isotropic by definition in every inertial frame in Einstein's view. After 1905, one had a relativity theory of the phenomena involved, made up of the codified formalism and the relativistic view of nature. This view implied the amazing reality of the relativity of simultaneity, of time dilation, etc. As is well known, it is impossible to infer uniquely the theory of relativity from the set of experiments on the effects in v/c. Thus alternative views of reality were admitted, although they generally faded away with time. This situation compares with that concerning quantum physics after the Aspect experiments: in much the same way as one had there different possible views of reality in connection with ether drift experiments, one has now different views of reality in connection with quantum facts, indeed as many as there are interpretations of QM [21].

We would like to suggest that, between 1905 and 1908, Einstein's relativity was a theory with little or no ontological interpretation (we think that the existence of a limit velocity is a strong ontological element, but we doubt that it was felt this way immediately after the appearance of Einstein's paper).

The Lorentz-Poincaré interpretation, which went along with an acceptable ontology, appeared, on the other hand, less acceptable on epistemological grounds. It seems to us that Minkowski's four-dimensional formulation of 1908 provided a strong ontological background (also in the technical sense specified in section 2) to relativity theory, in the sense that it made space-time a pseudo-Euclidean manifold with metric properties similar to those of ordinary three-dimensional Euclidean space. We do not claim that Minkowski was purposely looking for a new ontology; we simply would like to suggest that this was indeed the result. And, with the new ontology, great progress came with the transcription of electromagnetism and (relativistic) mechanics in the four-dimensional version. Let us, for further reference, point out that the new ontology conveyed by the four-dimensional version was bound to be considered as revolutionary in those days, and it sounds to most physicists perfectly accessible these days.

Before going to another less trivial example, a little digression on "crucial experiments" may reveal itself instructive. It has often been observed that crucial experiments are often such if looked at in retrospect [22]. It has however been pointed out that there have been experiments which were considered crucial when they were performed while they would not be considered as such anymore (Ben Dov 1993). Such is the case with the Foucault experiment comparing the speed of light in air and in a refractive medium. After Young's and Fresnel's experiments, the wave theory of light was prevailing. However, to the eyes of most physicists, the discriminating point was the different behaviour expected, according to the corpuscular and the wave theories, in refractive media, namely, according to the former conception the speed of light had to be larger in media with higher refractive indices.

Thus, Foucault experiment was considered decisive. But, of course, it would not be considered discriminating today, since the wave and corpuscular conceptions, which peacefully coexist, agree on predictions about the speed of light in refractive media. What has changed in the meanwhile are the basic ideas about what should be intended as a corpuscular theory of light, or, in other words, about the ontology of light: photons are not Newtonian corpuscles. We ignore if this will ever prove an instructive analogy: it only suggests, admittedly in a very vague way, that today's "crucial" experiments discriminating between QM and alternative theories may turn out to take on a different meaning in a future in which the ontology undergoing the concepts involved had deeply changed.

Le us now come to our last example: Newtonian gravitation. Borelli and Hooke had already made the essential step consisting in the assumption that an attractive force was necessary to keep a planet in its orbit. Hooke also speculated on the possibility that the force could go as the inverse of the squared distance, bu it was Newton who succeeded in proving that this law was necessary and sufficient for the elliptic orbits. As is universally known, Newton did not consider his force as a "cause" of the planets motions, but only as a "mathematical" force [23]. To this point, we might call his attitude "pragmatistic". But this does not mean that he did not set himself the problem of a "cause", or "physical agent" of his mathematical force, nor that he considered this question as irrelevant. However, he could not go any further than saying that he did not believe in an action at a distance, since he was not able to replace it with something else. In this way, "Newton had, to all appearances, made a future physics impossible, a physics acceptable to mechanical philosophers at least." (Mc Mullin 1989, p. 293). Mc Mullin draws in fact a comparison between this situation and that concerning quantum physics: "Though the escape routes were not as tightly sealed as they appear to be in the case of the Bell theorem and the associated experimental results, the challenge to assumptions about explanation must have seemed no less then than it once again does now." (Mc Mullin 1989, p. 293) And in some way, experiments on "non locality" [24], and the current views of them, "have made a future physics acceptable to mechanical philosophers impossible". General relativity has shown that action at a distance is unneccesary to explain the phenomena of gravitation. However, it is not the product of a search for a new ontology: rather, a new ontology of the phenomena and interphenomena of gravitation emerged together with it.

How long are we likely to wait for a new satisfactory quantum ontology? This question makes perhaps less sense than the pretension to learn from history. Let us nonetheless take the single circumstance that an "explanation" of Newton mathematical force has come after more than two centuries as a vague indication that it could take a long time indeed. It all depends, of course, on a very subjective feature, namely when an ontology is to be considered satisfactory. We have listed some of the reasons which make us propend for the idea that existing ontologies are not. Our view is authoritatively conforted by an analogous opinion expressed by D. Howard (1989).

ACKNOWLEDGMENTS

Some of the ideas presented in this paper have been debated within a study group with the participation of F. Bonsignori, F. Cannata, A. Desalvo, N. Guicciardini, V. Monzoni, and R. Rosa; we wish also to thank M. Roncadelli for a useful discussion.

Notes

[1] For example, Marshall, Santos, Selleri 1983, Marshall 1983, 1984, Marshall and Santos 1985, Ferrero and Santos 1985, Santos 1992, Garuccio 1994; we are here freely using the current terminology; for a thorough discussion of what accepting the conclusiveness of the Bohm-EPR class of experiments would actually imply, we refer the reader to the essays collected in Cushing and McMullin 1989.

[2] Cushing 1989, p.8: "The logical skeleton of the argument is that the assumptions of locality and determinism, plus the actual experimentally observed distributions of the real world, have produced the contradiction...". In agreement with D.Howard (1989), we tend rather to single out, as basic assumptions, those of separability and locality; we will come back to this point in the following.

[3] Or whatever is implied by the experimental results; see footnotes [1] and [2].

[4] "These challenges were, to some degree at least, implicit in the very first formulations of the quantum ideas in the early days of the century. But it was only with John Bell's formulation of his now celebrated theorem in 1964 that the full measure of the challenge came to be appreciated" (*Preface* of Cushing and McMullin 1989, p.XI).

[5] Referred to in Shimony (1989), p. 32, n..

[6] Bohr's view of the measurement process makes the description of a pysical (quantum) system independently of observations impossible; there are however views of quantum reality, like the de Broglie-Bohm causal interpretation, in which the measurement process can at least "be understood ontologically" (Bohm and Hiley 1993, p. 97 ff.).

[7] Herbert distinguishes between *quantum reality*, *quantum theory* and *quantum facts*. "Quantum reality doesn't show up directly in the quantum facts: it comes indirectly out of the quantum theory, which perfectly mirrors these facts." Hence, "we see quantum reality through a glass darkly..." (Herbert 1985, p. 57).

[8] See some of the essays collected in Cushing and McMullin 1989, in particular Howard, Stapp, and Teller; the expression "experimental metaphysics", used by Shimony (*ibidem*) does not seem to have a substantially different meaning. See also, of course, Bohm and Hiley 1993, and references to previous work by the same authors therein.

[9] I.Kant, *Kritik der reinen Vernunft*, B 303.

[10] The classical formulation of this distinction is presented in Mally 1904. See also Lambert 1983, p.126.

[11] In this sense, the search for a quantum ontology would not be considered a primary scientific issue.

[12] von Neumann 1932. It should be noted that von Neumann contented himself with showing that the frontier between the system observed and the measuring apparatus could be arbitrarily shifted, and that nowhere in his book is to be found an allusion to the collapse of the wave function as produced by the observer's conscience taking notice of the result of the measurement.

[13] Hereafter referred to, for simplicity, as NLBW theory. We reject the impulse to dismiss this view altogether, arising from considering, for instance, how measuring apparatuses record an event at one of the LEP experiments being performed at CERN with no intervention of human observers; or should one perhaps think that mental waves can make determinate what is yet undetermined in a magnetic tape or the like?

[14] This attitude is widely shared: see, for instance, Penrose 1987.

[15] Which we propose to name *degenerate ontology*.

[16] Everett (1957); De Witt (1973). As stressed by Bohm and Hiley, the formulations of the theory given in the two papers mentioned differ significantly [Bohm and Hiley (1993, p. 296)]; we are here explicitly referring to De Witt's version, the many-worlds interpretation, or MWI for short, since Everett's version, insofar as "it is not a theory of many universes, but a theory of many viewpoints about one universe" (Bohm and Hiley 1993, p. 303), seems to be subject, from our point of view, to a criticism similar to that outlined for the NLBW theory.

[17] We would like to observe the *independence* and the somewhat *complementary nature* of the "non-locality" and measurement problems: independence is witnessed by the circumstance that the measurement problem arises in connection with superposed but not necessarily entangled states; complementarity by the circumstance that experiment is the only way to disentangle the entangled states exhibiting "non-locality".

[18] To the extent that it is linked to a strictly corpuscular interpretation (Tarozzi 1988; we refer to this paper for an extended bibliography on the subject.)

[19] See, however, Accardi 1988.

[20] The theory is formulated, rather than in terms of "observables", in terms of what Bell (1987) has called "beables", "having a reality that is independent of being observed or known in any other way" (Bohm and Hiley 1993, p. 41).

[21] A comparison with the ether theories-relativity theory conundrum can also provide some indications which may prove useful for the purpose of selecting just one view of the quantum world. The first one is that apparently equivalent views may turn out to differ in a wider experimental context than the one at hand (this happened with Stokes's ether theory). Secondly, we may expect alternative views to survive for a long time. The Lorentz-Poincaré view has been recurrently, and even very recently, reproposed, although it can be maintained that it makes different predictions with respect to relativity theory in some istances. (See, for instance, Selleri, 1993.)

[22] See, in particular, Lakatos 1968.

[23] Mc Mullin 1989, p. 290 segg..

[24] The comparison between action at a distance and non locality is suggestive, and perhaps not only that.

References

1. Accardi, L. (1988). Foundations of Quantum Mechanics: a Quantum Probabilistic Approach, in *The Nature of Quantum Paradoxes*, edited by G.Tarozzi and A. van der Merwe, Kluwer, Dordrecht/Boston/London.

2. Bell, J.S. (1987). *Speakable and Unspeakable in Quantum Mechanics*, Cambridge University Press, Cambridge.
3. Ben-Dov, Yoav (1993). Local realism and crucial experiments, in *Frontiers of Fundamental Physics*, Proceedings of the Olympia Conference, September 1993, to be edited by M. Barone and F. Selleri.
4. Bohm, D., Hiley, B.J. (1988). Nonlocality and the Einstein-Podolsky-Rosen Experiment as Understood through the Quantum-Potential Approach, in *Quantum Mechanics versus Local Realism. The Einstein-Podolsky-Rosen Paradox*, edited by F. Selleri, Plenum Press, New York and London.
5. Bohm, D., Hiley, B.J. (1993). *The undivided universe - An ontological interpretation of quantum mechanics*, Routledge, London and New York.
6. Carnap, R. (1950). Empiricism, Semantics and Ontology, *Revue Internationale de philosophie*, **4**, pp. 20-40.
7. Cushing, J.T., McMullin, E. (1989). *Philosophical Consequences of Quantum Theory. Reflections on Bell's Theorem*, University of Notre Dame Press, Notre Dame, Indiana.
8. Cushing, J.T. (1989). *A background essay*, in Cushing and McMullin 1989, pp. 1-24.
9. De Witt, B.S. (1973). Quantum Mechanics and Reality, in *The Many-Worlds Interpretations of Quantum Mechanics*, edited by B.S. De Witt and N. Graham, Princeton University Press, Princeton, New Jersey, pp. 155-165.
10. Everett III, H. (1957). 'Relative state' formulation of quantum mechanics, *Reviews of Modern Physics* **29**, 454-62.
11. Ferrero, M., Santos, E. (1985). *Phys. Lett.* **108A**, 373.
12. Ferrero, M., Van der Merwe, eds. (1994). *Fundamental Problems in Quantum Physics*, Proceedings of the Oviedo Conference, September 1993, Kluwer, Dordrecht, 1994, in press.
13. Garuccio, A. (1994). "On the validity of Clauser and Horne factorizability", in Ferrero and van der Merwe 1994.
14. Ghirardi, G.C., Rimini, A., Weber, T. (1986). *Phys. Rev.*, **D34**, 470.
15. Herbert, N. (1985). *Quantum Reality*, Rider, London.
16. Howard, D. (1989). "Holism, separability, and the metaphysical implications of the Bell experiments", in Cushing and McMullin 1989, pp. 224-253.
17. Husserl, E. (1952). *Phänomenologische Untersuchungen zur Konstitution*, Husserliana IV, Martinus Nijhoff, The Hague.
18. Jarrett, J.P. (1989). "Bell's Theorem: A Guide to the Implications", in Cushing and McMullin 1989, pp. 60-79.
19. Kant, I. *Kritik der reinen Vernunft*, **B303**.
20. Kemp Smith, N. (1962). *A Commentary of Kant's "Critique of Pure Reason"*, Humanities Press, Cambridge.
21. Lakatos, I. (1968). "Criticism and the Methodology of Scientific Research Programmes", in *Proceedings of the Aristotelian Society*, **69** pp. 149-86.
22. Lambert, K. (1983). *Meinong and the Principle of Independence*, Cambridge University Press, Cambridge.
23. London, F., and Bauer, E. (1939). "La théorie de l'observation en mécanique quantique", No. 775 of *Actualités scientifiques et industrielles: Exposés de physique générale*, P. Langevin ed., English translation in Wheeler and Zurek 1983, pp. 217-259.
24. Mally, E. (1904). "Untersuchungen zur Gegenstandtheorie des Messens", in A. Meinong (ed.), *Untersuchungen zur Gegenstandtheorie und Psychologie*, Graz, pp. 121-262.
25. Marshall, T.W., Santos, E., Selleri, F. (1983). *Phys. Lett.*, **98A** (5).
26. Marshall, T.W. (1983). *Phys. Lett.*, **99A** (163); (1984) *Phys. Lett.*, **100A** (225).
27. Marshall, T.W., Santos, E. (1985). *Phys. Lett.*, **107A** (164).
28. McMullin, E. (1989). "The explanation of distant action: Historical notes", in Cushing and McMullin 1989, pp. 272-302.
29. Nelson, E. (1966). *Phys. Rev.*, **150** (1079).
30. Peat, F.D. (1990). *Einstein's Moon - Bell's Theorem and the Curious Quest for*

Quantum Reality, Contemporary Books, Chicago.

31. Penrose, R. (1987). "Newton, quantum theory and reality", in *Three hundred years of gravitation*, S.W.Hawking, W.Israel eds., Cambridge University Press, Cambridge.

32. Penrose, R. (1989). *The Emperor's New Mind*, Oxford University Press, New York/Oxford.

33. Reichenbach, H. (1944). *Philosophic Foundation of Quantum Mechanics*, Berkeley - Los Angeles.

34. Roncadelli, M. (1991). "Langevin Formulation of Quantum Dynamics", *Europhysics Letters* **16** pp. 609-615.

35. Roncadelli, M. (1992). "Langevin approach to quantum mechanics", preprint FN/T-92/43.

36. Roncadelli, M., Defendi, A. (1992). *I cammini di Feynman*, Quaderni di Fisica Teorica, Università degli Studi di Pavia, Dipartimento di Fisica Nucleare e Teorica.

37. Roncadelli, M. (1993). "Approccio a cammini aleatori alla meccanica quantistica", this conference.

38. Santos, E. (1992). "Does quantum mechanics predict violations of the Bell inequalities?", in *Bell's Theorem and the Foundations of Modern Physics*, edited by A. van der Merwe, F. Selleri and G. Tarozzi, World Scientific, Singapore.

39. Selleri, F. (1993). "On the Meaning of Special Relativity. If a Fundamental Frame Exists", preprint.

40. Shimony, A. (1989). "Search for a worldview which can accomodate our knowledge of mycrophysics", in Cushing and Mc Mullin 1989, pp. 25-36.

41. Stapp, H.P. (1989). "Quantum nonlocality and the description of nature", in Cushing and McMullin 1989, pp. 154-174.

42. Tarozzi, G. (1988). "The Italian Debate on Quantum Paradoxes", in *The Nature of Quantum Paradoxes*, edited by G.Tarozzi and A. van der Merwe, Kluwer, Dordrecht/Boston/London.

43. Teller, P. (1989). "Relativity, Relational Holism, and the Bell inequalities", in Cushing and McMullin 1989, pp. 208-223.

44. Von Neumann, J. (1932). *Mathematische Grundlagen der Quantenmechanik*, Springer.

45. Wheeler, J.A. and Zurek, W.H., eds. (1983). *Quantum Theory and Measurement*, Princeton University Press, Princeton.

46. Wigner, E. (1961). "Remarks on the mind-body question", in *The Scientist Speculates*, I.J. Good, ed., pp. 284-302, Heinemann, London; Basis Books, New York. Reprinted in Wheeler and Zurek 1983, pp. 168-181.

MACROREALISM, NON-INVASIVITY

AND QUANTUM MECHANICS

A quantitative approach

T. CALARCO and R. ONOFRIO

Dip. di Fisica "G. Galilei", Università di Padova,

Via Marzolo 8 - 35131 - Padova, Italy

Abstract. We discuss a quantum measurement model which allows to test in a quantitative framework the concept of non-invasivity introduced by Leggett and Garg. A new technique which allows to speed-up the computations is described and applied to a bistable potential schematizing consecutive measurements of magnetic flux in radio-frequency superconducting quantum interferometer devices.

1. Introduction

The debate on realism, *i.e.* the way everybody thinks to the macroscopic world, and its apparent contrast with quantum mechanics is well known, being also at the basis of famous *Gedankenexperimente* as the Schrödinger's cat. In recent years, due to the impressive technological advances, some of them are becoming feasible or at least their practical feasibility starts to be discussed by the experimentalists (Braginsky *et al.*, 1992). This is the case of the experiment proposed by Leggett and Garg (1985), consisting in the observation of violations to temporal Bell inequalities in quantum limited devices if quantum mechanics holds (Tesche, 1990; Paz *et al.*, 1993). The idea is to observe the evolution of a superconducting ring having two junctions through which Josephson effect occurs. In an rf-Superconducting Quantum Interferometer Device (SQuID) a quantum state is in general a superposition of two macroscopic currents corresponding to the clockwise and counterclockwise senses. Correlation probabilities for the execution of more than two measurements can be obtained either in quantum mechanics or in a realistic theory in which the flux has a definite value regardless of its observation. Stimulated by this possibility Leggett and Garg deduced, in complete analogy to the well known Bell inequalities for spin correlations (Bell, 1987), inequalities which allow to test the validity of quantum me-

C. Garola and A. Rossi (eds.), The Foundations of Quantum Mechanics, 59-70.
© *1995 Kluwer Academic Publishers.*

chanics versus any local realistic theory. In addition to the usual test which
is already obtained through spatial Bell inequalities, this proposal allows to
study a macroscopic superconducting state, giving also hints on the validity
of quantum measurement theory at the macroscopic level. Unfortunately,
there is also the caveat of the repeated monitoring of a single degree
of freedom, and this requires to introduce the concept of non-invasive
measurement, *i.e.* one which allows to neglect the effect of the measurement
on the subsequent evolution. However, as already observed (Ballentine,
1987; Peres, 1988), the Heisenberg original interpretation of quantum mechanics forbids to imagine a quantum measurement as one which does not
influence the subsequent dynamics. A quantitative model capable to clarify
this point is therefore highly demanded. This can be obtained on the basis
of the path-integral approach plus numerical simulations of an effective
Schrödinger equation taking into account the effect of the measurement
process on the observed system. Here we report on the strategy we have
chosen to deal with this delicate topic and in particular two technical
aspects of it. Firstly, in section 2 we deal with the possibility to describe
with numerical simulation the dynamics of a bistable potential, in our
particular case a square well with a central delta-like potential which is
analitically solvable and allows to test the efficiency of the numerical simulation of a tunnelling process. Secondly, the inclusion of the effect of repeated
impulsive measurements through a filter centered around the measurement
result is described in section 3. The merging of the two aspects occurs in
the conclusions, where a merit factor indicating the degree of invasivity of
repeated quantum measurements is shown to depend upon the time interval
between two consecutive impulsive measurements. When this interval is
equal to the tunnelling period the maximal degree of noninvasivity compatible with the Heisenberg uncertainty principle is obtained.

2. Tunnelling in a bistable potential

A first approximation for a bistable potential describing the dynamics of
the flux in an rf-SQuID is obtainable by adding a delta-like singularity in
the central point of a square-well section. The overall potential is therefore
written as

$$V(x) = \left\{ \begin{array}{ll} +\infty & |x| \geq L \\ V_0\delta(x) & |x| < L \end{array} \right. \tag{1}$$

and originates two regions, one on the right and one on the left of the origin.
Magnetic flux can oscillate between the two regions via tunnelling.

2.1. ANALYTICAL CALCULATION

The interval $[-L, L]$ is divided in three parts: the two halves of the well and a small region centered in the origin having radius $\epsilon \ll L$. The wave function can be expressed as a linear combination of plane waves in the two finite regions:

$$u(x) = \begin{cases} \text{I} & [-L, -\epsilon) & : & Ae^{ikx} + Be^{-ikx} \\ \text{III} & (+\epsilon, +L] & : & Ce^{ikx} + De^{-ikx} \end{cases} \tag{2}$$

By integrating in the region II we get

$$\int_{-\epsilon}^{\epsilon} \left[-\frac{\hbar^2}{2m} \partial_x^2 + V_0 \delta(x) \right] u(x) dx = \int_{-\epsilon}^{\epsilon} Eu(x) dx \equiv 0,$$

which implies

$$V_0 u(0) = \frac{\hbar^2}{2m} \partial_x u \Big|_{-\epsilon}^{\epsilon}. \tag{3}$$

Thus, by putting $\hbar = 1, m = \frac{1}{2}$, we get

$$\partial_x u = \begin{cases} \text{I}: & ikAe^{ikx} - ikBe^{-ikx} & \Longrightarrow & \lim_{\epsilon \to 0^+} \partial_x u(x)\big|_{x=-\epsilon} = ikA - ikB \\ \text{III}: & ikCe^{ikx} - ikDe^{-ikx} & \Longrightarrow & \lim_{\epsilon \to 0^+} \partial_x u(x)\big|_{x=+\epsilon} = ikC - ikD \end{cases}$$

which allows to write (3) as

$$V_0(C + D) = ik(C - D) - ik(A - B). \tag{4}$$

By including also the boundary conditions and the continuity condition in the origin we get, for the momentum eigenvalues, the equation

$$V_0 \sin^2(kL) = -k \sin(2kL), \tag{5}$$

whose solutions are:

1. the values for which both sides of (5) are zero,

$$k_m = \frac{m\pi}{L}, \qquad m = 1, 2, 3, \ldots; \tag{6}$$

2. the values determined by the condition

$$2k \cot(kL) = -V_0, \tag{7}$$

which comes from (5) for $k \neq \frac{m\pi}{L}$.

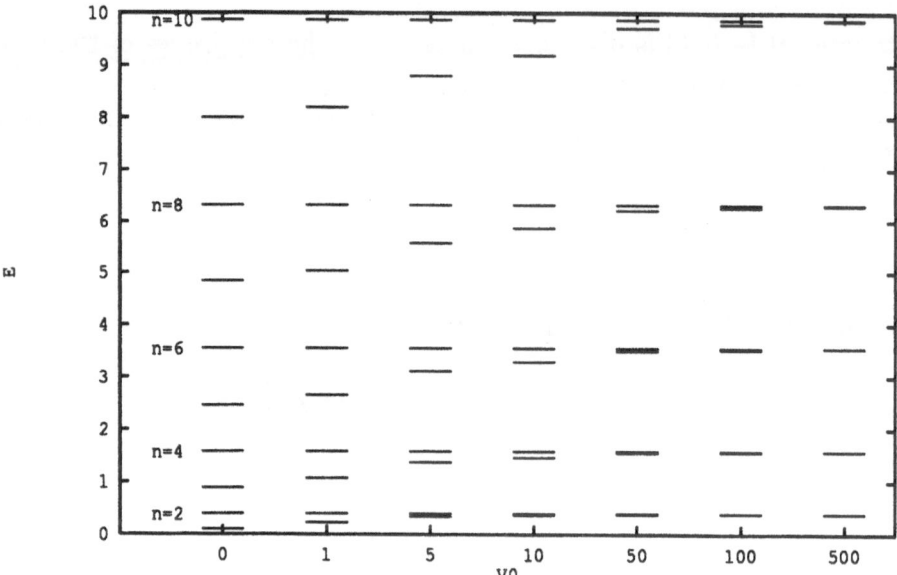

Figure 1.   Energy levels E_n, $1 \leq n \leq 10$, for various values of the delta-like potential.

The set of all the solution of (5) coincides, for $V_0 = 0$, with the well known sequence of the momentum eigenvalues in the square well without central barrier,

$$k_n = \frac{n\pi}{2L}, \qquad n = 1, 2, 3, \ldots$$

The values which fulfill (7) in this case coincide with the k_n of odd order; By increasing V_0 they approach from the left the integer multiples of π, the even eigenvalues, independent from V_0 as described in (6), as shown in figure 1.

For an impenetrable barrier ($V_0 = +\infty$), the energy levels should coincide with the square of these last, with a double degeneracy: in absence of tunnelling the eigenstates will become symmetrical or antisymmetrical combinations of the identical eigenstates of each half-well. The tunnelling probability different from zero which is obtained for a finite value of V_0 breaks the degeneracy and gives rise to doublets.

Let us now compute the eigenfunctions.

1. With $k = k_n \equiv \frac{n\pi}{2L}$, n even, from (2) and the boundary and continuity conditions we deduce

$$u_n(x) = \begin{cases} \text{I} & : \quad Ae^{ik_n x} - Ae^{-ik_n x} = \\ & = Ae^{ik_n L}\left(e^{ik_n(L+x)} - e^{-ik_n(L+x)}\right) \equiv A'\sin[k_n(L+x)] \\ \text{III} & : \quad Ce^{ik_n x} - Ce^{-ik_n x} = \\ & = Ce^{ik_n L}\left(e^{ik_n(L+x)} - e^{-ik_n(L+x)}\right) \equiv C'\sin[k_n(L+x)] \end{cases}$$

Therefore $u_n(0) \equiv \lim_{\epsilon \to 0} u_n(\epsilon) = 0$. This means that the condition of continuity of the derivative in the origin gives $A' = C'$. By also imposing the normalization

$$1 = \int_{-1}^{1} \{A' \sin [k_n(L+x)]\}^2 \, dx,$$

we get

$$u_n(x) = \sin[k_n(L+x)], \qquad n = 2, 4, 6, \ldots$$

on all the interval. These eigenstates are independent upon V_0 because they are null in the origin, where the delta-like potential is acting.

2. We have, still from the boundary and continuity conditions,

$$\left. \begin{array}{l} A = -Be^{2ik_nL} \\ C = -De^{-2ik_nL} \end{array} \right\} \implies -Be^{2ik_nL} + B = -De^{-2ik_nL} + D$$

(with n odd), from which we get

$$B = -De^{-2ik_nL}, A = D.$$

Therefore, reminding (2),

$$u_n(x) = \begin{cases} \text{I}: & De^{ik_nx} - De^{-ik_n(x+2L)} & = De^{-ik_nL}2i \sin[k_n(L+x)] \\ \text{III}: & -De^{ik_n(x-2L)} + De^{-ik_nx} & = De^{-ik_nL}2i \sin[k_n(L-x)] \end{cases}$$

By putting $N_n \equiv 2iDe^{-ik_nL}$, we have

$$u_n(x) = N_n \sin [k_n(L - |x|)], \qquad n = 1, 3, 5, \ldots$$

From the normalization condition we deduce

$$N_n = \frac{1}{\sqrt{L - \frac{\sin(2k_nL)}{2k_n}}}. \tag{8}$$

2.2. NUMERICAL CALCULATION

The time-dependent Schrödinger equation can be solved numerically by discretizing it on a proper space-time lattice (Press *et al.*, 1986). This requires to discretize also the potential (1) to get

$$V(x_j) = \begin{cases} 0 & x_j \neq 0 \\ \gamma & x_j = 0 \end{cases} \qquad (j = J+1),$$

with finite γ. This can be calculated by expressing the integral

$$\int_{-\epsilon}^{\epsilon} V_0 \delta(x) \psi(x) dx = V_0 \psi(0) \equiv V_0 \psi_{J+1} \tag{9}$$

in discrete form, as

$$\sum_{j=(J+1)-l}^{(J+1)+l} \gamma \delta_{j,J+1} \psi_j \Delta x = \gamma \Delta x \psi_{J+1}. \tag{10}$$

By comparing (9) and (10) it is easy to derive

$$\gamma = \frac{V_0}{\Delta x}.$$

In order to display the evolution of the wavefunction in a well of half-width $L = 1$, we have chosen the initial state

$$\psi(x, t = 0) = \sum_{n=1}^{\infty} A(k_n) u_n(x),$$

where

$$A(k_n) = \begin{cases} e^{-\frac{1}{8}(n-\bar{n})^2 \pi^2 \sigma^2} e^{-\frac{ik_n}{3}} & 80 \leq n \leq 120 \\ 0 & n < 80, n > 120 \end{cases} \tag{11}$$

with $\bar{n} = 100$ and $\sigma = 0.06$. We have also put $V_0 = 320$ to get a transmission coefficient almost equal to the reflection one.

The figure 2 allows to compare numerical and analytical (Segre *et al.*, 1976) results. The agreement is good for the greater part of the profile of the wavefunction besides some peaks not completely under control in correspondence of very fast oscillations. These do not influence the evaluation of the figure of merit indicating the non-invasivity in the quantum regime of measurement which we will introduce in the next section.

3. Impulsive quantum measurements

In the framework of the quantum measurement theory with the Feynman-Mensky path-integral approach (Mensky, 1979; 1993), the propagator of a system undergoing to a continuous measurement of its position between the times 0 and τ, with result $a(t)$ and instrumental uncertainty Δa, is written as a weighted path-integral:

$$K_{[a]}(x'', \tau; x', 0) = \int_{x(0) \equiv x'}^{x(\tau) \equiv x''} \mathcal{D}[x(t)] \exp\left\{ \frac{i}{\hbar} \int_0^{\tau} \mathcal{L}(x(t), \dot{x}(t), t) dt \right\} w_{[a]}[x],$$

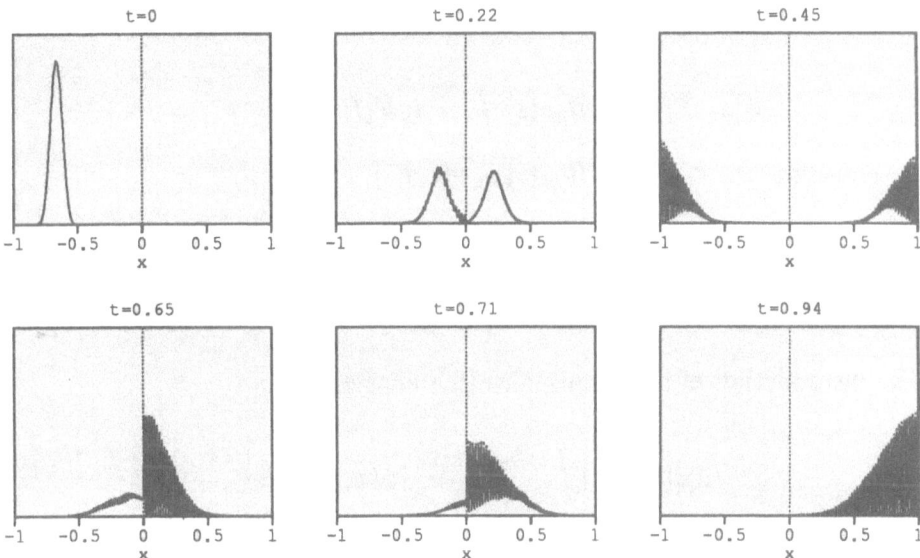

Figure 2. Time evolution of an initial Gaussian wavefunction in the square well with the delta-potential in the middle. Both the analytical and the numerical square moduli of the wavefuction are depicted, and they are almost completely overlapping.

where

$$w_{[a]}[x] = \exp\left\{-\frac{1}{2\Delta a^2 \tau}\int_0^\tau [x(t) - a(t)]^2 dt\right\}. \qquad (12)$$

Th most natural way to represent in this framework an impulsive measurement of the position at time 0 is as limit, for infinitesimal time intervals, of a continuous one. In this approximation,

$$K_\varepsilon(x'', x') \cong \lim_{\tau \to 0} K_{[a(t)\equiv\varepsilon]}(x'', \tau; x', 0) =$$

$$= \lim_{\tau \to 0} \int_{x(0)\equiv x'}^{x(\tau)\equiv x''} \mathcal{D}[x(t)] \exp\left\{\int_0^\tau \left(\frac{i}{\hbar}\mathcal{L} - \frac{[x(t) - \varepsilon]^2}{2\Delta a^2 \tau}\right) dt\right\}.$$

The imaginary (Lagrangian) part of the integrand at the exponent can be neglected, in the limit, with respect to the measurement term. If we apply to the time integration the mean value theorem, it follows

$$K_\varepsilon(x'', x') \cong \lim_{\tau \to 0} \int_{x(0)\equiv x'}^{x(\tau)\equiv x''} \mathcal{D}[x(t)] \exp\left\{-\frac{[x(0) - \varepsilon]^2 \not\tau}{2\Delta a^2 \not\tau}\right\} =$$

$$= e^{-\frac{(x'-\varepsilon)^2}{2\Delta a^2}} K(x'', 0; x', 0) =$$

$$= e^{-\frac{(x'-\varepsilon)^2}{2\Delta a^2}} \delta(x'' - x'). \qquad (13)$$

Let $\psi(x,t)$ be the wavefunction of a system subjected to an impulsive measurement, at the time t, with result ε. From (13) it follows

$$\begin{aligned}
\psi(x,t^+) &= R_{[\psi]}(\varepsilon) \int_{-\infty}^{+\infty} dy K_\varepsilon(x,y)\psi(y,t^-) = \\
&= R_{[\psi]}(\varepsilon)\hat{w}_\varepsilon\psi(x,t^-),
\end{aligned}$$

with

$$\hat{w}_\varepsilon \equiv \exp\left\{-\frac{(\hat{X}-\varepsilon)^2}{2\Delta a^2}\right\}.$$

The introduction of the renormalization constant

$$R_{[\psi]}(\varepsilon) \overset{\text{def}}{=} \left[\int_{-\infty}^{+\infty} e^{-\frac{(x-\varepsilon)^2}{\Delta a^2}}|\psi(x,t^-)|^2 dx\right]^{-\frac{1}{2}}$$

is due to the nonunitarity of the time evolution introduced by the measurement term. This is similar to the usual measurement theory of von Neumann (1955). The latter is simply recovered by choosing for the measurement operator the form (discontinuous and therefore less realistic)

$$\hat{w}_\varepsilon^{v.N.} \propto \theta(x-[\varepsilon-\Delta a])\theta([\varepsilon+\Delta a]-x). \tag{14}$$

For a generic system with discrete energy levels,

$$H|l\rangle = E_l|l\rangle,$$

any initial state can be expressed in terms of the energy eigenstates:

$$|\psi(t_0)\rangle = \sum_{l=1}^{\infty} c_l^{(0)}|l\rangle. \tag{15}$$

If the magnetic flux is measured as a generalized coordinate with results Φ_n, $n = 0, 1, \ldots, N$ at each of the times $t_n \equiv n\Delta T$, and always with the same fixed uncertainty $\Delta\Phi$, it results

$$|\psi_{\{\Phi_n\}_{n=0,\ldots,N}}(t_N^+)\rangle = \hat{w}_{\Phi_N}\left(\prod_{j=1}^{N} e^{-\frac{i}{\hbar}\hat{H}\Delta T}\hat{w}_{\Phi_{N-j}}\right)|\psi(t_0^-)\rangle. \tag{16}$$

The renormalization constants relative to each measurement factorize and then can be neglected in the calculation, since we are interested only in normalized probabilities.

By using (15) and inserting the unity operator $1 \equiv \sum_{m=1}^{\infty} |m\rangle\langle m|$, eq. 16 can be rewritten

$$|\psi_{\{\Phi_n\}_{n=0,\ldots,N}}(t_N^+)\rangle = \sum_{l,m=1}^{\infty} c_l^{(0)} |m\rangle\langle m|\hat{w}_{\Phi_N} \left(\prod_{j=1}^{N} e^{-\frac{i}{\hbar}\hat{H}\Delta T} \hat{w}_{\Phi_{N-j}} \right) |l\rangle$$

$$= \sum_{m=1}^{\infty} c_m^{(N)} |m\rangle, \tag{17}$$

where

$$c_m^{(N)} \stackrel{\text{def}}{=} \sum_{l=1}^{\infty} B_{ml}^N c_l^{(0)}, \tag{18}$$

$$B_{ml}^N(\Delta T, \Delta\Phi, \{\Phi_n\}) \stackrel{\text{def}}{=} \langle m|\hat{w}_{\Phi_N} \left(\prod_{j=1}^{N} e^{-\frac{i}{\hbar}\hat{H}\Delta T} \hat{w}_{\Phi_{N-j}} \right) |l\rangle.$$

By inserting N times $1 \equiv \sum_{n_i=1}^{\infty} |n_i\rangle\langle n_i|$ $(i = 1, 2, \ldots, N)$, it results

$$B_{ml}^N = \sum_{n_1,n_2,\ldots,n_N=1}^{\infty} W_{mn_1}^{\Phi_N} \left(\prod_{j=1}^{N-1} W_{n_j n_{j+1}}^{\Phi_{N-j}} \right) W_{n_N l}^{\Phi_0} \exp\left\{ -\frac{i\Delta T}{\hbar} \sum_{i=1}^{N} E_{n_i} \right\}, \tag{19}$$

$$W_{ij}^{\Phi_n}(\Delta\Phi) \stackrel{\text{def}}{=} \int_{-\infty}^{+\infty} u_i^*(\varphi) e^{-\frac{(\varphi-\Phi_n)^2}{2\Delta\Phi^2}} u_j(\varphi) d\varphi, \tag{20}$$

where the $u_i(\varphi) \equiv \langle\varphi|i\rangle$ are the energy eigenfunctions.

The measurement becomes noninvasive, in the sense defined by Leggett and Garg, in the limit for $\Delta\Phi \to \infty$, i.e. when one substantially does not make any measurement. It is indeed immediate to see that

$$\lim_{\Delta\Phi\to\infty} W_{ij}^{\Phi_n}(\Delta\Phi) \stackrel{(20)}{=} \delta_{ij}, \tag{21}$$

and then

$$\lim_{\Delta\Phi\to\infty} B_{ml}^N \stackrel{(19),(21)}{=} \sum_{\{n_i\}}^{\infty} \delta_{mn_1} \delta_{n_1 n_2} \ldots \delta_{n_N l} \exp\left\{ -\frac{i\Delta T}{\hbar} \sum_{i=1}^{N} E_{n_i} \right\} =$$

$$= \delta_{ml} \exp\left\{ -\frac{i\Delta T}{\hbar} \sum_{i=1}^{N} E_l \right\}; \tag{22}$$

now, because $N\Delta T \equiv t_N$,

$$\lim_{\Delta\Phi\to\infty} |\psi_{\{\Phi_n\}}(t_N^+)\rangle \stackrel{(17),(18)}{=} \sum_{l,m=1}^{\infty} c_l^{(0)} B_{ml}^N |m\rangle =$$

$$\overset{(22)}{=} \sum_{l=1}^{\infty} c_l^{(0)} e^{-\frac{i}{\hbar} E_l t_N} |l\rangle \equiv \tag{23}$$

$$\equiv e^{-\frac{i}{\hbar} \hat{H} t_N} |\psi_{\{\Phi_n\}}(t_0^-)\rangle.$$

This is equivalent to say that in the considered limit the measurement does not modify the subsequent evolution of the system. This shows also the consistency with the Mensky theory. When $\Delta\Phi \to \infty$ the weight functional (12) becomes unity, and the propagator becomes the usual one without the measurement effect.

The probability that the N-th measurement gives a result Φ_N, when the results of the previous $N-1$ ones are known, is expressed through

$$P_N(\Phi_N) \equiv \frac{\left|\left\langle \psi_{\Phi_N}(t_N^+) \middle| \psi_{\Phi_N}(t_N^+)\right\rangle\right|^2}{\int_{-\infty}^{+\infty} \left|\left\langle \psi_{\Phi_N}(t_N^+) \middle| \psi_{\Phi_N}(t_N^+)\right\rangle\right|^2 d\Phi_N}. \tag{24}$$

The denominator can be rewritten as

$$\left|\left\langle \psi_{\Phi_N}(t_N^+) \middle| \psi_{\Phi_N}(t_N^+)\right\rangle\right|^2 \overset{(17)}{=} \left(\sum_{m=1}^{\infty} \left|c_m^{(N)}\right|^2\right)^2$$

Thus, once the energy eigenvalues and eigenstates of a generic quantum system with discrete spectrum are known, from the energy eigenstates expansion of the initial wavefunction it is possible to derive the effective uncertainty value

$$\Delta\Phi_{\text{eff}}^2 = 2 \frac{\int_{-\infty}^{+\infty} (\Phi - \Phi_N)^2 \left(\sum_{m=1}^{\infty} \left|\sum_{l=1}^{\infty} B_{ml}^N(\Phi_0, \ldots, \Phi_{N-1}, \Phi) c_l^{(0)}\right|^2\right)^2 d\Phi}{\int_{-\infty}^{+\infty} \left(\sum_{m=1}^{\infty} \left|\sum_{l=1}^{\infty} B_{ml}^N(\Phi_0, \ldots, \Phi_{N-1}, \Phi) c_l^{(0)}\right|^2\right)^2 d\Phi},$$

$$\tag{25}$$

which expresses the spreading of the possible measurement results (Mensky et al., 1993) and is equal to $\Delta\Phi$, the instrumental error, in the classical limit. The complete computation seems to be quite difficult, due to the presence of multiple sums on the numerable set of the energy eigenstates. A direct numerical computation is, for the same reason, also unfeasible in practice. It is required a reasonable approximation based upon truncation of sums and integrals to finite values. Concerning the integrals it is evident that the probability $P(\Phi)$ that a measurement will give a result Φ will be zero outside a finite region of constant size (which is roughly given by the distance between the extrema of inversion of the corresponding classical motion for the maximal energy scale), unless the system is excited to very high energy. This last case does not occur because after a certain number of measurements the system should approach a collapse on an asymptotic state which depends upon ΔT. In the cases dealt in literature (Mensky *et al.*, 1993) this state, after each measurement, evolves according to the dynamics dictated by the particular potential and tends to reform itself almost identically in correspondance of the next measurement. Therefore in each instant the state of the system has non-negligible projection only on a finite number of energy eigenstates. Thus the sums in (25) can be truncated to a convenient maximal value N_{MAX}. The accuracy of the approximations has been tested through comparison with results already analytically known, as in the case of the harmonic oscillator (Mensky *et al.*, 1993).

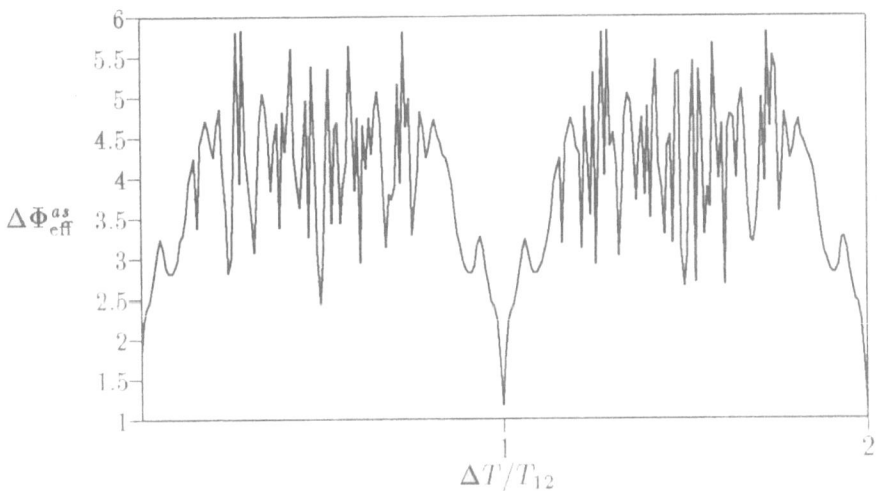

Figure 3. Effective magnetic flux uncertainty versus the quiescent time for the potential analyzed in section 2.

4. Conclusions

If the effect of the measurement discussed in section 3 is included in the dynamical description of the tunnelling dynamics of section 2, a model for repeated quantum measurements of magnetic flux in a bistable potential is obtained. The result is summarized in fig. (3) where the dependence of the asymptotic $\Delta\Phi_{\text{eff}}$ upon the time interval between consecutive measurements is depicted. When the quiescent time is a multiple of the tunnelling period each measurement, although it is a quantum one, does not alter the predictability of the next one, as if it were perfectly non-invasive. This represents a so-called stroboscopic quantum nondemolition measurement (Braginsky *et al.*, 1992; Caves *et al.*, 1981) of magnetic flux in the rf-SQuID. Work is in progress to apply this strategy to more realistic potentials to understand the role of the uncertainty principle in the temporal Bell inequalities (Calarco *et al.*, 1994).

This research has been supported by INFN, Laboratori Nazionali di Legnaro, Italy.

References

Ballentine, L.E. (1987). Realism and Quantum Flux Tunnelling, *Phys. Rev. Lett.*, **59** pp. 1493–1495.

Bell, J.S. (1987). *Speakable and unspeakable in quantum mechanics: collected papers in quantum mechanics*, Cambridge University Press, Cambridge, pp. 14–21.

Braginsky, V.B. and Khalili, F.Ya. (1992). *Quantum Measurement*. K.S. Thorne editor. Cambridge University Press, Cambridge.

Caves C.V., Thorne K.S., Drever R.W., Sandberg V., and Zimmermann M. (1980). On the Measurement of a Weak Classical Force Coupled to a Quantum Mechanical Oscillator I. Issues of Principle, *Rev. Mod. Phys.*, **52** pp. 341–392.

Calarco, T. and Onofrio, R. (1994). Optimal Measurements of Magnetic Flux in Superconducting Circuits and Macroscopic Quantum Mechanics, in preparation.

Leggett, A.J. and Garg, A. (1985), Quantum Mechanics versus Macroscopic Realism: Is the Flux There when Nobody Looks?, *Phys. Rev. Lett.*, **54** pp. 857–860; see also Leggett, A.J. (1980), Macroscopic Quantum Systems and the Quantum Theory of Measurement, *Suppl. Prog. Theor. Phys.*, **69** pp. 80–100.

Mensky M.B. (1979), Quantum Restrictions for Continuous Observation of an Oscillator *Phys. Rev. D*, **20** pp. 384–387.

Mensky M.B. (1993), *Continuous Quantum Measurements and Path-Integrals*, mailIOP Publishers, Bristol and Philadelphia.

Mensky, M.B., Onofrio, R., and Presilla, C. (1993). Optimal Monitoring of Position in Nonlinear Quantum System, *Phys. Rev. Lett.*, **70** pp. 2828–2831.

Paz, J.P. and Mahler, G. (1993). Proposed Test for Temporal Bell Inequalities, *Phys. Rev. Lett.*, **71** pp. 3235–3239.

Peres, A. (1988). Quantum Limitations on Measurement of Magnetic Flux, *Phys. Rev. Lett.*, **61** pp. 2019–2021.

Press, W.H., Flannery, B.P., Teukolsky, S.A., and Vetterling, W.T. (1986). *Numerical Recipes: the Art of Scientific Computing*, Cambridge University Press, Cambridge.

Segre, C.U. and Sullivan, J.D. (1976). Bound-state wave packets, *Am. Journ. Phys.*, **44** pp. 729–732.

Tesche, C.D. (1990). Can a Noninvasive Measurement of Magnetic Flux be Performed with Superconducting Circuits?, *Phys. Rev. Lett.*, **64** pp. 2358–2361.

von Neumann, J.V. (1955). *Mathematical Foundations of Quantum Mechanics*, Princeton University Press, Princeton.

PLANCK'S THEORY (1898-1906) AND THE BIRTH OF QUANTUM PHYSICS

P. CAMPOGALLIANI

Dip. di Fisica "G. Galilei", Università di Padova
Via Marzolo 8 - 35131 - Padova, Italy

Abstract. Generally historians of science studied Planck's theory of black body radiation (1900–1906) with focus of their analysis in the question whether resonator energy is continuous or discontinuous. Following an alternative historiographical approach, we consider this question incorrect and secondary and we think that the central point for the comprehension of Planck's theoretical work consists in his evolutive thought about irreversibility.

The main steps of this evolution are essentially the hypothesis of natural radiation, the elementary disorder principle, the combinatorial entropy and the phase–plane subdivision: focusing our attention on this conceptual development it is possible to understand how this theory is able to highlight some quantistic aspects without strict incompatibility with classical physics, particularly the appearance of energy elements, phase–space elementary regions and microscopical indetermination.

1. Introduction

In the second half of the nineteenth century, the succesful evolution of the kinetic theories of gases highlights the existence of two differents aims: on the one hand the intention of explaining thermodynamical laws, phenomenological laws and macroscopical magnitudes through mechanical laws, on the other hand the hope of achieving some basic information about the microscopical structures of the physical world.

Nevertheless, in the last decades of the century, this ambitious programme met with some consistent difficulties of two different kinds:

a) *specific difficulties;*
b) *basic difficulties.*

The specific difficulties were previously inherent in the disagreement between forecasts and the experimental response: e.g. the relation between the

C. Garola and A. Rossi (eds.), The Foundations of Quantum Mechanics, 71-84.
© 1995 *Kluwer Academic Publishers.*

specific heat of gases and the degree of freedom of molecules, the unexpected dependence of the specific heat of gases on temperature, the violation of the Dulong–Petit law by some solid specific heats etc.

The second kind of difficulty is substantially conceptual: it is inherent in the problem of explaining mechanically the irreversibility of all natural phenomena stated by the second law of thermodynamics. However molecular collisions follow the reversible laws of mechanics: this was the source of the paradox of reversibility (Loschmidt) and the recurrence paradox (Zermelo).

The reversibility paradox, formulated by Loschmidt in a series of papers in the years 1876–77 against Boltzmann's H–theorem, keeps its critical strength even after Boltzmann's successive theoretical constructions of the year 1877 containing the famous probabilistic interpretation of entropy.

The recurrence paradox, formulated by the German mathematician E. Zermelo between 1896–97 suggested by Poincaré's quasi–periodicity theorem, brings out the incompatibility between a phenomenological view of irreversibility and a mechanical view of the world grounded on reversible deterministic motion of molecules. Boltzmann's reply is unsatisfactory and furthermore his probabilistic entropy principle formulated in 1877 is still without any empirical confirmation or theoretical application.

Only in the first decade of the twentieth century, was Boltzmann's principle finally succesful in many cases while Planck's theory of black body radiation is undoubtedly the most famous but not the only one.

In those years Boltzmann's principle showed its validity and basic role in the birth of the quantum hypothesis both in Planck's and in Einstein's works. Therefore an understanding of the beginning of quantum physics needs to follow the evolution of the conception of irreversibility.

2. Planck's theory of black body radiation

Recently historians of science have studied Planck's work in depth but generally the focus of their detailed analysis has been this question: is Planck's theory of black body radiation formulated between 1900–1906 classical or quantistic? Their answer to this question was classical if resonator energy was valued continuous, and quantistic if valued discontinuous[1].

A final consequence of this historiographical approach is a particular research emphasis on the explicit statementes contained and the implicit persuasions suspected in Planck's works. These statements and persuasions relate specifically to resonator energy and here arises a great disparity of conclusions.

An alternative historiographical approach is desiderable which considers this question sterile and incorrect[2] and, on the contrary, thinks these following statements to be fundamental:

- the central point for the comprehension of Planck's theoretical work rests on his evolutive thought about irreversibility;
- quantistic doesn't necessarily imply resonator energy to be discontinuous;
- classical and quantistic are not in dichotomic opposition, however Planck's theory can be valued simultaneously classical and quantistic;
- in Planck's theories between 1900–1906 some quantistic peculiarities emerged without a clear incompatibility with previous classical physics.

I) PRESTATISTICAL PERIOD

Planck doesn't think in the same manner as his assistent Zermelo: the mechanical view of the world is not in conflict with the second law of thermodynamics. Instead, the incompatibility consists in the atomistic view of the world because it gives rise to Maxwell's demon experiment, to the statistical interpretation of irreversibility and finally to reversibility and recurrence paradoxes.

"The second law of thermodynamics, logically developed, is incompatible with the assumption of finite atoms.... yet there seem to be at present many kinds of indications that in spite of the great successes of atomic theory up to now, it will finally have to be given up and one will have to decide in favour of the assumption of a continuos matter"[3].

The irreversibility is not of a statistical nature but of an absolute nature and it needs to look for its explanation through continuum mechanics such as, for example, the electromagnetic field.

That conviction led the German scientist to study the physical processes leading to the black body distribution of the stationary radiation and led him to write a series of five papers *"Über irreversible Strahlungsvorgänge"* in the years between 1897–1899 culminating in the birth of the quantum hypothesis[4].

Planck initially hoped to obtain irreversibility through reversible processes only, through processes in which resonators absorb and emit electromagnetic radiation under conservative strength.

This hope was quickly abandoned in face of Boltzmann's definite criticism and consequently, in 1898, Planck was forced to introduce the statistical hypothesis of natural radiation, in analogy with the hypothesis of molecular disorder for the kinetic theories.

II) FIRST STATISTICAL PERIOD: NATURAL RADIATION HYPOTHESIS AND WIEN'S LAW

The introduction of the natural radiation hypothesis in 1898, involves the appearance of statistics in Planck's programm, in similar manner as molecular chaos appears in the kinetic theory of gases.

The mechanical laws are reversible but the molecular collisions aren't reversible because the statistical hypothesis of molecular chaos exists; in a similar way the electromagnetic laws are reversible but the processes of radiation emission and absorption aren't reversible because the statistical hypothesis of natural radiation exists.

With regards to this assumption, Planck explains:

"If it is stated that an electromagnetic ray possesses the properties of natural radiation, that in short means the following: the energy of radiation is distributed in a completely irregular manner among the individual partial vibrations, of which the ray can be thought to be composed.... The hypothesis of natural radiation, if assumed to be valid for all space points at all times, implies at its case the second law of thermodynamics when applied to the processes of radiation; that is, it is another expression of the same law"[5].

The most important papers where we can find Planck's considerations about this hypothesis, are the following:

- "Über irreversible Strahlungsvorgänge", Vierte Mitteilung, Berl. Ber, (1898), pp.449–476.
- "Über irreversible Strahlungsvorgänge", Ann. d. Phys. 1 (1900), pp.69–122.
- "Entropie und Temperatur strahlender Wärme", Ann. der Phys. 1 (1900), pp.719–37.
- "Zur Theorie des Gesetzes der Energieverteilung in Normalspectrum", Verhandlungen der Deutschen Physikalischen Gesellschaft 2 (1900), pp.237–245.
- *Vorlesungen über die Theorie der Wärmestrahlung* (1906), Vierter Abschnitt, Kap. 1, Fünfter Abschnitt, Kap. 2.

From the electromagnetic theory and the hypothesis of natural radiation, in May 1899, Planck found a basic equation for the evolution of resonator energy U_0:

$$\frac{dU_0}{dt} + 2\nu\sigma U_0 = \frac{3c^3\sigma}{16\pi^2\nu_0} J_0 \qquad (2.1)$$

U_0 resonator energy;
ν_0 resonator frequency;
σ damping constant;

J_0 field intensity.

From this equation one immediately gets a basic relation at equilibrium between U_ν, the average energy of a resonator having frequency ν, and ρ_ν, the spectral energy distribution of incident radiation having the same frequency:

$$\rho_\nu = \frac{8\pi^2\nu^2}{c^3}U_\nu \,, \quad \text{where} \quad \rho_\nu = \frac{3}{4\pi}J_\nu \,. \tag{2.2}$$

Defining the entropy of a resonator through the expression:

$$S = -\frac{U}{a\nu}\log\frac{U}{b\nu}. \tag{2.3}$$

Through the thermodynamical relation $\frac{\partial S}{\partial U} = \frac{1}{T}$ Planck can finally obtain a rigorous derivation of Wien's law.

At this point it is important to observe that the same basic differences exist between the gas hypothesis of molecular disorder and Planck's hypothesis of natural radiation.

Essentially:

- natural radiation has absolute validity and not statistical validity as for gases;
- it is an expression of temporal disorder and not of spatial disorder as for gases (however one must remember that in the electromagnetic field a Fourier component corresponds to a molecule).

On this purpose, Planck explains: "*Kurz gesagt: bei den Wärmeschwingungen eines Resonators ist die Unordnung eine zeitliche, während sie bei den Molekularbewegungen eines Gases eine räumliche ist*"[6].

III) SECOND STATISTICAL PERIOD: PLANCK'S LAW AND THE COMBINATORIAL APPROACH

However, a few months later, the experimentalists reported the first remarkable deviations from Wien's law for long wavelength of spectrum.

In front of the new difficulties, in October 1900[7], Planck interpolated the new experimental data in this way:

$$\frac{\partial^2 S}{\partial U^2} \sim \frac{1}{U}$$

for high frequency, that is in the region of validity of Wiens' law;

$$\frac{\partial^2 S}{\partial U^2} \sim \frac{1}{U^2}$$

for low frequency, that is in the new region in which results $U \sim T$ and therefore $\frac{\partial S}{\partial U} = \frac{1}{T} = \frac{1}{U}$.

Globally it will be:

$$\frac{\partial^2 S}{\partial U^2} = \frac{\alpha}{U(\beta + U)}$$

and integrating

$$\frac{\partial S}{\partial U} = \frac{\alpha}{\beta}\left[\log U - \log(\beta + U)\right] = \frac{1}{T}.$$

Then this expression immediately follows for the entropy of a resonator:

$$S = \alpha\left[\frac{U}{\beta}\log\frac{U}{\beta} - \left(1 + \frac{U}{\beta}\right)\log\left(1 + \frac{U}{\beta}\right)\right]. \qquad (2.4)$$

If one considers that Wien's displacement law, written by Planck in the form $S = \Psi\left(\frac{U}{\nu}\right)$, implies $\beta = A'\nu$, this expression, with (2.2), provides Planck's famous formula of black body radiation distribution.

In order to give a theoretical guise to this bare interpolating formula, a few weeks later, on 14 December 1900[8] Planck introduced the Boltzmann relation $S = k\log W$, the calculus of complexiones and consequently the unexpected discrete energy elements $\epsilon = h\nu$.

Let us now recall briefly the way Planck proceeded in his interpretation of the black body formula.

Let us consider a large number of linear monochromatically vibrating cavity resonators:

N resonators of frequency ν
N' resonators of frequency ν'
N'' resonators of frequency ν''

....

The total energy E_T is divided partly in the medium and partly E_0 among the resonators as vibrational energy.

We assign to them certain arbitrary energies, for instance, an energy E to the N resonators ν, E' to the N' resonators ν' and so on, where $E + E' + E'' + ... = E_0 < E_T$. Now we must look for the various possible distributions over the different colours and also for the various possible distributions over the separate resonators of each monochromatic group.

If we put $E = P\epsilon = NU_\nu$ where P is an integer, Boltzmann's relation provides:

$$S_N = k\log W = k\log\frac{(N + P - 1)!}{(N - 1)!P!} \simeq k\log\frac{(N + P)^{N+P}}{N^N P^P} \qquad (2.5)$$

S_N entropy of N resonators.

Then after some obvious calculus, we obtain:

$$S = k\left[\left(1 + \frac{P}{N}\right)\log(N + P) - \log N - \frac{P}{N}\log P\right] \qquad (2.6)$$

and therefore

$$S = k\left[\left(1 + \frac{U}{\epsilon}\right)\log\left(1 + \frac{U}{\epsilon}\right) - \frac{U}{\epsilon}\log\frac{U}{\epsilon}\right]. \qquad (2.7)$$

That is the expression for the entropy of a resonator obtained in the theoretical way; in comparison with expression (2.4) obtained by interpolation, the existence of the energy element $\epsilon = h\nu$ follows immediately.

A more elegant and rational way for deducing the formula and the energy element exists. One can find it in §150 of *Vorlesungen* where, for the first time in the history of physics, the subdivision of phase–space in elementary regions (Elementargebiete), in cells of size h, appears.

Briefly the energy U of a resonator is given by the expression

$$U = \frac{1}{2}Kf^2 + \frac{1}{2}\frac{g^2}{L}$$

f, g, canonically coniugate variables;

K, L constants of the system.

And the relation

$$\frac{1}{2}Kf^2 + \frac{1}{2}\frac{g^2}{L} = const.$$

is represented, in the plane (f,g), by elliptical curves having area

$$\pi\sqrt{\frac{2U}{K}} \cdot \sqrt{2UL} = 2\pi U\sqrt{\frac{L}{K}} = \frac{U}{\nu}.$$

Planck says of this:

"*If we assume that the entire phase–plane is divided by a large number of these ellipses into separate sections, such that the ring–shaped areas bounded by two successive ellipses become equal in magnitude to one another, such that $\frac{\Delta U}{\nu}$ = const., then we obtain a method of determining those sections ΔU of energy which correspond to equal probability and which we may therefore call energy elements. If we put the magnitude of an energy element ΔU equal to ϵ, and put the constant in the previous equation equal to h, then we arrive exactly at the former equation ($\epsilon = h\nu$) without having made use of Wien's displacement law. At the same time the elementary quantum of action h acquires a new meaning, namely it gives the area of an elementary region in the phase–plane of a resonator, no matter what its frequency*".

Here is the meaning of h as a quantum of action *(Wirkungsquantum)* and this apriori statement about this phase–space subdivision in elementary regions, for the theory of radiation, is what we can call Planck's principle.

3. Planck's conceptual evolution about irreversibility

The evolution of Planck's thought about irreversibility is crucial for understanding how this theory is able to highligth some quantistic aspects without imcompatibility with classical physics.

From this point of view one can say that this theory is both classical and quantistic or, if we like, that it is a classical theory suitable to admit quantistic aspects without any conflict.

The main steps of this evolution are essentially the following:

- hypothesis of natural radiation;
- elementary disorder principle;
- combinatorial probability and entropy;
- Planck's principle (phase–plane subdivision).

The papers where one can find Planck's thought about these points, are the following

- "Zur Theorie des Gesetzes....", 14 Dec. 1900 cit.
- *Vorlesungen* 1906, §§133–146 cit.
- "Atomistische Theorie der Materie", 30 Apr. 1909[9].
- "On the Unity of the Physical World Picture", Leiden Lecture 1908[10].
- *Vorlesungen*, 2nd ed., 1913, §§115,116,117[11].

Very important in this evolution is the wellknown reflection from the famous December 1900 paper, where are related the hypothesis of natural radiation, the elementary disorder principle, and the combinatorial probabilistic interpretation of entropy:

"Entropy means disorder and I thought that one should find this disorder in the irregularity with which even in a completely stationary radiation field the vibrations of the resonator change their amplitude and phase, as long as one considers time intervals long compared to the period of one vibration, but short compared to the duration of a measurement. The constant energy of the stationary vibrating resonator can thus only be considered to be a time average or, put differently, to be an instantaneous average of the energies of a large number of identical resonators which are in the same stationary radiation field........ Since the entropy of a resonator is thus determined by the way in which energy is distributed at one time over many resonators, I suspected that one should evaluate this quantity by introducing probability considerations into the electromagnetic theory.... to do this it was only necessary to extend somewhat the interpretation of the hypothesis of natural radiation".

In *Vorlesungen* these questions are examined again in greater width and depth, particularly in §135 and subsequent.

In Mechanics of gases there exists the hypothesis of molecular disorder, in Elektrodynamics the hypothesis of natural radiation: both are included in the general hypothesis we call "of elementary disorder". That means that all states and all processes are not absolutely definite but only roughly definite. There are always "unkontrollierbare Bestandteile" and the absolute state is always an ideal state and not a physical one: here, in the observable state, one meets a reality in which our microscopical knowledge is not infinitely exact:

"Therefore to define a state we don't need to know the detail inside an elementary region of a molecule; however the hypothesis of natural radiation gives the univocity of temporal evolution in spite of the lack of deterministic knowledge"[12].

Therefore the principle of elementary disorder implies both microscopical indetermination and macroscopical determination.

About the evolution of the hypothesis of natural radiation, in the paper of Dec. 1900 we can read:

"The probability of any state is porportional to the number of corresponding complexions...this proposition is the core of the whole of the theory presented here....it can also be understood as a more detailed definition of the hypothesis of natural radiation".

Briefly summing up the natural radiation, becoming elementar disorder, leads finally to Boltzmann's principle of probabilistic entropy. Nevertheless Boltzmann's principle, for obtaining entropy of radiation, is not enough: it also needs Planck's principle:

"In point of fact, the validity of this theorem (of the equipartition) requires the assumption that the distribution over all possible states with given total energy, is an ergodic one, or briefly speaking, that the probability that the state of the system lies in a definite little cell is simply proportional to the volume of this cell no matter how small the latter is chosen to be. This hypothesis is however not fulfilled in the case of black body radiation; for elementary regions, or cells, may not be chosen to be arbitrarily small, but their volume is finite, determined by the value of the elementary quantum of action"[13].

Planck's principle, defining the size of an elementary region of phase–plane, provides the way for calculating the entropy of resonators and then the entropy of radiation.

We must remember that Einstein in his 1906 paper[14] calls Planck's principle a different thing, that is the assumption that oscillator energy can assume only discrete values.

4. Quantistic aspects in Planck's theory 1900–1906

Summarizing we can say that in this theory some essentially new aspects are present which we can rightly take into account as quantistic.

Schematically these are the following:

1) That is a question: are there oscillator's energy levels?
2) Energy elements;
3) Phase–space elementary regions (quantum of action);
4) The constant h;
5) The microscopical indetermination.

1. ARE THERE OSCILLATOR'S ENERGY LEVELS?

As is well known, some historians of science think that the oscillator's energy in Planck's theory 1900–1906 is a discontinuous variable, that is the energy is quantized in discrete values (e.g. Klein), other historians of science think that the oscillator's energy is a continuous variable, that is the energy is still a classical magnitude (e.g. Kuhn): therefore, on this point, Planck's statement are often ambiguous and secondary. Nevertheless it seems more correct to think that the oscillator's energy, in those papers, is not considered instantaneusly definite; magnitude U is rather thought of as a time average and really the temporal equation for its evolution has only an approximated validity.

In §146 of *Vorlesungen* Planck explicitly says:

"We can define the energy U of one oscillator in a stationary field, only as a time average... also entropy S is not defined for an instant but only for a time interval".

There are continuous microscopical fluctuations of resonator energy, but that is beyond observability.

2. ENERGY'S ELEMENTS

A very important feature in Planck's combinatorial theory of radiation in comparison with Boltzmann's combinatorial theory of gases, is the appearence of the element of energy $\epsilon = h\nu$. Also in Boltzmann's theory it remains a non vanishing finite quantity ϵ after the division of energy continuum; but now differently from Boltzmann ϵ is not only a non vanishing quantity, ϵ is also an elementary not arbitrary quantity.

Furthermore in Boltzmann's theory this finite energy quantity disappears from the final expression of entropy; in Planck's theory the energy element $\epsilon = h\nu$ appears in all final expressions.

In §149 of *Vorlesungen*, Planck says:

"Here places an essential difference in comparison to the entropy expression of a gas, where the size of the elementary region we call dσ, disappears from the final result and remains only in the additive constant lacking any physical meaning".

Therefore the energy elements in Planck's theory of radiation are not a simple mathematical device as they are for Boltzmann, but a new physical feature requiring a corresponding interpretation, not yet easy to find!

3. PHASE–SPACE ELEMENTARY REGIONS

As we said before, in §150 of Vorlesungen appears, for the first time in the history of physics, the subdivision of the phase–plane of a resonator in elementary regions (Elementargebiete), that is in cells of size h. Then Planck introduced the word quantum of action (Wirkungsquantum) for appointing the constant h and this new feature of radiation physics will become very important in future quantum physics (e.g. the quantization of action and Bohr-Sommerfeld conditions). This granular constitution of phase–space, in Planck's theory, becomes an apriori statement comparable to a principle (Planck's principle) which we need to put near Boltzmann's principle, in order to be able to evaluate radiation entropy.

Indeed this elementary subdivision in cells of size h, leads to the energy element $\epsilon = h\nu$ as a natural consequence, explaining, at least partially, its existence and consequently illustrating the way to make combinatorial calculus.

Furthermore it expresses a quantitative measurement of elementary disorder, because the size, h, of these cells is present in the final formula of resonator entropy, a physical magnitude which measures the quantity of disorder.

4. THE NATURAL CONSTANT h

In the work of December 1900, one meets the first appearance of the famous constant h, which will be the basis of the whole of future quantum physics. Planck is aware of its fundamental importance even if its meaning is as yet unknown.

"There can be no doubt whatsoever that the constant h plays a definite role in the basic vibrational processes in emission; our theories up to now do not, however, provide any foothold for an investigation of this role from the electrodynamics side. And yet the thermodynamics of radiation will not have come to a fully satisfactory conclusions until the constant h is perceived in its full, universal meaning"[15].

The h constant has a fundamental role in radiation thermodynamics and in its small but finite value, consists the relation–difference with the (classical) case of the kinetic theories:

"The occasion that h is introduced with a finite value, is a basic feature of the whole developed theory. If one put h infinitely small, we obtain a radiation law (Rayleigh's law) as a special case of the general law"[16].

and later:

"Only if one could assume that the quantum of action h were infinitely small the general distribution law goes over into the special one.... Then the energies of all oscillators would be equal, corresponding to the equipartition theorem, which in general is not the case"[17].

5. MICROSCOPICAL INDETERMINATION

Obviously one can't mistake Planck's elementary disorder principle for Heisenberg's uncertainty principle, but at that time in Planck's papers one can show the beginning of the conceptual separation between the macroscopical state and the microscopical state. Really the microscopical state is never completely definite, it is no longer comparable to a point but to a grain because it always contains uncontrollable and not measurable elements:

"......... is not necessary to give more details about a molecule inside an elementary region..... is not necessary to know the amplitude and the phase for every partial component vibration of the ray [in a little interval]...."[18].

Then the evolution of a microscopical state is instantaneously unknown; one can't state that the temporal evolution of resonator is both continuous and deterministic.

This particular view of the microscopical world led Planck to conceive, withouth any conflict, new quantistic aspects near macroscopical classical physics.

5. Conclusion

At the end of this lecture it could be profitable to summarize schematically the rational connection among the different statements of Planck's conception about irreversibility.

It is necessary to keep in mind this general philosophical conception of the physical world, if we want to understand correctly the genesis and the meaning of this theoretical work.

I) Planck was dissatisfied with the residual antropormophical peculiarity still present in every formulation of the second law of thermodynamics: e.g. impossibility of perpetuum mobile, mental experiment of Maxwell's demon where a reference to human ability is still involved.

II) Consequently Planck shows that Boltzmann's principle relating entropy and probability, is a great achievement in emancipation.

III) Differently from Boltzmanns persuasion, Planck believes in the absolute validity of the second law; this basic point is guaranteed by the hypothesis of elementary disorder forbiding any antientropic evolution.

IV) The hypothesis of elementary disorder implies on one hand microscopical indetermination, on the other hand macroscopical univocal thermodynamical evolution.

It is precisely in this physical and metaphysical view of the world, that the ability of Planck's theory to receive, or better to catch some quantistic aspects of the emerging physical reality in the beginning of this century, consists.

Notes and references

(1) We refer here particularly to the basic works of T. Kuhn and J.M. Klein: T.S. Kuhn (1978), *Black-Body theory and the quantum discontinuity, 1894-1912*, Oxford University Press; T.S. Kuhn (1984), "Revisiting Planck", H.S.P.S., **14**, 2 pp.231-252; J.M. Klein (1962), "Max Planck and the beginnings of quantum theory", Archive for history of Exact Sciences, **1**, pp.459-79; J.M. Klein (1963), "Planck, Entropy and Quanta, 1901-1906", The Natural Philosopher, **1**, pp.83-108; J.M. Klein (1966), "Thermodynamics and Quanta in Planck's work", Physics Today, **19**, n. 11, pp.23-32.

(2) For example: Allan A. Needell (1988), Introduction to M. Planck *The Theory of Heat Radiation*, Tomash Publishers, AIP; Peter Galison (1981), "Khun and the Quantum Controversy", B.J. Ph. Sc., **32**, pp.71-85.

(3) Max Planck (1882), "Verdampfen, Schmelzen und Sublimieren", Ann. d. Physik, **15**, pp.446-475.

(4) Max Planck (1897-1899), "Über irreversible Strahlungsvorgänge", Berl. Ber. 1897, pp.57-68, Berl. Ber., 1897, pp.715-717, Berl. Ber., 1897, pp.1122-1145, Berl. Ber., 1898, pp.449-476, Berl. Ber., 1899, pp.440-480.

(5) Max Planck (1900), "Über irreversible Strahlungsvorgänge", Ann. d. Phys., **1**, pp.69-122, cit.

(6) Max Planck (1906), *Vorlesungen über die Theorie der Wärmestrahlung*, Leipzig, Barth, §146 in *The Theory of Heat Radiation* cit.

(7) Max Planck (1900), "Über eine Verbesserung der Wienschen Spectralgleichung", Verh. d. Phys. Ges., **2**, pp.202-204.

(8) Max Planck (1900), "Zur Theorie des Gesetzes der Energieverteilung in Normalspectrum", Verh. d.D. Phys., **2**, pp.237-245.

(9) Max Planck (1910), "Acht Vorlesungen über Theoretische Physik", Leipzig, Hirzel.

(10) Max Planck (1993), *La conoscenza del mondo fisico* (introd. E. Bellone), Bollati Boringhieri.

(11) Max Planck (1913), *Vorlesungen über die Theorie der Wärmestrahlung*, 2nd ed.; English translation by Morton Masius, Philadelphia, Blakiston, 1914, in *The Theory of Heat Radiation* cit.

(12) Max Planck (1906), op. cit., §135.

(13) Max Planck (1906), op. cit., §166.

(14) Albert Einstein (1906), "Zur Theorie der Lichterzeugung und Lichtabsorption", Ann. d. Phys., **20**, pp.199–206.

(15) Max Planck (1906), op. cit., §149.

(16) Max Planck (1906), op. cit., §150.

(17) Max Planck (1906), op. cit., §166

(18) Max Planck (1906), op. cit., §135.

AN ANALOGUE OF THE TUNNEL EFFECT IN CLASSICAL ELECTRODYNAMICS

A. CARATI and L. GALGANI
Dip. di Matematica, Università di Milano
Via Saldini 50 - 20133 - Milano, Italy

and

J. SASSARINI
Dip. di Fisica, Università di Milano
Via Celoria 16 - 20133 - Milano, Italy

Abstract. We refer on some recent studies on classical electrodynamics of point particles, as described by the Abraham–Lorentz–Dirac equation. Such studies exploit positively the well known existence of generic runaway solutions. Indeed the additional requirement that has to be imposed, namely the restriction to initial data giving rise to nonrunaway behaviour, turns out to allow for unexpected phenomena, for example a behaviour qualitatively similar to that occurring in the quantum tunnel effect. It is pointed out how this fact might be relevant for the problem of hidden parameters.

1. Introduction

Much attention has always been paid since the introduction of quantum mechanics, to the problem whether it is possible to produce a classical model for quantum mechanics; this is often called the hidden variable problem. In this connection, the model of classical electrodynamics appears to be particularly suited, because it naturally contains an unit of action, namely e^2/c, where e is the electron charge and c the speed of light, and so it can in principle lead to results where the characteristic action of quantum mechanics, namely Planck's constant $\hbar \simeq 137 e^2/c$, appears. Considering for simplicity the nonrelativistic version of the model, one is thus led to the so called Maxwell–Lorentz system, namely Maxwell equations for the field with current due to a particle, and Newton equation for the particle with Lorentz force due to the field. In such a model, the state of the particle,

85

namely position and velocity, has to be thought of as the observable state, while the the field plays the role of the (infinitely dimensional) hidden parameter. From the mathematical point of view both the mechanical state and the hidden parameter are to be considered at the same level as producing deterministic motion through solutions of the Cauchy problem, but then one concentrate one's attention on the evolution of the mechanical state.

In a certain approximation (namely the so–called dipole approximation) it is well known that one can deduce a closed equation for the particle, namely the famous Abraham–Lorentz–Dirac equation, in which the field has explicitly disappeared. However the equation turns out to be of third order instead of second order, so that the deterministic evolution is obtained when one fixes position velocity and acceleration, i.e. the ordinary mechanical state plus acceleration. Thus in this reduced motion the acceleration plays the role of the hidden parameter played in the complete system by the field, and is in fact to be thought of as determined by the initial field for the complete system. The equation reads

$$\epsilon \dddot{\vec{x}} = \ddot{\vec{x}} - \frac{1}{m}\vec{F}(\vec{x}) \,, \tag{1}$$

where $\vec{x}$ is the position vector of the electron, m its (renormalized) mass, $\vec{F}(\vec{x})$ an external force field, and

$$\epsilon = \frac{2}{3}\frac{e^2}{mc^3} \tag{2}$$

is the "small parameter". In the present paper we refer on a recent study made on such an equation, where a phenomenon was discovered having a certain qualitative resemblance to the quantum tunnel effect.

2. The problem of the runaways and of their elimination

Since the famous work of Dirac[2] of the year 1938 it is very well known that the Abraham–Lorentz–Dirac equation presents the problem of the runaway solutions. The simplest case is just that of the free particle, where equation 1 takes the form $\epsilon \dot{\vec{a}} = \vec{a}$, in terms of the acceleration $\vec{a} = \ddot{\vec{x}}$, and its solution is $\vec{a}(t) = \vec{a}_0 \exp t/\epsilon$. So the generic solution diverges exponentially (for positive times), and the only meaningful (or physical) solution is obtained when one takes the exceptional initial datum $\vec{a}_0 = 0$.

However, the existence of runaways as a generic fact can be understood by the following qualitative argument. Considering for simplicity the case of one degree of freedom, equation 1 can be written in the form

$$\dot{x} = v \,, \quad \dot{v} = a \,, \quad \dot{a} = \frac{1}{\epsilon}(a - F(x)/m) \,.$$

So, in the "extended phase space" x, v, a, outside a "small" layer about the "slow manifold" defined by $a - F(x)/m = 0$, the vector field defining the differential equation is "practically infinite" and essentially parallel to the a–axis, being directed away from the slow manifold. Thus, for initial data not too near the "slow manifold" it is clesr that the solution will go exponentially fast to infinity along the direction of the a–axis.

Now, the generic appearence of runaways does not exclude the possibility of exceptional solutions with bounded acceleration, as we have seen for the case of the free particle, and the proposal of Dirac was just to add such a requirement on the solutions. In particular, for scattering problems nonrunaway solutions of 1 are selected by imposing subsidiary conditions of mixed type, namely conditions of ordinary Cauchy type for position and velocity (so defining a unique solution for the corresponding "reduced" or "mechanical" problem $m\ddot{\vec{x}} = \vec{F}(\vec{x})$), and an asymptotic outgoing condition for the acceleration, precisely $\ddot{x}(t) \rightarrow 0$ for $t \rightarrow |\infty$. From the mathematical point of view, the condition of Dirac, together with conditions of a Cauchy type on the initial position and velocity, leads to a problem of Sturm–Liouville type, which in general might admit no solution. However, with ordinary potentials vanishing at infinity the reduced mechanical problem does admit scattering states for suitable initial data, and so the same might be expected to hold also for the Lorentz-Dirac equation 1 with subsidiary conditions of Dirac type, at least for suitable initial data. For example, for the free particle this occurs for $\vec{a}_0 = 0$.

3. The uniqueness problem, and the folding of the physical manifold in the case of a potential barrier

The uniqueness problem for the nonrunaway (or physical) solutions of the Abraham–Lorentz–Dirac equation can be formulated in the following way. Given an initial datum $(\vec{x}_0, \vec{v}_0)$ in the mechanical phase space, one asks whether there exists a value $\vec{a}_0$ of the acceleration such that the corresponding initial datum $(\vec{x}_0, \vec{v}_0, \vec{a}_0)$ in the extended phase space gives rise to a motion $\vec{x}(t)$ satisfying the Dirac prescription $\vec{a}(t) \rightarrow 0$ for $t \rightarrow +\infty$. Notice that neither existence nor uniqueness is obvious, because this somehow resembles a problem of Sturm–Liouville type. In the paper [9] a generic mechanism was exhibited causing nonuniqueness for scattering by barriers for large eneough ϵ.

The phenomenon of nonuniqueness can be described geometrically in the following way. Consider the subset of the extended phase space (position, velocity and acceleration) corresponding to motions satisfying the Dirac condition, and call it the "physical (or Dirac) manifold": uniqueness would correspond to the physical manifold being a graph, say $\vec{a} = \vec{g}(\vec{x}, \vec{v})$, while nonuniqueness means that it is folded; in the work [9] it was shown instead that it has in general infinitely many foldings. Moreover, it turns out that, in the case of scattering by a barrier, the initial data belonging to different branches of the folded physical manifold and having the same position and velocity give rise alternatively to motions of mechanical and of nonmechanical type, which are transmitted and reflected respectively[1].

Everything can be understood by making reference to the particularly simple case of a force vanishing outside a compact domain and linear inside it; consider for example the case with potential energy $V(x)$ given by

$$V(x) = 1 - x^2 \quad \text{for} \quad |x| < 1, \quad V(x) = 0 \quad \text{for} \quad |x| > 1,$$

with continuity conditions for the acceleration at the points $x = \pm 1$. Indeed, by general arguments (i.e. the so called stable manifold theorem), the qualitative results can then be transported to the case of a generic smooth barrier having a single maximum.

Take any initial datum z_0 in the extended phase space with coordinates $z = (x, v, a)$, in the region with vanishing force, for definiteness with $x_0 < 1$. Thus the particle will proceed by the well known solution of the free particle problem and, depending on v_0 and a_0, some solutions will be able to reach the point $x = -1$. Then one has to solve the linear problem $\dot{z} = Az$ where A is the 3 by 3 matrix

$$A = \begin{pmatrix} 0 & 1 & 0 \\ 0 & 0 & 1 \\ -2/\epsilon & 0 & 1/\epsilon \end{pmatrix}.$$

The eigenvalues of A turn out to be real for small ϵ, while an interesting bifurcation occurs at $\epsilon = \sqrt{2/27}$, because for larger values of ϵ the matrix A has a real negative eigenvalue and two complex conjugate eigenvalues with positive real part. So, in the case of "large ϵ" to which we now concentrate our attention, for the corresponding linear system the phase space turns out to be the direct sum of a one dimensional "stable linear space" E^s (corresponding to the real negative eigenvalue) and a two dimensional "unstable linear space" E^u, the restriction of the system to E^u being an unstable focus. The dispositions of such linear spaces are easily determined; to fix ideas, for $\epsilon = 1$ one has the eigenvalue -1 with eigenvector $(1, -1, 1)$ and the eigenvalues $1 \pm i$ with eigenspace spanned by $(1, 1, 0)$ and $(0, 1, 2)$. Now the exceptional solution which at $x = -1$ hits exactly the point of intersection with the straight line E^s will tend to the origin (which corresponds to the maximum of the potential) in an infinite time (case of the separatrix). But consider now any other solution coming from left. By linearity, it will have a component tending to the origin and another component spiraling out. Thus there exists a certain time at which it will reach either the plane $x = -1$ or the plane $x = 1$, and after such a time it will become a solution of the free particle problem again. So we have now to impose the limiting Dirac condition, which in this simple case reduces to $a = 0$. In other terms, the Dirac or physical solutions will be the exceptional ones which, by spiraling around the separatrix, will reach one of the planes $x = -1$, $x = 1$ exactly with $a = 0$; the first ones will be reflected solutions, and the other ones transmitted solutions. Among the solutions of the free particle problem reaching the plane $x = -1$ from left, those nearer and nearer to the point corresponding to the separatrix will perform a larger and larger number of turns before reaching again one of the planes with $|x| = 1$, and so there exist motions making any number of turns. In such a way the surface of the physical solutions (i.e. satysfying the Dirac condition $a(t) \to 0$ for $t \to +\infty$) will be folded, and with infinitely many foldings; moreover, to the "branch" related to a number n of turns, for a given initial mechanical state there will correspond both one reflected and one transmitted solution.

Thus the state of affairs has some similarity to the weak reflection effect of quantum mechanics, according to which a particle with energy slightly larger that the maximum of the barrier can be either transmitted or reflected by the barrier. Here, we find that there exists an energy strip, about a value slightly above the maximum of the barrier, such that for any value of the energy in the strip there exist both transmitted and reflected motions; moreover, for ϵ large enough the energy strip turns out to extend up to values relevantly lower than the maximum of the barrier, so that one has even an effect which is similar to the tunnel effect.

The similarity with the weak reflection effect or the tunnel effect of quantum mechanics is certainly striking, but one has to make clear that it is only qualitative and not also quantitative. The difference comes about when one realizes that the width of the strip is too small for physical values of the parameters. Indeed, the effective "small parameter" in the problem is a pure number that is obtained just by a rescaling, namely the number ϵ' defined by

$$\epsilon' = \frac{\epsilon}{\sqrt{mL^2/V_0}} \, ,$$

where m is the mass of the particle, L a typical length of the potential and V_0 the height of the barrier. By considering typical cases of physical interest, such as that of the alpha–decay, one finds for ϵ' values of the order of 10^{-4}, in place of values of the order 1 which are required to exhibit the nonuniqueness phenomenon.

4. Conclusions

We have thus illustrated the role of the "hidden parameter", namely the acceleration, in the model corresponding to the Abraham–Lorentz–Dirac equation: having fixed the mechanical state, the particle is transmitted through or reflected from the barrier, depending on the value of the hidden parameter. Some remarks are in order. First of all, the domain of the values taken by the hidden parameter depends on the mechanical state, because the number of sheets of the physical manifold above a given mechanical state depends on the state itself, increasing up to infinity when the mechanical energy approaches the value corresponding to the separatrix. Furthermore, the various sheets collapse together for $|x| \to \infty$; thus the various sheets become effectively indistinguishable far away from the potential barrier, and this means that it would be actually impossible to control the value of the acceleration, which thus plays the role of a really hidden parameter. All these facts seem to indicate that the hidden parameter entering the realistic model considered here has peculiar and unconventional properties which were never imagined before. This lets us hope that the consideration of the present model might prove useful for a discussion of the general aspects related to the problem of hidden parameters.

Notes

[1] From the historical point of view the situation is rather curious. Indeed, we were mainly referring to a review article by Plass,[7] to the book o Rohrlich[8] and to the mathematical paper of Hale and Stokes,[5] all written in the years 1960–1965. Now, in the paper of Hale and Stokes existence was proven, while the uniqueness problem was left open. On the other hand, through a review article by Erber (see [1], pag. 355) we found that two particular cases of nonuniqueness were already known. Everything goes back to Bopp who, in his beautiful paper[3] of the year 1943, in the middle of a discussion of a very general type, solved a very particular example (one–dimensional scattering by a potential step), showing that there exist cases with two solutions corresponding to the same initial mechanical data (position and velocity), one solution corresponding to reflection and the other to transmission. A little variant of that example (a potential step increasing linearly between the two constant values) was studied by R. Haag[4] in the year 1955, who showed in particular that nonuniqueness would occur only if the steepness were large enough. The same result of Bopp, namely existence of two solutions for a one–dimensional potential step, but for the relativistic Dirac equation, was rediscovered in the year 1976 by Baylis and Huschilt,[6] apparently unaware of the previous works of Bopp and Haag. All such examples were studied by the quoted authors in a very elementary way, suited to the particularly simple cases considerd. The thesis maintained by Erber is that, among the two "Dirac-type solutions" of Bopp and Haag, by some other physical reason only one

should be retained and the other discarded. On the other hand, in [9] it was shown that one has in general an unlimited number nonrunaway solutions, that appear all to be on the same footing, so that there is no way of choosing a privileged one among them, and some new interpretation is required.

References

1. Erber, T. (1961). "The classical theory of radiation reaction", *Fortschr. der Phys.*, **9** pp.343–392.
2. Dirac, P.A.M. (1938). *Proc. Royal Soc. (London)*, A**167**, pp.148–168.
3. Bopp, F. (1943). *Ann. der Phys.*, **42** pp.573–608.
4. Haag, R. (1955). *Z. Naturforsch.*, **10 A** (752).
5. Hale, J.K. and Stokes, A.P. (1962). *J. Math. Phys.*, **3** (70).
6. Baylis, W.E. and Huschilt, J. (1976). *Phys. Rev.*, **D13** pp.3237–3230.
7. Plass, G.N. (1961). *Rev. Mod. Phys.*, **33** (37).
8. Rohrlich, F. (1965). *Classical charged particles*, Addison-Wesley, Reading.
9. Carati, A., Delzanno, P., Galgani, L., Sassarini, J. (1994). "Nonuniqueness properties of the physical solutions of the Lorentz-Dirac equation", *Nonlinearity.*

A NOTE ON LIGHT DISPERSION FORMULA AND THE DEVELOPMENT OF MODERN PHYSICS

B. CARAZZA and G. GUIDETTI
Dip. di Fisica - Università di Parma
Viale delle Scienze - 43100 - Parma, Italy

and

N. ROBOTTI
Dip. di Fisica - Università di Genova
Via Dodecaneso 33 - 16100 - Genova, Italy

Abstract. This paper analyses the role of the dispersion formula in relation to the development of modern physics. After proving to be a decisive factor in the debate on the nature of X-rays, the dispersion problem highlighted the limits of the old quantum theory and played a vital role in the conceptual process that led to the formulation of Matrix Mechanics. The historical account presented in this paper enables us to comment on the meaning of Quantum Mechanics as it is normally interpreted.

1. Introduction

The refraction index behaviour as a function of frequency and the theoretical expression for the polarizability of a material may at first glance seem of secondary importance in the development of modem physics. However, the dispersion formula played a fundamental role on at least two occasions.

In the first instance it was used to establish definitively that X-rays are waves, or more precisely electromagnetic radiation.

When X-rays were discovered in 1895 (Roentgen, 1895) an interpretation of this type was by no means a foregone conclusion, because the phenomenology they demonstrated was so anomalous as compared with forms of electromagnetic radiation encountered up to that time.

93

C. Garola and A. Rossi (eds.), The Foundations of Quantum Mechanics, 93-104.
© 1995 *Kluwer Academic Publishers.*

The nature of X-rays was established only when reference was made to electromagnetic dispersion theory, and particularly to the formula obtained by Helmholtz (Helmholtz, 1893) for the refraction index.

Helmholtz's theory and above all Lorentz's model (which gained wider recognition only at a later date), of electrons elastically bound to their position of equilibrium (Lorentz, 1892), were used as a basis for the development of classical dispersion theory, to which Drude made a fundamental contribution (Drude, 1900).

If the damping term is ignored, the Lorentz-Drude formula for atomic or molecular polarizability as a function of the frequency of the incident radiation is given by:

$$\alpha = \frac{e^2}{4\pi \cdot m^2} \cdot \sum_i \frac{f_i}{v^2 - v_i^2}$$

where f_i is the number of electrons in an atom or molecule with a proper frequency v_i, and e,m are the charge and mass of the electron.

This expression predicts a rapid variation of the refraction index for frequency values at which absorption occurs, explains the so-called anomalous dispersion and does an excellent job of describing the phenomenological data.

At a later stage in the development of physics, the dispersion formula provided one of the most important opportunities to demonstrate the inadequacy of the so-called old quantum theory. Attempts to derive a similar result for polarizability on the basis of new quantum ideas prepared the ground for Matrix Mechanics, one of the two roots of Quantum Mechanics. It was the result of a particular approach involving the interpretation of classical expressions in quantum terms as we will go on to describe.

2. The discovery of X-Rays

In order to understand the role of the dispersion formula in determining the electromagnetic nature of X-rays, it is worthwhile trying to reconstruct the theoretical-experimental context regarding so-called "invisible" radiation in which X-rays were discovered.

The fact that the solar spectrum is not limited to the visible, but extends beyond visible violet and below visible red, began to be recognised at the beginning of the nineteenth century. Systematic studies of the matter were carried out around 1840 and led to the production of experimental evidence of infrared and ultraviolet regions thanks to Hershel, and Ritter and Wollaston respectively. Both regions were invisible and characterised by rays defined according to their properties as "calorific rays" and "chemical" or "phosphorogenic rays" (so-called because of their ability to cause florescent or phosphorescent effects).

In 1867 Hittorf (Hittorf, 1867) immediately made use of "phosphorogenic" rays in his interpretation of a new type of "invisible ray" that had just been discovered, and was connected with discharge processes in rarefied gas. On the basis of the power demonstrated by these new rays to produce fluorescent and phosphorescent effects, Hittorf proposed an interpretation in which they were identified as a form of ultraviolet rays emitted by the cathode. They were therefore called "cathodic rays".

A wide-ranging debate on the nature of these rays was immediately opened. It continued for about thirty years until the discovery of the electron. On one side were supporters of the so-called "wave theory", who, according to Hittorf, considered cathode rays as electromagnetic rays with a very short wave length. On the other were supporters of the "particle theory", who believed that the "rays" were negatively ionized atoms or molecules.

So far as concerns the study of "cathodic rays", we would like to mention an innovation introduced in 1894 by Lenard (Lenard, 1894) which, in our opinion, was decisive in opening the way to the discovery of X-rays, even if the process was not immediate and direct.

The innovation was strictly technical, but had important methodological implications and the object was to analyse the nature of cathode rays in the "cleanest" possible way.

Before the introduction of Lenard's innovative technique, cathodic rays were analysed in glass vacuum tubes, i.e. in the same place where they were generated.

Lenard changed all this. His starting point was Hertz's observation in 1892 (Hertz, 1892) that cathode rays were "capable of passing through thin metal films". He used this fact to design a new type of vacuum tube in which the wall opposite the cathode (which in normal conditions blocked the cathodic rays) was replaced with a thin metal film or "window".

Using this system and sufficiently low pressure (between 10^{-2} and 10^{-3} mmHg) the cathode rays, after passing through the window, were propagated outside the tube, enabling them to be studied in the "open air".

This innovation finally enabled the phenomenon to be studied, and sophisticated measurements to be made on the absorption of cathodic rays by various materials.

The use of Lenard's vacuum tube (with the metal window that with hindsight we know to be an excellent producer of X-rays) was, in our opinion, decisive in Roentgen's discovery of X-rays (Roentgen, 1896), as he himself admitted during an interview the following year in 1896.

When Roentgen made his measurements, a barium platinocyanide phosphorescent screen was placed about two meters from the tube. We now know this substance to be particularly sensitive to X-rays. In these conditions he found that at the moment of discharge the screen was illuminated by a "brilliant fluorescence".

As the tube was isolated from the outside, the effect observed could not be due to light rays or ultraviolet rays produced by the discharge. Likewise it was not possible to attribute the effect to a new type of cathodic ray, as Lenard's experiments had already shown that they were only propagated in air for a distance of about 10 centimetres. In the final analysis, the combination of all these conditions could not but suggest that a new previously unknown phenomenon was involved. This was the position adopted by Roentgen. After verifying that similar phosphorescent effects could be obtained (to a lesser degree) even by "shielding the screen with a 1,000 page book or a 15 mm thick aluminium plate", and that the same phenomena could be obtained using normal vacuum tubes, Roentgen attributed the effects observed to an as yet unknown type of "invisible radiation", which he appropriately called "X rays".

3. The dispersion formula and the nature of X-Rays

Following Roentgen's discovery, the attention of almost the entire scientific community focused on these new, spectacular "invisible rays". The first photographs made by Roentgen of the bones in a hand and other objects hidden from the human eye are famous, and were used to demonstrate science's ability to photograph the invisible.

The most pressing and important matter raised by Roentgen's discovery was that of establishing the nature of these rays, which were so different from those already encountered. On one hand they couldn't be identified with some sort of cathodic rays because, in addition to their high power of penetration, and unlike cathodic rays, they were not deviated by a magnetic field (as Roentgen immediately verified). On the other hand, their natural identification with electromagnetic waves such as ultraviolet light (which give rise to phenomena such as fluorescence and phosphorescence, leave traces on photographic plates and ionize gas) was hindered by the fact that, in addition to their penetrative power, and unlike all other known rays (ultraviolet, visible, infrared) they were not reflected, refracted or polarized.

To explain the "anomalous" behaviour of the new rays, Roentgen, in his first paper (Roentgen, 1896), made use of an old conception compatible exclusively with the electromagnetic theory of Weber (action at distance). This had been used in 1895 by Yauman to interpret a number of the properties of cathodic rays. Roentgen wrote:

"There seems at least some connection between the new rays and light rays in the shadow pictures, and in the fluorescing and chemical activity of both kinds of rays. Now, it has been long known that besides the transverse light vibrations, longitudinal vibrations might take place in the ether, and according to the view of different physicists must take place. Certainly their existence has not up till now been made evident, and their properties have not on that account been experimentally investigated.

May not the new rays be due to longitudinal vibrations in the ether?"

This and other even more imaginative conjectures, based also on the existence of "vortices in the ether", were put forward immediately after Roentgen's discovery. They were quickly superseded when, in the attempt to understand the optical behaviour of X-rays, scientists started examining what should have been the natural reference for these phenomena - dispersion theory. This was the only way to reach a solution to the problem of the nature of X-rays.

In this context, the electromagnetical formula proposed by Helmholtz in 1893 (Helmholtz, 1893) describing the refraction index as a function of frequency, played a fundamental role.

In the months that followed Roentgen's discovery, various physicists (including O. Lodge, J.J. Thomson, S. Thompson, P. Raveau, etc.) looked back over the dispersion theories so far proposed and discovered the theory Helmholz had developed three years earlier on the basis of Maxwell's equations (Helmholtz, 1893). As was widely publicised at the time in scientific journals, Helmholtz's theory predicted that short wave length electromagnetic waves had a refraction index equal to one whatever substance was used, and were therefore neither reflected, refracted nor polarized. And this was exactly what experiments on X-rays showed.

Refraction index formulae compatible with the behaviour of X-rays could also be deduced from other dispersion theories as Raveau observed for the first time at a meeting of the French Physics Society held immediately after Roentgen's discovery. Examples include Sellmeier's 1871 mechanic theory, the theory (once again based on mechanics) developed by Maxwell in 1869 and republished by Rayleigh in 1899 (Maxwell, 1869), Helmholtz's 1874 elastic theory and Ketteller's 1892 theory (Ketteler, 1892).

Unlike the others, and as was widely recognized, Helmholtz's 1893 theory had a fundamental advantage - almost thirty years after the publication of Maxwell's treatise this was the first dispersion theory to be based entirely on electromagnetic concepts.

Together with its ability to account for the anomalous optical properties of X-rays, this was an important point in favour of Helmholtz's theory. Its convincing prediction that at short wave lengths all materials behaved like a vacuum so far as dispersion is concerned led to the rapid and definitive solution of the problematic nature of X-rays. As was noted by a number of physicists, Helmholz's formula demonstrated that they were necessarily electromagnetic waves of extremely short wave length, and from then on were also referred to as "ultra-ultra violet rays".

In its 7 July 1896 edition the then prestigious English journal "Electrician", after defining the theory of Helmhotz as "a marvellous example of mathematical prediction", continued:

> "The great German physicist worked out the properties of electromagnetic waves of various lenghts, with the result that, three years and more ago, before even the peculiar properties of Roentgen rays, were known, Helmholtz accurately described, as regards reflection and refraction, the behaviour of the numerous varieties of X-rays now known".

After this contribution to the identification of the nature of X-rays, the dispersion formula was once again the focus of attention when the first quantum formulations encountered problems.

4. The crisis of the old Quantum Theory

With the formalization of general quantization rules and requirements for "correspondence" in order to assess the intensity of spectral lines the old quantum theory now seems relatively compact and rational. But all this, despite its successes, cannot hide the fact that the theory was limited to few phenomena; it was a kind of mathematical model that can describe the spectra of simple systems like hydrogenoid atoms, subjected eventually to electrical or magnetic perturbations, with a fair degree of accuracy. The theory was also flawed by internal contradictions, combining new quantum ideas with the notions and the language of classical physics (Jammer, 1966).

These defects were highlighted around 1921-22 when there was no rapid progress regarding complex atomic systems, and attempts to extend the first quantum methods to other phenomena involving the interaction of the electromagnetic field with matter had failed (Mehra, Rechenberg, 1982). The end result was a veritable "crisis" of the old quantum theory.

The most ambitious goal in the framework of the theory was probably the attempt to account for the chemical-physical properties of elements on the basis of their atomic structure. Bohr tried to achieve this goal, but the theory was not up to the task. We now know that a complete understanding of the structure and properties of atoms and molecules requires modern quantum mechanics, Pauli exclusion principle and electron spin, all of which were obviously missing at that time.

In addition to the difficulties encountered in dealing with complex atomic systems, a further difficulty was created by the assumption that stationary states were particular, "permitted" classical states. In some cases this led to manifest absurdities or physically unacceptable conclusions.

For example, in the case of degenerate systems, the choice of different systems of separable variables led to identical energy values for stationary states, but to different stationary orbits, implying that the physical description of the system was not unambiguous.

It is clear that the theory provided valid indications, as is demonstrated by the initial successes. But it became more and more evident that it could only be taken as an indication or bridge to a general, coherent quantum theory that could easily handle all microscopic processes.

In the meantime, to overcome the difficulties encountered, less and less use was made of quantization rules and more and more of the principle of correspondence. Reasoning "by correspondence" was raised to a form of art, which allowed the correct quantum expression to be "divined" starting from classical results.

5. The dispersion formula in the old Quantum Theory

The limits of the old quantum theory were particularly evident as regards the phenomenon of dispersion.

In a classical interpretation based on elastically bound electrons, the dispersion, emission and absorption of light by atoms or molecules are intrinsically connected, as it is verified empirically.

The charged harmonic oscillator model however, was in contrast with the atomic description given by the old quantum theory. In the Bohr-Sommerfeld atom, the electrons are not described as oscillators, but are subject to the coulomb potential of the central nucleus and follow keplerian orbits. And of course the classical analysis did not use the concept of "quanta", which was by then accepted. The new methods for studying atomic phenomena were therefore expected to provide in turn a quantum explanation of dispersion. The task in hand seemed a perfect occasion to apply perturbation methods.

The first physicist to embark on this endeavour was Epstein (Epstein, 1922), who immediately came across a major problem. The resonance phenomenon was found at the classical revolution frequencies of the electron in a stationary state and not, as was observed experimentally, at the absorption frequencies predicted by the theory,which are connected with the quantum jump between two stationary states.

The old quantum theory also did not provide a way of connecting emission, absorption and dispersion as was the case in classical theory and as was clearly demonstrated experimentally. A paper written by Einstein in 1917 (which had already thrown light on the process of interaction between matter and radiation) made an important contribution to obtaining a dispersion formula using the new quantum concept. (Einstein, 1917)

During his study of conditions for the thermodynamic equilibrium between the radiation and a system of molecules presumed to exist only in discrete energy states, Einstein introduced the coefficients of emission (both stimulated and spontaneous) and of absorption and derived their reciprocal relations.

Thanks to this Ladenburg was able to interpret the quantities f_i (dispersion electrons), which we have already seen in the classical expression for the polarizability, in the new quantum language (Ladenburg, 1921). Considering the classical expression for the intensity of an absorption line, and allowing himself to be guided by the correspondence principle, he obtained the relationship:

$$f_i = A_k^i \cdot \frac{p_k}{p_i} \cdot \frac{m \cdot c^3}{8\pi^2 \cdot e^2 \cdot v_{ik}^2}$$

by setting the classical absorption frequencies equal to quantum transition frequencies v_{ik}, and considering an atom in quantum state i.

In the above expression A_k^i are Einstein's coefficients for spontaneous transition from state k to state i, and p_i, p_k indicate the degeneration of the corresponding states.

Thanks to the work of Ladenburg, Drude's formula and similar expressions could be translated into quantum language by translating the classical terms they contained into the corresponding quantum terms.

Soon afterwards Ladenburg and Reiche explicitly proposed a quantum formula for polarizability (Ladenburg, Reiche, 1923). This was obtained simply by substituting the symbols f_i with the quantum expression in the Lorentz-Drude formula, and where the absorption frequencies are given by the transition frequencies.

Implicit in Ladenburg's work was a representation of the atom (so far as concerns interaction with the electromagnetic field) by a set of harmonic oscillators with proper frequencies egual to the atomic absorption frequencies. A similar idea is very close to Slater's "virtual radiation field" (Slater, 1924), later recovered in a paper by Bohr, Kramers and Slater (Bohr, Kramers, Slater, 1924).

At a later date the term "virtual orchestra" was coined to refer to the method of considering radiation-matter interaction by means of a virtual set of harmonic oscillators.

Another quantum formula for atomic polarizability which acknowledged Ladenburg's results was proposed by Kramers (Kramers, 1924), who developed it during the preparation of his paper with Bohr and Slater. Kramers' formula differs from that of Ladenburg by the addition of a second term to account for the possibility that there is a stimulated emission for an atom in a excited state, i.e. "negative absorption" in the terminology in use.

Kramers wrote:

"The reaction of an atom, so far as concerns incident radiation, can therefore be formally compared to the action of a set of virtual oscillators in the atom linked to every possible transition to other stationary states." (Kramers, 1924).

The perturbation method of dealing with a stationary atomic state under the influence of an external field (which proved problematic in the case of dispersion) was therefore abandoned. Satisfactory quantum formulae were on the contrary obtained by translating and interpreting the results of classical theory into quantum terms. It therefore became necessary to follow the principle of correspondence ever more faithfully, and find a general method for translating the solution of a classical problem into the language of quanta.

Shortly after the appearance of the formula of Kramers, Max Born provided one such general method (to which he gave the name "Quantum Mechanics") for translating classical perturbation theory into quantum expressions (Born,1924).

Following Born's method the classical quantities are translated into quantum ones by substituting derivatives (which appear in the perturbation method) with finite differences. This allowed Born to obtain Kramers' formula.

Using Born's indications it was still necessary to study the classical solution to the problem, which was then reinterpreted. True quantum mechanics could only be achieved by means of a theory that allowed problems to be posed in quantum terms, using quantum quantities and quantum concepts in the starting equations. It was Heisenberg who made this conceptual jump.

6. The Origins of matrix mechanics

Given the importance attributed to the principle of correspondence and the successful results obtained by applying it, the foundations for a general and coherent quantum mechanics had to be laid on that principle. It had to be formalized in such a way as to automatically provide results which at the time were generally obtained as plausible suppositions.

As we have seen the most sophisticated example of the use of the correspondance principle before Heisenberg is that of Born. But as we have already pointed out, Born's approach still meant that it was necessary to solve the classical problem first.

Heisenberg's radical approach was to do without the crutches provided by the classical conceptual framework from the outset. Why continue describing the movement of electrons using concepts such as orbit, and variables such as position and velocity (which could not be measured directly at a microscopic scale), when the final expressions contained macroscopic observable quantities?

Heisenberg decided to place the "virtual orchestra" at the center of the theory, and consider as basic elements of the physical description the observable quantum frequencies and the quantum transition amplitudes.

The final result of Heisenberg's work was the paper that marked the beginning of matrix mechanics. The paper was completed at the beginning of July in 1925 and was subsequently published with the title "Quantum - theoretical reinterpretation of kinematics and mechanical relations" (Heisenberg, 1925).

In his introduction Heisenberg states:

"It is well known that the formal rules used in quantum theory to calculate observable quantities such as the energy of the hydrogen atom can be seriously criticised for containing relationships between quantities that cannot, it appears, be observed in principle, such as the position and period of revolution of the electron. (...) It would seem more reasonable to attempt to establish a theory of quantum mechanics that is analogous to classical mechanics, but in which only relations between observable quantities appear." (Heisenberg, 1925).

As a new theory of quantum mechanics was close to being established, and the field was being cleared of non-observable classical terms, the question arose as to what to do with expressions belonging to classical physics.

As we have seen, the common aim was to translate each classical relation into the corresponding quantum expression, and in the work of Heisenberg they always appear in parallel. He observed that the basic equation of classical physics, i.e. Newton's law as well as the quantization rule of the old quantum theory (expressed as the famous "phase space integral") must continue to be regarded as valid as "dynamic" equations.

They must be reinterpreted by re-writing the terms they contain in the new language. What must be re-interpreted is therefore the description of microscopic objects (the "kinematics" of systems), as the title of the article indicates. Heisenberg therefore attributes a symbolic value to the variables that describe microscopic objects in classical mechanics, while their formal relations remain unchanged.

The final result is a set of algebraic equations that link quantum-mechanical quantities, i.e. "directly observable" quantities.

At this point we cannot fail to notice that this result can be presented or interpreted as a success for the methodological and epistemological petitions of neopositivism.

7. Final Remarks

It was not our intention to provide a detailed description of Heisenberg's original work but to demonstrate how it constitutes the final and most important of the attempts to apply the correspondence principle. These attempts were motivated by problems such as that of the quantum dispersion formula.

At last matrix mechanics was a successful way to translate in one fell swoop the formalism of classical mechanics into a new "quantum" formalism. This demonstrates the original nature of the new matrix mechanics, dictated by a decidedly positivistic attitude. And as the generally accepted interpretation of quantum mechanics is basically a product of the way of thinking which lead Heisenberg to the matrix mechanics, we can argue that Quantum Mechanics as it is commonly interpretated has the characteristics of a phenomenological theory related to the measurements of microscopic systems using macroscopic apparatus.

References

Bohr, N., Kramers, H.A. and Slater, J.C. (1924). 'The quantum theory of radiation', *Philosophical Magazine*, **47** (785).

Born, M. (1924). ' Uber Quantenmechanik', *Zeitschrift fur Physik*, **26** (379).

Drude, P. (1900). 'Zur Geschichte der elektromagnetischen Dispersionsgleichungen', *Annalen der. Physik*, **1** (437).

Einstein, A (1917). 'Zur Quantentheorie der Strahlung', *Physikalische Zeitschrisft*, **18** (121).

Epstein, P.S. (1922). 'Die Storungsrechnung im Dienste der Quantentheorie. I. Eine Methode der Storungsrechnung', *Zeitschrift fur Physik*, **8** (211); ' Die Storungsrechnung im Dienste der Quantentheorie. III. Krritische Bemerkungen zur Dispersionstheorie', *Zeitschrift fur Physik*, **9** (92).

Heisenberg, W. (1925) 'Uber die quantentheoretische Umdeutung kinematischer und mechanischer Beziehuungen', *Zeitschrift fur Physik*, **33** (879).

Helmholtz, H. (1893). 'Electromagnetische theorie der farbenzerstreuung', *Wiedemann's Annalen*, **XLVIII** (133).

Hittorf, M.W. (1867) 'Sur la conducibilité électrique des gaz', *Annales de Chimie et Physique*, **2** (268).

Jammer, M. (1966). *The Conceptual Development of Quantum Mechanics*, McGraw-Hill Book Company, New York.

Ketteler, E. (1892). 'Der Grenzbrechungsexponent fur unendlich lange Wellen; Transformation der Dispersions-gleichungen', *Wiedemann Annalen*, **46** (582).

Kramers, H.A. (1924). ' The law of dispersion and Bohr's theory of spectra', *Nature*, **113** (673).

Ladenburg, R. (1921). 'Die quantentheoretische Deutung der Zahl der Dispersionselektronen', *Zeitschrift fur Physik*, **4** (451).

Ladenburg , R. and Reiche, F. (1923). 'Absorption, Zerstreuung und Dispersion in der Bohrschen Atomtheorie', *Naturwissenschaften*, **11** (584).

Lenard, P. (1893). 'On the cathode rays in gases at the pressure of the atmosfere in the highest vacuum', *The Electrician*, **XXXI** (258).

Lorentz, H. A. (1892) 'La teorié électromagnétique de Maxwell et son application aux corps mouvants', *Arch. Neérl. d. Sc. Exact. et Nat.*, **25** (263).

Maxwell, C. (1869). *Mathematical Tripos Examination*, reported in: Lord Rayleigh (1899). 'The Theory of anomalous dispersion', *Philosophical Magazine*, **XLVIII** (151).

Mehra, J. and Rechenberg, H. (1982). *The Historical Development of Quantum Theory*, Springer-Verlag, New York.

Roentgen, K. (1895) 'On a new form of radiation', *The Electrician*, **XXXVI** (415). English translation from *Sitzungsber der Wurrzburger Physikal. Medic. Gesellschaft*, 28 Dec. 1895.

Slater, J.C. (1924) 'Radiation and atoms', *Nature*, **113** (307).

QUANTUM MECHANICS, OBJECTS AND OBJECTIVITY

E. CASTELLANI

Abstract. Turning from classical to quantum physics, new problems arise with regard to the traditional philosophical question of what a 'physical object' is. A recent 'group-theoretical' approach to the question as to whether it does make sense to speak of 'quantum objects' is illustrated, investigating the connection it affords with the traditional problem of the 'objectivity' of physical knowledge. The individuality issue for quantum particles is also taken into account.

> "Mais ce que nous appelons la réalité objective c'est, en dernière analyse, ce qui est commun à plusieurs êtres pensants, et pourrait être commun à tous."
> (H. POINCARÉ, *La valeur de la Science*, 1905)

1. Introduction

It is a rather common opinion that many aspects of the 'classical' concept of a physical object (when not the concept itself) must be given up in considering microphysical entities, i.e., the entities described by quantum physics. And it is largely agreed that, because of problems coming out specifically in a quantum context such as, for instance, the so-called 'measurement problem', new difficulties arise with regard to the traditional philosophical question of what can be taken as 'objective reality' in the world described by physics. Starting from this kind of assumptions, in what follows we shall take into consideration some aspects of the object problem with regard to contemporary physics. More precisely, we shall examine to which extent it is possible to speak of 'objects' in the case of microphysical entities and how this question can be related to the general problem of the objectivity of scientific knowledge.

Let us begin with what it is usually assumed to be a 'classical' physical object. This is, more or less, the idea of an object which is abstracted from

105

C. Garola and A. Rossi (eds.), The Foundations of Quantum Mechanics, 105-114.
© *1995 Kluwer Academic Publishers.*

our experience of everyday material 'things': that is, things having mass, a definite location in space and time and 're-identifiability' or persistence through time. How objects can in fact be identified through their mass and spatio-temporal determinations and whether we do need further identification criteria are questions typically debated in the philosophical literature on the object theme. Here, it will be sufficient to recall that a quite popular view on physical objects and their re-identifiability is that according to which, for identifying an object as that same object and no other, a prominent role is assigned to space-time properties: 'space-time location', 'space-time continuity' (i.e., a continuity condition upon the space-time path of an object) and the condition given by an 'impenetrability assumption' (objects which are distinct cannot occupy the same position at the same time) are the basic ingredients of such a view.

Now it is clear that a view as the one above cannot apply to the entities described by quantum physics. Microphysical entities — take, for instance, the so-called 'elementary particles' — are not always massive, to begin with. Their localizability at any instant of time is not a trivial question and 'spatio-temporal continuity' is no more at hand. Moreover, what can 'impenetrability' really mean in the case of particles presenting wave aspects?

Should we then totally dispense with the object concept at the microphysical level? Or should we try to generalize the concept in order to obtain such a comprehensive definition of a physical object that it could apply to everyday material things as well as to microscopic entities? This last way is that followed, for example, by W. V. O. Quine in his 'Whither Physical Objects?'. In this essay, what he obtains as a result of his attempt to arrive at a very general definition of physical objects is a progressive 'evaporation' of those same objects: from the material things or 'bodies' he begins with, to 'space-time regions', to end up with 'pure sets of numerical coordinates'.[1]

But a more 'relative' point of view can be adopted. This is the view on which what counts as a 'physical object' cannot be established once for all, for all kind of contexts. The basic assumption is that we don't dispose of a given notion of physical object: what the objects of a certain context are (for example, what the objects in the context of classical mechanics or quantum mechanics are) must be 'determined' and this process of determination largely depends on the characteristics of the context. We can put it also in this other way. What we usually do in defining something as an 'object' is pointing to given properties pertaining to this something which, we assume, are essential for conferring on their 'carrier' the dignity of an 'object'.[2] Conceiving of objects as 'carriers of properties', the question then becomes what kind of properties and prescriptions we do need in order to 'construct' or 'constitute' our objects, and this will depend, according to the above point of view, on the context in which we are dealing.

This is, in very short, the epistemological framework of an approach to the problem of physical objects which, although it is grounded on concepts such as those of *invariance* and *group of trasformations* that are not quite new, is just beginning to be systematically developed. To this approach, which can be called the *group-theoretical approach* to the object problem, will be devoted the following pages. We shall first describe in what the approach consists, recalling its historical background and underlining the possibility it affords of a connection with the objectivity issue. Then, in the last part of the paper, we shall briefly illustrate how, according to the group-theoretical approach, one can effectively proceed towards the determination of 'quantum particles'.

2. Invariance, symmetry groups and objectivity: the group - theoretical approach to the object question

As already said, the group-theoretical approach to the object problem is grounded on the notion of *invariance*. Now, invariance surely doesn't constitute any new argument in relation to the problem of physical objects. We have mentioned, for example, the importance attributed to the condition of 're-identifiability' in the philosophical debate on physical objects. Re-identifiability is nothing other than invariance through time or, more generally, invariance through change in time, which is also the notion ordinarily known as 'permanence' or 'persistence'. What is new, in the approach in question, is rather the exploitation of this traditional idea of 'permanence' by using the mathematical notion of 'invariance with respect to a group of transformation' (or 'symmetry'). This is basically motivated by the fundamental role which the notions of *invariance, group of transformations* and *symmetry* have acquired in contemporary physics. Let us shortly see in which sense this result, which is of purely mathematical and physical nature, can be used in relation to the philosophical issues of 'physical objects' and 'objectivity'.

To begin with, the possibility of using the theory of transformation group and their invariants in addressing the problem of determining objects is not limited to the case of physical objects. This possibility is grounded on the idea that individuating 'objects' in a given context is closely connected to individuating 'invariants' with respect to the symmetry group of the context. And, historically, this idea was first introduced with regard to 'geometrical objects', as a consequence of the new conception of geometry proposed by Felix Klein with his famous 1872 *Erlanger Programm*: that is, the conception according to which each 'geometry' is characterized by a given transformation group and the 'geometrical' properties of an object are those properties which are invariant with respect to the transformations of the group. With the application of group theory to other domains of science and in particular to physics, the above view could then be extended to other kind of objects and, first of all, to 'physical objects'.[3]

As regards physics, group theory revealed to be the appropriate mathematical tool for investigating the consequences of physical symmetries, that is, those symmetry properties of physical systems which are usually expressed in terms of *invariance principles* — first of all, the space-time symmetries postulated through the *principle of relativity*. A decisive support to the above conception of objects was in fact provided by the interpretation of the theory of relativity (special relativity) as a theory of physical invariance with respect to the group of transformations of spatiotemporal reference frames or 'observers'.[4] The common idea that what has 'objective value' should not depend upon the particular perspective under which it is taken into account could so be applied to physics in the following terms: *objective* is what is invariant with respect to the transformation group of reference frames, or, in the words of Hermann Weyl, "objectivity means invariance with respect to the group of automorphisms [of space-time]".[5] So understood, that of invariance with respect to the space-time symmetry group could therefore be taken as an *objectivity condition* for the physical description of the world.

The invariance postulated through the principle of relativity is that of physical laws, that is, of given relations between certain physical quantities. What is of interest, from the point of view of the object question, is how this objectivity condition for the laws of physics can be used with regard to the determination of 'objects' within a given physical domain.

The starting point for considering the significance of physical symmetries with respect to the constitution of objects is the well known 1939 work of Eugene Wigner on the representations of the inhomogeneous Lorentz group, the space-time symmetry group of special relativity, also known as 'Poincaré group'.[6] In determining the unitary representations of the space-time symmetry group within the context of relativistic quantum physics, Wigner's aim was at the same time to investigate the connection between such representations and the wave equations of quantum relativistic 'particles'. Among the far reaching consequences of his work, we find the possibility of obtaining a complete classification of (free) relativistic quantum 'elementary' systems in correspondence with the classification of the *irreducible representations* of the Poincaré group. It is precisely this possibility of associating 'elementary physical systems' with the irreducible representations of the symmetry group that provides the basic motivation for the group-theoretical way of approaching the object problem.

Since Wigner's work, it has indeed become quite usual to classify the so-called 'elementary particles' on the basis of their symmetry properties (to be more precise, the symmetries of the theory describing their behaviour). On this view, each 'elementary particle' is associated with an irreducible representation of the space-time symmetry group. This implies, in particular, that the particle in question has a given number of *invariant properties*, and that these are exactly the properties which characterize that sort of particles. It is important to note that what one obtains in this way is at most a *class* of particles, not a particle as an *individual object*. The invariant properties which are ascribed to a 'particle-object' on the basis of group-theoretical considerations — as, for example, definite properties of mass and spin are ascribed to a (quantum) particle which is associated with an irreducible representation of the Poincaré group — are necessary for determining that given particle (an electron couldn't be an *electron* without given properties of mass and spin), but they are not sufficient for distinguishing it from other similar particles. In addition to these 'necessary' properties (sometimes also called 'essential' properties), one does need further specifications in order to constitute a particle as an individual object. But before coming to the individuality issue, let us enter into some more details of the group-theoretical approach.

3. Constitution of 'particles'

Given a *physical system* (the 'object' to determine) and its description in terms of its *observables*, *states* and *dynamical law*, the point is to see what follows from the assumption of *space-time symmetry*, that is, the invariance of the theory with respect to the group of transformations of reference frames (the Galilei or Lorentz inhomogeneous group, depending on whether we have Galilean or Lorentz symmetry).

In general, if G is a symmetry group of a theory describing some physical system, this implies that the states of the system should transform into each other according to some *representation* of the group: the group operations are 'represented' in the states space by operations relating the states one to each other.

We can so begin to proceed towards the determination of a 'Galilean (Lorentz) particle' by defining as a 'Galilean (Lorentz) system' a physical system whose states form a representation space for the Galilei (Lorentz) group. In order to obtain 'particles', the strategy is then to select, among Galilean (Lorentz) systems, those which are *elementary*. That of 'elementarity' is a quite natural requirement for a physical system representing a single particle. In group-theoretical terms, an *elementary system* is a system whose set of states constitutes a representation space for an *irreducible representation* of the space-time symmetry group.[7] This means, in other words, that there is a correspondence between the 'elementarity' of the system and the 'irreducibility' of the representation associated with the system.

At this point, we can propose the following definition of a 'Galilean (Lorentz) particle':

> *Definition*: a Galilean (Lorentz) particle is an elementary Galilean (Lorentz) system, i.e., a physical system whose states form an irreducible representation space for the Galilei (Lorentz) group.[8]

Let us see what the above definition implies. For this purpose, we shall very shortly recall some aspects of the *Lie group formalism* appropriate for investigating the irreducible representations of space-time symmetry groups.

Space-time transformation groups such as the Galilei and Lorentz groups are *continuous groups*, and in particular *Lie groups*. This means that the group elements are functions of a certain number r of continuous parameters a_l ($l = 1, 2, ...r$) and that they can be written in terms of a corresponding number r of infinitesimal operators X_l, the *generators* of the group. The generators satisfy the 'multiplication law' represented by the 'Lie brackets' relations

$$[X_s, X_t] = c_{st}^q X_q$$

so forming what is called the *Lie algebra* of the group. The coefficients c_{st}^q are constants characterizing the structure of the group and therefore called the *structure constants* of the Lie group. For every Lie group, we can construct operators which are scalar quadratic in the infinitesimal operators X_l, the so-called *Casimir operators*

$$C = \sum_{s,t} (g^{-1})_{st} X_s X_t$$

with the matrix elements g_{st} given by $g_{st} = \sum_{pq} c_{sq}^p c_{tp}^q$. Casimir operators have the special property of commuting with all the infinitesimal operators X_l, that is, they are the fundamental *invariants* of the group. In the context of the theory of group representations, this implies, in particular, that in irreducible representations the Casimir operators are simple multiples of the unit operator, whence the possibility of labelling the representations directly in terms of the eigenvalues of these operators. The eigenvalue spectra of the invariants of the group therefore provide the labels for classifying the irreducible representations of the group. On this fact is grounded the possibility of associating the values of the invariant properties characterizing physical systems with the labels of the irreducible representations of symmetry groups.

This is the general scheme which can be abstracted from the usual way of proceeding in quantum relativistic physics (quantum field theory) for classifying elementary particles. Accordingly, in order to obtain 'Galilean (Lorentz) particles' one shall have to examine how the considered physical systems can respectively provide a representation space for the irreducible representations of the Galilei (Lorentz) group, and what kind of invariant properties can be consequently attributed to such systems. This can be done for 'classical' as well as 'quantum' systems.[9] In the quantum case, the representation space for the irreducible representations of the symmetry group will be provided by the Hilbert space of the states of the physical system.[10] For arriving at a *quantum particle*, one shall have therefore to examine the irreducible representations in this Hilbert space and the resulting invariant properties.

4. The individuality question

As already said, what we obtain, according to the above way of proceeding, are at most classes of particles, not particles as individual objects. We are so faced with the well known 'individuation problem', that is, in our case, the problem of how to obtain 'individuating' properties which can confer individuality on the particle in question, distinguishing it from the other similar particles.

We have mentioned how, for ordinary physical objects, a quite diffused view is that according to which individuality is conferred by means of 'accidental' spatio-temporal determinations such as, for example, the object's position and its spatio-temporal career or trajectory. As already noted, such kind of individuation theory cannot directly apply to microphysical entities. The situation is that which, more than forty years ago, was so summarized by Hans Reichenbach in his classical pages on the "genidentity of quantum particles": "The usual methods of identification break down in the atomic domain [...] we have to replace an individual examination of particles by inferences based on statistical properties of an assemblage of particles." [11]

The introduction of statistical considerations into the debate concerning microscopic objects raises a further problem with regard to the identity of quantum objects, the problem of the physical indistinguishability of particles of the same kind (often called, not very happily, 'identical particles'). The appearance of new statistics (the Bose-Einstein and Fermi-Dirac statistics) for aggregates of similar quantum particles and the connection which was established between those 'quantum statistics' and the *permutation invariance* principle for quantum particles of the same kind have started a very lively discussion on the significance of the notions of identity, individuality and indistinguishability in the case of quantum objects. In particular, the existence of collections of 'identical particles', that is, of particles which are *indistinguishable* (in the sense that they share all the same properties) but nevertheless *numerically* distinct, is at the ground of the much discussed so-called 'problem of identical particles', very often connected, in the literature, with the question as to whether the famous Leibniz's *Principle of the Identity of Indiscernibles* should be regarded as violated in quantum physics.[12]

Without entering, here, into the details of the individuality issue for quantum objects, let us very shortly mention a recent way of approaching the question from the group-theoretical point of view. The ground notion is that introduced by G. W. Mackey of a *system of imprimitivity* associated with a symmetry group. In the literature, the systematic use of the notion of imprimitivity systems with regard to the definition of physical particles is first due to C. Piron.[13] The idea is to arrive at a definition of a 'particle' as an individual object by determining 'individuating' observable quantities, such as for instance the position and momentum, with the help of the method of imprimitivity systems.

This is made by taking into account the *logico-algebraic viewpoint* for describing physical systems: that is, the viewpoint introduced in the 1930's by John von Neumann in the case of quantum mechanics, according to which to each physical system is associated a (orthocomplemented) lattice $\mathcal{L}$ of *propositions*, the *logic* of the system.

In such a framework, if $A\{E\}$ is the proposition that, for a given system S, the value a of the observable quantity A lies in the set $E \in \mathcal{B}(\mathcal{R})$ (where $\mathcal{B}(\mathcal{R})$ is the family of the Borel subsets of the real line $\mathcal{R}$), the observable A may be identified with the mapping

$$A : E \mapsto A\{E\}$$

of $\mathcal{B}(\mathcal{R})$ into the lattice $\mathcal{L}$.

Given the triplet $(A, \mathcal{B}, \mathcal{L})$ for a physical system, the method is then to consider what the action of the space-time symmetry group G implies for the elements of the triplet. To state it very shortly, we shall in particular have, for the observable A, a *condition of covariance* which can be expressed in terms of the following *imprimitivity condition*:

$$A\{\sigma(g)[\mathcal{B}]\} = S(g)[A\{E\}] \qquad (*)$$

where $\sigma(g)$ and $S(g)$ are representations of the group G in terms of the automorphisms of $\mathcal{B}$ and of $\mathcal{L}$, respectively. The triplet $(A, \mathcal{B}, \mathcal{L})$ satisfying the above condition is a *system of imprimitivity* for the group G.

Let us directly jump to the conclusion, that is the following resulting definition of a 'particle': a 'Galilean (Lorentz) particle' is a system of propositions $\mathcal{L}$, for which an irreducible representation of the Galilei (Lorentz) group is defined, such that the observable 'identifying' quantities A_i satisfy the imprimitivity condition (*).

The properties A_i are precisely those having an 'identifying' function for the particle to define. Let us just recall that, in the case of Galilean symmetry, this method of constructing 'particles' has been explicitly developed for either classical and quantum particles by taking as 'identifying' properties the observables position, momentum and time.[14] In the quantum case, a way of dealing with the well known problem of the incommensurability of quantities such as the position and momentum is provided by what has been called 'unsharp objectification', that is the formulation of the objectification problem in terms of 'unsharp' observables.[15] On such a possibility is grounded the recently proposed programme of a systematic (approximate) constitution of 'quantum objects' by means of the application of the method of imprimitivity systems to the case of unsharp observables.[16]

Notes and References

1. Quine, W.V.O. (1976). 'Whither Physical Objects?', in R. S. Cohen et al. (eds.), *Essays in Memory of Imre Lakatos*, Reidel Publishing Company, Dordrecht, pp. 497–504.

2. Assuming that the objectifying role is not attributed to some kind of 'substance' or 'essence' trascending the entity's set of properties.

3. On this point, of particular interest is the contribute of Ernst Cassirer, 'The Concept of Group and the Theory of Perception', *Philosophy and Phenomenological Research*, Vol. V, No. 1 (1944), pp. 1–35, where the significance of the developments of the theory of transformation groups for the object problem is taken into consideration with regard to 'objects' in the context of geometry, physics and perception theory.

4. Max Born, in his 'Physical Reality' (*Philosophical Quarterly* 3, 39 (1953), pp. 139-149), offers a clear example of such a way of thinking, connecting his conception of 'objects' as "what is permanent in the flux of phenomena, the invariants", with the view of the theory of relativity as "an extension of this programme [Klein's 'Erlanger Programm'] to the four-dimensional geometry of space-time".

5. Weyl, H. (1952). *Symmetry*, Princeton University press, Princeton, p. 132.

6. Wigner, E. (1939). 'On unitary representations of the inhomogeneous Lorentz group', *Annals of Mathematics* 40, No. 1, pp. 149–204.

7. Note that the concept of an 'elementary system'is indeed broader than the intuitive concept of an 'elementary particle'. The point was thoroughly discussed in Newton, T.D. and Wigner, E.P. (1949), Localized States for Elementary Systems, *Review of Modern Physics*, 21 pp. 400–406.

8. Or, to be more precise (if we want to stress the fact that what we get, in this way, is a *class* of equivalent elementary systems): a Galilean (Lorentz) particle is physical equivalence class of elementary Galilean (Lorentz) systems.

9. With regard to Galilean symmetry, a very thoroughly analysis of such programme of both classical and quantum particles can be found, in particular, in some contributes of J.M. Lévy-Leblond. See for instance his 'Galilei Group and Galilean Invariance', in Loebl, E.M. (ed.), *Group Theory and Its Applications*, Academic Press, New York (1971), Vol. II, pp. 221–299, and, for the quantum case, his 'Galilei Group and Nonrelativistic Quantum Mechanics', *Journal of Mathematical Physics*, 4 No. 6 (1963), pp. 776–788.

10. Taking into account, here, only the case of 'pure' states. A quite general treatment of representation spaces for the space-time symmetry groups including also the case of 'mixed states' is of course possible. See for example Currie, D.G., Jordan, T.F., Sudarshan, E.C.G. (1963), 'Relativistic Invariance and Hamiltonian Theories of Interacting Particles', *Review of Modern Physics*, **35** pp. 350–375.

11. Reichenbach, H. (1956). *The Direction of Time*, ed. by M. Reichenbach, University of California Press, Berkeley, p. 228.

12. With regard to this whole debate on identity, individuality and indistinguishability in quantum physics, see for example Van Fraassen, B., 'Statistical Behaviour of Indistinguishable Particles: Problems of Interpretation', in P. Mittelstaedt-E. W. Stachow (eds.), *Recent Developments in Quantum Logic*, Bibliographisches Institut, Mannheim (1985), pp. 189–202; which offers also a sort of revue of the related issues. For more recent contributes see, among others, French, S., 'Identity and Individuality in Classical and Quantum Physics', *Australasian Journal of Philosophy*, **67**, No. 4 (1989), pp. 432–446; and Van Fraassen, B., *Quantum Mechanics: An Empiricist View*, Clarendon Press, Oxford (1991), whose chapter 11 is entirely devoted to the problem of 'identical particles'; Redhead, M. and Teller, P., 'Particles Labels and the Theory of Indistinguishable Particles in Quantum Mechanics', *British Journal for the Philosophy of Science*, **43** (1992), pp. 201–218.

13. Piron, C. (1976). *Foundations of Quantum Physics*, W. A. Benjamin, Inc., Reading, Massachusetts, pp. 93 ff.

14. Piron, C. *op. cit.*, pp. 93 ff.

15. See, in particular, by Busch, P., 'Unsharp Reality and the Question of Quantum Systems', in P. Lahti-P. Mittelstaedt (eds.), *Symposium on the Foundations of Modern Physics 1987*, World Scientific, Singapore (1987), pp. 105–125, and 'Macroscopic Quantum Systems and the Objectification Problem', in P. Lahti-P. Mittelstaedt (eds.), *Symposium on the Foundations of Modern Physics 1990*, World Scientific, Singapore (1990), pp. 62–76.

16. Such a programme was recently illustrated in Mittelstaedt, P., 'The Constitution of Objects in Kant's Philosophy and in Modern Physics', in P. Parrini (ed.), *Kant and Contemporary Epistemology*, Kluwer Academic Publishers, Dordrecht (1994).

TOWARD A LOGIC OF UNSHARP QUANTUM MECHANICS

G. CATTANEO

Dip. Scienze dell'informazione - Università di Milano
Via Comelico 39/41 - 20135 - Milano, Italy

and

F. LAUDISA

Dip. di Filosofia - Università di Firenze
Via Bolognese 52 - 50139 - Firenze, Italy

Abstract. The standard logico - algebraic approach to quantum mechanics singled out a class of structures, called *state - event - probability* structures and underlying orthodox Hilbert space approach. These structures constitued the so called *logic of quantum mechanics.*

With the development of the unsharp formulation the corresponding structures underlying generalized quantum mechanics, and called *state - effect - probability* structures, have been introduced. Two possible axiomatic formulations for the logic of unsharp (i.e., generalized) quantum mechanics are presented; the mutual relationships are investigated and some steps are taken in the direction of proving their equivalence.

1. Quantum logic as axiomatic formulation of quantum mechanics (QM)

The logical approach to quantum mechanics is one of the several axiomatic formulations of quantum theory. The application of what has been called the *axiomatic method* to the quantum theory stems from the attempt of defining rigorously the assumptions underlying the several interpretations of quantum mechanics, although there is no general agreement on the effective usefulness of axiomatic approaches to the field of the foundations of quantum mechanics.

The 'philosophical' presupposition of the logical approach to quantum mechanics in the '60 has been the idea according to which the foundations of quantum theory are *logical*, or *logico-algebraic*, in nature. This idea was

C. Garola and A. Rossi (eds.), The Foundations of Quantum Mechanics, 115-126.
© 1995 *Kluwer Academic Publishers.*

clearly expressed in Finkelstein's papers of the sixties and early seventies (Finkelstein, [1969], [1972]). The analogy with non-Euclidean geometry actually was the starting point of the paper "Is Logic Empirical?", where Putnam, referring to the quantum-logical approach to QM, argues that some 'necessary truths' of logic might turn out to be false for empirical reasons: empirical data or results might force us to deny some logical laws and reinforce the idea that logic too is an empirical science (Putnam, [1969]).

Quite apart from the philosophical developments of Putnam's positions, this view of the fundamental role of logical structures in the investigations on the foundations of quantum theory is present in the main contributions to the quantum logical approach - starting from the seminal paper by Birkhoff and von Neumann up to Mackey, Jauch, Piron, Varadarajan and others. The standard quantum logical investigations have allowed to single out the logico-algebraic structures underlying the usual formulation of quantum mechanics, usually called *state-event-probability* structures. But with the development of a significant generalization of quantum mechanical formalism, namely the unsharp formulation of quantum mechanics, the question arises whether it is possible - in a quantum logical perspective - to single out the logico-algebraic structures that possibly underlry unsharp quantum mechanics. An affirmative answer to this question is provided by the family of *state-effect-probability* structures, a family of more general structures with respect to the state-event-probability ones. In quantum logical terms, the individuation of these structures would allow then to move from the logic of (standard) quantum mechanics to a *logic of unsharp quantum mechanics*, since from a foundational point of view, the state-effect-probability structures bear the same relation to unsharp quantum mechanics as state-event-probability structures do to standard quantum mechanics.

In the sequel we will present two possible axiomatic formulations for the set of effects and we will make a first attempt to investigate their mutual relationship, paralleling the equivalence between state-observable structures introduced by Mackey, [1963], and state-event-probability structures, as shown by Maczynski, [1973], and Beltrametti-Cassinelli, [1976].

2. The unsharp generalization of Hilbertian QM

The unsharp approach to QM has developed a generalization of conventional QM on grounds that were both theoretical and experimental, with reference - in particular - to measurement-theoretic issues (cp. Busch, Grabowski, Lahti, [1989]). A key point is represented by the generalization of the notion of observable, as developed on the basis of the statistical analysis of physical experiments. In general (Busch, Schroeck, [1989]) the possible results of a measurement of a given unsharp observable are elements Δ of a certain σ-algebra $\Sigma(\Omega)$ of subsets of a values set Ω. The observable is realized as a mapping associating to any possible result of the measurement Δ a bounded self-adjoint operator $F(\Delta)$ on the Hilbert space $\mathcal{H}$, whose spectrum is contained in the real unit interval $[0, 1]$ (effect operator); in this way, once denoted by $\mathcal{F}(\mathcal{H})$ the set of all such effect operators, the observable is a mapping $F : \Sigma(\Omega) \mapsto \mathcal{F}(\mathcal{H})$ satisfying the standard conditions of being an effect-valued measure. In QM the probability of obtaining the result Δ in the state ρ is described by the expectation value $P_\rho^F(\Delta) = Tr[\rho \circ F(\Delta)]$ which, as Δ runs in $\Sigma(\Omega)$, is a classical probability measure. If in this expression $F(\Delta)$ is meant to stand for a projection operator, the standard Born rule of orthodox QM is recovered; being $F(\Delta)$ a projection operator, however, is not a necessary condition for the consistency of the probabilistic framework just outlined; this justifies the generalization resulting from the above definitions.

3. Axiomatic formulation of unsharp QM

Since 'orthodox' quantum logic was introduced in order to provide conventional QM with an axiomatic foundation, the development of a unsharp generalization of QM naturally raises the question whether it is possible to single out the logico-algebraic structures that underly unsharp QM: an affirmative answer would allow - á la Birkhoff-von Neumann - to speak of the *logic* of the unsharp QM. In this section we will take into account two possible axiomatic formulations of the structure of the effects, and we shall give some results along the direction of showing their equivalence. The first formulation is a slight modification of the original approach by Cattaneo, Garola and Nisticò, [1989], (from now on CGN), whereas the second is due to Cattaneo and Laudisa, [1994], (in the sequel CL).

3.1. FROM STATE - EFFECT - PROBABILITY TO STATE - EVENT - PROBABILITY

In a 1989 paper Cattaneo, Garola e Nisticò have presented a particolar axiomatic approach to QM in the spirit of Ludwig's interpretation. In presenting the main elements of this approach, we will briefly provide the intended interpretation that is given to its primitive notions, and then we will formulate the axioms of the mathematical structure itself. In this approach two primitive sets and a mapping relating them to the real unit interval are introduced:

(1) the set S of the *states*, associated with the macroscopic apparatuses that prepare the individual samples and the ensembles of identical,

noninteracting physical systems;

(2) the set $\mathcal{F}$ of the *effects*, defined by dichotomic measurement apparatuses that interact with an individual sample and that can produce a
certain macroscopic yes-no alternative.

The probabilistic connection between the elements of these two sets is given
by a probability function $P : S \times \mathcal{F} \mapsto [0, 1]$:

(3) if $x \in S$ and $f \in \mathcal{F}$, the numerical quantity $P(x, f) \in [0, 1]$ represents
the probability of occurrence of the 'yes' alternative for f when the
systems are in the state x.

DEFINITION 3.1 – A state-effect-probability triple is a triple $(S, \mathcal{F}, P)$,
where S and $\mathcal{F}$ are two nonempty sets, called - respectively - the set of the
states and the set of the effects, and P is a function $P : S \times \mathcal{F} \mapsto [0, 1]$,
called probability function.

Given a state - effect - probability triple $(S, \mathcal{F}, P)$ and an effect $f \in \mathcal{F}$,

$$S_1(f) = \{x \in S : P(x, f) = 1\}, \quad S_0(f) = \{x \in S : P(x, f) = 0\}$$

are defined - respectively - as the *certainly-yes* and the *certainly-no domains*
of f.

DEFINITION 3.2 – Let $(S, \mathcal{F}, P)$ be a state-effect-probability triple; we
define the triple as a *state-effect-probability structure* (SEP) whenever the
following axioms hold.

Axiom 1. (i): for $x, y \in S$, if for every $f \in \mathcal{F}$ $P(x, f) = P(y, f)$, then
$x = y$.

Axiom 1. (ii): for $f, g \in \mathcal{F}$, if for every $x \in S$ $P(x, f) = P(x, g)$, then
$f = g$.

Axiom 2: an effect $\underline{1}$ exists, called the certain effect, such that for every
$x \in S$ $P(x, \underline{1}) = 1$.

Axiom 3: for every effect f there exists another effect f' such that for
every $x \in S$

$$P(x, f) + P(x, f') = 1.$$

Axiom 4: for every countable family of numbers in the real unit interval
$\{\lambda_n \in [0, 1] : n \in \mathbb{N}\}$ such that $\sum_{n \in \mathbb{N}} \lambda_n = 1$ and every corresponding
countable family of states $\{x_n \in S : n \in \mathbb{N}\}$, there exists a state $\sum_{n \in \mathbb{N}} \lambda_n x_n$
$\in S$, called the σ-convex combination of the x_n with weights λ_n, such that
for every $f \in \mathcal{F}$

$$P\left(\sum_{n \in \mathbb{N}} \lambda_n x_n, f\right) = \sum_{n \in \mathbb{N}} \lambda_n P(x_n, f).$$

Axiom 5. (i): for every number $\lambda \in [0, 1]$ and every effect $f \in \mathcal{F}$, there exists an effect, denoted by λf, such that for every $x \in \mathcal{S}$

$$P(x, \lambda f) = \lambda P(x, f).$$

Axiom 5. (ii): for every orthopair of effects $f, g \in \mathcal{F}$, i.e., such that for every $x \in \mathcal{S}$,

$$0 \leq P(x, f) + P(x, g) \leq 1$$

an effect, denoted by $f \oplus g$ and called the sum of f and g, exists such that for every $x \in \mathcal{S}$

$$P(x, f \oplus g) = P(x, f) + P(x, g).$$

Finally, we define the effect $\underline{0} = \underline{1}'$ the impossible or absurd effect.

From Axiom 5, in the present form (i)–(ii), it immediately follows the "orthodox" version of Axiom 5 in CGN, [1989]:

Axiom 5-CGN: for every n-tuple of effects $(f_1, \cdots f_n)$ and for every n-tuple of nonnegative real numbers $(\lambda_1 \cdots \lambda_n)$ such that $\sum_{j=1}^{n} \lambda_j = 1$, there exists an effect $\sum_{j=1}^{\oplus n} \lambda_j f_j$, called the convex combination of the effects f_j, with weights λ_j, such that for every $x \in \mathcal{S}$

$$P\left(x, \sum_{j=1}^{n} {}^{\oplus} \lambda_j f_j\right) = \sum_{j=1}^{n} \lambda_j P(x, f_j).$$

In general, the converse is not true.

DEFINITION 3.3 – Let $(\mathcal{S}, \mathcal{F}, P)$ be a state-effect-probability structure. Let us denote by $\leq$ the partial order relation on $\mathcal{F}$ defined in the following way: for $f, g \in \mathcal{F}$,

$$f \leq g \quad \text{iff} \quad \forall x \in \mathcal{S}, \ P(x, f) \leq P(x, g).$$

PROPOSITION 3.1 – *Let $(\mathcal{S}, \mathcal{F}, P)$ be a state-effect-probability structure. The structure $\langle \mathcal{F}, \underline{0}, \leq, ' \rangle$ is a Kleene poset, with maximum $(\underline{1})$ and minimum $(\underline{0})$, where the mapping ' is an involutive antiautomorphism satisfying the regularity (Kleene) condition, namely where the following properties hold:*
(1)   for every $f \in \mathcal{F}$ $f = f''$;
(2)   for every $f, g \in \mathcal{F}$, $f \leq g$ implies $g' \leq f'$;
(re)   let $f \leq f'$ and $g' \leq g$, then $f \leq g$.

DEFINITION 3.4 – Let $(\mathcal{S}, \mathcal{F}, P)$ be a state-effect-probability structure. For $f, g \in \mathcal{F}$, we say that f is orthogonal to g ($f \perp g$) whenever $f \leq g'$.

DEFINITION 3.5 – Let $(\mathcal{S}, \mathcal{F}, P)$ be a state-effect-probability structure. We define JP relation ($<_{JP}$) the following preorder relation: for $f, g \in \mathcal{F}$,

$f <_{JP} g$ iff $S_1(f) \subseteq S_1(g)$. We define then as JP equivalence ($\equiv_{JP}$) the equivalence relation induced by the JP relation: for $f, g \in \mathcal{F}$, $f \equiv_{JP} g$ iff $f <_{JP} g$ and $g <_{JP} f$, and we define JP proposition (for $f \in \mathcal{F}$, $[f]_{\equiv_{JP}}$) the equivalence class of effects with respect to the JP equivalence relation.

DEFINITION 3.6 – Let $(\mathcal{S}, \mathcal{F}, P)$ be a state-effect-probability structure and let $a, f \in \mathcal{F}$. Then a is an event iff the following two conditions hold:
   (1) $f \equiv_{JP} a$ implies $a \leq f$,   (2) $f \equiv_{JP} a'$ implies $a' \leq f$.

If we denote by $\mathcal{F}_e$ the set of the events it is possible to prove the following proposition (Cattaneo, Garola, Nisticò, [1989], Prop. 3.2).

PROPOSITION 3.2 – *Let $(\mathcal{S}, \mathcal{F}, P)$ be a state-effect-probability structure. If a JP proposition contains at least an event a, such event is unique and it has the greatest certainly no domain with respect to all other effects contained in the JP proposition (in this case it is said that a tests the JP proposition).*

PROPOSITION 3.3 – *For any σ-convex combination $\sum_{j\in\mathbb{N}} \lambda_j f_j$ of effects $\{f_j : j \in \mathbb{N}\}$ with weights $\{\lambda_j : j \in \mathbb{N}\}$, the following equations hold:*

$$S_1\left(\sum_{j\in\mathbb{N}} \lambda_j f_j\right) = \bigcap_{j\in\mathbb{N}} S_1(f_j) \qquad S_0\left(\sum_{j\in\mathbb{N}} \lambda_j f_j\right) = \bigcap_{j\in\mathbb{N}} S_0(f_j) .$$

A particular class of state-effect-probability structures is the one that we now introduce in which the first Axiom is reminiscent of the Jauch-Piron approach to axiomatic quantum theory and, anyway, it turns out to be a generalization of the above Proposition.

DEFINITION 3.7 – A state-effect-probability structure is said to be *complete* iff the following axioms hold:

Axiom JPc: for every set of effects $\mathcal{G} \subseteq \mathcal{F}$, there exists an effect $\Pi_{\mathcal{G}}$, called the product of the elements of $\mathcal{G}$, such that

$$S_1(\Pi_{\mathcal{G}}) = \cap_{\mathcal{G}} S_1(f), \qquad S_0(\Pi_{\mathcal{G}}) = \cap_{\mathcal{G}} S_0(f).$$

Axiom CT: for every effect $f \in \mathcal{F}$ there exists an event $a \equiv_{JP} f$ that tests the JP proposition $[f]_{\equiv_{JP}}$ generated by f.

In any complete state - effect - probability structure we denote by $\square$ the mapping from $\mathcal{F}$ into itself, associating with every $f \in \mathcal{F}$ the unique event in $[f]_{\equiv_{JP}}$, whose existence is guaranteed by the completeness Axiom CT and whose uniqueness is proved in Prop. 3.2: $\square : f \in \mathcal{F} \mapsto \square(f) \in \mathcal{F}_e$. Using this *"necessity"* mapping, the set of all effects turns out to be a BZ-poset (see CGN, [1989]); the set of BZ exact elements is an orthomodular complete lattice, coinciding with the set of all events.

Finally, a σ-orthogonality condition could be required [which is a countable extension of the restriction of Axiom 5. (ii) to the set of events].

Axiom σ-OG: for any countable set $\{a_n\}$ of pairwise orthogonal events, there exists an effect $f \in \mathcal{F}$ such that, for every $x \in \mathcal{S}$,

$$P(x, f) = \sum_{n \in \mathbb{N}} P(x, a_n).$$

In a complete state-effect-probability structure, this axiom guarantees that the sum of a countable family of pairwise orthogonal events coincides with the countable join of the family.

3.2. FROM STATE-EVENT-PROBABILITY TO UNSHARP QUANTUM LOGIC

The second possible formulation for the structure of effects is contained in a paper by Cattaneo and Laudisa, [1994]. The starting point is the well-known Mackey's approach to QM. Such approach envisages two primitive sets, the set $\mathcal{S}$ of the states of the physical entity under scrutiny and the set $\mathcal{O}$ of the observables that can be measured on the physical entity. The elements of the two sets are probabilistically connected in a standard way by a probability function $P : \mathcal{S} \times \mathcal{O} \times \mathcal{B}(\mathbb{R}) \mapsto [0, 1]$. If w denotes a state, A an observable and Δ a Borel set, the real number $P(w, A, \Delta) \in [0, 1]$ represents the probability that a measurement of A on the entity prepared in the state w will yield a result contained in Δ. The axioms of the standard Mackey's approach express some natural conditions to be required for the sets of states and observables and for the probability function connecting them. Owing to the axioms, for any Borel set Δ the mapping (characteristic function) $\chi_\Delta : \mathbb{R} \mapsto \mathbb{R}$ defines for any $A \in \mathcal{O}$ a unique observable $\chi_\Delta(A)$, called Δ-*event* generated by A. These observables are idempotent and are the Mackey's *questions*.

A description that is equivalent to Mackey's approach [Maczynski, (1973), Beltrametti and Cassinelli, (1976), see also Cattaneo, (1993)] is given by the *state-event-probability* structures. Before introducing this structure, we must premise some useful definitions.

DEFINITION 3.8 – Let $\mathcal{E}$ be a bounded orthomodular σ-orthoposet. A *probability measure* on $\mathcal{E}$ is any mapping $p : \mathcal{E} \mapsto [0, 1]$ such that:

(pm-1) $p(\mathbb{0}) = 0, \qquad p(\mathbb{1}) = 1$;

(pm-2) $\forall\{E_n : i \neq j, \ E_i \perp E_j\} \subseteq \mathcal{E}, \quad p(\vee E_n) = \sum_{n=0}^{\infty} p(E_n)$.

DEFINITION 3.9 – Let $\mathcal{E}$ be a bounded orthomodular σ-orthoposet. An $\mathcal{E}$-valued measure is any mapping $E : \mathcal{B}(\mathbb{R}) \mapsto \mathcal{E}$ such that:

(vm-1) $E(\emptyset) = \mathbb{0}, \qquad E(\mathbb{R}) = \mathbb{1}$;

(vm-2) $\Delta_1 \cap \Delta_2 = \emptyset$ implies $E(\Delta_1) \perp E(\Delta_2)$;

(vm-3) $\forall \{\Delta_n : i \neq j, \Delta_i \cap \Delta_j = \emptyset\} \subseteq \mathcal{B}(\mathbb{R})$, $E(\cup \Delta_n) = \vee E(\Delta_n)$.

We are now ready to introduce state-event-probability structures.

DEFINITION 3.10 – A state-event-probability structure is a triple $\langle \mathcal{E},$ $\mathcal{S}(\mathcal{E}), \mathcal{O}(\mathcal{E}) \rangle$ where:

(1) $\mathcal{E}$ is the set of all events, endowed with a structure of bounded orthomodular σ-orthoposet;

(2) $\mathcal{S}(\mathcal{E})$ is an indexed (by elements from $\mathcal{S}$) family of probability measures on $\mathcal{E}$, $\mathcal{S}(\mathcal{E}) := \{p_w : \mathcal{E} \mapsto [0,1] \mid w \in \mathcal{S}\}$, which is:

    (2-i) order-determining, i.e., for all $E_1, E_2 \in \mathcal{E}$, if $p_w(E_1) \leq p_w(E_2)$ for any $p_w \in \mathcal{S}(\mathcal{E})$, then $E_1 \leq E_2$,

    (2-ii) σ-convex, i.e., for any countable family of states $\{p_n : n \in \mathbb{N}\} \subseteq \mathcal{S}(\mathcal{E})$ and any corresponding family of positive numbers $\{\lambda_n : n \in \mathbb{N}\} \subseteq \mathbb{R}_+$ such that $\sum \lambda_n = 1$, we get that $\sum_{n \in \mathbb{N}} \lambda_n p_n \in \mathcal{S}(\mathcal{E})$;

(3) $\mathcal{O}(\mathcal{E})$ is an indexed (by elements from $\mathcal{O}$) family of event-valued measures $\mathcal{O}(\mathcal{E}) := \{E_A : \mathcal{B}(\mathbb{R}) \mapsto \mathcal{E} \mid A \in \mathcal{O}\}$ which is:

    (3-i) surjective, i.e., $\forall E \in \mathcal{E}$, $\exists A \in \mathcal{O}$, $\exists \Delta \in \mathcal{B}(\mathbb{R})$, s.t. $E_A(\Delta) = E$;

    (3-ii) closed with respect to the product for real Borel functions, i.e., for all $E_A \in \mathcal{O}(\mathcal{E})$ and all $f : \mathbb{R} \mapsto \mathbb{R}$, real Borel function, we have that $E_{f(A)} := (E_A \circ f^{-1}) \in \mathcal{O}(\mathcal{E})$.

The elements of $\mathcal{S}(\mathcal{E})$ and the elements of $\mathcal{O}(\mathcal{E})$ can be seen as the abstract counterparts - respectively - of the states and the observables of the physical entity (and often in the quantum logic literature they are themselves referred to as the states and the observables).

The set of all events $\mathcal{E}$ actually coincides with the set $\{\chi_\Delta(A) : A \in \mathcal{O}, \Delta \in \mathcal{B}(\mathbb{R})\}$ of Mackey's questions (Δ-events). In Mackey's structure, one passes from an observable A to the corresponding Δ-event by using the sharp Markoff kernel $\chi : \mathcal{B}(\mathbb{R}) \times \mathbb{R} \mapsto [0,1]$, defined as $\chi(\Delta, x) = 1$ if $x \in \Delta$ and 0 otherwise; this Markoff kernel was interpreted (Cattaneo-Laudisa, [1994]) as a mathematical description of a *macroscopic device* which *sharply localizes* windows [various χ_Δ as Δ running in $\mathcal{B}(\mathbb{R})$] in the reading scale of the apparatus testing the observable A. A generalization was obtained by introducing the *concrete macroscopic device* $\omega : \mathcal{B}(\mathbb{R}) \times \mathbb{R} \mapsto [0,1]$ where, for every $x \in \mathbb{R}$, $\omega_x : \mathcal{B}(\mathbb{R}) \mapsto [0,1]$ must be a Borel measure, and for every Borel set Δ, $\omega_\Delta : \mathbb{R} \mapsto [0,1]$ is the *membership measurable function* of Δ, describing an *unsharp localization* of the window which selects the region Δ in the reading scale of the apparatus testing the observable A (the use of measurable functions $\mathbb{R} \mapsto [0,1]$, but in a different semantical context, can be found in Garola - Solombrino [1983]).

Let $(\mathcal{E}, \mathcal{S}(\mathcal{E}), \mathcal{O}(\mathcal{E}))$ be a state-event-probability structure. Then, for every event-valued measure E_A, for every concrete device ω and for every Borel set Δ, the event-valued measure $E_{\omega_\Delta(A)} := (E_A \circ \omega_\Delta^{-1}) : \mathcal{B}(\mathbb{R}) \mapsto \mathcal{E}$ is said to be the *generalized event* (or Ludwig's *effect*) associated to the *generalized elementary statement* (or Piron's *question*)

(A, Δ, ω) : "the value of the observable A lies in Δ when the latter is tested by the concrete device ω."

Let $\mathcal{F}(\mathcal{E}) := \{E_{\omega_\Delta(A)} : A \in \mathcal{O}, \omega \in \mathcal{F}(\mathbb{R}), \Delta \in \mathcal{B}(\mathbb{R})\}$ be the set of all such event-valued measures, whose elements will be simply denoted by $F, G, \ldots$ in the following. In particular, an *exact* question is a question tested by the sharp localization device χ: (A, Δ, χ); this exact question is tested by the *sharp* event-valued measure of the form

$$E_{\chi_\Delta(A)} = (E_A \circ \chi_\Delta^{-1}) : \mathcal{B}(\mathbb{R}) \mapsto \mathcal{E}$$

characterized by the following rule ($\forall \tilde{\Delta} \in \mathcal{B}(\mathbb{R})$):

$$E_{\chi_\Delta(A)}(\tilde{\Delta}) = \begin{cases} E_A(\Delta) & \text{iff} \quad 1 \in \tilde{\Delta} \text{ and } 0 \notin \tilde{\Delta} \\ E_A(\Delta)' & \text{iff} \quad 0 \in \tilde{\Delta} \text{ and } 1 \notin \tilde{\Delta} \\ \mathbb{1} & \text{iff} \quad \{0,1\} \subseteq \tilde{\Delta} \\ \mathbb{0} & \text{iff} \quad \{0,1\} \subseteq \mathbb{R}/\tilde{\Delta} \end{cases}$$

The set of all sharp event-valued measures is denoted by $\mathcal{F}_e(\mathcal{E}) := \{E_{\chi_\Delta(A)} : A \in \mathcal{O}, \Delta \in \mathcal{B}(\mathbb{R})\}$ and it is easy to see that the (primitive) set of events $\mathcal{E}$ and the mathematically derived set of sharp event-valued measures $\mathcal{F}_e(\mathcal{E})$ are identifiable by the one-to-one and onto mapping

$$E_A(\Delta) \in \mathcal{E} \longleftrightarrow E_{\chi_\Delta(A)} : \mathcal{B}(\mathbb{R}) \mapsto \mathcal{E}.$$

For this reason, $\mathcal{F}_e(\mathcal{E})$ can be considered as the set of events, and $\mathcal{F}(\mathcal{E})$ as the set of *generalized* (unsharp) events. In order to provide $\mathcal{F}(\mathcal{E})$ with a structure, the following axioms are introduced.

DEFINITION   3.11 – The triple $(\mathcal{S}, \mathcal{F}(\mathcal{E}), p)$, where $\mathcal{S}$ is the nonempty set of all states, $\mathcal{F}(\mathcal{E})$ is the nonempty set of all generalized events and p is the probability function $p : \mathcal{S} \times \mathcal{F}(\mathcal{E}) \mapsto [0, 1]$ defined as

$$p(w, E_{\omega_\Delta(A)}) := \int \omega_\Delta d(p_w \circ E_A)$$

is an *usharp quantum logics* (UQL) iff the following axioms hold.
Axiom U: for $F_1, F_2 \in \mathcal{F}(\mathcal{E})$, and for every $w \in \mathcal{S}$,

$$p(w, F_1) = p(w, F_2) \quad \text{implies} \quad F_1 = F_2.$$

Axiom E: let $F_1, F_2 \in \mathcal{F}(\mathcal{E})$ be such that, for every $w \in \mathcal{S}$,

$$0 \leq p(w, F_1) + p(w, F_2) \leq 1;$$

then an effect $F \in \mathcal{F}$, called the sum, exists such that, for every $w \in \mathcal{S}$,

$$p(w, F) = p(w, F_1) + p(w, F_2).$$

It is then possible to prove the following proposition (Cattaneo - Laudisa, [1994], Prop. 6.2).

PROPOSITION 3.4 – *The set of all generalized events $\mathcal{F}(\mathcal{E})$ is a convex set endowed with a structure $\langle \mathcal{F}(\mathcal{E}), O, I, \leq, ' \rangle$ of partially ordered set w.r.t. the partial order relation*

$$F_1 \leq F_2 \quad \text{iff} \quad \forall w \in \mathcal{S}, \quad p(w, F_1) \leq p(w, F_2);$$

the poset is bounded by the minimum element $O := E_{\omega_\emptyset(A)}$ and by the maximum element $I := E_{\omega_{\mathbb{R}}(A)}$ and is endowed with a regular degenerate orthocomplementation $F = E_{\omega_\Delta(A)} \mapsto F' := E_{(1-\omega_\Delta)(A)}$.

Two particular mappings from $\mathcal{F}(\mathcal{E})$ into $\mathcal{F}_e(\mathcal{E})$ can be introduced:
(1) the **necessity** mapping $\nu : \mathcal{F}(\mathcal{E}) \mapsto \mathcal{F}_e(\mathcal{E})$, defined by

$$F = E_{\omega_\Delta(A)} \in \mathcal{F}(\mathcal{E}) \mapsto \nu(F) := E_{\chi_{\omega_\Delta^{-1}(\{1\})}(A)} \in \mathcal{F}_e(\mathcal{E});$$

(2) the **possibility** mapping $\mu : \mathcal{F} \mapsto \mathcal{F}_e$, defined by

$$F = E_{\omega_\Delta(A)} \in \mathcal{F}(\mathcal{E}) \mapsto \mu(F) := E_{\chi_{\omega_\Delta^{-1}(0,1]}(A)} \in \mathcal{F}_e(\mathcal{E}).$$

The necessity and the possibility mappings turn out to satisfy some relevant properties which are typical of the corresponding BZ modal operators (Cattaneo-Nisticò, [1989]). We can associate with any effect $F \in \mathcal{F}(\mathcal{E})$ the *certainly-yes* and the *certainly-no* domains

$$S_1(F) = \{w \in \mathcal{S} : p(w, F) = 1\}, \qquad S_0(F) = \{w \in \mathcal{S} : p(w, F) = 0\}.$$

The final (completeness) assumptions are the following (for the motivations, see Cattaneo, Laudisa, [1994]).

Axiom JPc: for any family $\mathcal{G}$ of generalized events, $\prod_{\mathcal{G}} F$ exists such that

$$S_1(\prod_{\mathcal{G}} F) = \bigcap_{\mathcal{G}} S_1(F), \qquad S_0(\prod_{\mathcal{G}} F) = \bigcap_{\mathcal{G}} S_0(F).$$

Axiom CC-1: For any generalized event F, its necessity $\nu(F)$ satisfies the condition $S_1(\nu(F)) = S_1(F)$.

Axiom CC-2: for any pair of exact events $\nu(F)$ and $\nu(G)$, $S_1(\nu(F)) = S_1(\nu(G))$ implies $\nu(F) = \nu(G)$.

4. The equivalence between SEP and UQL structures

In the previous section two different structures for the set of the effects have been considered. We will show in this section some links between them, along the direction of proving their strong equivalence.

PROPOSITION 4.1 – *Any SEP $(\mathcal{S}, \mathcal{F}, P)$ satisfies the axioms U, E, JPc, CC-1,2 of an UQL.*

Proof. The axiom U, E and JPc of UQL are validated due to the Axiom 1. (ii), 4 and 5. (ii) of SEP. As regards CC-1, the event $\square f$ is the event testing the JP proposition $[f]_{\equiv_{JP}}$ generated by f. Since $\square f$ belongs to $[f]_{\equiv_{JP}}$, it has the same certainly-yes domain and thus CC-1 is satisfied. As regards CC-2, let $f_1, f_2 \in \mathcal{F}$ and let $\square f_1, \square f_2$ be the associated events, for which, by hypothesis, $S_1(\square f_1) = S_1(\square f_2)$ and, according to CC-1, $S_1(f_1) = S_1(f_2)$. Then there exists an $f \in \mathcal{F}$ such that $f_1, f_2 \in [f]_{\equiv_{JP}}$. But also $\square f_1, \square f_2$ belong to $[f]_{\equiv_{JP}}$ and, due to the uniqueness of the event in a JP proposition, we have $\square f_1 = \square f_2$, i.e. CC-2.

PROPOSITION 4.2 – *Any UQL satisfies the axioms of a SEP.*

Proof. $\forall F$, $p(w_1, F) = p(w_2, F)$ implies, in particular, that $\forall A$, $\forall \Delta$ $p(w_1, E_{\chi_\Delta(A)}) = p(w_2, E_{\chi_\Delta(A)})$, i.e., $p_{w_1}(E_A(\Delta)) = p_{w_2}(E_A(\Delta))$; thus axiom 1. (i) follows from the (3-i) and the (2-1), definition 3.10. $\forall F := E_{\omega_\Delta(A)}$, $\forall \lambda \in [0.1]$ the generalized event $\lambda F = E_{(\lambda \omega_\Delta)(A)}$ satisfies axiom 5. (i). Axioms 1. (ii) and 5. (ii) of SEP are validated due to the axioms U and JPc of UQL, whereas the element I of proposition 3.4 satisfies axiom 2. Moreover, for any $F = E_{\omega_\Delta(A)}$, the element $I - F := E_{(1-\omega_\Delta)(A)}$ satisfies axiom 3. Axiom 4 trivially follows from the (2-ii), definition 3.10, applied to definition 3.11.

The proof that it is possible 'to close the circle' is still an open problem of some interest.

References

Beltrametti, E., Cassinelli, G. (1976). "Logical and mathematical structures of quantum mechanics", *Nuovo Cimento*, **6** pp. 321–404.

Busch, P., Grabowski, M., Lahti, P. (1989). "Some remarks on effects, operations and unsharp measurements", *Found. Phys. Lett.*, **2** pp. 331–345.

Busch, P., Schroeck, F. (1989). "On the reality of spin and helicity", *Found. Phys.*, **19** pp. 807–72.

Cattaneo, G. (1993). "The logical approach to axiomatic quantum theory", in G. Corsi et al. (eds.), *Bridging the Gap: Philosophy, Methematics, and Physics*, Kluwer Academic Publ., The Netherlands.

Cattaneo, G., Garola, C., Nisticò, G. (1989). "Preparation-effect versus question proposition structures", *J. Phys. Essays*, **2** pp. 197–216.

Cattaneo, G., Laudisa, F. (1994). "Unsharp axiomatic quantum theory", *Found. Phys.*, **24** pp. 631–681.

Cattaneo, G., Nisticò, G. (1989). "Brouwer-Zadeh posets and three valued Lukasiewicz posets", *Fuzzy Sets Syst.*, **33** pp. 165–90.

Cattaneo, G., Nisticò, G. (1994). "Coexistence of questions versus Jauch-Piron's compatibility", *Found. Phys.*, **24** pp. 1131–1152.

Finkelstein, D. (1969). "Matter, space and logic", *Boston Studies Philosophy of Science*, vol. **5**.

Finkelstein, D. (1972). "The physics of logic", in R. G. Colodny (ed.), *Paradigms and Paradoxes*, University of Pittsburgh Press.

Garola, C. and Solombrino, L. (1983). "Yes-No experiments and ordered structures in quantum physics", *Nuovo Cimento*, **77B** pp. 87–110.

Mackey, G. W. (1963). *The Mathematical Foundations of Quantum Mechanics*, Benjamin, NY.

Maczynski, M.J. (1973). "The orthogonality postulate in axiomatic quantum mechanics", *Int. J. Theor. Phys.*, **8** pp. 353–60.

Putnam, H. (1969). "Is logic empirical?", *Boston Studies Philosophy of Science*, vol. **5**.

INTERPRETATIVE REMARKS IN QUANTUM MECHANICS

G. CATTANEO

Dip. Scienze dell'Informazione - Università di Milano
Via Comelico 39/41 - 20135 - Milano, Italy

and

G. NISTICÒ

Dip. di Matematica - Università della Calabria
87036 - Arcavacata di Rende (CS), Italy

Abstract. The principle which states that non-commuting quantum observables cannot be measured together gives rise to interpretative questions in quantum mechanics. In this article we derive this principle, according to the Von Neumann approach, and some of these interpretative questions are discussed. In particular, the sense in which the mentioned principle may be 'outflanked' by means of correlations between outcomes of different observables, as in the Einstein, Podolsky and Rosen conceptual experiment, is clarified.

1. Introduction

A key tool for the interpretation of quantum physics is the principle according to which two observables $\mathcal{R}$ and $\mathcal{S}$ are simultaneously measurable, written $\mathcal{R}(SM)\mathcal{S}$, if and only if the corresponding operators R and S commute:

$$\mathcal{R}(SM)\mathcal{S} \quad \text{iff} \quad RS = SR. \tag{1}$$

In fact, this principle is a theorem of the theory formulated by J. Von Neumann in his book about the mathematical foundation of quantum mechanics (Von Neumann, 1955).

In this article we are concerned with some interpretative questions arising from this principle within the framework of Von Neumann approach to quantum mechanics.

First, in section 2, we outline Von Neumann theory; then a theorem stating principle (1) is easily proved. In section 3 we show how this principle

127

C. Garola and A. Rossi (eds.), The Foundations of Quantum Mechanics, 127-138.
© 1995 *Kluwer Academic Publishers.*

works within the 'interpreting engine' of quantum theory to translate mathematical relations of quantum formalism into factual statements.

A direct consequence of principle (1) is that two non-commuting quantum observables cannot be measured together. This fact, by itself, does not prevent the knowledge of the outcomes of both observables, if only one of them is measured. This happens, for instance, if the state of the system is simultaneously an eigenstate of both observables. Otherwise, this knowledge may be attained by means of correlations of the kind introduced by Einstein, Podolsky and Rosen in their thought experiment (Einstein, Podolsky and Rosen, 1935). More precisely, by measuring only the first of the two non-commuting observables, together with a third 'mirror' observable which commutes with both and whose outcomes are correlated, in the EPR sense, with the outcomes of the second observable; in such a way we have also indirect knowledge of the outcome of the second observable, besides the direct knowledge of the outcome of the first observable. In section 4 we argue how this kind of indirect knowledge of the outcomes of two non-commuting observables must be correctly interpreted according to quantum mechanics, taking into account the need for care noticed at the end of section 2.

In section 5 we show in which situations this indirect knowledge is possible and the way to obtain it is prescribed. We show, in particular, that there exist situations in which this knowledge is not possible, at least by means of a mirror observable.

2. Von Neumann theory

The primitive concepts of Von Neumann theory are essentially two.

- *Physical Magnitudes* (or *Observables*), whose set is denoted by $\mathcal{O}$. The set of the observables and their functional relations determine the nature of the physical system **S**.
- *Expectation Values* of observables (or $\mathcal{R}$-*functions*). These are a set of functions $Ev : \mathcal{O} \mapsto I\!R$ assigning a numerical value $Ev(\mathcal{R})$ to every observable $\mathcal{R}$, interpreted as the mean value of $\mathcal{R}$ over a certain ensemble of physical systems.

Basic assumptions, induced by the interpretation of the primitive concepts are the following.

B_1 Given an observable $\mathcal{R}$ and a function $f : I\!R \mapsto I\!R$, there is an observable, denoted by $f(\mathcal{R})$, whose outcomes are obtained by applying the function f to the outcomes of $\mathcal{R}$;

B_2 The set of $\mathcal{R}$-functions is endowed with a convex operation: if $\{Ev_1, Ev_2, ...\}$ is a countable set of $\mathcal{R}$-functions then $\sum_i \alpha_i Ev_i$, where $\alpha_i \geq 0$

and $\sum_i \alpha_i = 1$, is a $\mathcal{R}$-function too, corresponding to the physical operation of making a mixture with statistical weights α_i of the statistical ensembles of physical systems corresponding to the Ev_i.

In its turn, B_1 induces the definition of simultaneous measurability of observables.

DEFINITION 1. *Two observables $\mathcal{R}$ and $\mathcal{S}$ are measurable together if both are functions of a third observable $\mathcal{T}$:*

$$\mathcal{R}(SM)\mathcal{S} \quad \text{if} \quad \exists \mathcal{T} \in \mathcal{O} \quad \text{such that} \quad \mathcal{R} = f(\mathcal{T}) \quad \text{and} \quad \mathcal{S} = g(\mathcal{T}). \quad (2)$$

Von Neumann characterized quantum mechanics by means of the following axioms.

AXIOM 0. To every physical system, a complex and separable Hilbert space $\mathcal{H}$ can be associated such that to every observable $\mathcal{R}$ there corresponds one, and only one, selfadjoint operator R of $\mathcal{H}$. The correspondence is bijective.

AXIOM 1. Let $\mathcal{R}$ and $f : \mathbb{R} \mapsto \mathbb{R}$ be an observable and a function, respectively; if $\mathcal{R}$ is represented by the operator R, then $f(\mathcal{R})$ is represented by the operator $f(R) = \int_{\mathbf{R}} f(\lambda) dE^R(\lambda)$ (here $E^R(\lambda)$ denotes the spectral resolution of R).

AXIOM 2. Let $\mathcal{R}, \mathcal{S}, \ldots$ be observables represented by the operators $R, S, \ldots$. Then there exists a third observable $\mathcal{R} + \mathcal{S} + \ldots$, which is represented by the operator $R + S + \ldots$

AXIOM 3. If $\mathcal{R}$ is a non-negative quantity, then $Ev(\mathcal{R}) \geq 0$.

AXIOM 4. For every $\mathcal{R}$-function Ev, $Ev(a\mathcal{R} + b\mathcal{S} + \ldots) = aEv(\mathcal{R}) + bEv(\mathcal{S}) + \ldots$, where $\mathcal{R}, \mathcal{S}, \ldots$ are observables and $a, b, \ldots$ are numbers.

Axiom 1 establishes a strong link between the algebraic structures of the set of observables and that of the selfadjoint operators of $\mathcal{H}$, and this allows the proof of (1). Indeed, from axiom 1 it follows that $\mathcal{R}(SM)\mathcal{S}$ iff the operators R and S are functions of the same operator T. Now, a theorem of functional analysis states that two selfadjoint operators are functions of a third selfadjoint operator if and only if they commute. Thus, we get the following theorem.

THEOREM 1. *Two observables are simultaneously measurable if and only if the corresponding operators commute with each other.*

Before going on, we have to notice that one must be careful in interpreting the consequences of theorem 1. In particular troubles could arise in connection with interpretation rule B_1. For instance, it may happen that the same observable $\mathcal{R}$ is simultaneously function of two non-commuting observables $\mathcal{S}$ and $\mathcal{T}$:

$$\mathcal{R} = f(\mathcal{S}) = g(\mathcal{T}) \quad \text{and} \quad [S, T] \neq 0.$$

A superficial reading of B_1 may lead one to argue that, since the same outcome of $\mathcal{R}$ is simultaneously obtained from those of $\mathcal{S}$ and $\mathcal{T}$ by means of f and g, then $\mathcal{S}$ and $\mathcal{T}$ must have simultaneous outcomes, against theorem 1. Thus, in quantum mechanics it is not always correct to attribute outcomes to unperformed measurements. A detailed discussion of these problems may be found in (Readhead, 1990).

3. Interpreting Quantum formalism

For our purposes we need the following theorems.

THEOREM 2. (Von Neumann 55). *For every $\mathcal{R}$-function Ev satisfying axioms 0–4 there is a positive operator ρ, whose trace $Tr(\rho)$ is 1, such that $Ev(\mathcal{R}) = Tr(\rho R) \, \forall \mathcal{R} \in \mathcal{O}$.*

THEOREM 3. (Spectral representation). *For every selfadjoint operator R with spectrum $\sigma(R)$ there exists a unique family of projections $\{E^R(\lambda) \mid \lambda \in \sigma(R)\}$ (spectral resolution), increasing in λ with $E^R(-\infty) = 0$ and $E^R(\infty) = 1$, such that $R = \int_{\sigma(R)} \lambda dE^R(\lambda)^1$.*

These theorems, together with axioms 0–4, yield

PROPOSITION 1. *For every $\mathcal{R}$ - function Ev there exists ρ such that*

$$Ev(\mathcal{R}) = \int_{\sigma(R)} \lambda Tr(\rho dE^R(\lambda)).$$

By looking at proposition 1 we derive the following interpretative consequences:

D_1. $Tr(\rho dE^R(\lambda))$ *has to be interpreted as the probability that the outcome of $\mathcal{R}$ is a value in the interval $(\lambda, \lambda + d\lambda)$.*

D_2. *A number $\lambda \in \mathbb{R}$ is a possible outcome for $\mathcal{R}$ if and only if $\lambda \in \sigma(R)$.*

We call 0-1 observable any observable $\mathcal{P}$ having 0 and 1 as possible outcomes. Then, by D_1 and proposition 1 it follows:

D_3. *For a 0-1 observable $\mathcal{P}$ the expectation value $Ev(\mathcal{P})$ must be interpreted as the probability of occurrence of the outcome 1 for $\mathcal{P}$.*

By D_2 we derive the following proposition.

PROPOSITION 2. *If P is the selfadjoint operator associated to a 0-1 observable $\mathcal{P}$, then $\sigma(P) = \{0, 1\}$, i.e. P is a projection. The converse holds too.*

3.1. THE ORDERING RELATION

By $\mathcal{E}(\mathcal{H})$ we denote the set of all projections of $\mathcal{H}$. From now on, because of axiom 0, we identify every observable $\mathcal{R}$ with the corresponding selfadjoint operator R. $\mathcal{E}(\mathcal{H})$ is endowed with the following partial order relation:

$$P \leq Q \quad \text{if} \quad \langle \psi \mid P\psi \rangle \leq \langle \psi \mid Q\psi \rangle \quad \forall \psi \in \mathcal{H}. \tag{3}$$

What is the physical meaning of the formula $P \leq Q$? Now we can answer this question.

For every positive operator ρ with $Tr(\rho) = 1$ there is an orthonormal basis ψ_i of eigenvectors of ρ, so that

$$\rho = \sum_i \alpha_i |\psi_i\rangle\langle\psi_i| \quad \text{and} \quad \sum_i \alpha_i = 1, \tag{4}$$

where the α_i are the eigenvalues of ρ relative to the ψ_i. Hence $Ev(R) = \sum \alpha_i \langle \psi_i \mid R\psi_i \rangle$ and (3) yield

$$P \leq Q \quad \text{implies} \quad Ev(P) \leq Ev(Q) \quad \forall R\text{-function } Ev. \tag{5}$$

Since P and Q represent 0-1 observables, the formula $P \leq Q$ may be interpreted according to D$_3$.

I$_1$. *The probability of occurrence of outcome 1 for P is less or equal the probability of occurrence of outcome 1 for Q.*

Actually, the theory of Von Neumann provides a reinforcement of the interpretation I$_1$ of $P \leq Q$.

I$_2$. *Whenever P and Q are measured together, if the outcome of P is 1, then the outcome of Q is 1 too.*

To prove I$_2$, notice that it makes sense only if P and Q are performable together, and this is true because if $P \leq Q$ then $PQ = QP = P$. Therefore P and Q must be functions of another operator C. Actually, if we define the operator $S = \sum_{n=1}^{4} nS_n$, where $S_1 = PQ$, $S_2 = PQ'$, $S_3 = P'Q$, $S_4 = P'Q'$, and the functions $f(n) = \delta_{1n} + \delta_{2n}$ and $g(n) = \delta_{3n} + \delta_{4n}$, then $P = f(S)$ and $Q = g(S)$ hold. Therefore, by interpretation rule B$_1$, the outcome of S univocally determines the outcomes of both P and Q, when S, P and Q are measured together, according to the following Table 1.

S	P	Q
1	1	1
2	1	0
3	0	1
4	0	0

TABLE 1.

It becomes clear that interpretation I_2 holds if and only if outcome 2 for S is impossible. But if $P \leq Q$ this is true, since $PQ' = P(1-Q) = P - PQ = P - P = 0$.

This proves that in the Von Neumann quantum theory the order relation $\leq$ in $\mathcal{E}(\mathcal{H})$ has more than the merely probabilistic interpretation I_1.

3.2. CONDITIONAL PROBABILITIES

Let P and Q be two projections which commute with each other, so that they can be measured together. It makes sense to consider the probability that the outcome 1 for P occurs under the condition that the outcome 1 occurs for Q, i.e. the conditional probability $P(P \mid Q)$. We consider the observable S of the previous section. A look at table 1 makes it clear that

$$P(P \mid Q) = \frac{Tr(\rho S_1)}{Tr(\rho Q)} = \frac{Tr(\rho PQ)}{Tr(\rho Q)}. \tag{6}$$

A $\mathcal{R}$-function is said *pure* if it cannot be expressed as convex combination of other $\mathcal{R}$-functions, in the sense of B_2. $\mathcal{R}$-functions of the form $\langle \psi \mid R\psi \rangle$ are pure. Every $\mathcal{R}$-function admits the representation

$$Ev(R) = \sum_i \alpha_i Tr(|\psi_i\rangle\langle\psi_i|R) = \sum_i \alpha_i \langle \psi_i \mid R\psi_i \rangle;$$

then, from now on we restrict our arguments to pure $\mathcal{R}$-functions $\langle \psi \mid R\psi \rangle$. When ρ describes the *pure* expectation value $Ev(R) = \langle \psi \mid R\psi \rangle$, where $\psi \in \mathcal{H}$ and $\|\psi\| = 1$, (6) becomes

$$P(P \mid Q) = \frac{\langle \psi \mid PQ\psi \rangle}{\langle \psi \mid Q\psi \rangle}. \tag{6'}$$

The outcomes of P and Q are completely correlated when

$$P(P \mid T) = P(T \mid P) = 1 \quad \text{or} \tag{7.1}$$

$$P(P \mid T) = P(T \mid P) = 0 \tag{7.2}$$

Indeed, (7.1) (resp. (7.2)) means that the outcome for P is 1 if and only if the outcome for T is 1 (resp. 0). To be precise, the interpretation of (7.1) is the following.

D_4. *If T and P are measured together and (7.1) holds, then the outcome of T is 1 iff the outcome P is 1.*

4. Knowledge of outcomes of non-commuting observables

The wish to escape the rule according to which non-commuting observables cannot be measured together, in spite of its own status as a theorem, is perhaps as old as quantum physics itself. A way to show how this physical law could be 'outflanked' is to consider an experimental situation similar to that conceived by Einstein, Podolsky and Rosen in their famous article (Einstein, Podolsky and Rosen, 1935). Let P and Q be two projections such that $[P, Q] \neq 0$. Since P and Q do not commute, they cannot be measured together (theorem 1). Let the state of the system be described by the *pure* $\mathcal{R}$-function $Ev(R) = \langle \psi \mid R\psi \rangle$, so that it may be identified by the vector $\psi \in \mathcal{H}$ ($\|\psi\| = 1$). In such a case the conditional probabilities are, according to (6')

$$P(P \mid T) = \frac{\langle \psi \mid PT\psi \rangle}{\langle \psi \mid T\psi \rangle} \quad \text{and} \quad P(T \mid P) = \frac{\langle \psi \mid PT\psi \rangle}{\langle \psi \mid P\psi \rangle} \tag{8}$$

DEFINITION 2. *Let P and Q be two projections of the Hilbert space $\mathcal{H}$, and let ψ a vector in $\mathcal{H}$. A projection T is a mirror projection for $(P, Q; \psi)$ if:*

(i) $[T, P] = [T, Q] = 0$;
(ii) $\langle \psi \mid PT\psi \rangle = \langle \psi \mid T\psi \rangle = \langle \psi \mid P\psi \rangle$.

If T is a mirror projection for $(P, Q; \psi)$, then it is possible to measure together T and Q, by (i) in definition 2. On the other hand (ii) in definition 2 means that $P(P \mid T) = P(T \mid P) = 1$ when the state is described by the vector ψ; therefore the outcomes of P and T are completely correlated. Thus, by measuring together Q and T when the state is described by ψ we *know* simultaneously:

(W1) the outcome of Q and T by direct measurement;
(W2) that if P were measured instead of Q, the outcome would be the same of that of T.

In such a sense we get information, expressed by (W1) and (W2), about simultaneous outcomes of two noncommuting observables, namely P and Q. We should stress that the existence of a mirror projection does not mean that P and Q could directly measured together. The mirror projection provides an *indirect* knowledge of the outcome of P, as correctly expressed by (W2). Only in the sense expressed by (W2) one can assert that

> if a mirror projection for $(P, Q; \psi)$ exists, P and Q can be indirectly measured together.

Another remark seems necessary to us. We do not interpret the outcome of T quoted in (W2) as the outcome obtained in the simultaneous measurement with Q. If such a stronger interpretation were correct, then we could state that a simultaneous measurement of T and Q simultaneously provides

(W1') the outcome of Q;

(W2') the outcome of P, *if this last had been measured instead of T in the measurement actually performed.*

While (W1) and (W2) have been coherently derived within quantum theory by using the correct interpretation D_4, to get (W1') and (W2') one needs additional assumptions. For instance, they could be derived if D_4 were replaced by

D'_4. *If T is measured, then its outcome is the same of that of a, albeit unperformed, measurement of P.*

D_4 is a statement of complete correlation between the outcomes of P and T when they are measured together. D'_4 asserts that this complete correlation takes place, no matter whether both measurements are performed or not.

In fact, the compatibility of interpretation D'_4 with the formalism of quantum physics and its statistical interpretation has been questioned by several authors. We refer to the book of M. Redhead (Redehead, 1990) for a review and further references on this point.

5. When there is a mirror projection

In section 4 we saw that the existence of a mirror projection T allows knowledge, expressed by (W1) and (W2), about the outcomes of both P and Q, also in the case in which they do not commute. Elsewhere, this kind of knowledge has been called *weak knowledge of the outcomes of both P and Q* (Nisticò and Romania, 1994). In order to be a mirror projection, T must satisfy the following (necessary and sufficient) conditions

$$(M_1) \qquad\qquad [T, P] = [T, Q] = 0,$$

$$(M_2) \qquad\qquad PT\psi = T\psi = P\psi.$$

(M_2) is equivalent to condition (ii) in definition 2.

Let P and Q be two projections of $\mathcal{H}$. We have three cases in which there is obviously a mirror projection T.

$$P\psi_0 = \mathbf{0} \quad T = 0 \quad \text{for} \quad (P, Q; \psi_0); \tag{9.1}$$

$$P\psi_0 = \psi_0 \quad T = 1 \quad \text{for} \quad (P, Q; \psi_0); \tag{9.2}$$

$$[P, Q] = 0 \quad T = P \quad \text{for} \quad (P, Q; \psi), \forall \psi \in \mathcal{H}. \tag{9.3}$$

Since the set of projections which commute with both P and Q is always non-empty, the existence of a mirror projection depends on (M_2), i.e. on ψ. We found that, fixed P and Q, both cases are always realized: in $\mathcal{H}$ there are vectors ψ such that a mirror projection exists, but also vectors for which it does not exist. In the present section we outline the procedure singled out by one of us with Romania, to characterize those vectors which admit a mirror projection and to construct such a mirror, whenever it exists (Nisticò and Romania, 1994).

5.1. MIRROR PROJECTION IN THE COMMUTATION SUBSPACE

The *commutation subspace* of P and Q is the kernel of the commutator $[P, Q]$, i.e.

$$\mathcal{C}(P, Q) = \{\psi \in \mathcal{H} \mid PQ\psi = QP\psi\}.$$

The subspace $\mathcal{C}(P, Q)$ coincides with the subspace spanned by the common eigenvectors of P and Q. $\mathcal{H}$ may be decomposed into the direct sum $\mathcal{C}(P, Q)$ $\oplus \mathcal{R}(P, Q)$, where by $\mathcal{R}(P, Q)$ we denote the so called *rest*, i.e. the orthogonal complement of the commutation subspace $\mathcal{C}(P, Q)$. Let us denote the projections onto $\mathcal{C}(P, Q)$ and $\mathcal{R}(P, Q)$ by C and R. We have $[P, C] = [Q, C] = 0$, therefore the operators $P \mid_{\mathcal{C}(P,Q)} \equiv PC$, $Q \mid_{\mathcal{C}(P,Q)} \equiv QC$, $P \mid_{\mathcal{R}(P,Q)} \equiv PR$ and $Q \mid_{\mathcal{R}(P,Q)} \equiv QR$ are projections too. We proved the following proposition in (Nisticò and Romania, 1994).

PROPOSITION 3. *There exists a mirror projection for $(P, Q; \psi)$ iff in the Hilbert space $\mathcal{R}(P, Q)$ there exists a mirror projection T_r for $(P \mid_{\mathcal{R}(P,Q)}, Q \mid_{\mathcal{R}(P,Q)}; \psi_r \equiv R\psi)$.*
In such a case, $T = PC + T_r R$ is a mirror projection for $(P, Q; \psi)$.

Before facing the problem of finding the mirror projection for $(PR, QR; R\psi)$ in the Hilbert space $\mathcal{R}(P, Q)$, we can outline some consequences of prop. 3 on the interpretation of quantum mechanics.

First, let us consider the case in which $[P, Q] \neq 0$, but ψ is a common eigenvector of P and Q. This fact does not enable us to say that for such ψ the projections P and Q are measurable together. However we know the outcomes of a measurement of either P or Q before the performance of the measurement.

If ψ is a linear combination of common eigenvectors of P and Q, i.e. $\psi \in \mathcal{C}(P,Q)$, but it is not such an eigenvector, the outcomes of P or Q are no longer pre-determined. However, since $R\psi = 0$, by (9.1) T_r in prop.3 may be chosen equal to 0, and therefore there is a mirror projection $T = PC$ for $(P,Q;\psi)$; thus we get out the following consequence.

D_5. If $\psi \in \mathcal{C}(P,Q)$, it is possible to attain the weak knowledge of the outcomes of P and Q.

D_5 assigns a physical meaning to the fact that $[P,Q] \neq 0$ but $[P,Q]\psi = 0$, different from that, of a probabilistic nature[2], linked to uncertainty relations.

5.2. MIRROR PROJECTION IN THE 'REST'

Prop. 3 states that there is a mirror projection for $(P,Q;\psi)$ in the Hilbert space $\mathcal{H}$ if and only if there is one for $(PR, QR; R\psi)$ in the Hilbert space $\mathcal{R}(P,Q)$. To show our solution to this problem, we have to introduce further definitions.

First, to have a less heavy notation we put

$$PR = P_r, \quad QR = Q_r, \quad R\psi = \phi, \quad \phi_1 = P_r\phi, \quad \phi_0 = (1 - P_r)\phi \quad (10)$$

and we consider them as mathematical objects of the Hilbert space $\mathcal{R}(P,Q)$. The projections P_r and Q_r project onto the two subspaces M and N, respectively, of $\mathcal{R}(P,Q)$. One can prove (Halmos, 1969) that all four subspaces M, N, $M^\perp$ and $N^\perp$ have the same dimension. Given the linear mapping

$$E : M^\perp \mapsto M, \quad Ey = P_r Q_r y,$$

there is a unique *polar decomposition*[3] $E = |E|J$, and the isometric factor J turns out to be unitary.

Let $E(\lambda)$ be the spectral resolution (see theorem 3) of the selfadjoint operator $A^2 \equiv P_r Q_r P_r$ of the Hilbert space M.

By X_0 we denote the projection of M defined as

$$X_0 = \int_{\{\lambda \in \sigma_c(A^2) | dE(\lambda)\phi_1 \neq 0\}} \frac{|dE(\lambda)\phi_1\rangle\langle dE(\lambda)\phi_1|}{\langle dE(\lambda)\phi_1 \mid dE(\lambda)\phi_1\rangle}.$$

Once these necessary mahematical notions are introduced, we can state our main theorem.

THEOREM 4. *There is a mirror projection for* $(P_r, Q_r; \phi)$ *if and only if*

$$\begin{cases} X_0\phi_1 = \phi_1 \\ X_0 J\phi_0 = 0. \end{cases} \quad (11)$$

If (11) holds, then a mirror projection for $(P_r, Q_r; \phi)$ is

$$T_r = X_0 P_r + J^{-1} X_0 J (1 - P_r). \tag{12}$$

5.3. CONCLUDING REMARKS

The problem of the existence and of finding a mirror projection for $(P, Q; \psi)$ is solved by proposition 3 and theorem 4. We can proceed as follows.

(a) First we decompose the Hilbert space $\mathcal{H}$ into $\mathcal{C}(P, Q)$ and $\mathcal{R}(P, Q)$, then we define P_r, Q_r, ϕ, ϕ_1 and ϕ_0 according to (10).

(b.1) Second, if (11) in theorem 4 holds, then there exists a mirror projection T_r for $(P_r, Q_r; \phi)$ given by (12) and thus $T = PC + T_r R$ is a mirror projection for $(P, Q; \psi)$, by proposititon 3.

(b.0) If (11) in theorem 4 does not hold, then there is no mirror projection for $(P, Q; \psi)$.

If $\phi_1 \in M \setminus \{0\}$ then there is no mirror projection for $(P_r, Q_r; \phi_1 + J^{-1}\phi_1)$. Indeed in such a case $J\phi_0 = \phi_1$, so that conditions (11) in theorem 4 imply $\phi_1 = 0$. Therefore, for two projections in generic position there are always vectors which do not admit mirror projection. On the other hand, there are always vectors which do admit mirror projection; indeed if $\phi = \phi_1 \in M$, then the identity operator 1 is a mirror projection for $(P_r, Q_r; \phi)$.

Notice also that neither the set of vectors which admit mirror projection, nor the set of vectors which do not, are subspaces.

If the weak knowledge of the outcomes of (P, Q) were possible in every state $x \in \mathcal{H}$, then $\mathcal{C}(P, Q)$ should be the entire $\mathcal{H}$, i.e. $[P, Q] = O$. Now, in quantum theory there are always pairs of observables which do not commute; thus for every quantum system there are situations in which the weak knowlwdge cannot be attained.

Notes

[1] The integral must be understood as the Lebesgue-Stieltjes integral $\langle \psi \mid R\psi \rangle = \int \lambda d \langle \psi \mid E^R(\lambda)\psi \rangle$, which converges for every ψ in the domain of R. Full details relative to this and to the following note 3 may be found in specific textbooks, as for instance Reed, M. and Simon, B. (1975), "Methods of modern mathematical physics", **Vols. I-III**, New York, Academic Press.

[2] When $\psi \in \mathcal{C}(P, Q)$, the uncertainty relation for P and Q reads

$$\langle \Delta P \rangle_\psi \langle \Delta Q \rangle_\psi \geq 0,$$

which imposes no constraint between the probabilistic entities $\langle \Delta P \rangle_\psi$ and $\langle \Delta Q \rangle_\psi$, but says nothing about the outcomes of single measurement acts.

[3] For every bounded linear operator $E : \mathcal{H} \mapsto \mathcal{K}$ there is a unique pair $(|E|, J)$ of a positive operator $|E|$ of $\mathcal{H}$ and an isometric operator $J : \mathcal{H} \mapsto \mathcal{K}$ such that $E = |E|J$.

References

Einstein, A., Podolsky, B. and Rosen, N. (1935). 'Can quantum-mechanical description of physical reality be considered complete?', *Phys. Rev.*, **47** (777).

Halmos, P. (1969). 'Two subspaces', *Trans.Amer.Math.Soc.*, **144** (381).

Nisticò, G. and Romania, M.C. (1994). 'Knowledge about noncommuting quantum observables by means of Einstein-Podolsky-Rosen correlations', *J. Math. Phys.*, to appear.

Redhead, M. (1990). *Incompleteness nonlocality and realism – a prolegomenon to the philosophy of quantum mechanics*, Clarendon Press, London.

Von Neumann, J. (1955). *Mathematical foundations of quantum mechanics*, Princeton University Press, Princeton.

KUHN'S INTERPRETATION OF BOLTZMANN'S STATISTICAL HEREDITY IN PLANCK

P. CERRETA
Group of History of Physics, Dept. of Physical Sciences
University of Naples, Italy

Abstract. The heart of Kuhn's book *Black Body Theory and the quantum discontinuity 1894-1912* is the demonstration that Planck never intended in 1900-1 to introduce the quantum discontinuity into physics, with his distribution law, as many authoritative historians maintain it. Kuhn proves that Planck was very far from conceiving the necessity of quanta when he followed, more and more closely, Boltzmann's statistical concepts in order to find the distribution law he searched for.

We have analyzed the proofs of Kuhn's thesis and we have discovered that the various statistical concepts "inehrited" by Planck have their common foundation in Boltzmann's conception of infinite.

Kuhn, who is such a profound historian, could not escape from this fact. Kuhn, however, has not stressed this conceptual continuity either for giving a deeper explanation of the events of the birth of quanta, or for using that as a paradigm and saving, in this way, his work from the accusation that, in the black-body case, he was not able to use his famous interpretative scheme of the history of science.

Our conclusion on this matter is that Kuhn behaves like most of the physicists who consider mathematics as a constant of the history, and not a variable that plays a decisive role in the evolution of physics.

1. Kuhn's thesis of continuity between Boltzmann and Planck

In 1978 Kuhn published *Black-Body Theory and the Quantum Discontinuity 1894-1912* [1] (from this point on BBT), a book presenting the events regarding the origin of quanta in a radically new way.

In the author's intention, BBT was to point out the factors which proved to be crucial for the birth of quanta. For this reason he proposed to start a research program which, among other things, included the re-reading of all of Planck's works on black-body. This research allowed him to discover that even if Planck

139

C. Garola and A. Rossi (eds.), The Foundations of Quantum Mechanics, 139-146.

had started from very different positions, he had progressively come close to Boltzmann during the last years of 1800s.

The first part of BBT is essentially the narration of the stages of this process with the demonstration of the parallelism between their mathematics concepts. The topics confronted in those stages can be synthesized in order to put in evidence the implicit and explicit corrispondences presented by Kuhn:

BOLTZMANN'S GAS THEORY	PLANCK'S RADIATION THEORY
$f(u,v,w,t)$	$U(v,T)$
MOLECULAR DISORDER	NATURAL RADIATION
H-THEOREM	E.M. H-THEOREM
DIVISION OF CONTINUUM	DIVISION OF CONTINUUM
COMBINATORIALS	COMBINATORIALS
COMPLEXIONS	COMPLEXIONS
ENTROPY AND PERMUTABILITY	ENTROPY AND PERMUTABILITY
$S = K \operatorname{Ln} W$	$S = K \operatorname{Ln} R$

In the first line of the table, to Boltzmann's function $f(u,v,w,t)$ (of the gas molecule speed in H-Theorem), corresponds Planck's $U(v,T)$ function (distribution function of the resonator energy in the electromagnetic H-Theorem). In the following two lines, to the molecular disorder hypothesis which is at the base of Boltzmann's H-Theorem corresponds the natural radiation hypothesis, which Planck's electromagnetic H-Theorem rests on.

The other lines indicate the actual coincidence of Planck's steps on Boltzmann's, from the division of the continuum, to the use of combinatiorials up to the definition of entropy. The semplicity of this table mustn't make us think that Kuhn, in BBT, has espressed with so much linearity Planck's progress towards Boltzmann's mathematical concepts. But the outline shows very well Kuhn's purpose: to prove that Planck is Boltzmann's heir (the only one) and that, therefore, Planck's concept of discrete is the same as Boltzmann's.

In fact Planck's figure, which comes out from this re-reading of Kuhn, is that of a physicist who isn't at all in difficulty as regards the scientific tradition to which he belongs. On the contrary Planck is a scientist who works in complete continuity with the past, busy as he is in moulding his Radiation Theory on Boltzmann' s Gas Theory.

Kuhn notes that, until 1906, Planck had not asked himself the question of the quantization of energy, nor that of discontinuity in the processes of emission and absorbtion. So he concludes that, in 1900-1, Planck could not have consciously introduced such an important change as quantum discontinuity.

In fact, if the subdivision of energy in finite parts means to quantize, asserts Kuhn, the first quantization took place whith Boltzmann [2].

If, instead, to quantize means making a precise hypothesis in which it is affirmed that energy is a discrete quantity, then the first to do so, says Kuhn, was Einstein [3].

And what about Planck? To him is awarded the fact that he has introduced only implictly the fracture with the traditional physics [4].

The central thesis of the book, therefore, is that it was not Planck who introduced for the first time quanta in physics. This way Kuhn opposes the previous historians, according to whom Planck introduced them in 1900-1 with the blackbody distribution law.

As we can see, BBT is very interesting because it presents a thesis that puts the "standard" [5] version, maintained by very authoritative historians, in a critical position; but it is also very interesting for the scientific questions it debates, because his interpretation, with respect to that of other historians, deals in more profound way with the relations among mechanics, thermodynamics, electromagnetism and mathematics.

2. Boltzmann's mathematics inherited by Planck

But, returning to Planck, what does it mean for Kuhn that he had only an implicit role in the birth of the quantum theory?

What Kuhn means is that Planck did not reach quanta consciously, but he did not simply come across them either, after having found the probabilistic formula which resolved the question of the black-body distribution [6]. He has so deeply absorbed Boltzmann's thought as to make it "re-emerge" in a new scientific theory, that of radiation [7].

In order to understand what actually "re-emerges" in Planck let us quickly go over the group of Boltzmann's mathematical concepts which, according to Kuhn, is the nucleus of this heredity [8].

In his H-Theorem, Boltzmann "*introduces a function* f *of speed and time where* f(u,v,w,t)dudvdw *(usually abbreviated* fdω) *represents the number of molecules per volume unit which at time* t *have the speed components between* u *and* u+du, v *and* v+dv, w *and* w+dw. *This function* f(u,v,w,t) *specifies the distribution of the speed of molecules in the container in each instant* [9]. Kuhn specifies that Boltzmann requires that the volume dω be "*infinitesimal, but still containing a large number of molecules*" [10].

Through this function Boltzmann defines a new function H(t)= $\int$f logf dω and then demonstrates that dH/dt $\leq$ 0, that is H(t) can only, with passing of time, tend towards a minimum.

With this result Boltzmann later demonstrates that when H has reached the minimum, H_{min}, it can differ from the negative of the gas entropy (S) only for an arbitrary addictive constant that is $\Delta H_{min} = -\Delta S$. And so from the H-Theorem he derives the second law of thermodynamics [11].

But in order to obtain this thermodynamic result, observes Kuhn, Boltzmann *"makes the distribution function depend explicitly on time, he develops a differential equation for its dependence on time and treats its form at the instant to as an initial condition that the appropriate solution of the equation must satisfy. This topic is exactly what is needed for a function of the coordinates of a system in the classical - not statistical - mechanics . The form $f(u,v,w,t_0)$ describes the actual distribution of the molecular speeds at the instant t_0, and so it may appear as if it furnished, given an equation for $\partial f/\partial t$, a starting condition capable of working the same way as the starting conditions of classical mechanics. But, in reality, f is a sort of coarse grain function: an infinite number of different combinations of molecules within each small cell $d\omega$ is compatible with the same form of f at the instant t_0. Each of these combinations corresponds to a different starting condition for a mechanical system completely specified, and each of them brings to a different trajectory for that system in time. The actual form attribuited to f at the instant t_0 does not determine, however, the form of f at the following instant..."* [12].

The "coarse grain" that Kuhn sees in f is, actually, the finite number of molecules in $d\omega$ assumed by Boltzmann. Whose effect, still notes Kuhn, is that of leading Boltzmann to confuse the concept of <<molar>> (that is the group of molecules contained in the cell $d\omega$) with that of <<molecular>> [13]. This is the reason why, according to him, the coordinates of the individual molecule are more than once confused by Boltzmann with the coordinates of the cell that contains them. A further problem linked to such characteristic of f, Kuhn notes, are Boltzmann's passages from sums to integrals on this function. Kuhn affirms that their physical legitimacy is not evident in the case of the function Ω (= -H), also called measure of the permutability , where even if the quantities du, dv, dw are mathematical differentials, for Boltzmann they represent the dimension of cells *"still large enough to contain many molecules"* [14]. As regards this Kuhn comments that Boltzmann's opinions on the relationship between continuum and discrete haven't been closely examined [15]. And so he concludes that if Boltzmann's opinions are such, Planck's can't be different, being his heir. In fact, Kuhn affirms, Planck does not realize that his distribution function changes widely from one cell to another because the distribution of energy on the resonator is completely different from Boltzmann's distribution on the gas molecules [16].

3. The conception of infinite

From all of Kuhn's previous observations we can enucleate a fundamental theme: the question of infinite in Boltzmann. However we notice that he refrains from recognizing this element. He merely points out, as we have seen, that Boltzmann had superficial opinions as regards the relationship between continuum and discrete and refers to the reading of Dugas' book [17], but in a note. Almost to signify that the argument, even if important, had no other solution than Dugas'.

Dugas, from the beginning of *La Théorie physique* [18], concernes himself with the tipe of relationship that there is between Boltzmann and mathematics and defines Boltzmann's concepts of infinite and continuum, clarifying the particular way, that is atomistic, in which Boltzmann employed these concepts in the differential calculus (applied to physics). From him we learn that Boltzmann considered the purely formalist conception of calculus as methaphysical, that is separated from the atomistic representation of nature, but he admitted the use of automatisms of calculus, because he remebered that they couldn't be free from atomism.

In the same chapter, in order to complete the illustration of Boltzmann's mathematic finitism, Dugas presents the 1877 work on the probabilistic interpretetion of entropy. Precisely Boltzmann's work which, according to Kuhn [19], constituted Planck's chief model! Let us remember that this is the work in which Boltzmann considers a discrete number of molecules which can possess only discrete kinetics energies (that is: "quantizes"!). And Dugas quotes the words with which Boltzmann, with regard to possible objections, justifies that method by means of his conception of infinite. But we know that Boltzmann's finitist attitude isn't without ambiguity. His refusal of actual infinity is sometimes only apparent [20].

Therefore, the confusions observed by Kuhn as far as Boltzmann's mathematical methods, which we have recalled in the previous paragraph, can be reconducted to such ambiguity. But while Dugas shows to us that these contradictions derive from Boltzmann's refusal to use a mathematical-physics in which the use of actual infinite is admitted, for Kuhn they are simply superficial opinions. While Dugas reconducts the problem to the fundamentals and correlates it to the "*important consequence... in the quantum theory with Planck and Einstein*" [21], Kuhn skips it.

But, if Kuhn's purpose is to understand the genesis of quanta, that is the reason why from a certain period onward, in physics, very small quantities are assumed can't mathematically be sent to zero, we ask ourselves how he could have neglected Boltzmann's resolute finitism and, furthemore, how he could have done so after having gathered numerous proofs that the concept of infinite is the basic nucleus.

Kuhn could have specified the intellectual continuity between Boltzmann and Planck, after having carried on such a precise work as that of critically discussing all the conceptual dependence of the second from the first. Indeed, starting from Dugas' observations, he could have shown that both Boltzmann's quantization (that is his subdivision of continuum) and Planck's logically arise from a finitist conception, that is from a choice against the actual infinite. This way he would have really found the crucial factor that he resolved to look for. Kuhn's giving up in grasping this fact is probably due to his "blindness" as far as the role played by the foundations of mathematics in physical theories. For Kuhn mathematics is a constant of the history of science, the same as for all the physicists, and not a variable which determines the kind of theory that is being built up. Proof of this is the fact that among the examples of revolution given in *The Structure of Scientific Revolutions* [22] (from this point on SRS), Kuhn has not taken into consideration any historical examples in which the foundations of mathematics were at stake, at the contrary he has excluded that his scheme applies to mathematics [23].

On the other hand the questions relating to the actual infinite and to the potential infinite are very often considered as having a purely philosophical nature, that is not important for the understanding of how science has actually developed.

Let us remember besides that, in BBT, Kuhn has also given up interpreting the events with the categories of paradigm and revolution. In BBT there is, in fact, no trace of these concepts which were at the basis of the intepretative scheme that made him famous. This failure has been considered by scholars an event even more important than the new historical thesis on black-body maintained by the author [24]. In our opinion this failure is due both to the imprecise nature of his concept of paradigm [25] and because this time, contrarily to the historical examples presented in SRS, the facts analyzed invest the deeper relations existing between physics and mathematics. But the central thesis of the book - namely that the essence of Planck's work on radiation is limited to its continuity with the statistics that Boltzmann had developed for gases - could have at least outlined the existence of a paradigm wich guides the work of both Boltzmann and Planck: the paradigm of finitism.

It's not possible that this escaped Kuhn [26]! So it means that in this continuity Kuhn hasn't seen the characteristics of the "normal science". If the question of infinite had been chosen by Kuhn as a fundamental category [27] of paradigm, he could have probably used his interpretative scheme also in this case.

4. Conclusions

Destroyed the "standard" version of the birth of quanta, Kuhn had certainly the possibility of testing the vality of the interpretative scheme of SRS on BBT.

It is evident that the test has not been quite satisfactory. But Kuhn escluded that the relationship between BBT and SRS were necessary [28], when later on he was invited to clear it up. This way he followed his strange theory [29] according to which one can be a historian of science and a philosopher of science, but not at the same time!

We believe, on the contrary, that this theory is the rationalization of his failure. It is obvious in fact that this theory has no other justification than the fact that with the paradigm Kuhn hasn't been able to interpret his - so meticulous - work as historian!

Kuhn would have defended another thesis if, as we have proved, his interpretative tool had considered mathematics. He could have demonstrated that the birth of quanta in physics is the result of a contrast of mathematical paradigms of physics: Boltzmann's paradigm and the paradigm of the 19th century tradition. This way he could have saved his interpretative scheme!

References

[1] Kuhn, T.S. (1978). *Black-Body Theory and the Quantum Discontinuity 1894-1912*, Oxford, Clarendon Press - New York, Oxford U.P. It. tr.: *"Alle origini della fisica contemporanea. La teoria del corpo nero e la discontinuità quantica*, Il Mulino, Bologna, 1981.

[2] Kuhn, T.S. *BBT*, It.tr., p.222.

[3] *ibidem*, p.309.

[4] *ibidem*, p.238.

[5] Galison, P. (1981). "Kuhn and the quantum controversy", *British J. of Philos. of Science*, **32**.

[6] Kuhn, T.S. (1984). "Revisiting Planck", in *HSPS*, **14**:2, p. 236.

[7] Kuhn, T.S. *BBT*, op. cit., p.88.

[8] *ibidem*, 2nd chapter.

[9] *ibidem*, p.89.

[10] *ibidem*.

[11] *ibidem*, p.92.

[12] *ibidem*, p.96.

[13] *ibidem*, p.112.

[14] *ibidem*, p.115.

[15] *ibidem*.

[16] *ibidem*.

[17] Dugas, R. (1959). *La Théorie Physique au sense de Boltzmann*, Editions du Griffon, Neuchâtel.

[18] *ibidem*, pp.21--29.

[19] Kuhn, T.S. *BBT*, op. cit., p.130.

[20] Saiello, P. *Analisi storico-critica della Meccanica Statistica di Boltzmann*, Tesi di Laurea in Fisica, Università degli Studi di Napoli "Federico II", Anno accademico 1991-92, pp.162--171.

[21] Dugas, R. *La Théorie Physique au sense de Boltzmann*, op.cit., p.28.

[22] Kuhn, T.S. (1970). *The Structure of Scientific Revolutions*, Un. of Chicago. See Cerreta, P. and Drago, A., "Matematica e conoscenza storica. L' interpretazione di Kuhn della storia della scienza", in L. Magnani (ed.), *Conoscenza e Matematica*, Marcos Y Marcos, Milano, 1991, pp.353--364.

[23] Kuhn, T.S. *SRS*, p.34.

[24] Kuhn, T.S. "Revisiting Planck", op. cit., p.231.

[25] Mastermann, M. "The nature of Paradigm", in Lakatos, I. & Musgrave, A. (eds.), *Criticism and the Growth of Knowledge*, Cambridge Univ. Press, Cambridge, 170, pp.58--89.

[26] Kuhn, T.S. "Revisiting Planck", op. cit., p.245.

[27] Drago, A. (1990), "I quattro modelli di teoria scientifica", *Epistemologia*, **13** pp.303--324; "Il concetto di rivoluzione scientifica nella storia della scienza", in Bevilacqua, F. (ed.): *Atti del XII Congresso Nazionale di Storia della Fisica*, L'Aquila, 1991.

[28] Kuhn, T.S. "Revisiting Planck", op. cit., p.245.

[29] Kuhn, T.S. (1980). "The halt and the blind: philosophy and history of science", *Brit. J. Phil. Sci.*, **31,** p.185.

MACROSCOPIC QUANTUM COHERENCE AS A TEST
OF QUANTUM MECHANICS

L. CHIATTI
Laboratorio di Fisica Medica, CRS IRE, Roma

M. CINI
Dip. di Fisica Università "La Sapienza", Roma

M. SERVA
Dip. di Matematica and INFN, L'Aquila

Abstract. The proposal of using Macroscopic Quantum Coherence in a SQUID as a test of the validity of Quantum Mechanics (QM) for macroscopic systems (Leggett A.,1980; Leggett A. and Garg A. 1985) is considered. We note that if only Macroscopic Realism (MR) is assumed but the requirement that the flux measurement is non invasive (NIM) is dropped, only the measurement of the charge would discriminate between QM and MR. This discrimination however depends critically on the experimental parameters. There is a threshold above which QM is consistent with MR but the measurement is invasive as in QM.

PACS numbers: 03.65.Bz, 74.50.+r

1. Introduction

The main property of Quantum Mechanics (QM) is the superposition principle of states:

$$\psi = a_1\psi_1 + a_2\psi_2 \tag{1}$$

There is no question about its validity for microscopic systems. One should ask however whether it is also valid for macroscopic bodies. In fact, if the system is in the state (1) one has to accept that the variable G with eigenstates ψ_1 and ψ_2 is essentially undetermined until it acquires a definite value g_1 or g_2 in the act of

C. Garola and A. Rossi (eds.), The Foundations of Quantum Mechanics, 147-153.
© *1995 Kluwer Academic Publishers.*

measurement (Schrödinger's cat). Since we are accustomed to believe in macrorealism, namely to assume that the variables of macroscopic bodies always have a definite value (even if unknown to us) it is impossible to reconcile the validity of QM with our everyday experience.

It is therefore extremely interesting to investigate whether the possibility exists of making a macroscopic system such that the coherence effects of the linear superposition (1) may be detected. One of the most promising devices in this perspective is provided by the flux oscillations in a SQUID (Superconducting Quatum Interference Device)[Leggett A. 1980, 1986,1987].

The SQUID, as is well known, is a superconducting ring interrupted by a Josephson junction (JJ). The magnetic flux Φ through the ring reverses its sign as the current (of the order of 10^{22} Cooper pairs) oscillates in the ring. The junction is a thin non conducting barrier allowing the leakeadge of a current i_J. The SQUID is caracterized by a capacitance C and an inductance L. We neglect the resistance for the present considerations. The basic equations of the SQUID are the following.

If we denote by

$$\psi = \rho^{1/2} \, e^{\,i\theta} \qquad\qquad (2)$$

the common wave function of the Cooper pairs in the superconducting ring, the standard expression of the current $\mathbf{j}$

$$\mathbf{j} = (\rho/m)[(h/2\pi) \, \mathrm{grad}\theta - q\mathbf{A}] \qquad\qquad (3)$$

gives

$$(h/2\pi)(\theta_1 - \theta_2) = q \int \mathrm{rotA} \; dS = q \, \Phi \qquad\qquad (4)$$

where $\theta_1 - \theta_2$ is the phase difference on the two sides of the junction, because $\mathbf{j} = 0$ inside the superconductor.

Furthermore the current i_J through JJ is (q=2e):

$$i_J = i_o \sin(\theta_1 - \theta_2) = i_o \sin(4\pi e\Phi/h) \qquad\qquad (5)$$

where i_o is the critical current.

It is straightforward now to write down the classical energy of the SQUID:

$$H = (1/2) \, C \, V^2 + (1/2) \, L \, i^2 + \int V \, i_J \, dt \qquad\qquad (6)$$

namely, introducing the variable Φ and the elementary fluxon $\Phi_o = h/2e$

$$H_c = (1/2)C(d\Phi/dt)^2 + (1/2L)(\Phi-\Phi_{ext})^2 - (\Phi_o i_o/2\pi)\cos(2\pi\Phi/\Phi_o) \qquad (7)$$

We notice that the momentum p_Φ conjugated to Φ is

$$p_\Phi = C \, d\Phi/dt = Q \qquad (8)$$

namely the charge Q stored in the SQUID.

We now transform the classical Hamiltonian (7) into a quantum Hamiltonian by replacing p_Φ with the operator $(h/i)\partial/\partial\Phi$. We thus have:

$$H_q = - ((h/2\pi)^2/2C)\partial^2/\partial\Phi^2 + U(\Phi) \qquad (9)$$

with

$$U(\Phi) = (1/2L)(\Phi - \Phi_{ext})^2 - (\Phi_o i_o/2\pi) \cos(2\pi\Phi/\Phi_o) \qquad (10)$$

The external flux Φ_{ext} may be chosen equal to $\Phi_o/2$. Then U becomes symmetrical around $\Phi = \Phi_o/2$ for suitably chosen values of L, i_o and C. $U(\Phi)$ has the shape of a double well potential with the two minima Φ_+, Φ_- such that

$$\Phi_+ + \Phi_- = \Phi_o \qquad 0 < \delta\Phi = \Phi_+ - \Phi_- < \Phi_o \qquad (11)$$

Near each minimum the potential is approximately an harmonic oscillator potential of frequency ω. Denoting by $\psi_\sigma(\Phi)$ (with $\sigma=\pm1$) the ground state wave functions centered respectively at Φ_+,Φ_-, the Schrödinger equation

$$H_q\Psi(\Phi,t) = i \, (h/2\pi) \, \partial\Psi(\Phi,t)/\partial t \qquad (12)$$

admits solutions with initial value $\Psi(\Phi, t_o)= \psi_\sigma(\Phi)$ of the form

$$\Psi(\Phi,t) = \psi_\sigma(\Phi) \cos\Omega(t-t_o) - i \, \psi_{-\sigma}(\Phi) \sin\Omega(t-t_o) \qquad (13)$$

The frequency Ω may be calculated in terms of the potential parameters and is generally $<< \omega$. The probabilities of finding at time t the flux Φ in the states $\psi_\sigma(\Phi)$, $\psi_{-\sigma}(\Phi)$ are therefore

$$P(\sigma,t; \ \sigma,t_o) = \cos^2\Omega(t-t_o) \qquad P(-\sigma,t; \ \sigma,t_o) = \sin^2\Omega(t-t_o) \qquad (14)$$

2. Macroscopic Quantum Coherence as a test of Quantum Mechanics for macroscopic systems

A. Leggett and A. Garg have proposed [Leggett A., Garg A. 1985] to use the transitions of the magnetic flux trapped in a SQUID as a test of the validity of QM for macroscopic systems. More precisely they argue that the predictions of QM are incompatible with the following two assumptions which should characterize the behaviour of a macroscopic system:

(a) Macroscopic Realism (MR), namely that a macroscopic system will always be in either one or the other of two macroscopically distinct states;
(b) Non Invasive Measurability (NIM), namely that it is possible to determine the state of the system with arbitrarily small perturbation.

The proof goes as follows. Denote by σ_i (± 1) the value of Φ measured at time t_i. Then from (a) one derives

$$\sigma_1\sigma_2 + \sigma_2\sigma_3 + \sigma_3\sigma_4 - \sigma_1\sigma_4 \le 2 \qquad (15)$$

By averaging on a statistical ensemble and introducing the two times correlations K_{ij}

$$K_{ij} = < \sigma_i\sigma_j > \qquad (16)$$

one gets

$$K_{12} + K_{23} + K_{34} - K_{14} \le 2 \qquad (17)$$

Assumption (b) comes into play because if one wants to compare QM with MR one must assume that, by making flux measurements at t_1, t_2 or t_2, t_3 or t_3, t_4 or t_1, t_4 on *different* ensembles (since the QM predictions for K_{ij} are only valid if the evolution between t_i and t_j is not disturbed by intermediate measurements) the results are the same *as if* the measurements would be performed on a unique ensemble at all times.

On the other hand eq. (14) shows that QM gives:

$$K_{ij} = \sum_\sigma \sigma_i\sigma_j P(\sigma_i t_i; \ \sigma_j t_j) = \cos^2\Omega(t_i-t_j) - \sin^2\Omega(t_i-t_j) = \cos 2\Omega(t_i-t_j) \qquad (18)$$

This expression violates the Bell type inequality (17) (e.g. if we take $t_2-t_1 = t_3-t_2 = t_4-t_3 = \pi/8\Omega$ the l.h.s. is $2\sqrt{2}$). QM is therefore incompatible with MR+NIM.

3. Macroscopic Quantum Coherence and MR

It may be however interesting to investigate whether a separate test of MR would be possible without assuming NIM. An experimental proposal to test NIM independently has in fact been recently proposed [Cosmelli et al. 1993] and is currently in preparation. In fact, NIM is independent of MR and one may well conceive a theory in which the system will always have either one or the other of the two values of the flux, but the time evolution of the relevant probabilities will start again from the new initial condition corresponding to the measured value of Φ whenever a measurement is performed, as in QM.

A macrorealistic model of this type is obtained [Cini M. and Serva M. 1990, 1992] by describing the system by means of the density matrix

$$\underline{W}(t) = \psi_\sigma(\Phi)\, \psi^*_\sigma(\Phi)\, \cos^2 \Omega(t-t_o) + \psi_{-\sigma}(\Phi)\, \psi^*_{-\sigma}(\Phi)\, \sin^2 \Omega(t-t_o) \quad (19)$$

obtained from the density matrix W(t) of the pure state (13) by dropping the off-diagonal terms. The density matrix $\underline{W}(t)$ describes in fact a statistical mixture in which the flux is *either* in the state $\psi_\sigma(\Phi)$ *or* in the state $\psi_{-\sigma}(\Phi)$, representing the limit of the quantum density matrix when these states are macroscopically different [Cini M. 1983]. In fact when the distance between the two wells becomes macroscopically large, the overlap between $\psi_\sigma(\Phi)$ and $\psi_{-\sigma}(\Phi)$ vanishes exponentially. The same thing happens for the off-diagonal contributions to the mean value of any observable with a classical limit.

It is then clear that the predictions of the MR model for the flux measurements are identical to those of QM because both $\underline{W}(t)$ and W(t) give the same probabilities (20) of finding the flux in either the state $\psi_\sigma(\Phi)$ or $\psi_{-\sigma}(\Phi)$ at time t. It is therefore also clear that no experiment in which only the flux is measured can distinguish between this MR model and QM.

Only if one measures the variable Q conjugated to Φ can one test therefore the validity of the superposition principle for the macroscopic flux coherent oscillations in the double well. In fact, since

$$Q = (h/2\pi i)\, \partial/\partial\Phi \qquad (20)$$

one obtains, for the wave function $\Theta(Q, t)$ in Q-space

$$\Theta_\sigma(Q, t) = (h)^{-1/2} \int \Psi_\sigma(\Phi, t) \exp(-2\pi i Q\Phi/h) \, d\Phi \qquad (21)$$

By using (13) and taking into account that the ground state oscillator wave functions $\psi_\sigma(\Phi)$ are given by

$$\psi_{\pm 1}(\Phi) = (\beta/\pi)^{1/4} \exp[-\beta(\Phi-\Phi_+)^2/2], \qquad \beta = 2\pi C\omega/h \qquad (22)$$

we obtain the probability $P_\sigma(Q, t) = |\Theta_\sigma(Q, t)|^2$ of finding the value Q of the charge at time t

$$P_\sigma(Q, t) = 2(\pi/\beta h^2)^{1/2} \exp[-4\pi^2 Q^2/\beta h^2]\{1+\sigma \sin 2\Omega(t-t_o) \sin[2\pi Q\delta\Phi/h]\} \qquad (23)$$

The oscillating interference term is characteristic of QM. In the MR model (19) this term is instead missing. It is therefore possible, at least in principle, to discriminate between QM and MR by measuring the variable Q. The easiest thing to do might be to determine the sign of Q. Then the probability $P_\sigma(Q>0)$ is given by:

$$P_\sigma(Q>0, t) = (1/2) \{1 + \sigma \exp[-\beta (\delta\Phi)^2/4] \sin 2\Omega(t-t_o)\} \qquad (24)$$

The coefficient of the oscillating term, however, depends critically on the actual values of the experimental parameters. If one takes the values $C \approx 2.10^{-13}$F, $\omega \approx 3.10^9$ Hz, $\delta\Phi \approx \Phi_o/2$ from the available experimental proposals [Tesche C.D. 1987, Cosmelli et al. 1993], the absolute value of the argument of the exponential is of the order of one. This means that an increase of these parameters of a factor of some units is sufficient to reduce drastically the visibility of the quantum interference effect. If this happens QM is consistent with MR because there is no observable difference between the quantum mechanical density matrix W(t) and the density matrix $\underline{W}$(t) of the corresponding mixture. However, in this region, the measurement is still invasive as in QM.

It will therefore be of great interest to investigate experimentally the transition between the quantum and the classical region.

References

Cini, M. (1983). 'Quantum Theory of Measurement without Wave Packet Collapse', *Nuovo Cimento*, **73B** (27).

Cini, M. and Serva, M. (1990). 'Where is an object before you look at it?' *Found. of Phys. Lett.* 3,129

Cini M. and Serva M. (1992). 'Measurement in quantum mechanics and classical statistical mechanics', *Phys. Lett. A*, **167** (319).

Cosmelli, C., Diambrini-Palazzi, G., Di Cosimo, G., Di Domenico, A., Castellano, M.G., Leoni, R., Carelli, P., Cirillo, M., Chiatti, L., Scaramuzzi, F. (1993). 'Proposal for an Experiment for detecting Macroscopic Quantum Coherence with a System of SQUIDs'.

Leggett, A.J. (1980). 'Macroscopic Quantum Systems and the Quantum Theory of Measurement'. *Suppl. Prog. Theor. Phys.*, **69** (80).

Leggett, A.J. (1986). 'Quantum Mechanics at the Macroscopic Level', in *Directions in Condensed Matter Physics*, edited by Grinstein, G. and Mazenko, G. (World Scientific, Singapore), p. 189.

Leggett, A.J. (1987). 'Macroscopic Quantum Tunneling and Related Matters', in Proc.18th Int. Conf. on Low Temperature Physics, Kyoto, *Japan. Jour. Appl. Phys.*, **26**, 1986 Suppl. 26-3.

Leggett, A.J. and Garg, A. (1985). 'Quantum Mechanics versus Macroscopic Realism: Is the Flux There when Nobody Looks?', *Phys. Rev. Lett.*, **54** (857).

Tesche, C.D. (1987). 'Schrödinger's Cat: a Realization in Superconducting Devices', in Proc. 18th Int. Conf. on Low Temperature Physics, Kyoto, *Jap.Jour.Appl.Phys.*, **26**, Suppl. 26-3.

EXCHANGEABILITY AND INVARIANCE:
CLASSICAL ASPECTS OF QUANTUM CORRELATION

D. COSTANTINI
Ist. di Statistica, Università di Genova
Corso Paganini 3 - 16125 - Genova, Italy

and

U. GARIBALDI
CSSBT-CNR, Dip. di Fisica, Università di Genova
Via Dodecaneso 33 - 16146 - Genova, Italy

Abstract. In the paper some quantum correlations are determined using a classical probability function. These results are reached giving up independence and taking into account some suitable conditions of dependence among quantum particles.

1. Introduction

For several reasons since the twenties the elementary particle physics have neglected the classical probability and developed what is today called quantum probability. The official reason for this change is that whereas in classical mechanics propositions are described by sets, in quantum mechanics they are described by projection operators. Although never explicitly stated, for many authors the understood idea of this change is as follows: classical probability is suitable for dealing with situations characterized by subjective ignorance; quantum probability arises when ignorance becomes objective. To quote Feynman: "the laws of probabilities which are conventionally applied are quite satisfactory in analyzing the behavior of the roulette wheel but not the behavior of a single electron or photon of light" [1].

We are convinced that classical probability has been too hastily put aside. However, saying this we are not suggesting that we should return to classical probability theory. A lot of wonderful results has been reached with quantum probability. No reasonable person would make such a suggestion.

C. Garola and A. Rossi (eds.), The Foundations of Quantum Mechanics, 155-166.
© *1995 Kluwer Academic Publishers.*

What we mean to do is different. Physicists working in the twenties have constrained the notion of probability to one of its possible interpretations, namely that of relative frequency. As a matter of fact, this interpretation is strongly bound to the notion of stochastic independence. But the theory of probability, and classical probability too, is also suitable for handling conditional events, that is with dependence. Beyond any doubt, this has been shown by the history of this discipline since the seminal works of J. M. Keynes, B. de Finetti and H. Jeffreys. Taking this for granted, in the present paper we intend to show that an open-minded use of classical probability can lead to quantum results. In other words, taking dependence into account, classical probability can be used to derive some quantum correlations.

2. The elementary particles statistics

Let us consider a system of n particles and k states (of single particle) or oscillators all belonging the same energy level. $n_j, 1 \leq j \leq k$, are the occupation numbers of oscillators. $\mathbf{n} = (n_1, ...n_k)$, $\sum_{j=1}^{k} n_j = n$, is the state of the level. $\mathbf{N}^{n,k}$ is the set of all possible states of a level with total excitation equal to n. We want determine $P\{\mathbf{n}\}$, i.e. a probability distribution on $\mathbf{N}^{n,k}$, using a stochastic process that at each step describes the accommodation of a particle in an oscillator. To this aim we take into account the (predictive) probability function

$$P\{j|\mathbf{s}\}, \quad j = 1, ..., k, \quad 0 \leq s \leq n - 1 \tag{1}$$

i.e. the transition probability from $\mathbf{s}$ to $\mathbf{s}^j = (s_1, ..., s_j + 1, ..., s_k)$. Being a probability, for (1) the following (*basic*) conditions hold: **C1**, $P\{j|\mathbf{s}\} \geq 0$; **C2**, $\sum_j P\{j|\mathbf{s}\} = 1$; **C3**, $P\{j, g|\mathbf{s}\} = P\{j|\mathbf{s}\}P\{g|\mathbf{s}^j\}$. Moreover we also consider (*general*) conditions: **C4**, (*regularity*) for each j, $P\{j|\mathbf{s}\} > 0$; **C5**, (*exchangeability*) $P\{j|\mathbf{s}\}P\{g|\mathbf{s}^j\} = P\{g|\mathbf{s}\}P\{j|\mathbf{s}^g\}$; and for regular and exchangeable functions, **C6** (*invariance*) if $g \neq j, h \neq f, \mathbf{s}', \mathbf{s}'' \in \mathbf{N}^{s,k}$, then $Q_j^g(\mathbf{s}') = Q_h^f(\mathbf{s}'')$, where $Q_j^g(\mathbf{s}')$ is the heterorelevance quotient defined as

$$Q_j^g(\mathbf{s}) := \frac{P\{j|\mathbf{s}^g\}}{P\{j|\mathbf{s}\}}, \quad j \neq g.$$

We also consider the heterorelevance quotient at 0, i.e. $\eta := Q_j^g(\mathbf{0})$, and the autorelevance quotient at 0, i.e. $\rho := Q_j^j(\mathbf{0})$, $\mathbf{0} = (0, ..., 0)$..
For regular, exchangeable and invariant probability functions the following (*fundamental*) theorem holds: [2]

$$P\{j|\mathbf{s}\} = \frac{\lambda p_j + s_j}{\lambda + s}, \quad j = 1, ..., k, \tag{2}$$

where $\lambda := \eta/(1-\eta)$ and $p_j := P\{j|0\}$. This is the (initial) probability of the $j-th$ state (of single particle), that is the probability that an excitation will be created at the $j-th$ oscillator when the level is in the empty state. Hence $\mathbf{p} := (p_1, ..., p_k)$ is the (initial) distribution on the set of all possible states when the level is in the empty state. (2) holds for non regular functions too. If this is the case, λ is negative and the process must stop. As a consequence, the state of a single particle cannot increase indefinitely.

The last general condition we are considering is: **C7**,(*maximum entropy*) $H(\mathbf{p})$ is maximum. From this it follows that $\mathbf{p} = (k^{-1}, ..., k^{-1})$, and (2) becomes

$$P\{j|\mathbf{s}\} = \frac{\lambda/k + s_j}{\lambda + s}.$$

Finally let us consider the following (*special*) conditions:

$$\mathbf{CBE}, \ \eta = \frac{k}{k+1}; \quad \mathbf{CMB}, \ \eta = 1; \quad \mathbf{CFD}, \ \eta = \frac{k}{k+1}.$$

As a consequence, we have three values for the autorelevance quotient at 0:

$$\rho^{\mathbf{BE}} = \frac{2k}{k+1}; \quad \rho^{\mathbf{MB}} = 1; \quad \rho^{\mathbf{FD}} = 0,$$

and three transition probabilities:

$$P^{\mathbf{BE}}\{j|\mathbf{s}\} = \frac{s_j + 1}{k+s}; \quad P^{\mathbf{MB}}\{j|\mathbf{s}\} = k^{-1}; \quad P^{\mathbf{FD}}\{j|\mathbf{s}\} = \frac{1 - s_j}{k-s}$$

i.e. the Bose-Einstein transition; the Maxwell-Boltzmann transition and the Fermi-Dirac transition.

Using these transition probabilities, we can determine the three elementary particle statistics, more specifically:

$$P^{\mathbf{BE}}\{\mathbf{n}\} = \binom{n+k-1}{n}^{-1}, \quad P^{\mathbf{MB}}\{\mathbf{n}\} = \frac{n!}{\prod_{j=1}^{k} n_j!} k^{-n}, \quad P^{\mathbf{FD}}\{\mathbf{n}\} = \binom{k}{n}^{-1}$$

which are distributions on $\mathbf{N}^{n,k}$.

3. Average energy of an oscillator

From now on we shall pay more attention to bosons. [1] Using $P^{\mathbf{BE}}\{\mathbf{n}\}$, we can reach the marginal distribution of an oscillator with m excitations, that

---

[1] In the following we only give hints of the deductions when bosons are considered. For fermions and classical particles we give the final results. For a detailed discussion of the topic see [3] e [4].

is

$$P^{\mathbf{BE}}\{m||n,k\} = \frac{\left(\begin{array}{c} n-m+k-2 \\ n-m \end{array}\right)}{\left(\begin{array}{c} n+k-1 \\ n \end{array}\right)}, \quad 0 \le m \le n. \tag{3}$$

When $n \to \infty$ and $k \to \infty$ in such a way that $n/k \to \chi$ (we call this the thermodynamic limit), then

$$P^{\mathbf{BE}}\{m||n,k\} \to P^{\mathbf{BE}}\{m||\chi\} = \frac{\chi^m}{(1-\chi)^{m+1}}, \quad m = 0,1,2,...,$$

and putting $x = \chi/(\chi+1)$,

$$P^{\mathbf{BE}}\{m||x(\chi)\} = x^m(1-x) = \frac{x^m}{\sum_{m=0}^{\infty} x^m}.$$

Let the physical system be composed of K energy levels, each endowed with k_i oscillators. Hence the state of the system is represented by the table

$$\underline{\mathbf{n}} = \left[\begin{array}{c} \mathbf{n}_1 \\ ... \\ \mathbf{n}_i \\ ... \\ \mathbf{n}_K \end{array}\right]$$

where $\mathbf{n}_i$ is the state of the $i-th$ level. The horizontal margins of this table, that is $\mathbf{N} = (N_1,...,N_i,...,N_K)$, $N_i = \sum_i n_{ij}$, $\sum_i N_i = N$, describes the macrostate of the physical system.

We assume that, given $\mathbf{N}$, the K levels are statistical independent; on the contrary, the oscillators of the same level are correlated being fixed N_i. It follows that, given $\mathbf{N}$, the distribution on $\underline{\mathbf{n}}$ splits into K terms, that is

$$P\{\underline{\mathbf{n}}||\mathbf{N}\} = \prod_i P\{\mathbf{n}_i||N_i\} \tag{4}$$

where each value of $P\{\mathbf{n}_i||N_i\}$ can be determined as in section 2. Using the statistical jargon, what we are doing amount to the determination of a "sampling distribution" - a term that we shall use in what follows - for a stratified sample.

Supposing the validity of the thermodynamic limit for the whole system, the correlation among oscillators of the same level disappear, and the distribution on the table $\underline{\mathbf{n}}$ is the product of the marginal distribution of each entry. Moreover if we consider the $K-tuple$ $(\chi_1,...,\chi_K)$, the sampling distribution becomes

$$P\{\underline{\mathbf{n}}||(\chi_1,...,\chi_K)\} = \prod_i \prod_j P\{n_{ij}||\chi_i\} \tag{5}$$

where χ_i is the average occupation number of the $i - th$ level. A very natural way to look at χ_i is to consider it as the total number of excitation per oscillator in the assembly "system+thermostat".

Given the spectrum $(\epsilon_1, ..., \epsilon_K)$, the total number of excitations N in (4) is fixed as well as the total energy $E = \sum_i N_i \epsilon_i$. Due to the one-to-one correspondence between E and N, we call (4) the microcanonical distribution. On the contrary, (5) only fixes the average values of these quantities, that is $< N > = \sum_i < N_i > = \sum_i k_i \chi_i$ and $< E > = \sum_i k_i \chi_i \epsilon_i$. When these quantities or, what is better, their intensive counterparts, i.e. the chemical potential μ and the temperature T, are fixed by the thermostat, for the maximum entropy sampling distribution (the equilibrium distribution)

$$\frac{\delta S_i}{\delta \chi_i} = \frac{\epsilon_i - \mu}{T}$$

holds, where S_i is the entropy of one oscillator of the $i - th$ level and $S = \sum_i k_i S_i$. Thus we have

$$P^{\mathbf{BE}}\{m||\chi_i\} = \frac{\chi_i^m}{(1 + \chi_i)^{m+1}}, \quad P^{\mathbf{FD}}\{m||\chi_i\} = \chi_i^m (1 - \chi_i)^{1-m}, \quad (6)$$

obviously, in the first formula $m = 0, 1, 2, ...$ while in the second one $m = 0, 1$.

The entropy of the distribution on the possible states of an oscillator are

$$S_i^{\mathbf{BE}} = -\chi_i \ln \chi_i + (1 + \chi_i) \ln(1 + \chi_i), \quad S_i^{\mathbf{FD}} = -\chi_i \ln \chi_i - (1 - \chi_i) \ln(1 - \chi_i).$$

The values maximizing $S := \sum_{i=1}^{K} d_i S_i$, we call χ_i^*, are

$$\chi_i^{*\mathbf{BE}} = \frac{1}{\exp[(\epsilon_i - \mu)/T] - 1}, \quad \chi_i^{*\mathbf{FD}} = \frac{1}{\exp[(\epsilon_i - \mu)/T] + 1}. \quad (7)$$

In a sense these are the main results of our probability foundation of thermodynamic, but using our approach we can also show other interesting feature implicitly included in the transition probability of the process. In fact in (2) it is easy to recognize a classical term, $\frac{p_j \lambda}{\lambda+s}$, and a quantum one, $\frac{s_j}{\lambda+s}$. For bosons these terms become $\frac{1}{\lambda+s}$, respectively, $\frac{s_j}{k+s}$. With this in mind one immediately realizes the reason for which the probability of having m bosons in a state is $m!$ times greater than that of having m classical particles. In fact, using $P^{\mathbf{BE}}\{j|s\}$ and $P^{\mathbf{MB}}\{j|s\}$ we have

$$P^{\mathbf{BE}}\{m\} \propto \frac{m!}{\prod_{i=0}^{m-1}(k+i)}, \quad P^{\mathbf{MB}}\{m\} \propto k^{-m};$$

when $m \ll k$, the quotient of these probabilities gives the result.

4. Fluctuations: microcanonical values

Coming now to fluctuations, the usual way to calculate the variance of the marginal distribution of an oscillator makes use of distribution (6), that is the variance is determined after having performed the thermodynamic limit. We follow this way in the next section, but before doing this we show a way of determining microcanonical fluctuation directly. In order to calculate $Var(m_j||n)$, $m_j = m$ and $n = N_i$ we need $E(m_j^2||n)$ and $E^2(m_j||n)$ in the case in which the realization of the accommodation process ruled by (2) leads to m values $j_i = j$, that is when $m_j = \sum_i \delta(j_i - j)$, where $\delta(j_i - j)$ is the indicator of $j - th$ oscillator at the $i - th$ step of the process. If this is the case, we have

$$m_j^2 = \sum_{i,h} \delta(j_i - j)\delta(j_h - j) = \sum_i \delta^2(j_i - j) + \sum_{i \neq h} \delta(j_i - j)\delta(j_h - j)$$

and thus

$$E(m_j^2||n) = \sum_i E(\delta^2(j_i - j)||n) + \sum_{i \neq h} E(\delta(j_i - j)\delta(j_h - j)||n).$$

Moreover $E(\delta(j_i - j)\delta(j_h - j)||n) = P\{j_i = j, j_h = j\}, i \neq h$. Using (2) we have $E(\delta(j_i - j)) = k^{-1}$, $P\{j_i = j, j_h = j\} = \rho k^{-2}$, so that

$$E(m_j^2||n) = \frac{n}{k} + n(n-1)\frac{\rho}{k^2},$$

and finally

$$Var(m_j||n) = \frac{n}{k}\left(1 - \frac{1}{k}\right) + \frac{n(n-1)}{k^2}(\rho - 1)$$

Taking into account the values of ρ we have:

for **CBE**, $V(m_j) = \frac{n}{k}\left(1 - \frac{1}{k}\right) + \frac{n(n-1)}{k^2}\left(\frac{2k}{k+1} - 1\right) = \frac{n}{k}\left(\frac{k-1}{k} + \frac{n-1}{k}\frac{k-1}{k+1}\right)$

for **CMB**, $V(m_j) = \frac{n}{k}\left(1 - \frac{1}{k}\right)$

for **CFD**, $V(m_j) = \frac{n}{k}\left(1 - \frac{1}{k}\right) - \frac{n(n-1)}{k^2} = \frac{n}{k}\left(1 - \frac{n}{k}\right)$

In the same way we can calculate $Cov(m_j m_g||n) = -\frac{V(m_j||n)}{k-1}$, and thus the correlation coefficient $R(m_j m_g||n) = -\frac{1}{k-1}$. It is worth noting that through (2) we have been able to determine the values of the fluctuations also for classical particles when the margins are exactly fixed. The well-known values for classical and quantum particles are the thermodynamic limits of the previous formulae. The correlation coefficient approach to 0 when $k \to \infty$, that is, when this is the case, oscillators become independent. Moreover $R(m_j, m_g||n)$ does not depend on the quantum correlation, which is fixed by the ratio of the number of oscillators to λ, that is

$$c := \frac{k}{\lambda}.$$

It depends on the margins, that is on the macroscopic boundary conditions of the process.

5. Fluctuations: grandcanonical values

Coming back to (6) we note that these formulae are special cases of

$$P^c\{m|\chi_i\} = \binom{c+m-1}{m} \left(\frac{c}{c+\chi_i}\right)^c \left(\frac{\chi_i}{c+\chi_i}\right)^m \tag{8}$$

which summarizes all probability properties of oscillators. For $c > 0$, (8) is the negative binomial with $b := c^{-1}$ and $x := \frac{\chi_i}{b+\chi_i}$, i.e.

$$P^c\{m|\chi_i\} = \binom{c+m-1}{m} (1-x)^b x^m, \quad m = 0, 1, ...$$

For this distribution we have $E(m) = b\frac{x}{1-x} = \chi_i$ and

$$V(m) = \chi_i(1 - c\chi_i); \tag{9}$$

In a continuous way, c fixes the correlation term χ_i, which is 0 for independent (classical) particles and 1 for bosons.

For $c < 0$, in this case $b := |c^{-1}|$, putting $x := \frac{\chi_i}{b-\chi_i}$, (8) becomes

$$P^c\{m|\chi_i\} = \binom{b}{m} (1-x)^{-b} x^m, \quad m = 0, 1, ...$$

which for $p := \frac{x}{1-x} = \frac{\chi_i}{b}$ is the Bernoullian distribution

$$P^b\{m|\chi_i\} = \binom{b}{m} (1-p)^{b-m} p^m, \quad m = 0, 1, ..., b$$

with $E(m) = \chi_i$ and

$$V(m) = \chi_i(1 - |c|\chi_i). \tag{10}$$

(9) and (10) show the effect of the thermodynamic limit that amounts to annul the dependence among oscillators. Once this dependence vanishes, the grandcanonical values of the fluctuations appear.

Concluding the section we stress the way we have followed in order to derive the grandcanonical fluctuations. (9) and (10) has been deduced taking into account the whole system, that is: considering the distribution on $N^{s,k}$; marginalising this distribution to calculate the distribution on the possible states of an oscillator; and finally applying the thermodynamic limit.

6. Evolution matrices

Coming back to the conditional process, as we have said, the distribution (4) can be reached considering K exchangeable and invariant processes whose transition probabilities are (2) with the same value $c^{-1} = \lambda_i/k_i$ and initial uniform distributions. They are the maximum entropy distributions (or equilibrium distributions) given c and the external constraints. A physical interpretation of these K transition probabilities can be given considering them as K creation processes. Starting from the empty (initial) state, the following

$$P_i\{j||\mathbf{n}_i\} = \frac{c^{-1} + n_{ij}}{c^{-1} + N_i} = \frac{1 + cn_{ij}}{k_i + cN_i} \qquad (11)$$

is the probability of creating a particle in the $ij - th$ oscillator of the $i - th$ level. Considering this creation process, the macrostate N is a $K - tuple$ of external parameters. A process which randomizes N in order to simulate a thermal contact, is that in which one introduces a transition probability $P\{i|\mathbf{n}\}$ connecting the K disjoint processes ruled by (11). As a result we have a unique process whose transition probability is

$$P\{ij|\underline{\mathbf{n}}\} = P\{i|\underline{\mathbf{n}}\}P_i\{j|\mathbf{n}_i\}.$$

Obviously we need new assumptions for determining $P\{i|\mathbf{n}\}$.[2]

Besides these simple extensions of the conditional process, considering the transition probabilities (2) and (11) as creating probabilities, a new interpretation of the distribution (4), can be given. This intepretation allows to give up the statistical notion of sampling distribution as well as its physical counterpart, i.e. the ensemble as we shall see.

For the sake of simplicity let us consider a level with 2 oscillators supposing $c = 1$. Thus $k = 2$, $\mathbf{n} = (r, n - r)$, and we can write $P_n\{1|r\}$ instead of $P\{1|(r, n - r)\}$ which is the transition probability, and $P_n\{r\}$ instead of $P\{(r, n - r)\}$ which, on the contrary, is the probability of the state of the level. For the basic conditions the following equality

$$P_n\{r\} = P_{n-1}\{1|r - 1\}P_{n-1}\{r - 1\} + P_{n-1}\{2|r\}P_{n-1}\{r\}. \qquad (12)$$

---

[2] We have performed such an approach in [5]

holds. Obviously there are $n+1$ equations of this type. The conditions we have considered give $P\{1|r\} = \frac{r+1}{n+2}$. It follows that all probabilities (12) are summarized by a matrix with $n+1$ rows and n columns whose general term is

$$
C_{n-1} = \frac{1}{n+1}
\begin{bmatrix}
n & 0 & 0 & \cdots & 0 & 0 \\
1 & n-1 & 0 & \cdots & 0 & 0 \\
0 & 2 & n-2 & \cdots & 0 & 0 \\
\cdots & \cdots & \cdots & \cdots & \cdots & \cdots \\
0 & 0 & 0 & \cdots & n-1 & 1 \\
0 & 0 & 0 & \cdots & 0 & n
\end{bmatrix}
$$

in which the transition probabilities different from 0 are that from r to r and from r to $r+1$. In general, for each value of c, we have

$$
C_n = \frac{1}{cn+2}(c(n-r')+1)\delta(r'-r)) + (c(r'-1)+1)\delta(r'-r-1))
$$

$r = 0, 1, ..., n$, $r' = 0, 1, ..., n+1$, and $c = 2/\lambda$, which gives the transition probabilities from r to r'. It is easy to check that the sequence $C_{n-1}C_{n-2}...C_0$ applied to an initial distribution $P_0\{0\}$ gives the equilibrium distribution on a level.

We have just considered *creation* matrices, now we come to *destruction* matrices. These are matrices that starting from an equilibrium distribution on a level of n particles, transform it into an equilibrium distribution on a level with $n-1$ particles. The probabilities involved in a destruction are analogous to that of (11), and in order to determine the value we must make some assumptions. Let us suppose that the destruction of a particle is exchangeable. It follows that the destroying probability is n_j/n, and the corresponding matrix is

$$
D_n = \frac{1}{n}
\begin{bmatrix}
n & 1 & 0 & \cdots & 0 & 0 \\
0 & n-1 & 2 & \cdots & 0 & 0 \\
0 & 0 & n-2 & \cdots & 0 & 0 \\
\cdots & \cdots & \cdots & \cdots & \cdots & \cdots \\
0 & 0 & 0 & \cdots & 2 & n-1 \\
0 & 0 & 0 & \cdots & 1 & n
\end{bmatrix}
$$

which has n rows and $n+1$ columns. Contrary to what happens for the creation matrix, the destruction one does not depend on c. This means that the destroying process, given a $k-tuple$, does non depend on what particle we are considering. It is worth noting that we have the same result taking into account the principle of the detailed balance, that is

$$
P\{\mathbf{n}\} P\{j|\mathbf{n}\} = P\{\mathbf{n}^j\} P\{-j|\mathbf{n}^j\},
$$

where $-j$ means that we are destroying a particle in the $j-th$ oscillator.

What we have seen can be summarized as follows. The creation process is based on the correlation typical of the considered particle; on the contrary, for each type of particle, the destruction process is based on a hypergeometric (negative) correlation.

Based on C_n and D_n, we can define an *evolution* matrix

$$E_n := C_{n-1} D_n$$

which applied to $P\{\mathbf{n}\}$ shows the transformation of this distribution after the destruction and the creation of a particle. Applying E_n to a system of oscillators introduces into the system a dynamical evolution, that is, the possibility is given to two oscillators to exchange an excitation. Obviously these matrices can be applied to non equilibrium systems too. Due to the fact that E_n is a stochastic square matrix depending only on $\mathbf{n}$, the evolution process is Markovian homogeneous with equilibrium distributions as eigenstates. As a consequence, it can be shown that the elementary particle statistics are reached by the system whatever the initial distribution may be. Hence we have an evolution process which, for each value of the particle correlation, is a generalization of the celebrated Ehrenfest process.

7. Conclusion

A correct use of classical probability can explain some quantum correlations. In fact, what we have done is based on a classical probability function. The very difference between our function and that generally used in physics is that for the latter statistical independence holds whilst we have repeatedly used the condition **C3** without shortening. As we have said, the most important result of our work is the derivation of (7) without making use of any non probabilistic assumption. The way usually followed to derive (7), but also (9) and (10), is based on the notion of indistinguishability. As a consequence, one imputes these results to this notion (which is not probabilistic in character) supposed to be the marking property of quantum particles. We have reached the same results without any reference to indistinguishability showing that (7) can be derived in a purely probabilistic way.

But what is more illuminating, we have deduced the microcanonical values of the fluctuations only making reference to the proper correlation among particles. That is, in order to deduce the values of the fluctuations, we have avoided in section 4 any reference to the sampling distribution (4), which was the way followed by Einstein. What is essential in our approach is the correlation between two particles created in the same level at different steps i and h, that is

$$Cov(\delta(j_i - j)\delta(j_h - g)) = \left\{ \begin{array}{ll} (\eta - 1)k^{-2} & \text{if} \quad j \neq g \\ (\rho - 1)k^{-2} & \text{if} \quad j = g. \end{array} \right.$$

Obviously the condition of independence among particles, i.e. $\eta = \rho = 1$, makes this calculation quite unimportant. But this shortening characterizes classical particles (whether they exist or not) rather than classical probabilities. The calculation of $Cov(\delta(j_i - j)\delta(j_h - g))$ is the best way to illustrate what it is possible to do with a correct use of classical probability functions. In fact, to calculate the value of the covariance we can make use of **C3**, that is of the equality

$$P\{j, g|\mathbf{s}\} = P\{j|\mathbf{s}\}P\{g|\mathbf{s}^j\}.$$

But the complete formulation of the product axiom makes necessary the determination of $P\{g|\mathbf{s}^j\}$,. This means that one must take into account the dependence among quantum particles. This dependence, implicitly introduced via the indistinguishability, can be explicitly assumed via (2) and appropriate values of η, that is making clear the probability assumptions from which it follows.

To conclude this remark, we stress that classical particles are characterized by being $\eta = 1$ and $\rho = 1$. This is a consequence of putting

$$P\{j|\mathbf{s}\} = P\{j|\mathbf{s}^g\}.$$

Hence classical particles are characterized by the irrelevance in creating a particle on what oscillators have created the other particles. But this statistical independence is not a characteristic of classical probability which, at least in the case we have considered, is suitable for ruling non classical particles too.

A further remark deals with equilibrium and maximum entropy. In our approach, entropy is a property of probability distributions not of frequency distributions. The charcterization of equilibrium distribution, given the external constraints, is correct for any type of particle, if and only if, the interparticle correlation is taken into account from the beginning. On the contrary, the usual way to allot equal probabilities to each microstate allowed by the constraints, must change the notion or the support of the microstates according to the type of particles.

Finally, the conceptual scheme of the generalized Eherenfest process can, in principle, give a probability account of the equilibrium distribution and the attainment of it. This avoid the statistical notion of sampling distribution or its physical counterpart, i.e. the ensemble of independent replicas: useful conceptual tools to associate empirical meaning to the distributions $P\{\underline{\mathbf{n}}||\mathbf{N}\}$ and $P\{\underline{\mathbf{n}}||(\chi_1, ..., \chi_K)\}$ in an approach based on the relative frequency interpretation of probability.

References

1. Feynman, R.P. (1951). *The Concept of Probability in Quantum Mechanics*, Proc. 2nd Berkeley Symposium on Mathematical Statistics and Probability, University of California Press 1951, p. 533.
2. Costantini, D. and Garibaldi, U. (1989). "Classical and Quantum Statistics as Finite Random Processes", *Foundations of Physics*, 19 pp. 743–754.
3. Costantini, D. and Garibaldi, U. *Scambiabilità e correlazione nella meccanica statistica delle "particelle indistinguibili"*, to appear in D. Costantini and U. Garibaldi, *Probabilita' e fisica*, Bibliopolis, Napoli.
4. Costantini, D. and Garibaldi, U. *Scambiabilità e invarianza: aspetti classici della correlazione quantistica*, to appear in D. Costantini and U. Garibaldi, *Probabilita' e fisica*, Bibliopolis, Napoli.
5. Costantini, D. and Garibaldi, U. (1994). "Microcanonical and Canonical Distributions and Finite Exchangeable Random Processes", *Foundations of Physics*, 24 pp. 177–202.

EINSTEIN'S LIFE-LONG DOUBTS ON THE PHYSICAL FOUNDATIONS OF THE GENERAL RELATIVITY AND UNIFIED FIELD THEORIES[*]

S. D'AGOSTINO
Dip. di Fisica, Università di Roma "La Sapienza"
Roma, Italy

Abstract. Although Einstein frequently discussed the problem of the theory-experiment relationship, it has gone almost unnoticed by Einstein scholars that, in his papers, this problem is presented in the form of a search for criteria for the attribution of physical significance to the concepts of Relativity and Unified-Field theories, i.e. in the form of a stipulation of meaning.

Einstein confronted this problem beginning with his early approaches to Special Relativity in 1905 and, until his last years, he never ceased to search for possible solutions. Thus, problems concerning the "stipulation of meaning" interweave all of Einstein's methodological discussions on General Relativity and its generalisation into the Unitary-Field theories. In these discussions, Einstein often came in touch with the methodological views of mathematical physicists such as Weyl, Eddington, Levi-Civita, et al.

1. Introduction: a foundational problem in Einstein's Relativity

To the best of my present knowledge, it can safely be argued that the majority of recent studies on the foundation difficulties of Quantum Physics start from the assumption that the conceptual foundations of Classical Physics and Relativity theory were clear and unproblematic. Therefore, the present problems would concern only Quantum Physics. Jet, a simple inquiry into the literature and especially into Einstein's epistemological writings shows that, contrary to the view of a supposedly well-founded Classical Physics and Relativity, important foundation problems in these sciences are still in need of further analysis[1]. To discuss the foundation difficulties of Quantum Physics (henceforth QP) as if they alone existed against an ideal unproblematic background of Classical

167

C. Garola and A. Rossi (eds.), The Foundations of Quantum Mechanics, 167-178.

Physics (henceforth CP) and Relativity theories (RR), results in a limited approach to the historical documents.

It is known that Einstein's epistemological views were at times misinterpreted by physicists and philosophers. As an example, Einstein's Special Relativity (henceforth SR) was considered by Bridgmann in 1949 (Bridgmann 1959, p.335 ff.) as the "manifesto" of the fruitfulness of operational definitions for physical quantities. In his mature years Einstein rejected Bridgmann's interpretation (Einstein 1949b, p.679). As another example of this misunderstanding, let us take the case of Werner Heisenberg. In 1926, shortly after the publication of his paper on the presumed *Anschaulichkeit* of QT, Heisenberg confided to Einstein that he had actually taken the idea of observable quantities from Einstein's RR. Einstein quickly discredited Heisenberg's interpretation saying that, just to the contrary, he maintained that it is the theory which ultimately decides what can be observed and what cannot (AHQP 1963; Jammer 1966, p.198).

I argue that these misinterpretations were in part the result of a certain amount of ambiguity in Einstein's early epistemology, the hallmark of a conceptual situation which was properly named by Yehuda Helkana as "concepts in flux".

I will show that Einstein's approach to the problem of the meaning of the space-time interval in his SR easily lent itself to Bridgmann's and Heisenberg's criticism.

2. Einstein's problems with the stipulation of meaning for the Riemannian space-time continuum

In his 1923 Gothenburg lecture Einstein laid down explicitly a *stipulation of meaning* (SM) for the concepts of physics:

> Concepts and distinctions are only admissible to the extent that observable facts can be assigned to them without ambiguity (Einstein 1923, p.482).

He found that in Classical Mechanics (CM) the definitions of concepts, such as inertial system and free body, are circular; hence CM transgresses SM:

> Note in passing that the logical weakness of this exposition {i. e. the exposition of CM} from the point of view of the stipulation of meaning is the lack of an experimental criterium for whether a material point is force-free or not; therefore the concept of the inertial frame remains rather problematic (Einstein 1923, p.483. Parentheses { } mine).

Contrary to what might be expected, SR faces the same difficulty as does CP. The concept of inertial reference frame is also a fundamental foundation problem for that theory, because, from the SM point of view, a reference frame is just a combination of rigid rods. But, as known, rigidity would allow instantaneous signal transmission, thus contradicting a fundamental SR postulate. This justifies Einstein's conclusion in his Gothenburg lecture:

> I am mentioning these deficiencies of method because in the same sense they are also a feature of the SR in the schematic exposition which I am advocating here.

Concerning Einstein's problem with rigidity, it is worth mentioning that, in 1909, Max Born (Born 1909) called attention to this problem and proposed an original solution in various essays (Maltese & Orlando 1994).

Because SR, like CP, is not able to find a satisfactory SM-observant physical meaning for rigidity, it was logical for Einstein, in the same Gothenburg lecture, to explore GR as a possible basis for the solution of his problem. He then hinted at an indirect criterium as a guarantee for the physical meaning of the concepts: the *simplicity criterium* for theory validation (Einstein 1923, pp.485--489).

However, in spite of the above criterium, which appeared to be a way of circumventing the SM problem, he shortly thereafter in the same paper (Einstein 1923, p.487) presented GR as a radical solution to the rigidity problem: the abolition of finite rigid inertial frames and their substitution with local inertial frames. In a gravitation-free space the infinitesimal space-time interval of GR: $ds^2 = g^{mn} dx_m dx_n$, *has* to coincide with:

$$ds^2 = c^2 dt^2 - dx^2 - dy^2 - dz^2,$$

its correspondent in the pseudo-Euclidean space of SR.

He initially adopted this solution, i.e. the solution via the so-called Correspondence criterium or method, in his 1912 attempt to generalise (*Allgemeinen*) SR (Einstein 1934, pp.307--308). However, he was not satisfied with it. In his own words:

> It {the solution} was inevitably fatal to the simple physical interpretation of the coordinates, because it could no longer be required that coordinates differences should signify direct results of measurements with ideal scales or clocks (Einstein 1934, p.307. Parentheses { } mine).

3. "Correspondence" as a logically asymmetric method for correlating concepts and perceptions. Einstein's longing for a purer method

Given the premises above, it is not surprising that Einstein, in his 1936 essay *Physics and Reality* (Einstein 1950, pp.63--65), returned to the problem of a SM for *ds*, but, this time, through a more general approach founded on the *"Stratification of the Scientific System"*. A physical theory consists of a stratification comprising various levels: the lower level, which is also the most primitive in a diacronical sense, comprises concepts that are more directly related to perceptions (*Empfindungen*) and to the connecting theorems.

Although the upper level concepts are more distant from perceptions, this defect is balanced by an advantage: they gain in simplicity, i.e. in the clearness and distinctiveness of their axiomatic foundation, what they loose in empiricism. However, in order to make contact with the empirical level, the upper level concepts need to be reduced to their correspondents in the lower level. This is achieved by way of a mapping process. The mapping is achieved in the so-called Correspondence area, by relating the upper level concepts to their lower level correspondents (D'Agostino, Orlando 1994), but not viceversa. In this sense there is an asymmetry in the Correspondence relationship, the correspondence is univocal, not biunivocal:

> The relation is not analogous to that of a soup to beef but rather of wardrobe number to overcoat (Einstein 1950, p.64).

Once the Correspondence is established, the upper level concepts receive their physical meaning from their correspondent lower level concepts (Einstein 1950, p.81). In his view, Einstein confirms and generalises his Correspondence criterium of 1912 into a general feature of theories.

If one concedes that the above stratification and the related hjerarchy of concepts somehow absolve the upper level concepts from their SM-transgression, one should also admit that, concerning SM, this new method amounts to a transfer of meaning from the lower to the higher level through Correspondence rules which possess many degrees of freedom. For this reason, this transfer has been sometimes nicknamed "a transfer of meaning by decree" (D'Agostino, Orlando 1944). It risks to endanger the physical foundation of the higher level theory by depriving it of its autonomy with respect to the lower level counterpart. In this sense it appears as an hybrid method.

In 1923 Einstein already mentioned (Einstein 1923, p.484) an alternative method, which he considered a *purer method* (I argue: in the sense of being less hybrid) and attributed it to the contributions of Levi-Civita, Weyl and Eddington. In these discussions, Einstein often came in touch with the methodological views of these mathematical-physicists, all intrested in various formulations of the Unitary-Field theories (thereafter UFT).

As is known, Levi-Civita introduced (Levi-Civita 1917) the notion of parallel displacement into differential geometry in 1917. Through his contribution, he provided what seemed to be an indispensible tool for casting GR into a coordinate-free geometrical form, thus overcoming Einstein's problem with the rigid rod for the coordinates' physical definition. The Levi-Civita theory apparently had a great and immediate impact on Weyl's influential *Raum, Zeit, Materie* of 1918 (Janssen 1988, p.351). Weyl's theory made a considerable impression upon theoreticians and on Einstein himself, who wrote that its depth and boldness must charm every reader (Einstein 1918, p.480; Vizkin 1986, p.303). In Weyl's method, the meaning of concepts in a theory at level B > A should be founded without any recourse to the correspondent concept at level A. In 1918-19, this recourse was avoided by Levi-Civita, followed by Weyl, through the choice of an *affine geometry*. As is known, this geometry assumed that the *parallel transport*, i.e. the parallel displacement of a vector, is accompanied not only by a change in the vector orientation, as in Riemannian geometry, but also by a change in the vector's length[2].

Therefore it is not surprising that Einstein took this theory as an example of a *purer theory*, not committed to a more or less direct operational definition of the coordinates, hence more suitable for being transformed into an over-determined theory.

In 1923, Einstein introduced the new argument of a *purer method* after complaining that it is methodologically unjustifiable to base all physical considerations on the rigid or solid body and then finally to reconstruct that body atomically by means of elementary physical laws which in turn have been determined by means of the rigid measuring body (Einstein 1923, p.483).

This argument continues in his request of a new requirement for a *complete physical theory*: *the over-determination of physics equations*. In fact, the continuation of the passage above precises the features of the *purer method* which supposedly would have avoided the SM-transgression:

> Certainly it would be logically more correct to begin with the whole of the laws and to apply the <stipulation of meaning> to the whole first, i. e. to put the unambiguous relation to the world of experience last *instead of already fulfilling it in an imperfect form for an artificially isolated part, namely the space-time metric*. At the close of our considerations we will see that in the most recent studies there is an attempt, based on the ideas by Levi-Civita, Weyl, and Eddington, to implement that *logically purer metohod* (Italics mine).

I will show that the problems concerning SM also interweaved Einstein's generalisation of GR into a UFT.

4. Over-determined theories avoid the SM transgression

Einstein sketchily drew up in his 1936 "Physics and Reality" a possible mode of meeting the requirement for an over determined theory. He exemplified how a field theory can account for particles in the form of a Schwarzschild-type singularity-free solution for a modified differential equation: $g^2 R_{ik}=0$, in place of the former equation for empty space, $R_{ik}=0$. Einstein mentioned this example in connection with the purer theories of Levi-Civita, Weyl and Eddington, thus presenting it as a return to his 1923 proposal above for an over-determined theory (Einstein 1950, p.94).

In 1945, Einstein again took up his 1925 asymmetric theory, remaining faithful to this approach until the end of his life (Bergia 1991). As is known, a remarkable feature of this approach is the non-linearity of the resulting electromagnetic equations. It is this non-linearity which fulfils Einstein's request for over determined field equations whose spherical-symmetric singularity-free solutions can be interpreted as elementary particles. It can be reasonably argued that the complete fulfilment of the requirement above would correspond to Einstein's ideal of a *purer theory*.

It is well known that this ideal theory was never achieved in the span of Einstein's life and that it was considered hardly realisable by the majority of physicists in 1955; the more so after the expansion of the QM approach to particles in our modern theories.

Given the premises above, it can be argued that Einstein's 1949 discussion with Hans Reichenbach has to be understood as a further clarification of his 1923 and 1936 search for a *purer method*. Einstein's last "discussion" with Reichenbach was presented in Einstein's 1949 "Reply to Criticism" (Einstein 1949b), a part of his contributions to Schilpp's memorial work *Albert Einstein Philosopher-Scientist*. It offers an example of Einstein's mature thought on the difficulties of the meaning problem in RR and UFT.

Einstein's aim was to confute Reichenbach's argument on the possibility of deciding what is the real geometry of the world through an experimental check on the Euclidean congruence of physical rods (Einstein 1949b, p.677 ff.). Einstein's confutation runs as follows: in order to check an Euclidean congruence one needs rigid rods, but to control rigidity one actually has to resort to physical laws - such as those for the control of temperature-constancy, elasticity-coefficient, etc. - laws that, in their turn, need a previous assumption of rigidity for their foundation. In short, Reichenbach's proof of rigidity is considered by Einstein logically circular. (It is worth noting that Einstein's charge of circularity is, at bottom, an argument that Einstein also presented in his 1923 search for a SM-observant definition of rigidity).

In 1949, Einstein's own thesis is that *<Meaning> can be attributed to the individual concepts and assertions of a physical theory and to the entire system only insofar as it makes what is given in experience <intelligible>* (1949b, p.678).

It is evident that this thesis echoes the famous metaphor of the Kantian Copernican revolution, i.e. Kant's assertion that rationality is a precondition for reality and not viceversa. In fact, in an ensuing passage, Einstein admits his late adhesion to the basic tenets of Kant's philosophy (1949b, p.680).

5. Einstein's last views: the field-theory's incompleteness seen as a lack of a satisfactory criterium for postulation of meaning

The Einsteinian method of giving meaning through CR transfer did not satisfy Einstein's requirements for theory in 1949. His mature reactions to this method are expressed in this passage:

> One is struck [by the fact] that the theory (except for the four-dimensional space)[3] introduces two kinds of physical things, i.e., (1) measuring rods and clocks, (2) all other things, e.g., the electro-magnetic field, the material point, etc. *This, in a certain sense, is inconsistent;* strictly speaking measuring rods and clocks would have to be represented as solutions of the basic equations (objects consisting of moving atomic configurations), not, as it were, as theoretically self-sufficient entities. However, the procedure justifies itself because it was clear from the very beginning that *the postulates of the theory are not strong enough to deduce from them sufficiently complete equations for physical events* sufficiently free from arbitrariness, in order to base upon such a foundation a theory of measuring rods and clocks (Einstein 1959c, p.59) (Italics added).

In this passage, the 1923 SM problem - the introduction into theory of measuring rods and clocks - is joined to the 1936 under-determination problem (the postulates of the theory are not strong enough). I interpret this fact as evidence that, in 1949, Einstein matured the view that the two problems are parts of the more general problem of the physical basis of GR and UFT.

In fact, in the same year, he criticised the current theory of relativity (i.e. GR) for not meeting the requirements for a physical meaning of ds:

> For the construction of the present theory of relativity the following is essential:
> (1) Physical things are described by continuous functions, field-variables of four co-ordinates. As long as the topological connection is preserved, these latter can be freely chosen.
> (2) The field-variables are tensor components; among the tensors is a symmetrical tensor gik for the description of the gravitational field.
> (3) There are physical objects, which (in the macroscopic field) measure the invariant ds.
> *If (1) and (2) are accepted, (3) is plausible, but not necessary. The construction of mathematical theory rests exclusively upon (1) and (2).*
> A *complete* theory of physics as a totality, in accordance with (1) and (2) does not yet exist. If it did exist, there would be no room for the supposition (3). For the objects used as tools for measurement do not lead an independent existence along-side of the objects implicated by the field-equations (Einstein 1949b, p. 685. Italics added).

By connecting the latter statement on Completeness with Einstein's former passage (Einstein 1959c, p.59), one is led to the following definition: a theory is said to be complete if *the postulates of the theory are strong enough to deduce from them sufficiently complete equations for physical events* sufficiently *free from arbitrariness*. A physical theory of GR, i. e. a theory that gives physical meaning to the concepts of measuring rods and clocks by representing them as solutions of the basic equations, has to be based on postulates of the stronger type above. The impossibility of deducing from the foundational postulates or axioms of GR a physical meaning for *ds, this impossibility qualifies GR as an incomplete theory*.

The above definition neither excludes nor contradicts other known definitions of Einstein's incompleteness found in the literature, such as those presented in recent valuable studies (Fine 1986; Howard, 1989, 1990). I present it as a definition which conforms to Einstein's 1949 views.

An interesting perspective on these views is offered by Einstein's consideration that, due to the theory's *incompleteness,* postulates 1) and 2) suffice for the time to characterize only a mathematical but not a physical theory. This consideration implies that a future ideal *complete theory* should also represent a synthesis between pure mathematics and physics. In Einstein's ideal, this synthesis would consequently abolish the distinction between mathematical-physics and physics, the two traditions which were often counterposed in the historical development of western physics.

This *type of synthesis* was not accomplished by Einstein himself nor by subsequent Quantum physicists. They simply took another direction, somehow bypassing, rather than solving, the problems that Einstein had explored[4].

I do not consider Einstein's failure in reaching this synthesis as a shortcoming of his science, a missing exhaustiveness, as it were, of his methodological discourse. Such a reductive view would contradict, among the others, the largely high valuation of his theories. Rather, I believe that this missing exhaustiveness is to be valuated as an important contribution given by E. to our understanding of the deep philosophical implications of theoretical physics. In this connection, I like to refer to a passage by Gerald Holton:

> ..by always stating forthrightly and with eloquence his redefined position, Einstein not only helped us to define our own, but also gave us a virtually unique case study of the interaction of science and epistemology (Holton 1973, p. 246).

In support of his thesis, Holton quotes a Max Planck's statement which amazingly confirms the view above:

> ...a science is never in a position completely and exhaustively to solve the problem it has to face. We must accept that as a hard and fast, irrefutable fact,

and this fact cannot be removed by a theory which restricts the scope of science at its very start (Planck 1931, pp.15–17; Holton 1973, pp.244–245).

It is noteworthy that Planck's statement captured the complete adhesion of Einstein (Einstein, Introduction to Planck 1931; Holton 1973, p.244).

6. Conclusions

It has been rightly remarked (Vizgin 1986, p.310) that the scientific program underlying Weyl's theory and the companion geometrical UFTs was closely connected with the Göttingen tradition in mathematical physics. It represented a new form of interaction between mathematics and physics that was characteristic of the non classical theories of the twentieth century. Until the emergence of QP, the great theoretical penetration and the mathematical perfection of the geometric UFT program were seen as genuine advantages, notwithstanding their exceedingly weak connection with experience (Vizgin 1986, p.309).

However, the UFT program increasingly brought to light not only the great heuristic possibility of its new form of connection between physics and mathematics, but also certain dangers and difficulties, such as an overemphasis on the role of mathematical structure and an under-evaluation of the experimental and empirical aspects of theory. The latter difficulties should have appeared as grave defects especially when compared to the theoretical eclecticism of the QP program, and the predominance within it of the empirical over the theoretical (Vizgin 1986, p.309).

Given this situation, it is understandable that Einstein considered the meaning problem as a central problem, whose solution would have contributed to bringing the geometrized UFT to a more acceptable physical basis (Bergia 1993, p.188).

Clearly, Einstein's Completeness is also related to the Bohr-Einstein counter-position concerning the foundation of QP. Recent studies have interpreted this counter-position in the context of the known split between two Schools of the post-Kantian philosophers (Chevalley 1989a). In these studies, Einstein's and Bohr's different views on the foundation problems of modern physics are seen as a contrast between the two trends in post-Kantian philosophy, usually paraphrased in the key-words: *Anschauung* and *Symbol* (Chevalley 1989b, 1993). In short, in Bohr's view, the formalism of QP would represent a non directly visualizable (*anschaulich*) purely symbolic scheme. In contrast, the classical and the Einsteinian formalisms aimed at a direct intuitive interpretation; i.e. this formalism aims to be *anschaulich*, in the classical Kantian terminology.

In agreement with the thesis above, I argue that E's request for Completeness, as a form of direct connection between the theory's concepts and the perceptive experience, can be assimilated to a request for a Kantian synthesis. Moreover, the failure of Einstein's method to reach the SM requirement in its various forms was a failure of his request for a Kantian synthesis.

This assimilation would better clarify Einstein's adhesion to Kantianism. The fact that the synthesis he was aspiring to was not reached in his life, nor was it reached by Quantum physicists, opens a new area for the historical and critical research on the foundation problem of both GR an QP (Chevalley 1989b, p.151).

Notes

* *Due to space limitation, this paper is an abridged version of a study which will be published elsewhere.*

[1] Luckily, recent historical-epistemological studies on the foundation problems of Quantum Physics have adopted this viewpoint (Cattaneo & Rossi, 1991). It was also shared by many lecturers in the last Lecce Congress.

[2] It is not the components of the metric g_{ik}, but rather their ratios that have physical significance. This further leads to an extension of the general covariance group, in which scale transformation of the metric g_{ik} are also admissible (Vizgin 1986, p.302).

[3] I intend: a four dimensional non affine theory does not introduce electromagnetic fields.

[4] It may be taken as a peculiar aspect of the often winding paths of the history of physics that Bohr actually transfered the symbolic character of his QM to the Einsteinian theory of GR, in as much as it made use of a not directly visualizable symbolism (Chevalley 1993). More than a further misunderstanding, I think that Bohr's assignement may be properly understood as his realization that E.'s ideal Anschaulichkeit had actually failed.

References

AHQP (1963), "Interview with Einstein", February 15. I wish to thank the Accademia dei Quaranta, Rome, for permission to consult their copy of AHQP.

Bergia, S. (1991), "Attempts at Unified Field Theories (1918-1955): alleged failure and intrinsic validation/refutation criteria", to appear in vol. 4 of the *Einstein Studies Series,* Earmann, Janssen, Norton (eds.) (1993), "The fate of Weil's unified theory of 1918", in Bevilacqua, F. (ed.), *First Europ. Phys. Soc. Conference on History of Physics in Europe in the 19th and 20th Centuries,* Ed. Compositori, Bologna, pp.185–193.

Bridgmann, P.W. (1959), "Einstein's Theory and the Operational Point of View", in Schilpp, A. (ed). *Albert Einstein, Philosopher-Scientist,* Harper & Brothers Pbl., pp.333–354.

Born, M. (1909). "Die Theorie der stärren Elektrons in der Kinematik des Relativitätsprinzip", *Annalen der Physik,* **30** pp.1–56.

Cattaneo, G., Rossi, A. (eds.) (1991). *I Fondamenti della Meccanica Quantistica. Analisi Storica e problemi aperti*, Proceedings of the Camerino Congress, Camerino, Italy, EditEl, Commenda di Rende.

Chevalley, C. (1989a), "Histoire et Philosophie de la Mecanique Quantique. Traveaux Recent", *Revue de Synthèse*, IV, N.3-4, juil.-dèc. (1989b), "De Bohr et Von Neumann à Kant; L'Ecole allemand de logique quantique", in: *L'Age de la Science*, n.2, *Epistemologie*, ed. O.Jacobs. (1993), "Niels Bohr's Words and the Atlantis of Kantianism", in Faye, J. and Folse, H. (eds.), *Niels Bohr and Contemporary Philosophy*, Reidel.

D'Agostino, S. and Orlando, L. (1993). "Il principio di corrispondenza e la genesi della teoria gravitazionale einsteiniana", *Rivista di Storia della Scienza*, Hoepli, s. II, vol. 2, 1993, pp.51–74.

D'Agostino, S. (1993), "A consideration of the rise of theoretical physics in Europe and of its interaction with the philosophical tradition", in *First Eur. Physical Soc. Conference on History of Physics in Europe in the Nineteenth and Twentieth Centuries*, Ed.Compositori, Bologna, pp.5–28.

Einstein., A. (1923), "Fundamental Ideas and Problems of the Theory of Relativity", lecture delivered to the Nordic Assembly of Naturalists at Gothenburg, July 11, 1923, in *Nobel Lectures. Physics 1901-1921*, Elsevier 1967. (1933), "On the Method of Theoretical Physics". The Herbert Spencer Lecture delivered at Oxford, June 10, 1933. Oxford Clarendon Press, 15pp. Repr in *Phil. of Science*, vol.1, 1934, pp.162–169. The German Text is in *Mein Weltbild*, pp.176–187, Amsterdam, Querido, 1934, pp.269. Transl. *The World as I see It*, NY, Cocivi-Friede, 1934, pp.290. The Phil. Library, NY 1949. (1936a), "Physik und Realität", *Zeitschr. für freie deutsche Forschung*, Paris, vol. 1, n. 1,2. (1936b) "Physics and Reality", transl. *Franklin Int. Journal*, vol.221, N 3, 1936, pp.313–347. (1936d), "Physics and Reality", *Essays in Physics*, Phil. Libr., 1936. (1936e), "Che cos'è la teoria della Relatività", London Times, 28-11-1919; in A. Einstein, *Idee ed opinioni*, Schwarz, 1957, p.216, trad. it. da: A. Einstein, *Mein, Weltbild*. (1949a), "Reply to Criticism", in (P.A. Schilpp ed.), *Albert Einstein Philosopher-Scientist*, Library of Living Philosophers. (1949b) "Reply to Criticism", repr.in: *A. Einstein Philosopher-Scientist*, Harper & Brothers Publ., NY, 1959 2 vols, vol. 2, pp.665–688. (1949c), "Autobiographical Notes", in *A. Einstein Philosopher-Scientist*, Harper & Brothers Publ., NY, 1959, 2 vols, vol. 1, pp.1–95. (1950), "Physics and Reality", repr. in *Out of My Later Years*, Phil. Libr. Repr., NY, pp.59–97.

Fine, A. (1986). *The Shaky Game: Einstein, Realism and the Quantum Theory*, The University of Chicago Press.

Holton, G. (1973). *Thematic Origins of Scientific Thought. Kepler to Einstein*, Harvard Uiversity Press.

Howard, D. (1988), "Einstein and *Eindeutigkeit:* a neglected theme in the philosophical background to GR", in Eisestaedt, J. and Kox, A.J. (eds.), *Studies in the History of General Relativity*, Birkhäuser, 1992, pp.154--243; (1990), "Nicht Sein Kann Was Nicht Sein Darf", in Miller, A. (ed.), *Sixty-Two Years of Uncertainty: Hist., Phil. and Phys. Inquiries into the Foundations of Quantum Mechanics*, Plenum Press, pp.61–112.

Levi-Civita, T. (1917). "Nozione di Parallelismo in una varietà qualunque", *Rend. del Circolo Mat. di Palermo*, 42, pp.137--205.

Jammer, M. (1966). *The Conceptual Development of Quantum Mechanics*, Mc Graw Hill.

Janssen, M. (1992), "H.A. Lorentz's Attempt to give a Coordinate-Free Formulation of the Gen. Th. of Relativity", in Eisenstaedt, J. and Kox, A.J. (eds.), *Studies in the History of GR*, Birkhäuser, pp.334–365.

Maltese, G. and Orlando, L. (1994), "La condizione di rigidità in Relatività Ristretta e la genesi della Relatività Generale", in print. Kindly forwarded by the authors.

Planck, M. (1931). "Positivism and External Reality", *International Forum*, 1, No. 1, pp.12–16; 1, No. 2, pp.14–19.

Vizgin Vladimir, P. (1986), "Einstein, Hilbert, and Weyl: the Genesis of the Geometrical Unified Field Theory Program", in *Einstein and the History of General Relativity*, Birhäuser, pp.300–314.

PHYSICAL INTERPRETATIONS OF THE ŁUKASIEWICZ QUANTUM LOGICAL CONNECTIVES

M.L. DALLA CHIARA and R. GIUNTINI
Dip. di Filosofia, Università di Firenze
Via Bolognese 52 - 50139 - Firenze, Italy

Abstract. *Łukasiewicz quantum logic* is semantically characterized by the class of all *quantum MV algebras*. The standard model of this logic is based on *effects* in a Hilbert space. We discuss the physical interpretation of different kinds of conjunction that arise in this framework.

1. Introduction

Łukasiewicz many-valued logics ([1], [3]) represent generalizations of classical logic, where the basic connectives *and* and *or* have been split into two forms of conjunction and disjunction. The first kind of conjunction is non idempotent: generally, a repeated assertion "A and A" is not equivalent to a simple "A" (*Repetita iuvant!*). Similarly for the *or*. The second kind of conjunction and disjunction have a lattice-behaviour. Hence they are idempotent, commutative and associative. The standard semantic model of Łukasiewicz infinite many valued logic is based on the [0,1]-interval:

$$\mathcal{M}_{[0,1]} = \langle [0,1], \oplus, {}^*, \mathbf{0}, \mathbf{1} \rangle,$$

where

i) $\mathbf{1} = 1$; $\mathbf{0} = 0$.

ii) $a^* = \mathbf{1} - a$.

($*$ corresponds to the negation).

iii) $a \oplus b = \begin{cases} a + b, & \text{if } a + b \le 1 \\ 1, & \text{otherwise} \end{cases}$

(in other words, $\oplus$ represents the *truncated sum*).

C. Garola and A. Rossi (eds.), The Foundations of Quantum Mechanics, 179-185.
© 1995 Kluwer Academic Publishers.

On this basis the following operations and relation are defined:

$$a \odot b = (a^* \oplus b^*)^*$$

$$a \cap b = (a \oplus b^*) \odot b$$

$$a \cup b = (a \odot b^*) \oplus b$$

$$a \sqsubseteq b \text{ iff } a \cap b = a.$$

One can easily verify that $\oplus$, $\odot$ are not idempotent, whereas $\cap$, $\cup$ represent respectively the *inf* and the *sup* operations in the linearly ordered lattice $\langle [0, 1], \sqsubseteq \rangle$ (where $\sqsubseteq$ turns out to coincide with the natural order for the reals).

Our $[0, 1]$-structure represents a particular example of an MV algebra: it is also called the *standard MV algebra* [1]. Differently from the standard case, a generic MV algebra $\mathcal{M} = \langle M, \oplus, ^*, 0, 1 \rangle$ is not necessarily linearly ordered: generally, the relation $\sqsubseteq$ is only a partial order. A Boolean algebra turns out to be a particular case of an MV algebra, where $\oplus$ is idempotent. As a consequence, the two disjunctions $\oplus$ and $\cup$ collapse into one and the same operation (similarly, the two conjunctions).

An interesting intuitive interpretation of the Łukasiewicz connectives has been proposed by Mundici [4]. The basic idea is the use of *Ulam games*, where players are supposed to lie a certain number of times (the number may be either determined or undetermined). Instead of lying players, one may also think of information sources that are disturbed by a certain noise.

Quantum MV algebras (introduced in Giuntini [5] [2]) represent weakenings of MV algebras, where the second kind of conjunction and disjunction ($\cap$ and $\cup$) do not have a lattice behaviour: in particular, $\cap$ and $\cup$ are, generally, non commutative. *Łukasiewicz quantum logic* (LQL) is then defined as the logic that is semantically characterized by the class of all quantum MV algebras. The standard semantic model of LQL is the structure

$$\mathcal{E}(\mathcal{H}) = \langle E(\mathcal{H}), \oplus, ^*, 1, 0 \rangle,$$

based on the set of all *effects* in a Hilbert space $\mathcal{H}$. As is well known, effects represent a kind of maximal mathematical representative of the notion of *physical property*, that is compatible with the statistical rules of Hilbert-space quantum mechanics. A linear bounded operator E in a Hilbert space $\mathcal{H}$ is called an *effect* iff for any density operator W: $\text{Tr}(WE) \in [0, 1]$ (in other words, E admits a Born probability).

Following the $[0, 1]$-analogy, the operations $\oplus, ^*, 1, 0$ of $\mathcal{E}(\mathcal{H})$ are defined as follows:

i) $1 = 1\!\text{I}$
   (where $1\!\text{I}$ is the identity operator).

ii) $\mathbf{0} = \mathbb{O}$

   (where $\mathbb{O}$ is the null operator).

iii) $E \oplus F = \begin{cases} E + F, & \text{if } E + F \in E(\mathcal{H}); \\ 1\!1, & \text{otherwise} \end{cases}$

   (where $+$ is the usual operator-sum).

iv) $E^* = 1\!1 - E$.

The operations $\odot$, $\sqcap\!\!\!\!\!\;$, $\mathbb{U}$ and the relation $\sqsubseteq$ are defined like in the $[0, 1]$-case. The relation $\sqsubseteq$ coincides with the usual effect-order relation $\leq$ (where, $E \leq F$ iff for any density operator W: $\text{Tr}(WE) \leq \text{Tr}(WF)$).

Our effect structure turns out to be "very close" to an MV algebra. However one of the MV axioms (the so called Lukasiewicz axiom: $(a \odot b^*) \oplus b = (b \odot a^*) \oplus a$) is here violated. As a consequence, the conjunction $\sqcap\!\!\!\!\!\;$ and the disjunction $\mathbb{U}$ are no more commutative.

A quantum MV structure can be similarly induced also on the set $P(\mathcal{H})$ of all the orthogonal projections of $\mathcal{H}$. It is sufficient to put:

$$P \oplus Q := \begin{cases} P + Q, & \text{if } P + Q \in P(\mathcal{H}); \\ 1\!1, & \text{otherwise.} \end{cases}$$

The operations *, $\mathbf{1}$, $\mathbf{0}$ will be defined like in the effect-case. Also here, the relation $\sqsubseteq$ turns out to coincide with the natural projection-order.

As is well known, differently from effects, projections have a lattice-structure (with respect to the natural order). However, the *infimum* and the *supremum* operations do not coincide with $\sqcap\!\!\!\!\!\;$ and $\mathbb{U}$, which generally preserve their non commutativity.

The lattice structure of the projections also permits us to induce a quantum MV algebra on $P(\mathcal{H})$, according to an alternative method: it is sufficient to identify the quantum MV sum $\oplus$ with the supremum operation ($\sqcup$) of the lattice. As a consequence, one obtains:

i) $P \sqcap\!\!\!\!\!\; Q = (P \sqcup Q^*) \sqcap Q$

ii) $P \mathbb{U} Q = (P \sqcap Q^*) \sqcup Q$

We will discuss the following problem: are there any interesting physical interpretations for the different conjunctions, that arise in our Hilbertian models of Lukasiewicz quantum logic?

2. Physical interpretations

According to the standard interpretation of Hilbert space quantum mechanics, projections represent possible *sharp properties* of the physical system under investigation. At the same time, proper effects (which are not projections) can be generally regarded as *unsharp properties*, that may be disturbed by some noise.

Let us first consider the case of the sharp quantum MV algebra

$$\langle P(\mathcal{H}), \oplus, ^*, \mathbf{1}, \mathbf{0}\rangle,$$

where $\oplus$ coincides with the supremum operation $\sqcup$. In such a situation, $\oplus$ and $\odot$ will represent the usual quantum logical disjunction and conjunction, respectively, whose physical meaning has been largely investigated in the literature.

Owing to the well known correspondence between orthogonal projections and closed subspaces in a Hilbert space, we may equivalently refer to the following isomorphic structure

$$\langle C(\mathcal{H}), \oplus, ^*, \mathbf{1}, \mathbf{0}\rangle,$$

where $C(\mathcal{H})$ is the set of all closed subspaces of $\mathcal{H}$ and the operations are defined in the expected way.

Let $X, Y \in C(\mathcal{H})$. We will have:

i) $X \cap\!\!\!m\, Y = (X \sqcup Y^*) \sqcap Y$

ii) $X \cup\!\!\!w\, Y = (X \sqcap Y^*) \sqcup Y$

In other words, $\cap\!\!\!m$ corresponds to the so called *Sasaki projection* of X into Y. The following theorems hold:

Theorem 2.1   Let $\mathcal{H}$ be a Hilbert space. For any vector $\psi \in \mathcal{H}$ and for any two closed subspaces X, Y of $\mathcal{H}$:   if $\psi \in Y$ and $\exists \phi \in X$ s.t. $\psi = P_Y \phi$, then $\psi \in X \cap\!\!\!m\, Y$.

Proof. Suppose that $\psi \in Y$ and $\exists \phi \in X$ s.t. $\psi = P_Y \phi$. We have to prove that $\psi \in X \cap\!\!\!m\, Y$. By the Projection Theorem, $\phi = \phi_1 + \phi_2$, where $\phi_1 \in Y$ and $\phi_2 \in Y^*$. Thus, $\psi = P_Y \phi = \phi_1$. Hence, $\psi = \phi - \phi_2 \in X + Y^* \subseteq X \sqcup Y^*$. $\square$

Theorem 2.2   Let $\mathcal{H}$ be a Hilbert space s.t. $Dim(\mathcal{H}) < \infty$. For any vector $\psi \in \mathcal{H}$ and for any two closed subspaces X, Y of $\mathcal{H}$:   if $\psi \in X \cap\!\!\!m\, Y$, then $\psi \in Y$ and $\exists \phi \in X$ and $\psi = P_Y \phi$.

Proof. Suppose that $\psi \in X \cap\!\!\!m\, Y$. Then, $\psi \in Y$ and $\psi \in X \sqcup Y^*$. Since $\mathcal{H}$ has finite dimension, $X \sqcup Y^* = X + Y^*$. Thus, $\psi = \psi_1 + \psi_2$, where $\psi_1 \in X$ and $\psi_2 \in Y^*$. Let $\phi := \psi_1 = \psi - \psi_2$. By construction, $\phi \in X$. It remains to prove that $\psi = P_Y \phi$. Now, $P_Y \phi = P_Y \psi - P_Y \psi_2$. By hypothesis, $\psi \in Y$. Further, $\psi_2 \in Y^*$. Thus, $P_Y \psi - P_Y \psi_2 = \psi - \mathbf{0} = \psi$. $\square$

One can easily check that Theorem 2.2 (and trivially Theorem 2.1) hold even if one refers to unitary vectors ψ and ϕ (representing pure states). Theorem 2.2 cannot be generalized to infinite-dimensional Hilbert spaces as the following result shows.

Theorem 2.3   Let $\mathcal{H}$ be a separable Hilbert space s.t. $Dim(\mathcal{H}) = \infty$. Then, there are two closed subspaces X, Y of $\mathcal{H}$ and a vector ψ s.t. the following conditions are satisfied:

i) $\psi \in (X \sqcap Y^*)$.

ii) $\forall \gamma \in X: \psi \neq P_Y \cdot \gamma$.

Proof. Let $B = \{\phi_n\}$ and $C = \{\psi_n\}$ be two *infinite orthonormal sequences* of vectors of $\mathcal{H}$ s.t. $\forall \phi \in B, \forall \psi \in C: (\phi, \psi) = 0$.
Let X be the closed subspace generated by

$$\left\{ \delta_n = \cos\left(\frac{1}{n}\right) \phi_n + \sin\left(\frac{1}{n}\right) \psi_n \mid n = 1, 2, \cdots \right\}$$

and let Y be the closed subspace generated by B.

i) Let $\psi := \sum_{n=1}^{\infty} \sin\left(\frac{1}{n}\right) \psi_n$. By Halmos [6], $\psi \in X \sqcup Y$ and $\psi \notin X + Y$.
We want to show that $\psi \in Y^*$. Let ϕ be any vector of Y. We have to prove that $(\psi, \phi) = 0$.
First, we show that $\forall n: (\psi_n, \phi) = 0$.
$(\psi_n, \phi) = (\psi_n, \sum_{m=1}^{\infty}(\phi, \phi_m)\phi_m) = \sum_{m=1}^{\infty} (\psi_n, (\phi, \phi_m)\phi_m) = (\psi_n, (\phi, \phi_n)\phi_n)$
$= 0$.
Thus, $(\psi, \phi) = \left(\sum_{n=1}^{\infty} \sin\left(\frac{1}{n}\right) \psi_n, \phi\right) = \sum_{n=1}^{\infty} \left(\sin\left(\frac{1}{n}\right) \psi_n, \phi\right) = 0$. Consequently, $\psi \in (X \sqcup Y) \sqcap Y^*$.

ii) Let $\gamma \in X$. Let us suppose, by contradiction, that $\psi = P_Y \cdot \gamma$. By the Projection Theorem, $\gamma = \gamma_1 + \gamma_2$, where $\gamma_1 \in Y$ and $\gamma_2 \in Y^*$. Hence, $\psi = P_Y \cdot \gamma = \gamma_2$. By hypothesis, $\gamma \in X$. Thus, $\psi = \gamma - \gamma_1 \in X + Y$, contradiction.   □

Theorems 2.1-2.3 guarantee a natural physical interpretation for the non commutative conjunction $\sqcap$. Suppose a physical system σ, with the associated Hilbert space $\mathcal{H}^{\sigma}$, and suppose $Dim(\mathcal{H}^{\sigma}) < \infty$. Let the vector ψ represent a pure state of σ in $\mathcal{H}^{\sigma}$. The following relation holds:

> ψ *verifies* the conjunctive property $X \sqcap Y$ ($\psi \in X \sqcap Y$) iff there exists a pure state ϕ of σ such that:
>
> a) ψ is the result of a transformation of ϕ after the performance of a Y-measurement (by application of the projection postulate);
> b) ϕ *verifies* X ($\phi \in X$);
> c) ψ *verifies* Y ($\psi \in Y$).

In other words, the system in state ψ is $X \sqcap Y$ iff the system *was* X and it *is now* Y, after the performance of a Y-measurement.

Owing to Theorem 2.3, this equivalence breaks down in the infinite dimensional case. However, Theorem 2.1 guarantees that $\sqcap$ may still represent a *temporal* non commutative conjunction, in a weaker sense.

Let us now consider our second way of inducing a quantum MV structure on the set $P(\mathcal{H})$ of the sharp properties of σ. As we already know, the disjunction $\oplus$ is, in this case, defined as follows:

$$P \oplus Q = \begin{cases} P + Q, & \text{if } P + Q \in P(\mathcal{H}); \\ 1\!1, & \text{otherwise.} \end{cases}$$

As a consequence one obtains:

$$P \odot Q = (P^* \oplus Q^*)^* = \begin{cases} P \sqcap Q, & \text{if } P^* \sqsubseteq Q; \\ \mathbb{O}, & \text{otherwise.} \end{cases}$$

This justifies the following intuitive interpretation:
   a state ψ *verifies* the property $P \odot Q$ iff ψ *verifies* both P
   and Q and further the pair P, Q gives rise to a strongly *stable*
   alternative (whenever not-P then Q).

A similar interpretation can be assumed also in the case of the standard quantum MV algebra, based on $E(\mathcal{H})$. For, there holds:

$$E \odot F = \begin{cases} (E^* \oplus F^*)^*, & \text{if } E^* \sqsubseteq F; \\ \mathbb{O}, & \text{otherwise.} \end{cases}$$

Whence, by additivity of the probability measure μ_ψ (determined by the state ψ):

$$\mu_\psi(E \odot F) = 1 \text{ iff } E^* \sqsubseteq F \text{ and } \mu_\psi(E^* \oplus F^*)^* = 1 \text{ iff } E^* \sqsubseteq F \quad \text{and}$$

$$\mu_\psi(E) = \mu_\psi(F) = 1.$$

In other words, ψ verifies $E \odot F$ iff ψ verifies both E and F and further the pair E, F gives rise to a strongly stable alternative (whenever not-E then F).

Finally, let us discuss the question concerning the intuitive meaning of ⓜ. One can easily show that both projections and effects satisfy the following relation:

$$E \text{ ⓜ } F = \begin{cases} E, & \text{if } E \sqsubseteq F; \\ F, & \text{otherwise.} \end{cases}$$

As a consequence, we will have:
   a state ψ verifies $E \text{ ⓜ } F$ iff ψ verifies the second member F,
   and also the first member E (provided that E implies F).

In other words, generally, a conjunction $E \between F$ "looses the memory" of the first member. This memory is preserved, just in the case where the first member implies the second one. This is the intuitive reason why *associativity* (which generally fails for $\between$ [3]) is preserved only in the following weaker form

$$(E \between F) \between G = (E \between F) \between (F \between G).$$

Namely, the repetition of the middle element F guarantees the "conservation of memory".

Notes

(1) For a precise definition of MV algebra, see Giuntini, this volume.
(2) See also Dalla Chiara, Giuntini [2].
(3) See Giuntini [5].

References

1. Chang, C.C. (1957). "Algebraic analysis of many valued logics", *Transactions of the American Mathematical Society*, **88** pp.467–490.
2. Dalla Chiara, M.L. and Giuntini, R. (1994). "Unsharp quantum logics", *Foundations of Physics*, **24** pp.1161–1177.
3. Mundici, D. (1986). "Interpretation of AF C^*-algebras in Lukasiewicz sentential calculus", *Journal of Functional Analysis*, **65** pp.15–63.
4. Mundici, D. (1992). "The logic of Ulam's game with lies", C. Bicchieri and M.L. Dalla Chiara (eds.), *Knowledge, Belief and Strategic Interaction*, Cambridge University Press, New York.
5. Giuntini, R. "Quantum MV algebras", (preprint).
6. Halmos, P. (1957). *Introduction to Hilbert Space and Theory of Spectral Multiplicity*, Van Nostrand, New York.

RELATIVISTIC QUANTUM MECHANICS AND PATH INTEGRAL FOR KLEIN-GORDON EQUATION

G.F. DE ANGELIS
Dip. di Matematica, Università di Roma "La Sapienza"
Piazzale A.Moro 2 - 00185 - Roma, Italy

and

M. SERVA
Dip. di Matematica and INFN, Università dell'Aquila
67010 - Coppito (L'Aquila), Italy

Abstract. The subject of the paper is a path integral rapresentation for the semigroup $\{e^{-tH_1}\}_{t\geq 0}$ generated by the quantum Hamiltonian H_1 of a relativistic spinless particle in an external electromagnetic field. The result is compared with the "Feynman-Kac" formula which holds for relativistic Schrödinger operators.

1. Introduction

A sensible physical model depicts ordinary bodies as collections of pointlike nuclei and electrons attracting or repelling each other by Coulomb forces. According to classical physics atoms should be unstable against collapse because their classical energy is not bounded from below but according to quantum mechanics they have a ground state energy $E_{\min} > -\infty$. This result doesn't solve the stability problem for matter in bulk. Due to the long range of Coulomb interaction why two separated pieces of matter do not feel each other's influence ? The point is that even if bodies are globally neutral it is not obvious that they cannot become polarized in such a way as to attract or repel each other. An important facet of stability is the saturation property of the binding energy per particle i.e. the requirement that for a system of N bodies not only the ground state energy $E_{\min}(N)$ is finite (stability of the first kind) but also that $E_{\min}(N) \sim -C\,N$ for some positive constant C (stability of the second kind). In order to appreciate this point let suppose for instance that $E_{\min}(N) \sim -C\,N^r$ with $r > 1$. Then bringing

187

together two large pieces of matter each containing say 10^{23} particles, would release a huge amount of energy $\Delta E = 2C(2^{r-1} - 1)10^{23r}$ therefore bodies should collapse under Coulomb interaction! The problem of the stability for a system of point charges was raised about forty years after the birth of quantum mechanics by Fisher and Ruelle[10] and the answer was given only in the late sixties and early seventies beginning with two papers of Dyson and Lenard[8,9,14]. One can simplify the theory of N electrons interacting with K nuclei by assuming (Born-Oppenheimer approximation) that nuclei have an infinite mass and are located at fixed but arbitrary space points $\mathbf{R}_1, \ldots, \mathbf{R}_K$. Due to the large mass ratio between nuclei and electrons such an assumption is not a bad one, moreover it gives an exact lower bound because allowing a finite mass can only raise the energy. Now the system consists of N (Coulomb interacting) electrons in the static electric field

$$\mathbf{E}(\mathbf{r}) = \sum_{i=1}^{K} \frac{Z_i|e|}{|\mathbf{r} - \mathbf{R}_i|^3}(\mathbf{r} - \mathbf{R}_i) \tag{1.1}$$

generated by K positive charges $Z_1|e|, \ldots, Z_K|e|$ and its Hamiltonian H_N is given by

$$H_N = -\frac{\hbar^2}{2m} \sum_{i=1}^{N} \Delta_i - \sum_{i=1}^{N} \sum_{j=1}^{K} \frac{Z_j e^2}{|\mathbf{r}_i - \mathbf{R}_j|} + \sum_{1 \le i < j \le N} \frac{e^2}{|\mathbf{r}_i - \mathbf{r}_j|} + \tag{1.2}$$

$$\sum_{1 \le i < j \le K} \frac{Z_i Z_j e^2}{|\mathbf{R}_i - \mathbf{R}_j|}$$

Dyson and Lenard demonstrated that the ground state energy of (1.2) satisfies the inequality $E_{\min}(N) \ge -C N$ for all $\mathbf{R}_1, \ldots, \mathbf{R}_K$ provided that electrons behave as fermions because if they obeyed Bose-Einstein statistics matter definitely wouldn't be stable as $E_{\min}(N) \sim -C' N^{7/5}$ in that case. The key point for the stability[14] rests on the fact that the kinetic energy $K(\rho)$ of a system of N fermions grows with its density $\rho(\mathbf{r}) = N \int |\psi(\mathbf{r}, \mathbf{r}_2, \ldots, \mathbf{r}_N)|^2 d\mathbf{r}_2 \ldots d\mathbf{r}_N$ according to the law $K(\rho) \ge A \int_{\mathbb{R}^3} \rho(\mathbf{r})^{5/3} d^3\mathbf{r}$ for each antisymmetric N-particles wave function $\psi(\mathbf{r}_1, \ldots, \mathbf{r}_N)$, a result which depends in a critical way on the nonrelativistic relation $K(\mathbf{p}) = |\mathbf{p}|^2/2m$ between kinetic energy and momentum embodied in Schrödinger's theory which is notoriously unphysical in presence of nuclei with high atomic number because relativistic corrections are not negligible in that case. In the spirit of Dyson and Lenard, stability of matter should proceed from Quantum Electrodynamics which, however, is beyond an effective mathematical control in the sense of constructive field theory. The main present approach to the stability of relativistic matter[15] rests upon the replacement of the operator $-\frac{\hbar^2}{2m} \sum_{i=1}^{N} \Delta_i$ by its relativistic version $\sum_{i=1}^{N} \sqrt{-\hbar^2 c^2 \Delta_i + m^2 c^4}$

which defines the "relativistic Schrödinger operators" and, in particular, the one-body Hamiltonian $\tilde{H}_1 = \sqrt{-\hbar^2 c^2 \Delta + m^2 c^4} + e\, A^0$ for just one "electron" in the electric field $\mathbf{E} = -\nabla A^0$. We remark that the relevant relativistic corrections to one- electron energy levels in a Coulomb field proceed not only from kinematics but also from spin-orbit coupling and Darwin terms, therefore the adopted solution is physically suspect. Relativistic Schrödinger operators differ in many aspects from usual Schrödinger Hamiltonians. For instance, in the Coulomb field of a nucleus, the ground state energy of $\tilde{H}_1 = \sqrt{-\hbar^2 c^2 \Delta + m^2 c^4} - Ze^2/r$ collapses[12] to $-\infty$ when Z exceeds a critical value Z_c close to the inverse fine constant α^{-1} but there is[15] a theorem which relates the (second kind) stability of the many body Hamiltonian

$$\tilde{H}_N = \sum_{i=1}^{N} \sqrt{-\hbar^2 c^2 \Delta_i + m^2 c^4} - \sum_{i=1}^{N} \sum_{j=1}^{K} \frac{Z_j e^2}{|\mathbf{r}_i - \mathbf{R}_j|} +$$

$$\tag{1.3}$$

$$\sum_{1 \leq i < j \leq N} \frac{e^2}{|\mathbf{r}_i - \mathbf{r}_j|} + \sum_{1 \leq i < j \leq k} \frac{Z_i Z_j e^2}{|\mathbf{R}_i - \mathbf{R}_j|}$$

to the (first kind) stability of one "electron" in the Coulomb field of one atomic nucleus. In the latter case, the spectrum of $\tilde{H}_1$ is bounded from below[12] provided that $Z < Z_c = 2/\pi\alpha$ and the theorem says that if $Z_j < Z_c$ for all $j = 1, \ldots, K$, then $\tilde{H}_N$ exibits stability of second kind for fermions of spin 1/2 when $2\alpha \leq 1/47$. Since the actual value of α is $\approx 1/137$, at least in our corner of Universe, it would seem that the theory of relativistic Schrödinger operators is in a good shape but what about their physical status? In other words is there any physically reasonable roughening of Q.E.D. leading to Hamiltonians (1.2)? In the Furry picture of Q.E.D. in presence of an external field A_{ext}^{μ} (for instance the Coulomb field of one or several infinite massive nuclei), the unperturbed Hamiltonian $H_{0,M}$ of "matter" is that of the quantized Dirac field or (for spinless particles) of the Klein-Gordon one interacting with A_{ext}^{μ}. Such "external field approximation" of Q.E.D. is a rough picture since accounts only for the interaction of each particle with A_{ext}^{μ} but not with the quantized electromagnetic field which is responsible for their mutual Coulomb repulsion. Besides that, the unperturbed theory may not one which conserves the number of bodies as an external electric field could create pairs[16] at non-zero rate in the static limit but when the field strenght is below some critical value $E_c \approx \frac{m^2 c^3}{e\hbar}$ the "Klein Paradox" can't occur and the Dirac (or Klein-Gordon) equation provides a well defined linear quantum field theory[17,18] and the only relevant operators are the one-particle and one-antiparticle Hamiltonians $H_1, \overline{H}_1$. They are not trivial as include the interaction with the external field and one can construct from them the "unperturbed" matter Hamiltonian $H_{0,M}$ by second quantization. We want to compare one-body relativistic Schrödinger operators $\tilde{H}_1 = \sqrt{-\hbar^2 c^2 \Delta + m^2 c^4} + V$ with one-particle Hamiltonians in the external field approximation of Q.E.D. We choose to do that precisely in the Klein-Gordon case because, for trivial reasons, the Klein-Gordon H_1 doesn't include any spin-orbit coupling exactly as $\tilde{H}_1$ and therefore one may believe that the two relativistic theories coincide at one-body level. Besides that, Klein-Gordon quantum mechanics in an external field is far less known than the Dirac one which, of course, is much more interesting from a physical point of view. What is the Hamiltonian of a relativistic spinless particle in an external electromagnetic field? Is it true that $H_1 = \sqrt{c^2(-i\hbar\nabla - e/c\,\mathbf{A})^2 + m^2 c^4} + e\,A^0$ when $\mathbf{E} = -\nabla A^0 \neq \mathbf{0}$? A convenient approach to the problem is to study the semigroup $\{e^{-t\tilde{H}_1}\}_{t\geq 0}$, indeed we shall be able to give an explicit path integral representation of Euclidean Klein-Gordon propagators which, moreover, exploits ideas and techniques useful in the Dirac case[7]. We remark that relativistic Schrödinger semigroups $\{e^{-t\tilde{H}_1}\}_{t\geq 0}$ admit[2,3] the path-integral ($\hbar = c = m = 1$)

$$(e^{-t\tilde{H}_1}\psi)(\mathbf{x}) = e^{-t}\,\mathbb{E}_\mathbf{x}\left(\psi(\xi_t)\exp - \int_0^t V(\xi_s)\,ds\right) \tag{1.4}$$

where e^{-t} accounts for the rest energy while $t \mapsto \xi_t$ is a Markov process in the three-dimensional space with generator $L = 1 - \sqrt{-\Delta + 1}$, a jump process at variance with the well known Feynman-Kac formula of nonrelativistic quantum mechanics where paths are those of a three-dimensional Brownian motion $t \mapsto \mathbf{X}_t$. It may be interesting to observe that $t \mapsto \xi_t$ can be constructed[2,4] in terms of space-time diffusions. Indeed $\xi_t = \mathbf{X}_{\tau_t}$, $\{\tau_t\}_{t\geq 0}$ being the family of Markov times defined by

$$\tau_t = \inf\{s \geq 0 : s + W_s^0 = t\} \tag{1.5}$$

where $s \mapsto W_s^0$ is an extra one-dimensional Brownian motion independent from $s \mapsto \mathbf{X}_s$ and starting from 0. In other words, if $X_s^0 = s + W_s^0$ and $s \mapsto X_s^\mu$ is the four-dimensional diffusion $X_s^\mu = (X_s^0, \mathbf{X}_s)$ (which starts from the space-time point $(0, \mathbf{x})$ when $s \mapsto \mathbf{X}_s$ starts from $\mathbf{x} \in \mathbb{R}^3$), then τ_t is the first hitting time of the hyperplane $\Sigma_t = \{(x^0, \mathbf{x}) \in \mathbb{R}^4 : x^0 = t\}$ by $s \mapsto X_s^\mu$ while the three-dimensional random variable ξ_t represents the space-coordinates of the point $X_{\tau_t}^\mu$ on the wordline $s \mapsto X_s^\mu$. The different physical status of H_1 and $\tilde{H}_1$ can be grasped by inspecting the path integrals of the corresponding semigroups which, as we shall see in the next section, are subtly different when the electric field is not trivial. We add the fact that (1.4) can be generalized[5] in a gauge-invariant way when a magnetic field $\mathbf{B} - \nabla \times \mathbf{A}$ is superimposed to the electric one $\mathbf{E} = -\nabla A^0$. By minimal coupling, $\tilde{H}_1 = \sqrt{c^2(-i\hbar\nabla - e/c\mathbf{A})^2 + m^2 c^4} + e\,A^0$ and (1.4) becomes

$$(e^{-t\tilde{H}_1}\psi)(\mathbf{x}) = e^{-t}\,\mathbb{E}_{(0,\mathbf{x})}\left(\psi(\xi_t)exp - \left\{e \int_0^t A^0(\xi_s)\,ds + \right.\right.$$

$$(1.6)$$

$$\left.\left. ie \int_0^{\tau_t} \mathbf{A}(\mathbf{X}_s)\cdot d\mathbf{X}_s\right\}\right)$$

if the vector potential $\mathbf{A}$ satifies $\nabla \cdot \mathbf{A} = 0$. One can notice at once that it displays a remarkable asimmetry between electric and magnetic contributions which doesn't fit relativistic covariance and may be a case against relativistic Schrödinger operators.

2. Klein-Gordon semigroups and their path integrals

According to the best tradition of constructive field theory in this section we take any $d \geq 2$ space-time dimensions as $d = 4$ plays no critical role. The corresponding dimension $d - 1$ of space will be $n \geq 1$. We assume natural units $\hbar = c = m = 1$ and we consider the Klein- Gordon equation

$$\Box_d\varphi + 2ie\,A_\mu^{ext}\partial^\mu\varphi + (1 - e^2 A_\mu^{ext}A^{ext\mu})\varphi + ie(\partial_\mu A_{ext}^\mu)\varphi = 0 \qquad (2.1)$$

in an external electromagnetic field $F_{ext}^{\mu\nu} = \partial^\mu A_{ext}^\nu - \partial^\nu A_{ext}^\mu$ which we always take static in order to preserve invariance under time-translation. As it is well known, (2.1) cannot be interpreted as a simple relativistic version of the Schrödinger equation because the conserved current

$$J_\varphi^\mu = i\left(\overline{\varphi}(\partial^\mu + ieA_{ext}^\mu)\varphi - \varphi(\partial^\mu - ieA_{ext}^\mu)\overline{\varphi}\right) \qquad (2.2)$$

has no probabilistic meaning since the charge density

$$\rho_\varphi = i\left(\overline{\varphi}(\partial^0 + ieA_{ext}^0)\varphi - \varphi(\partial^0 - ieA_{ext}^0)\overline{\varphi}\right) \qquad (2.3)$$

is not positive. It is better to look at (2.1) as a field equation which, together with the canonical commutation relations for φ, defines an "external field problem"[17,18], the accepted name[18] for a general class of theories in which one or more quantized fields interact with classical "external sources" or "external fields". As they are given c-number functions of space-time coordinates, one neglects the back reaction of the quantum field upon the source itself, a sensible assumption for macroscopic sources but not too bad for a microscopic one much larger than field quanta, for instance an atomic nucleus with its surrounding Coulomb field. The linear quantum field theory (2.1) breaks[17] when the external electric field has a strength above the critical threshold for pair creation because a static field (which acts from $t = -\infty$) has been created an infinite number of particles at any time and the

Fock space formalism can't cope with that. Nevertheless, (2.1) makes sense for weak fields and there will be a (symmetric) Fock space $\mathcal{F}_s(\mathcal{D}_1) \otimes \mathcal{F}_s(\overline{\mathcal{D}}_1)$ built upon the one-particle and one-antiparticle Hilbert spaces $\mathcal{D}_1, \overline{\mathcal{D}}_1$. We shall give an explicit construction of $\mathcal{D}_1$ and of the one-particle Hamiltonian H_1 acting on $\mathcal{D}_1$ through a path-integral representation of its semigroup $P^t = e^{-tH_1}$. We begin with the free equation

$$(\Box_d + 1)\varphi = 0 \tag{2.4}$$

in order to make clear its quantum meaning. In that case $\mathcal{D}_1$ can be identified in a standard way[17] with the linear space of normalizable positive frequency solutions. They are the tempered distributions

$$\varphi(x) = (2\pi)^{-\frac{n}{2}} \int_{H^+} \psi(p) e^{-ip_\mu x^\mu} \mu(dp) \tag{2.5}$$

where $H^+ = \{ p = (p^0, \mathbf{p}) \in \mathbb{R}^d : p^\mu p_\mu = 1, p^0 \geq 1 \}$ is the positive mass shell equipped with the standard Lorentz-invariant measure $\mu(dp) \equiv d^n\mathbf{p}/2p^0$ and $\int_{H^+} |\psi(p)|^2 \mu(dp) < +\infty$. For such solutions the conserved charge $\int_{\{x^0=t\}} \rho_\varphi \, d^n\mathbf{x}$ defines a Lorentz-invariant Hilbert norm since

$$\|\varphi\|_{\mathcal{D}_1}^2 = \int_{\{x^0=t\}} \rho_\varphi \, d^n\mathbf{x} = \int_{H^+} |\psi(p)|^2 \, \mu(dp) \tag{2.6}$$

The resulting Hilbert space carries an irreducible unitary representation of the Poincaré group with spin $s = 0$, therefore it describes[19] the quantum states of a relativistic spinless particle on which the free one-particle Hamiltonian group $\{e^{-itH_1}\}_{t\in\mathbb{R}}$ acts by time-translation

$$(e^{-itH_1}\varphi)(x^0, \mathbf{x}) = \varphi(x^0 + t, \mathbf{x})$$

We want to exibit another representation of $\mathcal{D}_1$ which is especially fit for the free semigroup $\{e^{-tH_1}\}_{t\geq0}$ obtained from $\{e^{-itH_1}\}_{t\in\mathbb{R}}$ by the Wick rotation $t \to -it, t \geq 0$. We remark that positive frequency solutions are boundary values of holomorphic functions because, by looking at (2.5), it is easy to see that for $z = (z^0, z^1, \ldots, z^n) \in \mathbb{C}^d$, $\tilde{\varphi}(z) = (2\pi)^{-n/2} \int_{H^+} \psi(p) e^{-ip_\mu z^\mu} \mu(dp)$ is holomorphic when Im z belongs to the past open light cone $V_- = \{\eta \in \mathbb{R}^d : \eta_\mu \eta^\mu > 0, \eta^0 < 0\}$. The complex domain $\Omega = \{ z \in \mathbb{C}^d : \text{Im } z \in V_- \}$ contains all points $z = (ix_d, x_1, \ldots, x_n)$ where $(x_1, x_2, \ldots, x_d) \in \mathbb{R}^d$ and $x_d < 0$. From now on $D \subset \mathbb{R}^d$ will be the domain $D = \{ (x_1, \ldots, x_d) = (\mathbf{x}, x_d) \in \mathbb{R}^d : x_d < 0 \}$, with a boundary $\partial D = \{(\mathbf{x}, 0), \mathbf{x} \in \mathbb{R}^n\}$ which can be identified with the n- dimensional Euclidean space $\mathbb{R}^n$ and there is a one to one map between positive frequency solutions φ and Euclidean wave functions u defined through

$$u(\mathbf{x}, x_d) = (2\pi)^{-\frac{n}{2}} \int_{H^+} \psi(p) e^{(p^0 x_d + i\mathbf{p}\cdot\mathbf{x})} \mu(dp) \tag{2.7}$$

by restricting $\tilde{\varphi}(z)$ to the Schwinger points inside D. They are exponentially vanishing when $x_d \downarrow -\infty$ and satisfy the Euclidean free equation

$$(-\Delta_d + 1)u = 0 \tag{2.8}$$

by transition from the hyperbolic to the elliptic case through complex domains[13]. It easy to check that Euclidean wave functions belong to the Sobolev space $H^1(D)$ of square-integrable functions on D with square-integrable first derivatives which is the natural one to be considered[1] in searching for weak solutions of (2.8). By Plancherel's theorem and elementary integrations, it turns out that

$$\|\varphi\|_{\mathcal{D}_1}^2 = \int_D (\Sigma_{i=1}^d |\partial_i u|^2 + |u|^2)\,dx \tag{2.9}$$

Since the left hand side of (2.9) is the standard squared norm in $H^1(D)$, we got a new representation of free one-particle states in which $\mathcal{D}_1$ is identified with the closed linear subspace of $H^1(D)$ consisting of all weak solutions of (2.8) normed according to

$$\|u\|_{\mathcal{D}_1}^2 = \int_D (\Sigma_{i=1}^d |\partial_i u|^2 + |u|^2)\,dx \tag{2.10}$$

The free semigroup $\{P^t\}_{t\geq 0} = \{\exp -tH_1\}_{t\geq 0}$ acts on $\mathcal{D}_1$ by

$$(P^t u)(\mathbf{x}, x_d) = u^t(\mathbf{x}, x_d) = u(\mathbf{x}, x_d - t) \tag{2.11}$$

as it follows from Wick rotation. Formula (2.11) defines a reality preserving contractive semigroup on the whole $H^1(D)$ which, however, is *not* a self-adjoint one. It becomes self-adjoint only when *restricted* to the subspace of weak solutions which is invariant under translations. In order to check this important point we bound ourselves to real functions as $\{P^t\}_{t\geq 0}$ is reality preserving. Let u^t and v^τ be the time-translated of u and v by t and τ and

$$F(t,\tau) = (P^t u, P^\tau v)_{H^1} = \int_D (\Sigma_{i=1}^d \partial_i u^t \partial_i v^\tau + u^t v^\tau)\,dx \tag{2.12}$$

One can check that

$$\frac{\partial F}{\partial t} - \frac{\partial F}{\partial \tau} = \int_D \left((-\Delta_d u + u)^t \partial_d v^\tau - (-\Delta_d v + v)^\tau \partial_d u^t \right) dx$$

therefore

$$\frac{\partial F}{\partial t} - \frac{\partial F}{\partial \tau} = 0 \tag{2.13}$$

when u and v solve (2.8). It follows that $F(t,\tau) = G(t+\tau) \Rightarrow (P^t u, P^\tau v)_{H^1} = (P^\tau u, P^t v)_{H^1}$ *for all* $t, \tau > 0$ namely that $\{P^t\}_{t\geq 0}$ *is self-adjoint.* Because

$\{P^t\}_{t \geq 0}$ is a strongly continuous contractive semigroup, it has a positive self-adjoint generator H_1 which is the free one- particle Hamiltonian by definition. Now it is important to remark that when u_0 belongs to the half-integer Sobolev space $H^{1/2}(\mathbb{R}^n)$, an existence and uniqueness theorem holds[1] for the Dirichlet problem

$$\begin{cases} (-\triangle_d + 1)u = 0 & \text{in } D \\ u|_{\partial D} = u_0 \end{cases} \qquad (2.14)$$

namely for each $u_0 \in H^{1/2}(\partial D)$ there exists one and only one weak solution of (2.14). By exploiting this one to one correspondence, finally we identify the one-particle space $\mathcal{D}_1$ with the Hilbert $H^{1/2}(\mathbb{R}^n)$ of boundary data equipped with the norm $\|u_0\|_{\mathcal{D}_1} = \|u\|_{H^1(D)}$ where u is the unique solution corresponding to u_0 and the free semigroup acts on $H^{1/2}(\mathbb{R}^n)$ by the obvious rule

$$P^t : u_0 \in H^{1/2}(\partial D) \mapsto u^t|_{\partial D}$$

In a more explicit way

$$(P^t u_0)(\mathbf{x}) = u(\mathbf{x}, -t) \qquad (2.15)$$

By the previous discussion, it follows that $\{P^t\}_{t \geq 0}$ is a strongly continuous, reality preserving, contractive self-adjoint semigroup therefore its generator H_1 is self-adjoint, reality preserving and positive, indeed $H_1 = \sqrt{-\triangle_n + 1}$. Since there exist path-integral formulas for solving Dirichlet problems, this approach provide a path integral representation of the free Euclidean propagator but it is better to consider the more interesting case of an external field where positive frequency solutions are not explicitly given and it is expedient to define them through the Euclidean strategy. Therefore we consider the imaginary-time version of (2.1) namely

$$-\triangle_d u + 2e\mathcal{A}^\beta \partial_\beta u + (1 - e^2 \mathcal{A}_\beta \mathcal{A}^\beta)u + e(\partial_\beta \mathcal{A}^\beta)u = 0 \qquad (2.16)$$

where $\mathcal{A}_\beta$ is the Euclidean electromagnetic "four" potential defined by $\mathcal{A}_d = \mathcal{A}^d = A^0_{ext}$, $\mathcal{A}_\beta = \mathcal{A}^\beta = iA^\beta_{ext}$ for $\beta = 1, \ldots, n$. Of course $\partial_d \mathcal{A}^\beta = 0$ for $\beta = 1, \ldots, d$ by our assumption of a static external field. The elliptic equation (2.16) provides a Dirichlet problem in the domain D from which we construct the one-particle Hilbert space $\mathcal{D}_1$ and the one-particle Hamiltonian H_1. The analogous Dirichlet problem in the upper half-space $\overline{D}$ is related to the one- antiparticle structure but it is clear (by time-reversal and charge conjugation) that it is enough to make the change $e\mathcal{A}_\beta \rightarrow -e\mathcal{A}_\beta$. For the sake of simplicity, we shall consider purely electric external fields as the effect of magnetic ones is quite trivial. If $V = eA^0_{ext} = eA^d$, our Dirichlet problem looks

$$\begin{cases} -\Delta_d u + 2V\partial_d u + (1-V^2)u = 0 & \text{in } D \\ u|_{\partial D} = u_0 \end{cases} \tag{2.17}$$

but now we must explain when an external electric field $\mathbf{E}_{ext}$ is "weak". In general the strenght of $\mathbf{E}_{ext}$ should be tested[6] by inspecting the quadratic form

$$\phi \in H^1(\mathbb{R}^n) \mapsto E(\phi, \phi) = \int_{\mathbb{R}^n} (|\nabla \phi|^2 + (1-V^2)|\phi|^2)\, d^n\mathbf{x} \tag{2.18}$$

which is related to the classical energy functional of the Klein-Gordon field. Roughly speaking, a weak field is one for which (2.18) is positive definite because the fullfillement of this condition provides a positive energy gap between vacuum and one-pair states. In $d = 4$, when $V(\mathbf{x}) = Ze^2/|\mathbf{x}|$, by exploiting Hardy's inequality

$$\int_{\mathbb{R}^3} \frac{|\phi(\mathbf{x})|^2}{|\mathbf{x}|^2}\, d^3\mathbf{x} \leq 4 \int_{\mathbb{R}^3} |\nabla \phi|^2 d^3\mathbf{x}$$

one can see that the positivity condition holds if $Z < 1/2\alpha$ but it is violated when $Z \geq 1/2\alpha$. In order to simplify the matter, we shall assume that V be bounded with the supremum $\|V\|_\infty$ of $|V|$ strictly lesser than one. Accordingly, we shall equip $H^1(D)$ with the new norm

$$\|u\|^2_{H^1} = \int_D (\Sigma^d_{i=1}|\partial_i u|^2 + (1-V^2)|u|^2)\, dx \tag{2.19}$$

which is equivalent to (2.10) by $\|V\|_\infty < 1$. We start by defining the one-particle space as the closed subspace of $H^1(D)$ consisting of all its elements which are weak solutions of

$$-\Delta_d u + 2V\partial_d u + (1-V^2)u = 0 \tag{2.20}$$

Of course, $\mathcal{D}_1$ is invariant under translation in the time direction and will be normed according to (2.19). In order to check the soundness of the new V-depending Hilbert structure, we remark that in a purely electric field the charge density ρ_φ of a Klein-Gordon wave function is

$$\rho_\varphi = i(\bar\varphi\partial_0\varphi - \varphi\partial_0\bar\varphi + 2iV\bar\varphi\varphi) \tag{2.21}$$

therefore the squared norm of a positive frequency solution should be

$$\|\varphi\|^2 = \int_{\{x^0=t\}} (i\bar\varphi\partial_0\varphi - i\varphi\partial_0\bar\varphi - 2V|\varphi|^2)\, d^n\mathbf{x} \tag{2.22}$$

The right hand side of (2.22) doesn't depend on time and it can be evaluated for $t = 0$. On the other hand, by the theorem of divergence applied to the domain $D \subset \mathbb{R}^d$, it easy to check that for Euclidean wave functions u

$$\|u\|_{H^1}^2 = \lim_{x_d \uparrow 0} \int_{\mathbb{R}^n} (\bar{u}\partial_d u + u\partial_d \bar{u} - 2V |u|^2) \, d^n\mathbf{x} \qquad (2.23)$$

a formula which coincide with (2.22) by formal analytic continuation. Having got our one-particle space, we pick up the right one-particle Hamiltonian by defining its semigroup $P^t = \exp -tH_1$ still through (2.11). By the same procedure of the free case, it turns out that $\{P^t\}_{t\geq 0}$ is a self-adjoint, strongly continuous contractive semigroup on $\mathcal{D}_1$. Indeed, when u and v are weak real solutions of (2.20), the continous function $F : (0, +\infty) \times (0, +\infty) \mapsto \mathbb{R}$ defined by

$$F(t, \tau) = \int_D (\Sigma_{i=1}^d (\partial_i u)^t (\partial v)^\tau + (1 - V^2) u^t v^\tau) dx$$

satisfies the partial differential equation (2.13) in the distributional sense. Therefore $(P^t u, P^\tau v)_{H^1} = (P^\tau u, P^t v)_{H^1}$, namely $\{P^t\}_{t\geq 0}$ is self-adjoint and has a self-adjoint generator H_1. Because $\{P^t\}_{t\geq 0}$ is still contractive, H_1 is positive. Physically this means that a positive mass-gap survives the perturbation of the free theory by the external field. Now we come to the path integral representation of $P^t = \exp -tH_1$. In order to do that, we exploit once again the one to one correspondence between boundary data and solutions of (2.17) because, by standard theorems[1], it turns out that for each $u_0 \in H^{1/2}(\partial D)$ there is one and only one weak solution $u \in H^1(D)$ of (2.17) when $\|V\|_\infty < 1$. Finally we identify $\mathcal{D}_1$ with the space $H^{1/2}(\mathbb{R}^n)$ endowed with the V-depending norm $\|u_0\|_{\mathcal{D}_1} = \|u\|_{H^1}$ where u is the unique solution of (2.20) issuing from u_0. The semigroup $\{P^t\}_{t\geq 0}$ is carried on $H^{1/2}(\partial D)$ in the same way as in the free case, namely $P^t u_0$ is the trace of u^t on the boundary. Usually semigroups come out from the Cauchy problem for parabolic partial differential equations (as the imaginary-time Schrödinger one) but they can be generated by an elliptic operator L in some half d-dimensional space provided that L has "time-independent" coefficients in which case the semigroup act on boundary data for the corresponding Dirichlet problem. There are well known path-integral formulas[11] for solving such elliptic problems. Let D be a domain of the Euclidean space $\mathbb{R}^d$ with boundary ∂D and $L = -a_j^i \partial_i \partial^j - 2b^i \partial_i$ a second order partial differential elliptic operator, namely one for which the real and symmetric matrix valued function $x \in D \mapsto (a_j^i)(x)$ is positive definite in each point of the domain and, therefore, may be represented as $a_j^i = (\sigma \sigma^T)_j^i$. Let us consider the d-dimensional diffusion $s \mapsto Y_s =$

$(Y_s^1, \ldots, Y_s^d) = (\mathbf{Y}_s, Y_s^d)$ defined by Itô's stochastic differential equations

$$dY_s^i = b^i(Y_s)\, ds + \sum_{j=1}^{d} \sigma_j^i(Y_s)\, dW_s^j \quad i = 1, \ldots, d \tag{2.24}$$

where $s \to (W_s^1, \ldots, W_s^d)$ is a d-dimensional Brownian motion. Let τ_D be the first hitting time of the boundary ∂D by the diffusion $s \mapsto Y_s$, then the solution u of the Dirichlet problem

$$\begin{cases} Lu + cu = 0 & \text{in the domain } D \\ u\big|_{\partial D} = u_0 \end{cases} \tag{2.25}$$

is given by[11]

$$u(x) = \mathbb{E}_x \left(u_0(Y_{\tau_D}) \exp -\frac{1}{2} \int_0^{\tau_D} c(Y_s)\, ds \right) \tag{2.26}$$

where $\mathbb{E}_x(\cdot)$ means the expectation value when the process starts from the point $x \in D$. Therefore

$$\mathbb{E}_x(F[Y_.]) = \int_{\Omega_x} F[Y_.]\, \mathcal{D}\mu_Y^x$$

if $F[Y_.]$ is some functional on the space Ω_x of continous paths $s \geq 0 \mapsto Y_s \in \mathbb{R}^d$ starting from x and $\mathcal{D}\mu_Y^x$ the (functional) probability measure which weights such paths according to (2.24). When D is unbounded, it is important[11] that $\tau_D^x < +\infty$ a.s. for all $x \in D$ which, unfortunately, is not automatically guaranteed by the drift b^i in (2.17). In order to pick up a more convenient Dirichlet problem, it is better to define auxiliary Euclidean wave functions v by

$$u(\mathbf{x}, x_d) = e^{x_d} v(\mathbf{x}, x_d) \tag{2.27}$$

They have the same traces as the old ones on the boundary $x_d = 0$ of the half-space $x_d < 0$ but they satisfy the new equation

$$-\Delta_d v - 2(1-V)\partial_d v + V(2-V)v = 0 \tag{2.28}$$

The advantage of (2.28) over (2.20) is due to the fact that now the drift $b^i = (0, \ldots, 0, 1-V)$ points towards ∂D because we assumed $\|V\|_\infty < 1$. Since $b^d \geq 1 - \|V\|_\infty > 0$, by comparing the hitting time τ_D of $s \mapsto Y_s$ with the corresponding $\tilde{\tau}_D$ of the diffusion $s \mapsto \tilde{Y}_s$ defined by $d\tilde{Y}_s^d = (1 - \|V\|_\infty)\, ds + dW_s^d$, $d\tilde{Y}_s^i = dW_s^i$, $i = 1, \ldots, d-1$, it is easy to see, by standard result, that $\tau_D^x < +\infty$ a.s. $\forall x \in D$ (moreover $\mathbb{E}_x(\tau_D) < +\infty$). Therefore we can confidently apply (2.26) and we get

$$u(\mathbf{x}, x_d) =$$
$$e^{x_d}\mathbb{E}_{(\mathbf{x},x_d)}\left(u_0(\mathbf{Y}_{\tau_D})\exp -\tfrac{1}{2}\int_0^{\tau_D} V(\mathbf{Y}_s)(2 - V(\mathbf{Y}_s))\,ds\right) \qquad (2.29)$$

Now we make a "change of variables" inside the path integral (2.29) by Girsanov's formula[11]. When two diffusions in $\mathbb{R}^d$, $s \in [0,\tau] \mapsto X_s^i = (\mathbf{X}_s, X_s^d)$ and $s \in [0,\tau] \mapsto Y_s^i = (\mathbf{Y}_s, Y_s^d)$ share the same Brownian noise and start from the same point x but have different drifts

$$dY_s^i = (b^i(Y_s) + \delta b^i(Y_s))\,ds + \sum_{j=1}^d \sigma_j^i(Y_s)\,dW_s^j \quad i = 1,\dots,d \qquad (2.30)$$

$$dX_s^i = b^i(X_s)\,ds + \sum_{j=1}^d \sigma_j^i(X_s)\,dW_s^j \quad i = 1,\dots,d \qquad (2.31)$$

then the functional measure $\mathcal{D}\mu_Y$ which weights the paths of the first process is absolutely continous with respect to the corresponding measure $\mathcal{D}\mu_X$, moreover

$$\mathcal{D}\mu_Y = e^{\{\Sigma_{i=1}^d \int_0^\tau \phi^i(X_s)\,dW_s^i \frac{1}{2}\Sigma_{i=1}^d \int_0^\tau (\phi^i(X_s))^2\,ds\}}\,\mathcal{D}\mu_X \qquad (2.32)$$

where ϕ^i and δb^i are related by

$$\delta b^i(x) = \sum_{j=1}^d \sigma_j^i(x)\phi^j(x) \qquad (2.33)$$

As reference diffusion we choose the free one $s \mapsto X_s = (\mathbf{X}_s, X_s^d)$ defined by

$$dX_s^i = dW_s^i \quad i = 1,\dots,n$$
$$dX_s^d = ds + dW_s^d \qquad (2.34)$$

and we obtain the new path integral

$$u(\mathbf{x}, x_d) = e^{x_d}\mathbb{E}_{(\mathbf{x},x_d)}\left(u_0(\mathbf{X}_{\tau_D})\exp -\int_0^{\tau_D} V(\mathbf{X}_s)\,dX_s^d\right) \qquad (2.35)$$

where τ_D is now the first hitting time of ∂D by $s \mapsto X_s$. By its definition

$$(P^t u_0)(\mathbf{x}) = u(\mathbf{x}, -t) = e^{-t}\mathbb{E}_{(\mathbf{x}, -t)}\left(u_0(\mathbf{X}_{\tau_D})\exp -\int_0^{\tau_D} V(\mathbf{X}_s)\,dX_s^d\right)$$

It is expedient to translate the free process forward in time in order to display the dependence of P^t on t in a more explicit way. By this device we get

$$(e^{-tH_1} u_0)(\mathbf{x}) = e^{-t} \mathbb{E}_{(\mathbf{x},0)} \left(u_0(\mathbf{X}_{\tau_t}) \exp - \int_0^{\tau_t} V(\mathbf{X}_s) \, dX_s^d \right) \qquad (2.36)$$

where $\tau_t = \inf\{ s \geq 0 : s + W_s^d = t \}$ is the first hitting time of the hyperplane $\Sigma_t = \{x \in \mathbb{R}^d : x^d = t\}$. Formula (2.36) remind us of the usual Feymann- Kac one but with the difference that the path-dependent ordinary integral $\int_0^t V(\mathbf{X}_s) \, ds$ is replaced by the stochastic integral with a random upper limit $\int_0^{\tau_t} V(\mathbf{X}_s) dX_s^d$. The jump Markov process $t \mapsto \xi_t$ (appearing in the path integral of the one- body relativistic Schrödinger semigroup $\{e^{-t\tilde{H}_1}\}_{t \geq 0}$) can be represented[2,4] as $\xi_t = \mathbf{X}_{\tau_t}$ and therefore (2.36) looks

$$(e^{-tH_1} u_0)(\mathbf{x}) = e^{-t} \mathbb{E}_{(\mathbf{x},0)} \left(u_0(\xi_t) \exp - \int_0^{\tau_t} V(\mathbf{X}_s) \, dX_s^d \right) \qquad (2.37)$$

which is subtly different from

$$(e^{-t\tilde{H}_1} \psi)(\mathbf{x}) = e^{-t} \mathbb{E}_{(\mathbf{x},0)} \left(\psi(\xi_t) \exp - \int_0^t V(\xi_s) \, ds \right) \qquad (2.38)$$

except, of course, in the free case $V = 0$. In presence of a magnetic field, (2.37) must be modified according to

$$(e^{-tH_1} u_0)(\mathbf{x}) = e^{-t} \mathbb{E}_{(\mathbf{x},0)} \left(u_0(\xi_t) exp - \left\{ e \int_0^{\tau_t} A_\beta(\mathbf{X}_s) \, dX_s^\beta + \right. \right.$$

$$(2.39)$$

$$\left. \left. \tfrac{e}{2} \int_0^{\tau_t} (\partial_\beta A^\beta)(\mathbf{X}_s) \, ds \right\} \right)$$

a more elegant and covariant looking path-integral than that one which holds[5] for relativistic Schrödinger operators in a magnetic field. By dimensional analysis

$$\tau_t = \inf \left\{ s \geq 0 : s + \sqrt{\frac{\hbar}{mc^2}} \, W_s^d = t \right\}$$

in c.g.s. units and it is not difficult to check[4] that $t \mapsto \tau_t$ converges in probability (uniformly on bounded intervals) to the deterministic time t in the formal nonrelativistic limit $c \uparrow +\infty$. Therefore $t \mapsto \xi_t$ approaches the $\tilde{n}$-dimensional Brownian motion in the same limit. By subtracting the rest energy mc^2, it is clear that the nonrelativistic limit of (2.37) is the right one, namely the Feynman-Kac formula of nonrelativistic quantum mechanics in n-space dimensions.

References

1) Brezis, H. (1987). *Analyse Fonctionelle, Théorie et Applications*, Masson.
2) Carmona, R. (1988). "Path integrals for relativistic Schrödinger operators", *Lectures notes in Physics*, **345**, Springer-Verlag.
3) Carmona, R., Masters, W.C. and Simon, B. (1990). "Relativistic Schrödinger operators: asymptotic behaviour of the eigenfunctions", *J. Funct. Anal.*, **91** (117).
4) De Angelis, G.F. and Serva, M. (1990). "Jump process and diffusion in relativistic stochastic mechanics", *Ann. Inst. Henry Poincaré, Phys. Théor.*, **53** (301).
5) De Angelis, G.F. and Serva, M. (1990). "On the relativistic Feymann-Kac-Itô formula", *J. Phys. A: Math. Gen.*, **23** L965.
6) De Angelis, G.F. and Serva, M. (1992). "Imaginary-time path integrals from Klein-Gordon equation", *Europhys. Lett.*, **18** (477).
7) De Angelis, G.F. and Serva, M. (1992). "Brownian path integral from Dirac equation: A probabilistic approach to the Foldy-Wouthuysen transformation", *J. Phys. A: Math. Gen.*, **25** (6539).
8) Dyson, F.J. and Lenard, A. (1967). "Stability of matter". I, *J. Math. Phys.*, **8** (423).
9) Dyson, F.J. and Lenard, A. (1968). "Stability of matter". II, *J. Math. Phys.*, **9** (698).
10) Fisher, M.E. and Ruelle, D. (1966). "The stability of many- particles systems", *J. Math. Phys.*, **7** (260).
11) Freidlin, M. (1985). *Functional Integration and Partial Differential Equations*, Princeton University Press.
12) Herbst, I.W. (1977). "Spectral theory of the operator $(p^2+m^2)^{1/2} - Ze^2/r$", *Commun. Math. Phys.*, **53** (285).
13) Lewy, H. (1929). "Neuer beweis des analytischen charakters der lësungen elliptischer differential gleichungen", *Math. Ann.*, **101** (609).
14) Lieb, E.H. (1976). "Stability of matter", *Rev. Mod. Phys.*, **48** (553).
15) Lieb, E.H. and Yau, H.T. (1988). "The stability and instability of relativistic matter", *Commun. Math. Phys.*, **118** (177).
16) Nenciu, G. (1987). "Existence of the spontaneous pair creation in the external field approximation of Q.E.D.", *Commun. Math. Phys.*, **109** (303).
17) Wightman, A.S. (1973). "Relativistic wave equations as singular hyperbolic systems", in *Partial Differential Equations*, Proceedings of Symposia in Pure Mathematics, **23**, *Amer. Math. Soc.*.
18) Wightman, A.S. (1978). "Invariant wave equations: general theory and applications to the external field problem", *Lectures Notes in Physics*, **73**, Springer- Verlag.
19) Wigner, E.P. (1939). "On unitary representations of the inhomogeneous Lorentz group", *Ann. Math.*, **40** (149).

ONTOLOGICAL DETERMINATENESS IN QUANTUM MECHANICS AND SPECIAL RELATIVITY

M. DORATO
Via Biferno 4 - 00199 - Roma, Italy
Ph-Fax (396) 8621-3555. E-mail: Dorato@itcaspur.caspur.it

Abstract. The conceptual tension between the special theory of relativity and quantum mechanics is examined from the point of view of the ontological determinateness of events in Minkowski spacetime. While the standard interpretation of quantum mechanics requires an "open", indeterminate future in which events have "fuzzy" attributes before measurement, special relativity has often been thought to require that the future lobes of light-cones at each point contain only fully determinate events. Recent attempts at introducing indeterminateness in Minkowski spacetime are discussed in light of the philosophical implications of experiments violating Bell's inequalities.

1. The tension between Special Relativity and Quantum Mechanics

In the history of the relationships between the special theory of relativity (STR) and quantum mechanics (QM), the principles of the two theories have often been contrasted, especially by those who don't accept the so-called "Copenhagen interpretation" and argue that quantum mechanics is incomplete[1]. For the enemies of this interpretation, which is still the orthodox one in the physicists' community, the *complementarity* referred to by Bohr between the spatio-temporal description of continuous phenomena, typical of STR, and the description of discontinuous changes of states in atomic systems, typical of QM, may become a *contradiction*.

After the celebrated Bell's theorem (1964), the discussion on the relationship between the two theories has been focussing on the principle of *locality*, a core tenet of STR, which seems violated by those correlations at a distance predicted by QM and confirmed by experiments designed to test Bell's inequality (Aspect *et al.* 1982). More precisely, if by 'locality' we mean, as I

201

C. Garola and A. Rossi (eds.), The Foundations of Quantum Mechanics, 201-211.
© 1995 *Kluwer Academic Publishers.*

propose, *lack of action at a distance*, one can assume with confidence that between STR and QM there is "peaceful coexistence", since there is no correlation between the settings of the 'spin meter' in one wing of the experiment and the outcomes obtained in the other wing[2]. If instead by 'locality' we mean *separability* — the idea, that is, that the properties of a compound system $A \oplus B$ are just the sum of those of its spatially separated components A and B, or, alternatively, that the relational properties of A and B are supervenient with respect to the intrinsic ones — then the compatibility appears much less evident. The two spacelike-related measurement outcomes are in fact not statistically independent[3]. If every field theory presupposes the independence and the distinct identity of spacelike related space-time regions (separability), it has been plausibly argued that the conflict between STR and QM is indeed deep and unavoidable (Howard 1989).

There is, however, another aspect of the conflict between the two theories that has received less attention, and which is connected with the possibility of introducing objectively indeterminate 'quantum events' in Minkowski spacetime. To be clear about what is at stake, let us stipulate that the magnitudes of microentities characterizing a quantum system (in short, quantum events) are *ontologically determinate* just in case the numerical values that such magnitudes assume are always definite (sharp) independently of, and immediately before, a measurement[4]. It is well known that, according to the anti-hidden-variable interpretations, QM violates such requisite of determinateness: when the predictions of the theory are referred not to *ensembles* but to individual systems, the wave function Ψ does not represent our ignorance of hidden parameters (deterministic or stochastic), but rather a complete physical description of magnitudes that before measurement are objectively indeterminate. Moreover, at least within the environment provided by Hilbert space, the Kochen and Specker's Paradox (1967) seems to imply that it is not possible to assign in a non-contextual way values to all observables in all states, which implies an indeterminateness of some kind for at least some quantum events.

In order to understand the nature of the above mentioned conflict with the ontology presupposed in STR, it is enough to recall that the major contributors to the theory of relativity[5] have held the view that after the demise of absolute simultaneity, all events in Minkowski spacetime have to be regarded as ontologically determinate, since the distinction between past and future is merely relational or even conventional. Even though the problem of the determinateness of events in Minkowski spacetime has been the subject of a recent debate concerning the possibility of making room for objective becoming within STR[6], the implications of Bell-type experiments on this issue have not been taken into account. This problem seems particularly urgent, since we don't have as yet a clear idea of the violation of separability, of the acausality of the correlations of compound systems, and, what is most important, of the ontological status of quantum microentities.

2. Determinateness in Minkowski spacetime

Let us recall the three interpretations of QM listed by Redhead (1987: 45) with respect to a quantum system in which some state is not in an eigenstate of an observable Q. The first interpretation (A) claims that the value of Q is always sharp though unknown (hidden variables). The second interpretation (B) holds that the value of Q is indeterminate ("fuzzy") independently of our knowledge (potentiality view of QM). The third is the standard Copenhagen interpretation (C), according to which the value of Q is meaningless or undefined until a measurement or an observation occurs. Since a particle having a certain sharp magnitude at a time-place can be regarded in short as *a determinate event*, arguments claiming that STR implies that all events in Minkowski spacetime are determinate independently of our knowledge would be compatible only with the first of these interpretations[7].

In other words, if the argument put forward by Putnam (1967), Reitdijk (1966) and Maxwell (1985) in favor of the universal determinateness of all events in STR could be defended by the recent attack of Stein (1968, 1991)[8], we would have demonstrated that the structure of STR alone requires a hidden-variable completion of the theory. If such completion could not be provided as a consequence of the Bell experiments[9] or of other no-hidden variable theorems, we would have to conclude that STR and QM are indeed incompatible. It will therefore be appropriate to briefly re-examine such arguments in light of Bell experiments.

Let us assume that O is a spin meter centered in an inertial reference system (x, t). O's state here-now concides with event a. Since e represents the value of the spin of a particle shortly before measurement, it is located in the absolute future of a (see Fig. 1 below). Let us also introduce a dyadic relation of determinateness δ holding between any two events a and b, which here are treated as non-vanishing regions of Minkowski spacetime: $a\delta b$ means that b is ontologically determinate as of a. Since the values of spins *already registered* by the meter O are determinate as of a, any event in the causal *past* of a (the past lobe of the light cone at a) is determinate as of a. The question that I will try to answer concerns the status of determinateness of events belonging to a coupled system of Bell type. Such events are spacelike related to a since they are measured by the other spin meter O', whose here-now-state is event b, which is simultaneous with a in the system (x, t). Suppose that, according to interpretations B or C, for every spacetime region coinciding with a measurement apparatus a, there is some event e later than a some of whose magnitudes are, respectively, either indeterminate with respect to *a* independently of our knowledge, or undefined[10]. In symbols, ¬*aδe*.

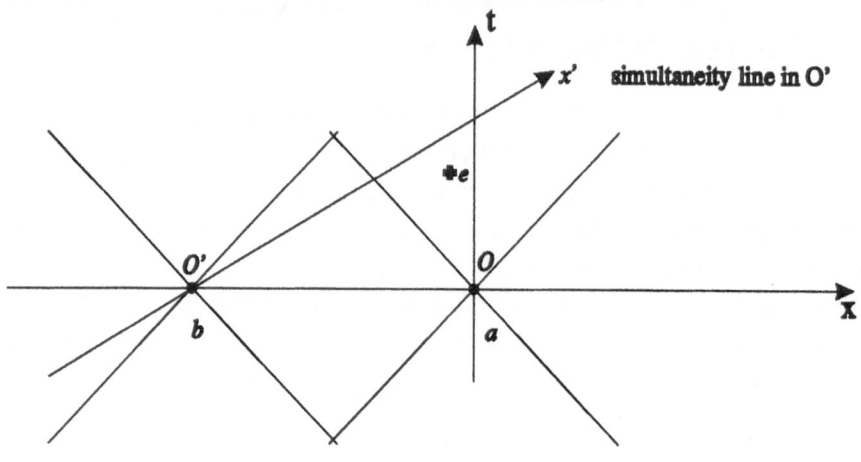

Fig. 1

Since, following Stein (1991), it is plausible to require that δ be *transitive and reflexive*, a crucial hypothesis that is needed to prove that the (anti-hidden variable) assumption ¬*aδe* leads to a contradiction with STR is *symmetry* of δ. If δ is an equivalence relation, and if it is not trivial in the sense that its holds for at least two distinct points, it can be easily proved that it is the *universal* relation, that is, that it holds for all points[11].

To study the problem of symmetry for δ, it is sufficient to restrict our attention to the case of spacelike related events. If we could establish that any two events belonging to an arbitrary hyperplane of simultaneity are co-determinate, i.e., determinate as of each other, we would have proved that symmetry holds for spacelike related events. Since any timelike vector can be written as the sum of two spacelike-related vectors, it is easy to see that symmetry and transitivity of δ allows us to conclude that any event lying in the future lobe of the light cone at each point is determinate with respect to the "here-now". If *aδf* and *fδa* holds for any couple of spacelike-related events *a* and *f*, and (*a, e*) is a timelike vector, we can always write (*a, e*) = (*a, f*) + (*f, e*), where the vectors on the right side of the equality sign are appropriate spacelike vectors. Since *aδf* and *fδe* holds by hypothesis, transitivity gives us *aδe* and symmetry gives us *eδa*. Independently of whether *e* in the future or in the causal past of *a*, we have both *aδe* and *eδa*, which gives us a contradiction with the hypothesis ¬*aδe* postulated by *B* or *C*. All events would be determinate.

The crucial hypothesis underlying the simple proof above is that any two events belonging to any hypersurface of simultaneity are co-determinate. On the one hand, this hypothesis seems reasonable given the statistical correlation of the values of the two measurement outcomes in Bell experiments (non-separability of the events). On the other hand, it seems to contrast with the relativistic injunction that 'the present moment, strictly speaking, does not extend beyond itself' (Robb 1921). The question we should try to answer is whether the experimental evidence of non-separability is sufficient to overcome the operationally derived unrealism about the spatially distant typical of STR.

3. Determinateness of events in spatial hypersurfaces

In order to shed light on the alternative above, it is necessary to analyze more precisely the meaning of determinateness. This can't be done without the concept of 'possible world'. Since we said intuitively that a future event e is determinate with respect to the here-now a if and only if e's observables O are sharp-valued before measurement, one can alternatively express the same concept by requiring that a future event is determinate just in case it has the same value in all possible future worlds (worlds compatible with the here-now and its causal past). An indeterminate event will then be one which has different values for its observables in different possible worlds, corresponding to different possible measurement outcomes.

Granted that the most important result of the violation of Bell's inequality is the statistical dependence of the outcomes obtained in one wing of the experimental set-up with the outcomes obtained in the other, it seems reasonable to require that the spacelike-related events whose magnitudes these outcomes represent be regarded as co-determinate[12]. The *non-separability* of the spacelike-related measurement outcomes might be considered sufficient for holding that events in any of the hypersurfaces intersecting a, b or both (see Fig. 1) are co-determinate, despite the fact that the statistical correlation between the outcomes may have no common cause.

Note that here I am not claiming that the collapse of the wave packet associated with the compound system is a physical process happening instantaneously on some privileged hypersurface. This alleged physical process would be relativistically non-invariant. Rather, I am just claiming that, as a matter of conceptual analysis, it is counterintuitive to hold that two events at a distance are indeterminate as of each other — as follows from Stein's definition of the relation of determinateness in terms of the relation of causal precedence (1991) — if their properties are statistically correlated. Since in its turn, such co-determinateness ensures symmetry and therefore the universality of δ, this argument implies that the only interpretation of QM that is compatible with STR is A (the hidden variable theory).

The main argument suggesting to take the other horn of the dilemma invokes the verificationist, operationalist roots of STR. Einstein's original critique to the

Newtonian absolute simultaneity was based on the impossibility in principle of verifying simultaneity at a distance, that is, the synchronicity of clocks that are far away from each other. To presuppose that events belonging to a hypersurface of simultaneity are co-determinate even at a distance might be regarded as a violation of the fundamental methodological rule of STR, according to which information or influence on event a can only be transmitted by events in the causal past of a. Any event which is spacelike related to a, as are events belonging to any hypersurface crossing a, cannot be connected by any non-tachionic causal signal to a, and should accordingly be regarded as indeterminate with respect to it.

In practice, the limitation above is due to the finiteness of the velocity of light: if we had infinitely fast signals, the co-determinateness of events at a distance would be ensured by the possibility that events causally interact with each other instantaneously. Suppose a sufficient criterion of determinate reality is the physical possibility that events interact causally at a distance. Then the limiting character of the velocity of light excludes from the class of determinateness all events lying outside the light-cone. It is this observation that led Robb (1921) and Capek (1961) among others to assert that the notion of the universe at an instant, is strictly speaking, meaningless. If the region outside the light cone is indeterminate with respect to the intersection of the two lobes (the origin or the here-now), then the region of determinateness for each point a in Minkowski spacetime would coincide with the causal past of a. In this case, interpretations A and B of QM would peacefully coexist with STR.

How can we evaluate the relative weight of these two arguments, respectively in favor and against a hidden variable completion of QM? Rather than offering a knock-down argument, the following observations are meant to convince us of the difficulty of the problem, and to bring us nearer to its future solution. The question of the status of determinateness of spacelike-related events is linked to deep problems in the foundations of physics, and has no easy solution of the kind prospected by Putnam (1967).

(i) Events lying outside the light-cone at a must be either regarded as all determinate or as all indeterminate, according to the two alternatives examined above. In other words, we must exclude the possibility that the whole space-time be divided in a region of determinateness, corresponding to events below and on a hypersurface of simultaneity, and in a region of indeterminateness, corresponding to events above the hypersurface. *Pace* certain theories of measurement, according to which such hypersurface might be *de facto* privileged by the instantaneous collapse of the wave packet in Bell experiments (Popper 1982, Maxwell 1985), the partition given by a hypersurface of simultaneity has no objective significance in Minkowski space-time, since a hypersurface is not invariant with respect to automorphisms preserving the causal structure of the space-time. A hypersurface is determined by a direction of motion: for a given inertial world-line, there is a unique hypersurface which is orthogonal to that line. If we assumed that the events belonging to such hypersurface were determinate, we would make an assertion of ontological determinateness depend on *a state of motion*, which is too fragile a concept to bear the weight of a distinction between what is determinate and what isn't. By a slight acceleration in fact, we would change inertial reference system, and therefore the relative hypersurface of simultaneity: the ontological status of an event which is far away with respect to me would change by taking an airplane! If the collapse of the wave packet is treated as a physical process occurring instantaneously in a hypersurface orthogonal to the world-line of the observer, it becomes a highly dubious assumption.

(ii) It has not been noted that EPR's sufficient criterion of reality presupposes quite explicitly the co-determinateness of events at a distance, an hypothesis that is prohibited by the interpretation of STR presented above: "if, without in any way disturbing the system, we can predict with certainty (...) the value of a physical quantity, then there exists an element of reality corresponding to that quantity" (Einstein, Podolski, Rosen 1935: 778). Supposing that the outcomes were indeterminate as of each other, one could not even talk about predictability, and the latter would apply only to events of coupled systems that are in mutual causal contact or at least quasi-coincident spatially.

This observation seems to guide us toward a possible solution of the problem presented above. Exactly as in the predictability argument of EPR one must presuppose the simultaneous co-existence of events in hypersurfaces of simultaneity, it could be alleged that even to talk about the non-separability of the *two* measurement outcomes in Bell experiments, one must presuppose, against Robb and Capek, that the 'now' extends beyond the 'here', in the sense that the notion of the determinate existence of a world-wide instant is not meaningless. This could be backed up by a distinction between the spatio-temporal properties of events in the spacelike region, which are indeterminate because they are conventional or relative to a frame, from the other, more "intrinsic" properties (charge, spin, mass, etc), which could be nonetheless definite with respect to events belonging to a hypersurface. While this relationalist distinction would be problematic in the general theory of relativity, in STR it can be made to look much more plausible, since the Minkowski spacetime "arena" is independent from the physical occurrences.

In any case, this peculiar contrast between the EPR argument, which accords determinate existence to the spatially distant, and the verificationist spirit of STR, which denies it in the name of the finiteness of the velocity of light, seems to confirm the truth of Einstein's description of the physicist as an incurable epistemological opportunist. The following letter of the father of STR to Born, dated 24 March, 1948, further illustrates this fact:

> I just want to explain what I mean when I say that we should try to hold on to physical reality... If a physical system stretches over the parts of space a and b, then what is present in b should somehow have an existence independent of what is present in a [separability]. What is actually present in b should not thus depend upon the type of measurement carried out in the part of space, a [locality]; it should also be independent of whether or not, after all, a measurement is made in a...However, if one renounces the assumption that what is present in different parts of space has an independent, *real existence,* [co-determinateness of events in an hypersurface] then I do not know what physics is supposed to describe. For what is thought to be a "system" is, after all, just conventional, and I do not see how one is supposed to divide up the world objectively so that one can make statements about the parts. (Born 1969: 223–234, transl. by D. Howard, my italics).

The three interpolations show with clarity that according to Einstein the hypothesis of realism does not just consist in locality and separability, but also in the co-determinateness of events at a spatial distance. From this passage it is not clear whether Einstein distinguishes the independent existence of two events that are simultaneous at a distance (separability) from their determinate existence (co-determinateness). However, if it is granted that to talk about two non-separable but reciprocally indeterminate spacelike-related events is to talk nonsense, from the well-know relativistic injunction of not extending the present beyond itself, we should not conclude that spacelike-separated events are ontologically indeterminate as of each other.

This observation should make us propend for the first horn of the dilemma presented above. The problem of non-separability is not, as Einstein claims, that it becomes impossible to cut out a system from the rest of the environment in a non-conventional way, but rather that the experimentally confirmed correlations at a distance presuppose the determinate existence of different parts of space, against the positivistic claim of irrealism of the spatially distant that lies at the foundation of STR.

(iii) However, before invoking the incompatibility between STR and interpretations B or C, which may even mean incompatibility between QM and STR if A is not available, there is another option that we should consider. Together with the full view of reality (all events are determinate) and the empty view of the future (only the events in the causal past of each event are determinate), there is the view according to which the past is unreal, that also seem to be backed up by the results of Bell experiments. This option claims that the causal past is unreal, because, in the absence of a direct causal link between the two statistically correlated outcomes a and b, we cannot find any event (hidden parameter) lying below the hypersurface to the past of a and b that can screen-off a from b. If there were such a screener-off, in fact, the Bell inequality would not be violated (see Butterfield 1989). The inclination to deny the existence of a common probabilistic cause of the correlations in the past of a and b means that the corresponding past region of space-time does not contain timelike curves connecting with a and b. This is another way of saying that as of a and b, the past is fully indeterminate, a result that Wheeler has independently made popular with his interpretation of the delayed-choice experiment. In this case, if one insisted on denying the simultaneous determinateness of events at a distance and refused to believe in a hidden variable interpretation of QM, one could hold a *hic et nunc* view of reality, according to which for each event a, it and it alone is determinate. While such a "solipsistic" view (Stein 1968) would hardly be incompatible with STR, it could be distinguished from its opposite (the full view) only with difficulties. If every event in STR can be here-now, and each here-now is determinate with respect to itself, every event in STR would turn out determinate. The only difference with the full view would be that in the latter events would be determinate also as of all other events, an option that in view of the non-separability of QM appears much more plausible.

Notes

[1] In this context, and according to the point of view of Einstein-Podolsky-Rosen (1935: 777), (EPR), a complete theory is such that "every element of the physical reality must have a counterpart in the physical theory". EPR take this condition as a necessary one for every acceptable physical theory.

[2] In short, QM does not violate what in the literature is known as 'parameter independence'. See for instance Ghirardi-Rimini-Weber (1980), Shimony (1984), Cushing and McMullin (1989). For various senses of 'locality', see Redhead (1987).

[3] Selleri and Tarozzi (1986) define Einstein locality as entailing also separability, besides other physical principle, and conclude in favor of an opposition between STR and QM. It is however advisable to distinguish between locality as absence of action at a distance and separability in the sense defined here.

[4] The definiteness in question takes into account the indeterminacies due to experimental errors. Here I also assume the Principle of Faithful Measurement, according to which the result of measurement reveals the value possessed by the observable immediately before measurement (see Redhead 1987: 89).

[5] Among such scientists, we find Einstein, Eddington, Weyl and Gödel. See M. Dorato, (1992). *The Reality of the Future*, Ph.D. Thesis, The Johns Hopkins University, ch.1.

[6] See Costa de Beauregard (1966), Putnam (1967), Rietdijk (1966), Stein (1968, 1991), Maxwell (1985), Sklar (1985, ch. 11).

[7] While this assertion is obvious for B, it requires only a little reflection to see that the ontological determinateness of the magnitudes of quantum events *before* measurement allegedly ensured by STR would count as decisive evidence against the positivistic injunction, typical of C, of regarding the question about the value of those magnitudes as devoid of meaning.

[8] See also Clifton R. and Hogarth M. "The definability of objective becoming in Minkowski spacetime", forthcoming in *Synthese*.

[9] By 'Bell experiments' I obviously mean experiments realized to test the Bell's inequalities.

[10] Here it is important to distinguish 'indeterminate' from 'undetermined'. An event is 'indeterminate' just in case (some of) its magnitudes are unsharp. An event in undetermined if laws of nature don't fix its occurrence given present conditions. A certain amount of indeterminateness can be compatible with determinism (see note 6).

[11] This can be obtained as a result of Malament's third proposition in his 1977 paper. The additional hypothesis that is needed is implicit definability in terms of the relation of causal connectibility, which also holds of δ.

[12] 'The particle x having spin up or down at a certain time-place' can be regarded as an event having a certain magnitude at that time-place.

References

Aspect, A. *et al.* (1982). "Experimental test of Bell's inequalities using time-varying analyzers", *Physical Review Letters*, **49** pp.1804--7.

Bell, J. (1964). "On the Einstein-Podolsky-Rosen Paradox," *Physics*, **1** pp.195--200.

Born, M. (ed). (1969). *Albert Einstein—Hedwig and Max Born. Briefwechsel, 1916-1955.* Munich, Nymphenburger.

Butterfield, J. (1989). "A space-time approach to the Bell inequality", in Cushing McMullin 1989, pp.114--144.

Capek, M. (1961). *The philosophical impact of contemporary physics.* Princeton, New Jersey.

Costa de Beauregard, O. (1966). "Time in relativity theory. Arguments for a philosophy of being", in Fraser, J., Haber, F. and Müller (1972) (eds). *The study of time.* Berlin, Springer Verlag, pp.417--433.

Cushing, J. and McMullin, E. (1989). *Philosophical Consequences of Quantum Theory.* Notre Dame, Indiana, University of Notre Dame Press.

Einstein, A., Podolsky, B., Rosen, N. (1935). "Can quantum-mechanical description of reality be considered complete?", *Physical Review*, Ser. 2, **47** pp.777--80.

Ghirardi, G.C., Rimini, A. and Weber, T. (1980). "A general argument against superluminal transmission through the quantum-mechanical measurement process." *Lettere al Nuovo Cimento*, **27** pp.293--298.

Howard, D. (1989). "Holism and separability", in Cushing e McMullin 1989, pp.224--53.

Kochen, S. and Specker, E.P. (1967). "The problem of hidden variables in quantum mechanics", *Journal of Mathematics and Mechanics*, **17** pp.59--87.

Malament, D. (1977). "Causal Theories of Time and the Conventionality of Simultaneity", *Nous*, **11** pp.293--300.

Maxwell, N. (1985). "Are Probabilism and Special Relativity Incompatible?", *Philosophy of Science*, **52** pp.23--43.

Popper, K. (1982). *Quantum Theory and the Schism in Physics*. London, Hutchinson.

Putnam, H. (1967). "Time and Physical geometry", *Journal of Philosophy*, **64** pp.240--247.

Redhead, M. (1987). *Incompleteness, Non-locality and Realism*. Oxford, Oxford University Press.

Rietdijk, C. (1966). "A rigorous proof of determinism derived from the special theory of relativity", *Philosophy of Science*, **33** pp.341--344.

Robb, A. (1921). *The absolute relations of time and space*. Cambridge, Cambridge Unviersity Press.

Selleri, F. and Tarozzi, G. (1986). "Why quantum mechanics is incompatible with Einstein Locality", *Physics Letters A*, **119** pp.101--9.

Shimony, A. (1984), "Controllable and uncontrollable non-locality", in Kamefuchi *et al.*, *Foundations of Quantum Mechanics*. New York, Wiley, pp.225--230.

Sklar, L. (1985). *Philosophy and Spacetime Physics*. Berkeley, University of California Press.

Stein, H. (1968), "On Einstein-Minkowski spacetime", *Journal of Philosophy*, **65** pp.5--23; (1991), "On relativity theory and the openness of the future", *Philosophy of Science*, **58** pp.147--167.

Tarozzi, G. 1992a, *Filosofia della microfisica*. **Vol. 1**, Modena, Accademia Nazionale di Scienze, Lettere e Arti, Mucchi; (1992b) (a cura di). *Il paradosso della realtà fisica*. Modena, Accademia Nazionale di Scienze, Lettere e Arti, Mucchi.

Van Fraassen, B. (1991). *Quantum Mechanics. An empiricist view*. Clarendon Press, Oxford.

DUALISM AND INCOMPLETENESS OF QUANTUM MECHANICS. TOWARDS NEW CONSISTENT THEORIES

A. DRAGO
Group of History of Physics - Dept. of Physical Sciences
University of Naples

Abstract. In the '70s the research on foundations of science recognized two basic options - respectively, on the kind of mathematics and on the kind of logic. By means of these options, the foundations of quantum mechanics are scrutinized anew. It results an intrinsic dualism owing to which the measurement theory takes different choices on these options than the theory of the imperturbed evolution of the system. Moreover, the past two debates on wave-corpuscle dualism and incompleteness are reduced to two definite problems. The last one obtains a definite positive answer when it is analysed in the mathematics which is alternative to the classical one - i.e., constructive mathematics.

1. A new interpretative scheme for physics foundations

My viewpoint on the foundations of a physical theory is new since it takes in account the new results of the 60's on the foundations of both mathematics[1] and logic[2]: there are at least two ways of founding each theory.

By linking these results to the theories of classical physics, I concluded that the foundations of a physical theory include two options. The first option concerns the kind of mathematics (whether appealing to actual infinity (AI) - as classical mathematics does -, or bounded to the unlimited finite only or potential infinity (PI) - as constructive mathematics is). The second option concerns the kind of logic, i.e. either the classical logic or non-classical logic, where $\neg\neg A \rightarrow A$ does not hold true; in correspondence, we have either a deductive organization of the theory, as Aristotle first suggested (AO), or a theory centered upon a problem (PO), whose new method of solution

213

C. Garola and A. Rossi (eds.), The Foundations of Quantum Mechanics, 213-227.

constitutes the theory itself. As a result, I obtained the following general scheme of interpretation of classical physics[3].

TABLE 1

Some centuries **Effective science** (as scientific geniuses determined it through two fundamental options)	One century **Subjective science** (as scientists conceived it through surrogatory notions)	One generation **Objective science** (as teachers formalized it through tools of reasoning)
NEWTONIAN MoST (AO + AI)	'Dissolution of the finite cosmos and geometrization of space'	Classical logic *Analytic method* Infinitesimal analysis (main example: differential equations of the 2° order)
CARNOTIAN MoST (PO + PI)	'Evanescence of the force-cause and discretization of matter'	Non-classical logic *Synthetic method* Symmetry or a cycle (main example: S. Carnot's cycle in thermodynamics)

Legenda: MoST=Model of Scientific Theory; AO=Aristotelian Organization; PO=Problematic Organization; AI=Actual infinity; PI=Potential infinity

The table represents the main two models of a scientific theory (MoST) - the Newtonian one (to be called a paradigm, owing to its great influence) and the Carnotian one - each one resulting from two opposite couples of choices: AO and AI, PO and PI. The table synthetises three kinds of representation, in correspondence to three different levels of awareness on the foundations of a theory. Note that the categories by which Koyré interpreted the birth of modern science are here recognized as pertaining to the subjective representation; they constitute a clever synthesis by means of intuitive notions of the Newtonian choices AO and AI. For the Carnotian MoST, I suggested corresponding categories which synthetizes by means of intuitive notions the choices PI and PO taken by the Carnotian theories of the 19th Century - i.e. L. Carnot's mechanics, S. Carnot's thermodynamics, classical chemistry. The discordance between these categories substantiates in subjective terms the incommensurability between the two MoSTs whose basic choices are different.

When applied to a given theory, the interpretative scheme offers one dichotomic parameter, i.e. the two MoSTs. Then, the fitting of a theory in a particular MoST has to be verified under definite criteria constituted by the characteristic features of the three representations.

2. Which basic choices by quantum mechanics?

Let us interpret quantum mechanics (QM). The following analysis on the foundations of QM will result to be more accurate than the occasional analyses which in the times of acute crises Einstein, Planck, Heisenberg have performed respectively on the basic notions of space and time, the continuous or discontinuous formalism of theoretical physics, some basic differences between the classical description and the quantum description of reality - i.e. always on partial aspects of the foundations here considered.

By the present interpretation the so-called "hidden variables" are meant as the two dichotomic options on mathematics and logic. Rather than physico-mathematical variables, they are here meant as two logico-mathematico-philosophical variables. Owing to their high degree of abstractness they stood unrecognized so long time. Moreover, the resulting interpretation will suggest more than new formulas and new notions - in fact, new theories.

But which basic choices? Let us inspect first the current subjective representation of QM[4]. It is apparent that it agrees with the two Carnotian statements: "discretization of matter", till to discretize even light; and "evanescence of force-cause", since this last notion disappears in quantum jumps for leaving place rather to the notion of potential energy. It is also apparent that QM does not share the Newtonian statements (the statement "geometrization of space" only can be viewed as substantiated - yet, in a very abstract way - by Hilbert's space).

Thus, the philosophical trends of the 19th Century physics qualify QM as a Carnotian theory. After the failure of energetists' program for overcoming by means of the subjective representation of thermodynamics the Newtonian theory, QM represented a new and thereafter decisive step for overcoming the Newtonian paradigm - i.e. that complex of basic notions, mathematical techniques, basic theoretical aims which constituted the estate of a traditional theoretical physicist -,. This fact gives reason for the surprising novelty of the new theory.

However, our analysis by means of the subjective representation refers to subjective notions, which lack any formalization. Let us apply then the effective representation. Here, we meet an ambiguous situation under two respects. Let us consider first the option on the infinity. It is apparent that quanta introduced PI in theoretical physics. But, next, Schrödinger equation (SE) as well as Hilbert space restored at all classical mathematics which appeals to AI. Furthermore, it was evidentiated in QM a non-classical logic, which corresponds to a PO choice by QM; of course, the main problem of QM is the uncertainty in the measurements. But a more accurate inspection shows that the theory privileges w.r. to the body of the theory an a priori part - the amplitude of probability as well as SE - , just as in an AO theory the part of the axioms is related to the part of theorems. Moreover, as if it was an AO theory, QM apparently dispenses itself with quantum logic. The further application of the objective representation gives no more insights in these ambiguities.

Two conclusions are allowed by the above application: Either to abandon the interpretative scheme as essentially inadequate to modern physics' theories; or to charge QM to be an ambiguous theory resulting from the juxtaposition of two distinct parts, i.e. SE and measurement theory. The well-known von Neumann's paradox together with the arguments by some authors[5] support the latter alternative, which I will choose in the following; that is, I conclude that QM is a PI and PO theory in its part of the measurement theory and it is an AI and AO theory in its part describing the imperturbed evolution of a quantum system. Under this light one gives reason for the great novelty of the new theory of QM; not only it followed the philosophy of the minoritarian group - the thermodynamicists' one - but also it was lacking consistency w.r. to the basic choices; i.e. not only it introduced different basic choices - i.e. incommensurability - w.r. to the Newtonian ones, but even inside the QM itself. As a result, that theory which definitely led the physicists to abandon the Newtonian paradigm, in fact introduced them into an ambiguous foundational situation which - by comparison with past classical physics - exhalted ironically the consistency of the old paradigm.

The way out of such an intricate situation is easily suggested: to reformulate QM as a theory consistently choosing PI and PO. This program may take advantage from Heisenberg's matrix mechanics, which among past formulations represents a good approximation of the wanted result. However, this theory has to be improved: i) by introducing constructive mathematics instead of merely finitist mathematics; ii) by stating the main problem of the PO theory - of course, the uncertainty principle - by means of a non-classical logic; iii) by introducing that mathematical technique which has been recognized as the typical technique of the PI & PO theories, i.e. symmetry[6]. A specific paper by A. Pirolo and myself is devoted to this suggestion[7].

3. The relevance of quantum logic

In the following I will give reason for the some more oddities QM intoduced. What I want to show is that when the basic choices of the theory are recognized in a clear-cut way, then the past philosophical difficulties either no more constitute real problems, or are translatable in sharply defined problems which can be suitably solved in an effective way.

Let us start from a revisitation of QM under the light of the choice OP. According to the above characterization of QM, a part only - i.e. the measurement theory, whose choice is PO - exhibits a non-classical logic, the remaining part following classical logic; as a consequence, any mixture of logical laws is allowed. No surprise if quantum logicians called "a labyrinth"[8] their field of research.

In any PO theory the main problem is represented by a double negated statement (DNS). Actually, we can say about measurement theory: "It is *not* true that two conjugate variables are *not* measurable at the same time". Although this statement $\neg\neg A$ holds true, it does not imply - as classical logic wants - A ($\neg A$ either). Under this light the classical logical law which fails in QM is not the distributivity law, as Birkhoff and von Neumann stated[9], but the double negation law. Actually several quantum logicians support independently this alternative[10].

However, according to the present interpretation, non-classical logic was effective in theoretical physics since the first PO theories - i.e. Lazare Carnot's and S. Carnot's theories[11]; then, what historical novelty by quantum logic? Really, in the old theories each DNS refers either to a single magnitude (for ex. the value of velocity in the inertia principle of L. Carnot's mechanics) or to two subordinate magnitudes (heat and work in S. Carnot's thermodynamics). Instead, in QM the DNS concerns the two variables defining together the state itself of the system; in other words, in QM the DNS cannot more be suppressed - as in mechanics -, or camouflaged by means of the word "equivalent" - as in thermodynamics, since this change would blur the core itself of the theory. Thus, quantum logic is at the same time a partial and yet decisive novelty in the development of theoretical physics.

One more phenomenon of QM theory can be explained, i.e. the staunch insistence by the Copenhagen school for supporting the mathematical interpretation of the physical state as a complex function. Really, a DNS which does not imply the corresponding affirmative sentence has not an easy counterpart in mathematical terms. For ex., in thermodynamics the DNS about heat and work was translated - owing of the lack of a mathematical symbol for "equivalency" - into an equality sign pertaining to the formula of energy conservation; that originates a paradox, since not $L = Q$ - as the first principle states -, rather $L = Q_1 - Q_2$ holds true - a statement justified by the subsequent second principle only. However, inasmuch as heat and work are two subordinate variables, the theory is not troubled by this false equality. But in QM the DNS concerns the definition of the state itself; hence, for representing a state QM has to invent a mathematical operation which - likely to logical negation in non-classical logic - is almost identity, in the sense that this operation, when performed two times, has to give as result a similar but not identical state than the original one. This requirement is not satisfied by ordinary operations in the real number field, where; yet, it is satisfied in the complex field, by making use of the idempotent operation of conjugation in association to the cycle of two operations, the evolution of a square number and the square root; in short, the algebraic property $(a+ib)(a-ib) = a^2+b^2$. In this way QM mechanics represents in fact the new DNS: "It is *not* true that the state is *not* represented by real quantities"; the translation from real numbers to complex numbers and back results in a similar but different situation than the original one. The commutator represents a further way of obtaining the same kind of translation. All that takes example, as all historians of QM well know[12], from the classical wave theory, Fourier's series, etc. which were considered as auxiliary tools of the differential techniques of wave theory; and not from the statistical mechanics which is unable to represent an almost identity in mathematical terms. Moreover, let us note that wave theory, being a PO theory could anticipate the above mathematical technique, whereas statistical mechanics, being an AO theory could not.

Moreover, one can give reason for both von Neumann's paradox and its nature. In any AO theory a double fault - as Berkeley first evidentiated in Newton's calculus - holds true[13]. The first fault is to introduce a Platonist notion. In QM, one puts SE together with the amplitude of probability as Platonist axioms; that constitutes a fault since whatsoever physical theory is instead obliged to be adequate to the empirical data. Then, for recuperating the lost reality the theory introduces - often, by its second axiom - a correction, which actually contradicts the previous hypothesis. In QM, measurement theory forbids through the uncertainty principle to speak at the same time of the absolutely accurate values of two conjugate variables which instead are involved by SE. One follows the quite not Platonist strategy of looking for the means value of the measurement. This one is obtained by the well-known technique of integrating the product of the state function - transformed by the operator associated to the measure - with its complex conjugate function. By such a second fault - just as Berkeley maintained - the correct result of the measurement is paradoxically obtained. However, this characteristic feature of any AO theory - i.e. the double fault - becomes in QM a true contradicion - truely, an antinomy - inasmuch as from the viewpoint of the IP and OP part of the theory, i.e. measurement theory, the part including the state function and SE is incommensurable. Hence, the formally correct result of a calculation does not clean the foundational trouble that the double fault induces in the other part of the theory which is of the OP type.

By inspecting classical theories, a similar double fault occurs in thermodynamics too, which claims to be an AO theory[14]. Here too, the first Platonist notion is so strong to generate the above-mentioned paradox. For eliminating this paradox the verbal statement of the first principle says that L and Q are "equivalent" magnitudes, without defining this word. However, the contradiction results to be ineffective - hence, it constitutes a paradox - inasmuch as the theory works on state variables only, which include neither L nor Q. In conclusion, the "von Neumann Paradox" is not specific of QM, being rather a common feature of AO theories; yet, in QM the nature of an antinomy is specific to the quantum von Neumann paradox.

4. The four ontological models of a physical system

In the next three sections we will inspect QM from the viewpoint of the PI choice, again an alternative choice to the Newtonian paradigm. Owing to the lack of a well-defined, constructive mathematics, in past times physicists supported their arguments by means of the physical imaginery of peculiar, subjective notions. In this way, they however explored more than the two MoSTs in the above. Really, when we consider any combination of the two dichotomic options, we obtain four MoSTs. Table 2 represents them and the theories that in the history of physics substantiated them.

TABLE 2

	ACTUAL INFINITY	*POTENTIAL INFINITY*
ARISTOTE-LIAN ORGA-NIZATION	*The Newtonian Paradigm* Newton's Mechanics 1687); Newton's Optics(~1700); Maxwell's electro-magnetism (1873) **mass-point**	Descartes' Optics (1630);Acoustics (1740); Classical Thermodynamics (1851); Physical-Chemistry (~1878) **wave**
PROBLEM-BASED ORGA-NIZATION	**many-particles** Lagrange's Mechanics (1788); Boltz-mann's Statistical Mechanics (~1900)	**system** L. Carnot's Mechanics (1782); S.Carnot's Thermodynamics(1824); Classical Chemistry (1866) *The Carnotian Model*

This wider potentiality of the four MoSTs was effective at the philosophical level since the exploration by theoretical physicists of the subjective representation of the theories led them study a physical system by referring their ideas to a typical entity, i.e. as a straightforward application o of the theory. Mathematics first suggested this kind of ontology inasmuch as it represented space as a set of the typical entity, a point; Descartes and then Newton added to any point three coordinates and a mass. The ontology of mass-point was very influential along two centuries. It met an alternative when thermodynamics suggested the notion of a system, which disregards its internal composition. However, owing to the lack of a support by an advanced mathematics, this alternative was not so effective as was then the electromagnetism' one; now, to any point is added rather than a mass, a field extending on the whole space. The change was emphasized by a very wide, philosophical debate on this subject. However, such a notion of the electromagnetic field gave theoretical dignity to the abstract notion of wave, previously known as the particular mechanical phenomenon of material wave. Next, the wave played the role of an alternative ontology to mass-point. As a consequence, few decades before the discovery of quanta, the mass-point ontology proliferated in four distinct ontologies, each surrogating in subjective terms the basic choices of the corresponding MoST. In particular, let us remark that what between two MoSTs is an incommensurability phenomenon originating from a difference in their objective choices, between two ontological models - inasmuch as they are no more than philosophical notions and intuitive images - it becomes a not well-defined phenomenon; it is called irreducibility or complementarity of the two corresponding ontologies, by referring to the radical variation in meaning that any notion may suffer when it is considered in two different MoSTs.

5. From the physico-philosophical debate to a definite problem of theoretical physics: The lack of an anti-Newtonian wave theory

All that give reason for the great emphasis received by two discoveries of QM. Photons introduced a surprising ontology, i.e. massless quanta; then, is born the dualistic ontology of a wave-particle for a quantum system. Really, any quantum system is also a system in the thermodynamic sense, i.e. a global system disregarding its internal constituents but for the resulting global magnitudes - for instance: the entropy of a quantum system. Furthermore, any quantum system is a statistical system too, since the measurement operations require an unlimited set of equivalent systems to be experimented. As a consequence, we recognize that in QM the true ontological conflict concerns not two models only - as for synthesis' sake physicists assumed - but all four models we listed in the above.

Moreover, the problem of four ontologies is a philosophical problem in nature. Hence, it does not pertain to theoretical physics, except for it represents in both a reductive and blurry way the very foundations of a scientific theory, i.e. the four MoSTs. Rather, both quanta and wave-particle complementarity has to be considered as an obsolescence of the ontologies inasmuch as evidence for the basic choices ever more appeared. No surprise if in the 10's and in the 20's most physicists despaired to overcome such a radical philosophical crisis.

However, the arguments of the previous section leave open a question; why quantum physicists emphasized so much the wave ontology among the remaining three ontologies,   till to hope to solve the crisis by a mere improvement of the theory of this basic notion? Surely, owing to the strong influence of the ontology of mass-point in theoretical physics, the ontology of the wave was for a long time disregarded. A further explanation concerns theoretical physics as a whole.

Let us inspect the history of wave theories. Here we found out a surprising fact. "Strongly enough, in the standard histories... the history of acoustic has been largely a neglected subject"[15], wrote Lindsay in a cursory review of this history. When one inspects history of light one sees an almost similar neglect. For ex., the great relevance of Huygens' theory of wave is mainly a historical artifact; Huygens never calculated the wave fronts. Instead, Newton's *Optiks* was capable of performing the first reductive operation in the development of physics; it included Descartes' geometrical optics - born just some decades before it - in the mechanical theory, since the nature of light was explained by luminous mass-points. This mechanical optics was so influential to cancel any alternative - although its mathematical formalism, i.e. differential equations, was unpracticable when applied beyond elementary phenomena. After the experimental evidences for a wave optics, the new theory was received by the academic world inasmuch as Fresnel was able to apply to waves the mathematical tools pertaining to the Newtonian tradition, i.e. differential equations. But just next, Maxwell's electromagnetism subsumed the whole field in a theory of a higher level. When one adds that acoustic theory was considered as subsumed by Newtonian mechanics since D'Alembert - and then Lagrange - solved the string problem, one is forced to conclude that in its entire history never, except for very short periods, wave theory gained relevance in classical theoretical physics. Moreover, a wave theory of the same theoretical relevance as thermodynamics, electromagnetism or statistical mechanics never existed; for ex., wave theory was unable to give relevance to Bernoulli's suggestion for the superposition principle; often it advanced incidentally, by means of improvements coming from different theories; for, ex., Fourier's series came from heat theory, transversal waves from electromagnetic theory, etc.. We may conclude that when physicists started modern physics they lacked a developed theory of waves, if not as a subordinate theory of Newtonian theories. As a consequence, the ontology of a wave still appeared as largely undeveloped. So, this quick history gives further reason for the emphasis on the wave theory by 20th Century physicists.

Actually, the birth itself of the theory in QM supported this emphasis. The novelty of Planck's theory is fully understood only when we consider the history of the foundations of the wave theory. The mechanistic influence led physicists to think - through continuum mathematics - of a wave as a phenomenon referring mainly to the whole space and time, both notions conceived as the absolute, basic notions of even wave theory[16]. Yet, the wave travelling in a void space is incompatible with physical measurements, since the case of the interaction of waves with matter only allows us to give experimental evidence of waves. Unfortunately, till Planck's time the study of wave interactions has been confined to few, singular cases (interference, dispersion, absorption), while resonance phenomena were considered in a very bounded way. The high abstractness of this traditional approach is shown by the following fact too. In order to explain the general framework of the interaction of waves with matter, the physicists guessed the existence of a new substance, ether. However, anyone knows that this kind of interaction wave light - ether led to one of the most impressive failure of past theoretical physics. Thus, the novelty of the black-body theory consists also in representing a study of a strong interaction waves-matter. In addition, let us note that in the same years, one more phenomenon of strong interaction wave-matter, i.e. spectroscopy produced an extraordinary result; it gave experimental confirmation to a whole theory, chemistry, built by one century efforts on merely inductive arguments. Moreover, let us remark that the study-case of the black-body constituted the paradigmatic case of this kind of interaction since it suggested an universal formula. Under this light, Planck's theory starts a new theory of waves as a theory of their strong interactions with matter. This essential novelty is confirmed by the revolutionary result, i.e. to dismiss differential equations for relying on finite mathematics (PI) only.

A revolution even result by arguing in terms of the options of Table 1. Except for Huygens' theory[17], wave theory always was conceived through the Newtonian choices AI and AO[18]. Planck's theory was rightly seen as a revolution inasmuch as it did at last two decisive moves; it introduced PI (quanta) in the study of light and moreover by missing axiom-principles it followed a PO theory, i.e. it puts a key, universal problem - just the interaction wave-matter. By referring to Table 1, one may state that truely Planck's theory started an anti-Newtonian theory.

In § 2 I suggested that a PI & PO theory has to be formulated by means of symmetries[19]. Unfortunately, in Planck's time this technique was almost ignored, owing to the long delay in the reception by theoretical physicists of group theory. Then, no surprise if twenty years after Planck's time, Schrödinger introduced a wave theory by restoring the old formalism of the classical continuum and moreover by putting Ψ as an a priori principle, i.e. by coming back to an AI & AO theory. I conclude that the resulting QM represents a merely first approximation w.r. to a formulation of the same field of phenomena by means of a theory of waves strongly interacting with matter.

Even the subsequent development of quantum mechanics required to overcome Schrödinger's formalism by introducing the "pest of physics", i.e. group theory. Thereafter, S-matrix, Feynmann diagrams, supersymmetry introduced radically new theories with respect to the old one.

The Schrödinger's coming back to an AO theory is emphasized by the well-known paradox of Schrödinger's cat. This paradox reveals its nature when the cat is substituted by the observer himself. Then this one says: "I will be either dead or alive"; which is the same as "I will be either not alive or alive"; in its turn, according to formal logic this sentence is equivalent to "If I will be alive, then I will be alive"; that constitutes the well-known Lapalisse's paradox. Whose nature is to mistake a problem - to know when life will be end - with what is suggested by the dominant attitude which sees a science as a deductive system (AO) only, i.e. an axiom from which to draw implications. As a general rule, any problem which is changed into an axiom generates obviously paradoxical deductions. This point is evidentiated at best in mathematical logic when we take as an example a DNS representing the crucial of a theory; then its change to an axiom is performed by forcing it to an affirmative sentence -, i.e. $\neg\neg A \to A$ - possibly by means of a peculiar word which cancel the negations from our attention; e.g. "die" instead of " not alive" ; just what does not hold true in both non-classical logic and in a PO theory - a contradiction. Clearly, in Schrödinger paradox the DNS is: "It is not absurd that the cat is either dead or alive", a trivial sentence; the paradox originates from its incorrect change to an affirmative sentence, that cancels the problematic nature of the theory and hence the uncertainty principle; the resulting mixture of classical and quantical aspects obviously gives the paradox.

6. From the physico-philosophical debate to a definite mathematical problem: Proof of the incompleteness of quantum mechanics

Now let us study QM from a formal IP viewpoint, i.e. by means of constructive mathematics. A strong opposition to QM came from Einstein. His celebrated paradox questions the completeness of quantum mechanical description of the reality[20]. After 1935 a colossal, inconclusive debate took place on this subject. Here, I will show that this philosophical problem represents actually a mathematical problem, provided that - as it is suggested in § 1 - one realizes that in QM we have beforehand to choose a kind of mathematics relying on PI only.

When a theory makes use of AI, no bound exists to its mathematical tools; rightly, Hilbert stated that "Ignorabimus" does not exist in this kind of mathematics. Being AI mathematics unable to decide the incompleteness of a physical theory, physicist were forced to argue on this problem by means of a physical, operationistic viewpoint only.but intermixing subjective notions as determinism, causality, locality, all notions which may include an appeal to AI. Instead, in a PI mathematics - whose tools are a priori bounded and well-known - the completeness problem of a theory is stated in a clear-cut way: do the bounded mathematical tools of the theory represent all physical situations? In particular, do the solutions of SE include all physical situations dealt with by QM? Or, in constructive terms: when SE is solvable by a general algorithm? Or, more precisely, are there physical situations which cannot be obtained by the general algorithm of solution of a differential equation like to SE?

There is no unique definition of a mathematics relying on PI, though all are called constructive mathematics (CoM). The most interesting one is Bishop's CoM. It was proved that counterparts of classical weak solutions - which may be physically relevant - cannot be obtained[21]. One can improve Bishop's CoM by adding Markoff's principle; Aberth's CoM results. In the latter one a general algorithm obtains all the strong solutions and in addition a subset of weak solutions too; however, not all weak solutions are obtained again[22]; in other terms, weak solutions always give rise to undecidable problems.

Rather than considering such a result as an evidence for the unadequacy of CoM to meet all needs of theoretical physics[23], I suggest that in a CoM the undecidable problems stress Platonist claims that incorrectly past theoretical physics presented as theoretical truths. In particular, the undecidable problems given by weak solutions stress that a differential equation - since it is unable to represent all relevant physical situations - cannot be considered as an universal axiom for the whole theory - that amounts to state the incompleteness of QM when it is based on the SE meant as an axiom in an AO theory[24]. Indeed, let us note that EPR paradox deals with physical situations including discontinuities; it involves a weak solution of wave equation. Thus, Einstein suggested a mathematical problem in a misleading way, as if it were a physical problem only.

In general, any AO theory whose axioms includes a differential equation - like Euler's mechanics, Lagrange's mechanics, Hamilton's mechanics - cannot be a complete theory because any differential equations is not solvable by a general algorithm for whatsoever physical situation. Let us remark that in classical theories this phenomenon was already considered, although in intuitive terms only. Since D'Alembert's time, many scientists distrusted on "general systems" like Newtonian mechanics. By rejecting its main notion, force-cause, they developed rather a mechanics of the shock of the bodies[25] - actually, a physical situation involving weak solutions of differential equations. Unfortunately, in past times the blind trust in infinitesimals (actually, AI) covered the problem of the incompleteness of classical theories; only this Century's wide inquiry - motivated by QM - on the foundations of physics led to stress such a phenomenon -, unfortunately not in the correct way.

Conclusions

According to the content of § 1, since the origin of QM - and special relativity too - theoretical physics included contrasting and fading ontologies, opposite choices, incommensurability phenomena. The lot of improvements of the theory, although resulting as satisfying from a merely technical viewpoint yet did not offered adequate solutions to the new fundamental problems.

M. Jammer, at the end of a relevant paper, wrote: "...Aristotelian and Newtonian physics... produced each a coherent and intellectually satisfying world picture, modern physics has not been able to do so. This is the reason... the physicist finds himself... lost in a universe which he does not comprehend"[26].

That agrees with the main result of this paper; QM represents an irreversible step out of the monopoly of theoretical physics by the Newtonian MoST; however, QM represents at the same time an inconclusive step inasmuch as its dualism qualifies it as an unachieved theory; the first step for achieving a more consistent theory is to look for a new formulation, where to recognize the crucial role of non-classical logic in QM by starting from the typical DNS of the uncertainty principle.

Notes and bibliography

[1] Bishop, E. (1967). *Foundations of Constructive Mathematics*, McGraw-Hill, New York.

[2] Prawitz, D. (1977). "Meaning and Proof", *Theoria*, 3 pp. 6--39.

[3] The potentialities of this interpretative scheme have been successfully explored in mathematical logic ("Is Gödel's incompleteness theorem a consequence of the two kinds of the organization of a scientific theory?", in Wolkowski, Z.W. (ed.) (1993), *First International Symposium on Gödel's Theorems*, World Scientific, London, pp. 107--135); in philosophy ("Leibniz' <Scientia Generalis> reinterpreted and accomplished by means of modern scientific theories", in Cellucci, C. (ed.) (1993), *Convegno SILFS*, (Lucca), in press) and in social sciences ("A paradigm-shift in conflict resolution. An approach suggested by history of science" submitted to *Gandhi Marg*). For an useful comparison of the present approach with recent, narrower approaches to the interpretation of QM, see Cartwright, N. (1990), "The Born-Einstein debate: where application and explanation separate", *Synthese*, **81** pp. 271--282; Combourieu, M.C and Rauch, H. (1992), "The wave-particle dualism in 1992: A summary", *Found. Phys.*, **22** pp. 1403--1434.

[4] This section synthestizes and improves the results presented in my paper: "Alle origini della meccanica quantistica: le sue opzioni fondamentali", in Cattaneo, G. and Rossi A. (eds.) (1991), *I fondamenti della meccanica quantistica. Analisi storica e problemi aperti*, Editel, Commenda di Rende, pp. 59--79.

[5] For ex., Maynard, L. (1984), *The Enigma of Probability and Physics*, Reidel, Boston; Dalla Chiara, M.I (1983), "Problemi logici della meccanica quantistica", *Sapere*, dic. pp. 42--46.

[6] Barut, A.O. (1986), "Dynamics and Symmetry: Two distinct methodologies from Kepler to supersymmetry" in Gruber, B. and Lenczewski, L. (eds.), *Symmetry in Science, II*, Plenum P., New York, pp. 37--50. I qualified Barut's claim by linking symmetry technique to the PI & PO choices: "Alternative mathematics and alternative physics: the method for linking them together", submitted to *Epistemologia*.

[7] Drago, A. and Pirolo, A., *this volume*.

[8] Van Frassen, B. (1974). "The Labyrinth of Quantum Logic", in Cohen, R.S. and Wartofsky, M.W. (eds.), *Logical and Epistemological Studies in Contemporary Physics*, BSPS XIII, Reidel , Dordrecht, pp. 196--209.

[9] Birkhoff, G. and von Neumann, J. (1936). "The Logic of Quantum Mechanics", *Ann. Math.*, 37 pp. 23--24.

[10] See for ex. Fine, A. (1972), "Some conceptual problems of Quantum Theory" in Colodny, R.G. (ed.), *Paradigms and Paradoxes*, U. Pittsbourgh P., pp. 3--31; Mittelstaedt, P. (1978), *Quantum Logic*, Reidel; Garden, R.W. (1984), *Modern Logic and Quantum Mechanics*, Hillger, Bristol.

[11] Drago, A. (1991). "Incommensurable scientific theories. The rejection of the double negation logical law", in Costantini, D. and Galavotti M.G. (eds): *Nuovi problemi della logica e della filosofia della scienza*, CLUEB, Bologna, vol. 1 pp. 195--202.

[12] For ex, Mehra, J. and Rechenberg, H. (1982), *The Historical development of Quantum Mechanics*, Springer, ch. 2.V and vol. 4.

[13] Drago, A. (1988). "A characterization of Newtonian paradigm", in Scheurer, P.B. and Debrock , G. (eds.): *Newton's scientific and Philosophical Legacy*, Kluwer Ac. P., pp. 239--252.

[14] One may add that even thermodynamics is composed by an AI and AO part (the first principle) and a PI and PO part (S. Carnot's theory). theory. Some authors already remarked that the first principle does not agree with the second one. Moreover, Rosenfeld, L. (1953), "Strife about Complementarity", *Science Progress*, July n°163, pp. 393--398, suggested in classical thermodynamics just the same complementarity as in QM; a rebuttal is the paper of the same title by Bunge M. (1955), *Br. J. Phil. Sci.*, 6 pp. 1--12, 141--154. The ambiguity of classical thermodynamics will appear also in Table 2, where - being meant as a Ao&PI theory - it is associated to theontology of "wave" and not to that of "system", as instead S. Carnot' thermodynamics correctly is.

[15] Linsday, R.B. (1945). "Historical Introduction" to Strutt, J.W., Baron Rayleigh, *The Theory of Sound*, Dover, New York, V-XXXII, p. XI.

[16] Beyond previous reference, see Kassler, J.C. (1980), "The <Science> of music to 1830", *Arch. Int. Hist. Sci.*, 30 pp. 111--136, exp. p. 131.

[17] Let us remark that in C. Huygens' book one finds out DSNs also: for ex. "L'on *ne* sauroit douter que la lumière *ne* consiste dans le mouvement d'une certaine matière"; *Traité de la lumière*, Leiden, 1690, p.2. Moreover, he opposed infinitesimals, as notions based on the actual infinity.

[18] Shapiro, R. (1984), in " Mathematics and experiment in Newton's theory of colour", *Phys. Today*, **37**, Sept., pp. 34--42, stressed that Newton's book is incomplete since he was unable to organize it as an AO theory.

[19] One can suggest some recent theories as suitable candidates for covering this role in black-body theory. For ex., the theory of crystals including phonons; or a acoustic black-body theorized as a set of at most 3N resonant dipoles. Let us remark that the position expressed in the above about the ontological models is close to Heidenberg's, Bohr's and Franck's views inasmuch as it denies a strong link between an ontological model and reality; but it is a realist one not only since it scrutinizes reality by models which are extracted from the most realistic theories of reality, but since it supports the "realistic" choices, IP and OP w.r. to the Platonist choices, AI and AO.

[20] Einstein, A., Podolsky, B. and Rosen, N. (1935). "Can Quantum mechanical description of physical reality be considered complete?, *Phys. Rev.*, **47** pp. 777--780.

[21] Bishop, E. and Bridges, D.S. (1985), *Constructive Analysis*, Springer, Berlin; Bridges, D.S. (1991), review of the book in the following note: *Bull. Am. Math. Soc.*, **24** 216-228, pp. 226--227.

[22] Pour-El, B. and Richards, I. (1989). *Computibility in Analysis and Physics*, Springer, Berlin.

[23] See the above reference and in addition: Hellmann, G. (1993), "Constructive Mathematics and Quantum Mechanics: Unbounded Operators and the Spectral Theorem", *J. Phil. Logic*, **22** pp. 221--248.

[24] Drago, A. (1982), "Carathéodory's Thermodynamics and Constructive Mathematics", *Lett. Nuovo Cim.*, **34** pp. 52--56; "Relevance of Constructive Mathematics to Theoretical Physics", in Agazzi, E. *et. al.* (eds.), (1986), *Logica e filosofia della scienza oggi*, CLUEB, Bologna, vol. 2, 267-272 (abstract in *J. Symb. Logic*, **52** (1987), p. 316; *J. Symb. Logic*, **58** (1993), pp. 1139-40; "Constructive mathematics and theoretical physics; From wave equation to a general method", *J. Symb. Logic*, **59** (1994) in press.

[25] Hankins, T.L. (1970). *Jean D'Alembert, Science and Enlightment*, Clarendon, Oxford, pp. 107--110.

[26] Jammer, M. (1979). "A consideration of the philosophical implications of the new physics" in Radnitzky, G. and Anderssen, G. (eds.), *The Structure and Development of Science*, Reidel, 1979, 41-61, p. 59.

QUANTUM MECHANICS REFORMULATED BY MEANS OF SYMMETRIES

A. DRAGO and A. PIROLO
Group of the History of Physics - Dep. of Physical Sciences
University of Naples

Abstract. Here we present a formulation of Quantum Mechanics that is founded on the fundamental choices - only potential infinity and problematic organization - that - according to previous results by A. Drago - are characteristic of the alternative theories to the Newtonian one; hence on the symmetries and without differential equations. In this new formulation, the fundamental problem of the theory is recognized in Heisenberg's uncertainty relations. We state the "Uncertainty Principle" by a double negated sentence that is not the same as an affirmative one, that is to say a characteristic sentence of non-classical logic which in this way is introduced from the beginning of the theory. Then, we state the commutation relations by means of the classical symmetries, that essentially follow Jordan's new version of Heisenberg's formulation.

1. A working-plan

According to the point of view of one of us (A.D.) every physical theory is characterized by two basic options(1), that are fundamental for the construction of the whole theory.

The first option concerns the organization of the theory: a theory can be organized by a deductive scheme, that is to say it is based on principles-axioms from which to draw, by the deductive method, the laws of that theory; this kind of organization is called Aristotelic Organization (AO), from the philosopher that theorized it as a necessary condition for a theory in order to belong to science. On the other hand, a theory can be based on a central problem whose resolutive method constitutes that theory; this kind of organization (2) is called Problematic Organization (PO).

C. Garola and A. Rossi (eds.), The Foundations of Quantum Mechanics, 229-237.
© 1995 *Kluwer Academic Publishers.*

The second option concerns the kind of relationship between Physics and Mathematics, that carries out the choice on the kind of infinity; whether <u>actual infinity</u> (AI) (that is to say, the infinity represents a new point on the line of the reals), or <u>potential infinity</u> (PI) (i.e. the possibility to go further than any given bound).

These two fundamental options work as parameters capable to characterize the theories of classical physics; for example, both L. Carnot's mechanics and S. Carnot's thermodynamics choose PO & PI, while both Newton's mechanics and Maxwell's electro-magnetism choose AO & AI.

What choices by Quantum Mechanics? A previous study(2) suggested an ambiguity of Quantum Mechanics w. r. to the two above pairs of choices (AI&AO, PI&PO). In particular, we remember an historical episode that certainly is revealing of an ambiguity in the foundations of Quantum Mechanics; for about a year the equation of Schroedinger was contrasted by the formulation of Heisenberg, that being deliberately operational in nature employed matrices and algebraical calculations (that is to say PI) instead of differential equations ; moreover, it put the uncertainty as the crucial problem of the theory. These features constitutes evidence for the choices PI & PO. The subsequent unification of the two theories is still under criticism(3). In fact, some scholars(4) assert the Quantum Mechanics is made up of two stumps -the one of Schroedinger equation and the other one of the theory of the measurement - that we cannot bring about to a unitarian theory.

From the point of view of the foundational option, the way for getting over the incongruities of Quantum Mechanics is to reformulate it by rigorously following a couple of choices. In fact, the dominant tradition in theoretical physics takes the AI & AO choices, which are also the choices of the Hilbert Space by giving a priori all possible functions-solutions (AI) and by putting as first principle the equation of Schroedinger (AO). Then the opposite PI & PO choices appear not so much explored; in fact, the formulation of Heisenberg has no more been developed; that would not have been easy, because an efficient mathematics that uses the PI only has been produced not before 1967(5). In the following we will formulate a Quantum Mechanics according to the choices PO & PI.

To obtain this new formulation we have to avoid differential equations, because they are historically and logically tied to the choices AO & AI. Instead, all PO&PI theories make use of the symmetries in place of the differential equations(6).

At glance we see that this program is more general and foundational than the well-known programs of Einstein, de Broglie, Bohm etc.., i.e. to reformulate Quantum Mechanics by means of some hidden variables. Our new variables are well known (organization, infinity, symmetries), but they are so foundational that they require a deeper investigation than to find some new equations of the motion or some new physical mechanism for conciliating the two concepts, the wave and the particle.

Howeever our task is difficult because the reference theory, the formulation of Heisenberg, hasn't been conceived by him by considering the two fundamental choices PO & PI, and above all he did not introduced the symmetries. On the other hand, if we look in the past history of the physics for taking example from PO & PI theories with symmetries, no one of them helps us: Carnot's mechanics is PO & PI with symmetries, but these ones depend on a 2° fundamental equation inferred in an incomprehensible way(7); S. Carnot's thermodynamics, too, is a PI & PO theory, but - beyond the fact to be based in an essential way upon the out-of-date theory of the caloric - by materializing the symmetries with the cycle of Carnot it lacks a formal technique representing them(8).

In alternative, we can formulate directly Quantum Mechanics as based on PI & PO without referring to previous models, since the historical novelty of Quantum Mechanics may free us from the characteristic prejudices of AO&AI theories and hence it may suggest by itself a model for this kind of theory. For that we will follow this second alternative.

2. Matrix Mechanics by T.F. Jordan

In 1985 T.F. Jordan wrote a book on Quantum Mechanics(9) that is close to the formulation we are looking for. He follows that approach of Heisenberg's formulation which, programmatically avoiding the differential equations, represents the measurable quantities of quantum systems by means of matrices. In fact, Jordan by giving a central role to the "Uncertainty Relations" of Heisenberg doesn't start from axioms.

In the first chapter, he recalls a characteristic experimental fact of Quantum Mechanics, pointed out by Heisenberg. Q and P magnitudes, representing respectively position and moment, can be measured simultaneously with a bounded accuracy (the principle of uncertainty). If by measuring Q we obtain an "exact" value, P measurement gives the probability only according to which it is possible to obtain one of the "exact" values of P. After some chapters on technicalities, he dedicates the third chapter to the study of the matrices in . If the physical quantities Q and P are represented by matrices then it is clear that QP can be not the same as PQ.

Jordan points a great newness out of Quantum Mechanics: "Quantum Mechanics makes a distinction beetwen a physical quantity and its values"(10). That is to say, the mathematical representation of a physical quantity is not only the range of values it can assume, but it is another mathematical object too.

He starts by illustrating the meaning of what he calls the "strange equation":

$$QP - PQ = i\hbar;$$

where the square of the difference of these two physical quantities can be a negative number and particularly $-\hbar^2$. The following chapters solve this problem by representing a physical quantity by a matrix. Then all matrices are chosen so to have the same mathematical dimension for a particular system, hence they can be summed up and multiplied between them; in this way the equations that join different quantities are written, in matrices terms, as algebraical equations. Then, Jordan applies these rules to the study of both spin and intrinsic magnetic moment and it is easy for him to obtain the states of a system of two particles with spin 1/2 using only Pauli's matrices with their commutation rules.

In seventeenth chapter, Jordan translates the principle of uncertainty of Heisenberg by means of matrices.

Last chapters of Jordan's book deal with to the symmetries and their role in Quantum Mechanics. The matrices that represent some physical quantities represent at the same time the infinitesimal generators of the symmetry transformations of an isolated physical system. Then the commutation properties of the physical quantities arise from the commutation properties of the transformations. As a conseguence, the matrix representing the infinitesimal generator of a transformation, represents also the physical quantity that remains unchanged under that transformation. This fact points out the fundamental role that symmetries play in Jordan's formulation.

Since the introduction to his book, Jordan declares he wants to give a formulation without the use of the differential equations. Moreover, by using elementary algebra only, Jordan's formulation doesn't need IA. Thus, Jordan chooses IP although he does not declares it. Moreover, Jordan does not organize the theory in an axiomatic way; in fact he considers the not commutativity problem of two quantum variables as the most important newness of Quantum Mechanics. Yet, his formulation does not present the characteristic features of a PO since it does not identify in a clear-cut way the crucial problem of Quantum Mechanics. Even if we suppose that implicitly Jordan considers the "uncertainty problem" as the crucial problem, he does not formulate this problem by means of a double negation statement, that is an essential characteristic of PO choice. Probably, Jordan solves the problem of outletting AO by following a merely "didactic" presentation, as it was suggested by some scholars which supported an alternative to AO in modern science(12).

In 1986 K. Jagannathan by reviewing this book(13) pointed out its weak points as well as its strong points. He holds that the method of Jordan's formulation constitutes a new method to approach Quantum Mechanics since it shows that the most relevant novelties of this theory can be represented by a very simple mathematics.

Jagannathan thinks Jordan's formulation is alternative to those based on Schroedinger's equation or on operators in Hilbert's spaces; it is alternative on the other hand to those based on Feynman's path integrals. But he remarks that Jordan's method for generalizing the results from the study of the spin to more general cases - for ex., to conyinuous spectrum operators - is not a rigorous one. By Jagannathan, some relations "appear casually and miraculously"(14). This weakness is attributed to the fact Jordan has written a book for laymen and thus he is obliged to neglect some important mathematical passages.

To overcome these problems, Jagannathan suggests to Jordan "to write a second book requiring to the reader some familiarity with Quantum Mechanics; (...) in such way he can show how this approach is organized and how it joins and completes the usual presentation"(15).

3. New formulation of Quantum mechanics by PO & PI

Jordan's book will be the starting point to reformulate Quantum Mechanics in a clearly problematical way without applying AI in the mathematics; the symmetries are already introduced by its formulation.

In the historical period coming before the birth of Quantum Mechanics a great discussion led physicists to recognize the problem of the simulataneous measurement of two physical quantities. The result was summarized in the so-called "Uncertainty Principle". It does not represent an axiom for a deductive system, rather a methodological principle on how to do simultaneous measurements. Then, in the new formulation we will consider the "Uncertainty Principle" as the crucial problem around which Quantum Mechanics has to be developed. It can be formulated in the following way, by means of a double negated sentence that is not the same as an affirmative one - the typical way of the beginning of a PO: "It is not true that two physical quantities are not measurable at the same time".

By removing the two double negations we obtain: "Two physical quantities are measurable at the same time", what in general is obviously false. On the other hand, we cannot uphold the negated sentence: "Two physical quantities are not measurable at same time", because these measurements are possible, provided that we allow the product on the uncertainties of these measurements get a minimum which is different from zero. Then, the formulation of the "Uncertainty Principle" by means of a double negated sentence represents a logical problem (because - -A is not the same as A) that is at the same time a physical problem.

This feature - a double negated sentence that is not the same as the affirmative sentence - if put as the starting point of Quantum Mechanics, shows that the logic of this theory is different from the classical logic - a fact that some scholars found out shortly afterwards the birth of the same theory(16) - since just the law -- A ≠ A holds true in non-classical logic.

According to Jordan, the Uncertainty Principle can be mathematically characterizated by representing the physical quantities by means of matrices in a complex field. So, if A and B are matrices that have real eigenvalues, then the mathematical properties of the matrices give:

$$\sqrt{\langle (A-\langle A \rangle)^2 \rangle}\sqrt{\langle (B-\langle B \rangle)^2 \rangle}=\frac{1}{2}|AB-BA|$$

where the symbol $\langle \; \rangle$ represents the average value and $|\;|$ the module.

We solve the above-mentioned central problem of Quantum Mechanics by identifying the magnitudes which don't commute and the value of their commutator. To do this, we use the symmetry properties of an isolated system and the combining properties of the transformations. Indeed, by using the changing properties of the physical quantities under the following operations: space and time translations, rotations and global changes of speed (Galilei's transformations) and, moreover, by using the combining properties of the above-mentioned operations, we obtain (by the same operations of Jordan's calculus of matrices (17)) the relations between the matrices representing physical magnitudes and the matrices representing the transformations. The commutation properties of the transformations agree with the commutation relations of the physical quantities linked to them, which are the well-known commutation relations between two variables out of position, linear moment and angular moment.

The commutation relations between physical quantities solve the uncertainty problem. In the case of commuting quantities these relations give zero, i.e. these quantities can be measured at the same time without uncertainty; instead, in the case of not commuting quantities they give the minimum value of the product of the uncertainties. Let us note that this result has been obtain by the essential and determinant use of the symmetries, as it has to be in a PO & PI theory.

Now we can apply this method to all problems of Quantum Mechanics: by studying the symmetries properties of the physical system we identify both the commuting quantities and the values of the commutators of the not commuting quantities; using again operations between matrices only, this knowledge leads to quantize the physical magnitudes and to define the possible states of the system. Some example of these calculations are given in Jordan's book(18).

The following flow-chart sums up this method.

(In each square it is shown the author who introduced that content of the same square).

Flow-chart of the new formulation of quantum mechanics founded on the simmetry

FUNDAMENTAL PROBLEM: The measurementsat the same time

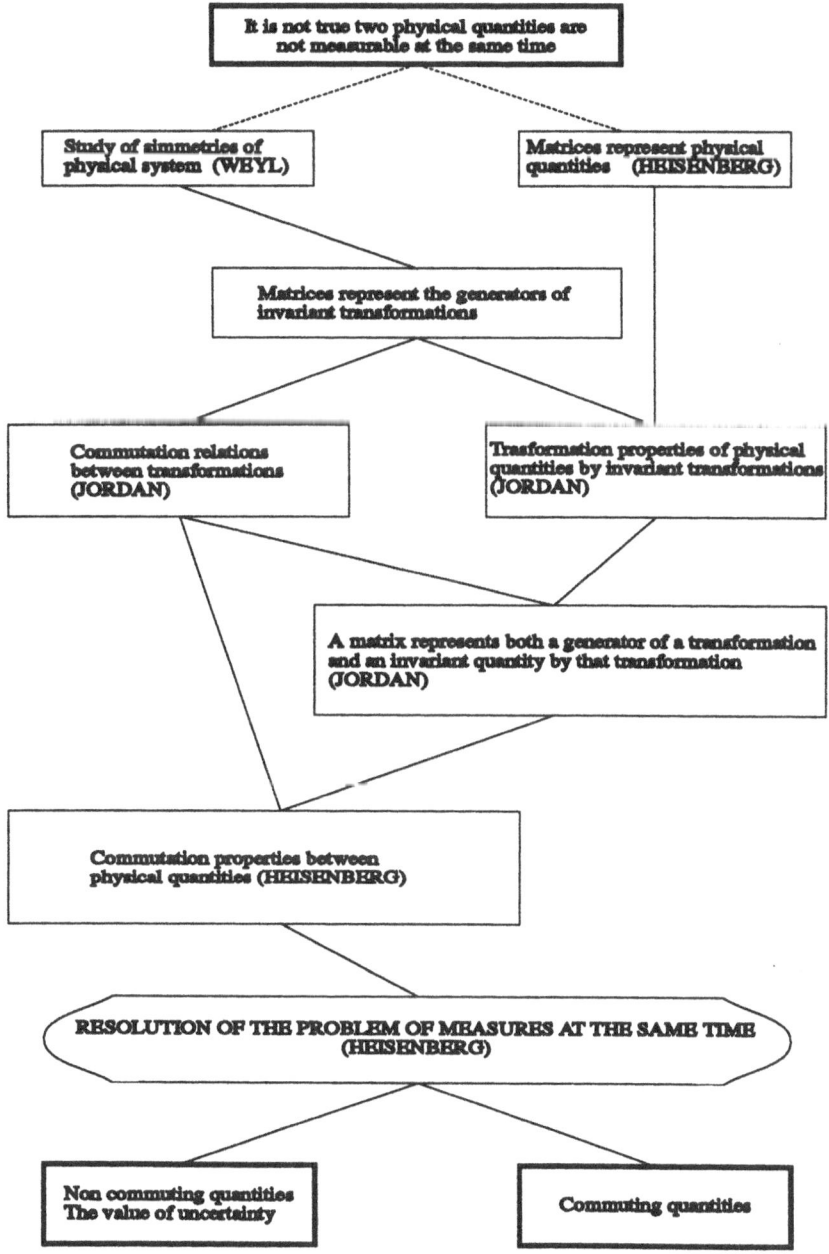

dot lines=methodological choices

full lines=deductions

4. An appraisal of the result

An argument pointing out the division of Quantum Mechanics in two stumps is Von Neuman's paradox on the measurement; the time evolution of the system is given by Schroedinger's equation, which, however, cannot represent in any way the quantum jump, or generally the reduction of the wave packet or the reduction of the mixture of states to one state. According to the interpretation of the fundamental options, the paradox is caused by the presence of contrary pairs of choices in the same theory, as we illustrated in the first section.

Paradoxes should not be yet in a formulation that is only PO & PI. Then by which way does the new formulation describe, in no paradoxical way, a measurement on the physical system? Merely by disregarding - just as the classical PO & PI theories do: for example thermodynamics - that description of the time evolution of every phenomenon which results from Schroedinger's equation -; then, there isn't the reduction of the wave packet here, because we don't ask how a system evolves beyond the measurements, just in agreement with Heisenberg's program. Then, we avoid here the typical paradox of Quantum Mechanics according to the Heisenberg indication, although at the cost of a reduction of our interest.

As noticed by Jagannathan when he reviewed Jordan's book, a weakness of this formulation concerns the study of the continuous spectrum operators. Since the use of matrices seems to be bounded to discrete case, in our formulation this problem could be solved giving up the previous intuitive method of attributing mathematics with AI to the differential equations and the mathematics with PI to the discrete, finite matrices. Really, since 1967 a PI mathematics had been formulated(5) which is able to solve differential equations also (but the set of their solutions is more reduced than the classical set). By means of this constructive mathematics, one can study whether the matrices which are required by Jordan's Mechanics are all constructive. If the answer were affirmative, the PI choice would be confirmed. But the solution of this problem is not easy, since the constructive algebra has been developed later than calculus and it presents cumbersome points(19); for that problem it is necessary a further work.

Another problem concerns Planck's constant h. Here, we introduced it in order to agree with the dimensions of the commutation relations. However, both facts that its value is constant w.r. to the values of a quantity and that its value is the same for all quantities are not justified here. Moreover, they are not clarefied by Jordan either, or in other books we know. However, the value of h can be inferred from some experiments, for example black-body radiation.

References

1) A. Drago: *Le due opzioni. Per una storia popolare della scienza*, La Meridiana, Molfetta (BA), 1991. An application of this interpretation to classical mechanics is A. Drago: "A characterization of the Newtonian Paradigm", in G. Debrock, G.B. Scheurer (eds.): *Newton's Philosophical and Scientific Legacy*, Kluwer Acad. P., 1988 239-252; to thermodynamics: A. Drago: "The alternative Content of Thermodynamics: The Constructive Mathematics and the Problematic Organization of the Theory" in K. Martinas et alii (eds.): *Thermodynamics. Facts, Trends, Debates*, World Scientific, Singapore, 1991, 329-344.

2) A. Drago: "All'origine della meccanica quantistica. Le sue opzioni fondamentali" in G. Cattaneo, A. Rossi (eds.): *I fondamenti della meccanica quantistica*, Editel, Cosenza, 1991, 59-79.

3) N.R. Hanson: "Are Wave and Matrix Mechanics Equivalent Theories?", in H. Feigl, S.Maxwell (eds.): *Current Issues in the Philosophy of Science*, Halt, New York, 1961, 401-424

4) L. Mayants: *The Enigma of Probability and Physics*, Reidel, Boston, 1984. M.L. Dalla Chiara, G. Toraldo di Francia: *La formalizzazione delle teorie fisiche*, Boringhieri, Torino, 1992. Dirac, too, asserts that the Heisenberg and Scroedinger pictures are not equivalent; in "Foundation of quantum mechanics", *Nature*, 4941 (1964), p. 115-116.

5) E. Bishop: *Foundations of Constructive Mathematics*, Mc Graw-Hill, New York,1967; O. Aberth: *Computable Analysis*, Mc Graw-Hill, New York, 1980.

6) A. Pirolo: "Analisi storico-critica delle simmetrie nella fisica teorica"; Tesi di laurea in Fisica, Univ. Napoli, a.a. 1992-93.

7) L. Carnot: *Essai sur les machines en général*, Defay, Dijon, 1782.

8) S. Carnot: *Réflexions sur la puissance motrice du feu*, Bachelier, Paris 1824, (Blanchard 1978). Probably, chemistry too substantiates symmetry by means of the periodic table of elements; however, that does not suggest any formalism. Moreover its PO is based on the problem of the existence of the indivisible of matter; but at present, after the experimental evidence for atoms and molecules, this is no more a problem.

9) T.F. Jordan: *Quantum Mechanics in Simple Matrix Form*, John Wiley & Sons, New York, 1985.

10) T.F. Jordan: op. cit. pag. 3.

11) T.F. Jordan: op. cit. pag. 187.

12) I. Lakatos: *Mathematics, Science and Epistemology*. Philosophical Papers, vol 2, pt. 1.2, Cambridge U.P., 1978.

13) K. Jagannathan: Review to of T.F.Jordan op. cit., *Am.J.Phys.* 54, (1986), p.1154.

14) K. Jagannathan: op. cit. p.1154, II column.

15) K. Jagannathan: op. cit. p.1155, II column.

16) G. Birkhoff, J. von Neumann: "The Logic of Quantum Mechanics", *Ann. Sci*, 37 (1936) 23-43.

17) T.F. Jordan: op. cit., cap. 23-27.

18) T.F. Jordan op. cit. cap. 18-22.

19) F. Richman, R. Mines: *Constructive Algebra*, Springer, Berlin, 1988.

A PRIORI SCHEMES IN QUANTUM MECHANICS

V. FANO
Ist. di Filosofia - Università di Urbino
Via Saffi 9 - 61029 - Urbino, Italy

Abstract. The present work explores the possibility that the standard formulation of quantum mechanics might be viewed within a gnoseological perspective of Kantian type. In particular it seems possible to maintain that the theory of knowledge described in the second edition of the *Critique of Pure Reason* is not essentially jeopardised by quantum theory, but for what concerns the structure of the *a priori* defined by Kant. On a number of grounds, however, one might argue, albeit not fully conclusively, that even the schemes chosen by Kant are justified by quantum physics.

1. Introduction

According to certain physicists, for instance Werner Heisenberg, quantum mechanics, by questioning causality and space-time continuity, brings to completion the break up of the Kant concept of *a priori*, which had originated in non-Euclidean geometry and in relativity.[1]

As is well known, Kant's thought is multifaceted and fluctuating, so it is not easy to decide whether his transcendental philosophy would have been able to encompass the conceptual novelties of quantum mechanics, or whether indeed the latter theory causes its refutation once for all.

As reference point we may consider the second edition of Kant's *Critique of Pure Reason,* in which, concerning the issue considered, a more critical perspective than the one contained in *Metaphysische Anfangsgründe der Naturwissenschaft,* 1786, is presented.[2] As far as the *Opus postumum* is concerned, it must be said that, notwithstanding the clarifying effort of Vittorio Mathieu,[3] who showed that it represents a step forward in the critical conception of physical knowledge, this work, being unfortunately unaccomplished, does not allow us to draw any conclusion from it.

C. Garola and A. Rossi (eds.), The Foundations of Quantum Mechanics, 239-254.
© 1995 *Kluwer Academic Publishers.*

Quantum mechanics on the other hand is not unequivocally defined either. Indeed there are different interpretations of the mathematical formalism of the theory, as well as different theories of measurement which provide its connection with experience. We are going to take into account, for our comparison, only the theory itself in its merely mathematical formulation, without further interpretation, such as Bohr's complementarity, Einstein's realism, or Heisenberg's potentia concept. We find this procedure viable, since the mathematical structure of the theory, no matter how many the different possible formulations, possesses, after von Neumann's *Mathematische Grundlagen der Quantenmechanik*,[4] a fairly consolidated core.

A more complicated situation prevails in the case of measurement, due to the multiplicity of available theories, none of which can be considered as fully accepted. However, to make our task simpler we proceed as if in quantum mechanics there existed a fairly clear-cut connection between the numbers to be introduced in the place of variables in the equations of the theory and those obtainable from experiments, without any need of reduction postulate, observer's consciousness or other physical interpretations of the measurement process, such as for instance the one recently proposed by Ghirardi, Rimini and Weber.[5] This assumption should not give us any problem, since, on one side, Kant doesn't deal with the problem of measurement; on the other, the conflict, if any, between Kant's philosophy and quantum mechanics is rather a problem of principles, than a problem of connection with experience.

2. The gnoseological irrelevance of reality in itself

What we want to do essentially is to focus our attention on selected characteristics of the theory of knowledge of the second *Critique*, which are amenable to a comparison with the conceptual structure of quantum mechanics.

First it seems that we can say the following: within Kant's perspective

A) reality as such does not play any role in the process of knowledge.

As is well known, any object can be considered under two view points: as thing in itself - as possible object of sensation -, or as appearance - as thing embedded in our receptive faculty -. The thing in itself, that is the object independent of our perception, does not have anything to do with knowledge. In other words all the knowing process is based on the appearance, which is moulded by the forms of *a priori* pure intuition, namely space and time. Thus, no experience which leaves aside the space-time framework of our intuition is possible.

The single structural characteristic of space and time bearing on our problem, is the infinite divisibility or continuity of both, explicitly stated by

Kant.[6] This is indeed the single property of space and time which quantum mechanics might question.

At any rate, the issue we want to underline in this first part of the theory of knowledge according to Kant is what we have called thesis A, namely the total gnoseological irrelevance of reality in itself.

3. The *a priori* schematism

Space and time as forms of intuition are essential to frame the manifold of sensibility; however physics is made up by structures a lot more complex than the purely space-time one. This science indeed is developed through a series of multifarious forms of judgements. Kant believes to detect twelve of them, which should cover exhaustively the field of possible judgements. Judgements are the product of the understanding; the same way as sensible intuitions were moulded by the pure intuition of space and time, all the judgements are moulded by one of the twelve categories which oversee their forms.

Physics is therefore constituted as the application of this possible forms of judgements to the sensible intuition already framed within space and time.

Summarising briefly, Kant, in proposing this solution to the problem of physics, encounters, among others, the following two obstacles: he must in the first place, demonstrate that the twelve forms of judgement are indeed correspondent to the twelve modalities of understanding synthesis. Namely, that the application of the categories of the understanding - quantity, quality, relation and modality - to the sensible manifold corresponds effectively to the twelve formal characteristics of judgement. If this is true, Kant has succeeded in anchoring the twelve forms of the understanding synthesis to the logic, thus ensuring their legitimacy. In the second place, Kant must show how it is possible to apply the forms of understanding synthesis to the sensible manifold; in other words, he must find a mediation between the categories of understanding and the space-time intuition. According to Kant, indeed, a profound gap separates sensibility and understanding: on the one side, the sensibility has an intuitive relationship with its object, although intermediated by pure intuition; on the other, the understanding cannot perceive its objects by intuition, so there is no such thing as an intuitive understanding juxtaposed to sensible intuition. In other words the understanding proceeds in a purely formal way, without any contribution whatsoever from intuition. One may also say that the understanding *thinks*, without being able to refer its thoughts to an object, since it cannot *know*; on the other side, the sensibility perceives by *intuition* without being able to formulate any judgement, therefore it is not any form of knowledge either. It follows that the application of the categories of understanding to the sensible manifold, that is the knowledge proper, is beset with difficulties, since it must connect what is heterogeneous, thought and intuition, *viz.* judgement and sensation.[7]

We must dwell upon these two problems a while more, since they are particularly relevant with respect to quantum mechanics.

The sensible intuition proposes to the understanding a content already framed within space and time. The understanding acts *a priori* with its unifying processes upon this content. Kant ensures us that the possible forms of understanding synthesis are exactly the twelve categories, which correspond to as many forms of judgement. Thus we can say:

B) that the understanding acts completely *a priori* with respect to sensible intuition;

C) The forms of synthesis overseeing the understanding activity are determined in absolutely defined manner and are justified by what Kant calls metaphysical deduction.[8]

We must keep in mind both theses *B)* and *C)*, since they will turn out to be useful in the following.

On the other hand, the problem of the applicability of the understanding categories to the sensible intuition remains unsolved, and to it Kant devotes its celebrated transcendental deduction.[9] An example might give an idea of the large gap opened between the sensible intuition and the judgement. On the one side the sensible intuition supplies us with a series of representations ordered in space and time; on the other, we must define the formal conditions which allow us to formulate judgements such as «two bodies attract one another with a force proportional to the inverse square of the distance and to the masses». How is it possible to bridge the gap between a series of space-time ordered representations and such a complex judgement? The problem here is not so much to establish the truth or falsehood of this judgement, but rather to understand how the forms of understanding synthesis have made it possible, by acting on the space-time manifold.

The answer Kant gives is as simple as it is disquieting. There exists beside understanding and sensibility, a further faculty namely the *productive imagination*, capable of setting up *schemes*, which allow a mediation between sensibility and understanding. While the understanding oversees these schemes, the material is supplied by pure intuition. In particular, the schematism of the *Critique of Pure Reason* is wholly founded on the pure intuition of time. In other terms, the understanding, before coming in contact with the sensible manifold, gets mixed, as it were, with *a priori* intuition, setting up forms which allow the sensibility to be framed, so as to bring about the mediation between sensibility and understanding.

Also in this case an example might help to comprehend this processes.

Let us start with the perception of a plate of circular form. Let's suppose that a moment after, when the plate is not present any more before our senses, we wish to *reproduce* it in our imagination. Such mental act is based on the use of a faculty, which Kant appropriately calls *reproductive imagination*.[10] This type of imagination is not capable of mediating between the categories of understanding and the sensible manifold, since it can only supply mere copies of the latter.

Let's go back to our example. The problem of going from sensible intuition to understanding has to do with the possibility of formulating the judgement «the plate is circular» on the basis of the plate perceived. On one side we have a sensible perception of the plate, or its reproduction in the imagination, on the other the concept of circle, which can be defined as «the ensemble of coplanar points equidistant from a given point». If considered in these terms, the concept and the sensation do not have anything in common, which might provide their mutual connection. On the one hand the border of the plate does not appear as an ensemble of points equidistant from a given point, on the other the concept as such does not have anything circular in the intuitive sense of the term.

In order to provide a connection between sensation and concept we must elicit a further faculty, which allows to *draw* in the pure *a priori* space the locus of points thus defined. This is only possible if the understanding concept is applied to the pure intuition of space. This brings to the *production* - as opposed to reproduction - of a scheme, to the formulation of which the productive imagination has taken part as well. The resulting schemes partake both of the characteristics of understanding and of intuition, insofar as they are the product of the spontaneity of the former applied to the form of receptivity of the latter, by means of the productive imagination. In the final result however, that is in the schemes produced, the imagination does not play any role. It only did contribute indeed, as we will see better in the following, to the formation of the scheme, but without entering in its effective structure. One might also say that the scheme is the procedure by which it is possible to construct the image starting from the concept; in our example the image of the circle from its concept.[11]

In the second edition of the *Critique* the imagination, albeit a crucial element of the process, plays a definitely subordinate role with respect to that of understanding, so that the schemes are unequivocally determined *a priori* no matter what.

We might conclude this partial and concise analysis of Kant's theory of knowledge, summarising the last issue in the following thesis:

D) The application of the categories of understanding to the sensible manifold occurs by means of the schemes, which are set up by the productive imagination - without the latter partaking of their final structure -, and which supply the procedure for the construction of images starting from the concepts.

The latter thesis must be taken into consideration along with the three previously stated:

A) reality as such does not play any role in the process of knowledge;

B) the understanding acts completely *a priori* with respect to sensible intuition;

C) the forms of synthesis overseeing the understanding activity are determined in absolutely defined manner and are justified by what Kant calls metaphysical deduction.

One cannot hold that these four theses can be exactly obtained from Kant's text, which, as is well known, is liable to a number of interpretations; nevertheless, we are going to utilise them for the comparison with quantum mechanics, so that eventually, even if we did not accomplish an analysis of the relationship between Kant's philosophy and quantum theory, at least we will have established to what extent this physical theory can either confirm or falsify this *a priori* perspective.

In the following paragraphs we will discuss one by one the four theses formulated and their relation with quantum physics.

4. Presentationalism and realism

The first thesis is very important, since it marks a distinction between two different gnoseological perspectives. On the one side we have what we could call «presentationalism»[12], according to which our perception are not directed toward the things, but rather toward the things inasmuch as perceived, that is toward the ideas of things. Thus following this perspective a clear distinction prevails between the two worlds, the one of our perceptions, the only one to which we have access, and the one beyond perception, on which nothing can be said. On the other hand, what we can call «realism» holds that we perceive things directly rather than the idea of things. On the basis of this viewpoint the object we perceive and the object in itself, although conceptually two different entities, since the first is embedded in our receptivity, while the second is the object independently of us, are absolutely equal. In other words we perceive things exactly as they are.

If, on the one hand realism ascribes a crucial gnoseological relevance to the thing in itself, on the other presentationalism tends to exclude the latter from the process of knowledge.

One cannot hold that thesis *A* is totally presentationalist; however one can say that it also invalidates the gnoseological role of the thing in itself.

Quantum mechanics, as well as every other physical theory since Galileo, does not approach reality in itself, but rather reality insofar as *measured*. Every physical theory indeed is concerned with numbers, which corresponds to the quantification of certain elements of perception. Now, we attribute to these numbers a certain objectivity, which on the contrary does not pertain to subjective sensations. We might say that in a sense these numbers should represent fairly well something which is substantially independent of ourselves.

The same does not occur in quantum mechanics. The equations of this theory concern exclusively *observations*. Quantum mechanics compel us to reduce its object to the one defined by observations in the moment in which they are observed; *vice versa*, if we do not observe, no guarantee of objectivity is given. That much we can easily infer, for instance, from the standard interpretation of the double slit experiment.[13]

In other terms, under this perspective, the physical theory is nothing but a connection of measurements, to which no objectivity, independent of measurements, corresponds.

This means that, whereas classical physics could yet pretend - perhaps unsoundly - to describe an objectivity independent of measurement, quantum mechanics must of necessity limit its object to measurements. In other words, quantum mechanics is not a theory of things in themselves, since it does not allow any correlation between the object perceived (measured) and the independent one; it is, on the contrary, a theory of the object inasmuch as perceived (measured). It follows that, in this theory, reality independent of its perception does not play any gnoseological role whatsoever.[14]

We might then say that quantum mechanics favours a presentationalist view rather than a realistic view of the process of knowledge, namely it tends to oust the thing in itself from any gnoseological role. It might well be that such an issue was already implicitly present in classical mechanics, but certainly this is made explicit in all its theoretical strength in quantum mechanics.[15]

On the basis of this discussion we can state that quantum theory leans toward confirmation of the thesis we called *A)*.

5. The object does not possess any structure preceding its conceptualisation

Thesis *B)* stresses the fact that understanding and pure intuition supply the whole conceptual frame necessary for the process of knowledge. Indeed, space, time and the twelve categories frame the object within their conceptual grid, while the object in itself does not supply any structural contribution, except its mere indistinct presence. One might say that this second thesis is in a way complementary to the previous one. If, in fact, reality in itself does not play any gnoseological role, then all the burden of conceptualisation must reside *a priori* in the subject.

Similarly to the first one, this thesis marks an important theoretical juxtaposition between those who hold the total determination of the object by the conceptual scheme in which it is embedded, and those who advocate on the contrary the presence of an essential contribution of the object in determining its own conceptualisation. Here again we might speak of presentationalism on one side, of realism on the other.

In classical mechanics there is undoubtedly a certain distance between the image of objects we have in daily life and that implicit in the theory. There is no such thing indeed as material points, or motion without friction, even less isotropic and homogeneous space-time as presupposed by the theory. Nevertheless to some extent the object of classical mechanics seems amenable to the object of first-hand experience. One might still claim that material points, frictionless motions etc. are abstractions and idealisations originated from daily experience. One might then argue that the direct object of experience has supplied elements to the conceptualisation of the theory.[16] Certainly classical mechanics is also so abstract, that it might be envisaged as completely detached from daily experience, so much so indeed that even Kant sees it in this way. To Kant indeed the object, independent of its reception, has no structure contributing to its conceptualisation. Nevertheless, the world of classical mechanics still allows visualisation, so that, perhaps wrongly, we were able to think of the theory in realistic terms, namely as an abstraction and idealisation of the structures already implicit in the object of daily experience.

No such possibility exists in quantum mechanics. The object of quantum mechanics is completely unrelated to the one of daily experience and is defined solely and exclusively by the equations of the theory. The object of quantum mechanics cannot therefore be viewed as a transformation of the object of daily experience. Rather, it is connected to experience through the numbers obtained from measurements. The overall conceptual frame of these numbers, not only that, even the choice of the quantities to be measured, is contained in the equations of the theory. In quantum mechanics therefore no possibility is left to attribute structure to the object before its embedding in the theory. The structure is completely determined by the theory.

We think we can say that the principal concept of the theory, namely the wave function, interpreted as probability wave, is the paradigm of this gap between perception and the object of the theory. Indeed no transformation of the direct object of perception can lead us to the structure of a probability wave.

It seems thus that also thesis *B)*, complementary to thesis *A)*, stating the total *a priori* quality of conceptualisation, is corroborated by quantum theory.

6. Causality and continuity of space and time

After analysing theses *A*) and *B*), which form the general framework of the theory of *a priori* knowledge, and were substantially confirmed by quantum physics, let us turn now to the last two theses, which specify more exactly the structure of the process of knowledge.

According to thesis *C*), the *a priori* structures, which preside over the synthetic activity of understanding, are determined unequivocally once for all and justified by the metaphysical deduction. This is doubtless the one among Kant's theses which encountered most mistrust, and also the one which was most strongly opposed on the basis of the new physical and mathematical theories.[17]

We do not want to delve in the details of a comparison between the *a priori* established by Kant and the conceptual structure of quantum theory. Let us only comment on a few points. The *a priori* principles held by Kant which are most strongly questioned by quantum physics are undoubtedly the principle of causality and the continuity of space and time. Indeed the uncertainty relationships connecting position and momentum, or time and energy, imply, as is well known, the impossibility of a description both causal and continuous in space-time.[18]

Actually the break up of the *a priori* principles of the *Critique of Pure Reason* is not so definitive. We must first underline that Bohr's complementary interpretation by juxtaposing the space-time description to the causal description, aims, by way of a ingenious trick, at saving the *a priori* principles of Kant's theory, denying exclusively their simultaneous applicability.

For what concerns in particular the continuity, we must say that the conceptual structure of quantum space-time is not so very different from the classical one. Time is reversible in quantum mechanics not unlike the one of classical mechanics; it is homogeneous; space is both homogeneous and isotropic. Finally the space-time discontinuity is only relative, since it can disappear the moment we give up expecting to know the future evolution of the system. Not only that, but no one can say that the structure of space and time quantification as continuous ensembles of real numbers is changed at all.

On the other hand, to Kant, causality means temporal succession in agreement with a given rule.[19] He does not say anything about the nature of this rule. As early as 1929 Hugo Bergmann stressed the point that, no matter what, in quantum mechanics a probabilistic causality principle holds.

Finally, we must remark that in his *Opus postumum* Kant presents us with a much more flexible and rich transcendental schematismus than the one of the *Critique*, so that, although he does not ever question the impossibility to modify the *a priori*, nevertheless its application can vary considerably in the new perspective.

We cannot take these arguments as conclusive; however, there is no doubts that, even if quantum mechanics had demonstrated the necessity to revise the *a priori* structures chosen by Kant, nevertheless the general *a priori* approach would remain incorporated with the theory. In other words: it might be that quantum mechanics has falsified thesis *C*); anyway this thesis is a specification of theses *A*) and *B*), which still constitute the general framework of Kant's theory of knowledge, and which are apparently not questioned by quantum mechanics.

7. The role of productive imagination

Thesis *D*) states that categories can only be applied to the sensible manifold through the schemes, which are set up by the productive imagination - without the latter being involved in their final structure -, and which supply the procedure for building up images starting from the concepts.

We already said that Kant, after discovering the *a priori* intuitions of space and time, as well as the twelve categories which preside over the formation of any kind of judgement, was not able to connect the synthetic activity of understanding with sensible intuition. This is the reason why he thought of a contribution of the productive imagination, capable to apply the categories of understanding to the pure intuition of space and time, so as to bridge the gap between understanding and sensible intuition. Already while discussing thesis *B*) we noticed that in quantum mechanics a gap not to be filled separates the object of immediate perception and the object of theory. Thus, the object is exclusively determined by the theory and no realistic interpretation, in the sense already out-lined, is allowed. Namely, we cannot think of the theory as a theory of the thing in itself, nor transfer any structural property from the thing in itself to the theory. Thus also quantum mechanics makes a complete distinction between the activity of understanding and the receptivity of sensibility.

Such a gap can only be bridged if we find a scheme in which to insert the quantum state description implicit in the mathematical formulation of the theory. In other words if we want to find a confirmation of Kant's gnoseology in quantum physics, we cannot satisfy ourselves with the thesis according to which the wave function is implicitly defined by mathematical axioms, rather we must demonstrate how it can be justified by application of the categories to the pure intuition of space and time mediated by productive imagination. If the wave function indeed were only the entity defined by mathematical formalism, its applicability to the sensible manifold would not be warranted, and it would remain only as an empty form suspended above the world of experience. Only if we were able to detect which categories of understanding are applied to the intuition of space and time and which is the contribution of the productive imagination, then we could conclude that the state description of quantum mechanics is not only consistent, but liable to be applied to the sensible manifold. In conclusion if we were able to demonstrate that the scheme of the wave function is the result of the application of categories to pure intuition, it would follow, being pure intuition the one which frames the sensible manifold, that the quantum state description would be justified when applied to experience.

Let us then investigate the schematismus of the wave function.

As is well known the wave function is a mathematical entity whose square of the absolute value represents the probability of finding a particle; that is, for every point of the space it has one value which, once squared, gives the probability to find in that point the particle described by the particular wave function under consideration. One might then say that the wave function is a *space distribution*, so that not only the pure intuition of time, but also the intuition of space are involved.

However the magnitude which is distributed over the space by the wave function does not have any *material* characteristic, namely doesn't have anything substantial such as could be detected through a physical effect. What is distributed over space is rather the square root of the probability of finding the particle, were we to effect a measurement.

Now, we know that the *a priori* schemes proposed in the *Critique of Pure Reason* are the result of applying the twelve categories of understanding to the pure intuition of time. The single category which might produce a scheme in which to insert the wave function is that of possibility, since the probability of finding a particle can be viewed as the *possibility* that it finds itself at a given instant in a given place. But the schematismus of the *Critique* is exclusively connected with the intuition of time. Anyway this is really no constraint. Time is no doubt the primary form of intuition, inasmuch as it comprehends all the sensible representations, both internal and external. But nothing really prevents us from extending the schematismus to space. The very examples Kant chooses are on occasion space-time schemes.[20]

According to Kant, the scheme of possibilities with respect to the pure intuition of time is that something occurs at least in one time, without such time be determined.[21] This scheme is juxtaposed to the scheme of reality (*Wirklichkeit*), according to which something is in a determined time, and to the scheme of necessity, according to which something is at every time. Nothing prevents us from applying modal categories to space, so that something *can* be at a given instant, in different places, or *is* in a determined place, or *is necessarily* in all places.[22]

This concept of possibility with respect to space-time might well supply the scheme in which to insert the notion of wave function.

It has often be stressed that imagination does not play any role in the structure of quantum mechanics.[23] This characteristic of the theory clarifies itself particularly in the concept of wave function. In classical physics, on the other hand, the concepts of wave and particle appeared to be able to be visually represented. This is no more true for the wave function, which, even if a spatial distribution, cannot be visualised any longer, being a probability distribution, or in terms of Kant's schematismus, a distribution of possibility of finding a particle.

The wave function thus does not have a structure amenable to imagination, even though space plays a very meaningful role in it.

The schemes on the other hand are not connected to imagination, from a structural point of view, since they depend on the application of categories of understanding to the *a priori* forms of intuition. In the schematismus indeed the productive imagination plays an important role in the *genesis* of the scheme, but does not contribute to its final structure. For instance the scheme of triangle, as stated in a celebrated excerpt by Kant,[24] is not an imagined triangle, since in this case it would have to be necessarily equilateral or isosceles or scalene, but rather the formal relation underlying all these types of triangles, which cannot be imagined. According to Kant, therefore, the productive imagination only plays a relevant role in the preparation of schemes, but is not a constitutive ingredient of their structure. Similarly, in quantum mechanics the wave image and the particle image are still relevant in the comprehension and application of the theory. But the wave function as such is neither wave nor particle, rather a scheme capable of generating both images, exactly in the same way as the scheme of the triangle in general is capable of generating equilateral, isosceles and scalene triangles.

We have thus demonstrated, on one side, that the wave function is amenable to the scheme of possibility within the *a priori* intuition of space; on the other, that both the wave function and Kant's schemes, although resulting from a contribution of imagination, nevertheless in their constitution do not allow for the presence of this faculty.

A comment is however necessary. Not even waves and particles would be able to be visualised within a philosophically more exact perspective. They are indeed the schemes of classical physics and as such do not allow for imagination within their structure. In fact, the difference between the notion of wave or particle and the notion of wave function lies in the *distance* from the images which contributed to the creation of these schemes. The triangle in general, although not imaginable, can nevertheless be decently approximated by an imagined triangle. The same way a classical wave or a material point while not images, can be approximated by means of visually imaginable objects. The wave function on the other hand not only cannot be imagined, but even can't easily be approximated with an image, being generated by two images incompatible with one another, that is the wave image and the particle image. But these two images, which contributed to the creation of the scheme, do not actually belong to it. Indeed we only find them in the subjective comprehension of the wave function concept, but not in its conceptual structure.

Summarising, the wave function is liable to be framed in the scheme of possibility with respect to space-time. It is no image, although it makes use of images in its subjective genesis. We may then say that the wave function is justified in its application to the sensible manifold by an adequate reconstruction of its schematismus.

In conclusion thesis *D*), which states the relevance of the schemes in the application of understanding to sensible intuition, as well as their structural independence of imagination, is confirmed by quantum mechanics.

8. Concluding remarks

We can now conclude by summarising the results of our brief investigation. Our feeling is that quantum mechanics finds a good collocation within a theory of knowledge of Kantian type, which, while maintaining experiment as fundamental connection between reality and theory, aims, on one side, at devaluating the gnoseological role of the thing in itself, and on the other, at loading *a priori* all the burden of conceptualisation on the subject, while confining to a lower level the structure of the object, if any.

Nevertheless we must say that the *a priori* structure determined by Kant once and for all might *possibly* be questioned by the theory. The last thesis does not contrast the fact that what determines the object is the *a priori* and not the other way around.[25]

At this point we must emphasise an important point. It is not quite clear to us whether quantum mechanics has really demolished causality and the continuity of space and time as *a priori* principles; however if this was the case, it would seem to provide a good reason to abandon an *a priori* approach in science and in the theory of knowledge. If indeed Kant's *a priori* schemes have been questioned, this can only mean that reality sneaked inside the *a priori* structure of the subject. Would it not be safer to think that such a thing as objectivity and objective structures independent of measurement exists, and not to reduce the object to its instantaneous measurements? Why should not one assume that in measurements actually an aspect of reality in itself is made apparent? and that also in reality in itself there exist structures with which the subject must reckon, while objective structures are not *solely* determined by the categorisation of the subject? An answer to these questions call for more investigations and goes beyond the limits of present discussion.

Notes

[1] Heisenberg, 1958, upholds actually the relative applicability of Kant's *a priori* schemes, such as space, time and causality. A similar viewpoint is expressed by C.F. von Weizsäcker, 1941 and 1979.

[2] See Mathieu, 1958, pp. 10-14 and Scaravelli, 1968, pp. 12 and 93n.. The choice of the second edition of the *Critique* rather than the first one has to do with the fact that in the former imagination plays second fiddles to understanding; this makes a comparison with quantum mechanics easier.

[3] See the partial Italian translation of *Opus postumum* by Mathieu, Kant, 1963, which to date can be considered to provide one of the best interpretation of the nearly two thousands pages of Kantian notes devoted to the passage from metaphysics of nature to physics.

[4] Von Neumann, 1932.

[5] The problem of measurement in quantum mechanics is no doubt one of the most controversial issues of the theory, see in this connection Tarozzi, 1992, pp. 101-197.

[6] *Critique of Pure Reason*, B 553 and B 256.

[7] *Critique of Pure Reason*, B 176.

[8] *Critique of Pure Reason*, B 159.

[9] *Critique of Pure Reason*, B 159.

[10] *Critique of Pure Reason*, B 151.

[11] This example is taken with some liberty from *Critique of Pure Reason*, B 176.

[12] The term was introduced by Gustav Bergmann, 1967. The gnoseological perspective to which it is referred, however, goes back at least as far as to Descartes' concept of *idea*, and it was seriously questioned by the Scottish philosopher Thomas Reid. In the realm of epistemology, a recent approach by Agazzi, 1969, expresses significantly critical arguments against it.

[13] See for instance Feynman, 1965, III, p. 1-12.

[14] This is the so called *restrictive* interpretation of quantum mechanics, as defined by Reichenbach, 1989, pp. 44ff., the only one in which no causal anomalies have to be allowed for.

[15] The fact that already classical mechanics favours a presentationalist interpretation is supported by the circumstance that Kant developed his partially presentationalist theory of knowledge as a side product of the foundations of classical mechanics; see for instance the recent collection edited by Butts, 1986, or Freedman, 1992, or the already classical work by Buchdahl, 1969.

[16] This is for instance Heisenberg's opinion, Heisenberg, 1930, p. 9.

[17] See for instance Reichenbach, 1977, p. 83.

[18] The thesis that quantum mechanics demands discontinuities in space and time (falsifying in such a way Kant's *a priori*) has been stated for instance by Çapek, 1961, pp. 240-1; an opposite point of view is expressed by Grünbaum, 1968, pp. 15f. The fact that quantum mechanics completely does away with causality is, for example, the thesis of Eddington, 1928, chap. 14 and Earman, 1986, chap. X; while Nagel, 1961, chap. 10, maintains a different opinion.

[19] *Critique of Pure Reason*, B 183.

[20] The example of circular dish, B 176, and the one of the triangle, B 180.

[21] *Critique of Pure Reason*, B 184.

[22] It remains to be seen what the conceptual impact of the new scheme will be.

[23] See for instance Nagel, 1961, p. 306. and Agazzi, 1969, pp. 292ff.

[24] *Critique of Pure Reason* B 180.

[25] Weizsäcker, 1979, maintains that the new physics, although it disproves the *a priori* structures chosen by Kant, can be thought of as an *a priori* scheme with respect to experience; an *a priori*, however, which has a historically determined validity.

References

Agazzi, E. (1969). *Temi e problemi di filosofia della fisica*, Milano.

Bergmann, G. (1967). *Realism. A Critique of Brentano and Meinong*, Madison.

Bergmann, H. (1929). *Der Kampf um das Kausalgesetz in der jüngsten Physik*, Braunschweig.

Buchdahl, G. (1969). *Metaphysics and the Philosophy of Science*, Oxford.

Butts, R.E. (1986). Edited by, Kant's Philosophy of Physical Science, Dordrecht.

Çapek, M. (1961). *Philosophical Impact of Contemporary Physics*, Princeton.

Earman, J. (1986). *A primer on Determinism*, Dordrecht.

Eddington, A.S. (1928). *The Nature of the Physical World*, Cambridge.

Feynman, R.P., Leighton, R.B. and Sands, M. (1965). *The Feynman Lectures on Physics*, Massachussets.

Friedman, M. (1992). *Kant and the Exact Science*, Cambridg Mass.

Grünbaum, A. (1968). *Modern Science and Zeno's Paradoxes*, Northampton.

Heisenberg, W. (1930). *Die physikalischen Prinzipien der Quantentheorie*, Leipzig.

Heisenberg, W. (1958). *Physics and Philosophy*, New York.

Kant, I. (1963). *Opus postumum*, it. tr. edited by V. Mathieu, Bologna.

Mathieu, V. (1958). *La filosofia trascendentale e l' Opus postumum di Kant*, Torino.

Nagel, E. (1961). *The Structure of Science*, New York.

Von Neumann, J. (1932). *Mathematische Grundlagen der Quantenmechanick*, Berlin.

Reichenbach, H. (1977). *Gesammelte Werke*, edited by A. Kamlah and M. Reichenbach, II, Braunschweig.

Reichenbach, H. (1989). *Gesammelte Werke*, V.

Scaravelli, L. (1968). *Scritti kantiani*, Firenze.

Tarozzi, G. (1992). *Filosofia della microfisica I*, Modena.

Von Weizsäcker, C.F. (1941) *Die Tatwelt*, 17 pp. 66--98.

Von Weizsäcker, C.F. (1979). In P. Bieri, R.P. Horstmann e L. Krüger, edited by, *Transcendental Arguments and Science*, Dordrecht, 1979, pp. 123--58.

AMONG QUANTUM MECHANICS, WAVE OPTICS, AND CHARGED-PARTICLE BEAM TRANSPORT: TOWARD A POSSIBLE UNIFIED FORMAL DESCRIPTION

R. FEDELE

Dip. di Scienze Fisiche, Università di Napoli "Federico II"
Mostra d'Oltremare Pad. 20 - 80125 - Napoli, Italy

Abstract. It is shown that the correspondence between non relativistic Quantum Mechanics and electromagnetic Wave Optics can be implemented in paraxial approximation with the Particle Beam Transport (optics and dynamics). This is done by introducing the recently proposed Thermal Wave Model (TWM) which assumes that the evolution of a charged-particle beam is governed by a Schrödinger-like equation for a complex function, the so-called beam wave function, whose squared modulus is proportional to the particle number density. This implemented correspondence suggests, at least in paraxial approximation, to develop a formal unified framework capable of describing together different *optical* and *dynamical* phenomena for better understanding, from a physical point of view, one subject by using the same language and similar concepts developed in the other ones. In addition this unified framework would be useful for transferring the quantum computational techniques into the other quantum-like theories in order to solve concrete physical problems. As an example, a recent relevant application of TWM to the accelerator physics is presented.

1. Introduction

It is well known that, starting from Hamilton-Jacobi equation, wave mechanics (WM) has been constructed as a possible counterpart of electromagnetic (e.m.) wave optics (WO) in extending, to a wave context, the analogy between classical mechanics (CM) and geometrical optics (GO) to a wave context [1]. This way, Schrödinger was able to introduce his wave equation and fund quantum mechanics (QM) [2]. In particular, e.m. WO and non relativistic (n.r.) WM formally correspond each other when we consider the Helmholtz equation in comparison with the stationary Schrödinger equation. But, if we want to take into account non-stationary effects in

255

C. Garola and A. Rossi (eds.), The Foundations of Quantum Mechanics, 255-271.

e.m. wave optics, we have to consider the d'Alembert equation (e.m. wave equation) which is a hyperbolic partial differential equation containing also a second-order time-derivative. At this step, the correspondence with nonrelativistic Quantum Mechanics is no longer valid, because the non-stationary Schrödinger equation is a complex parabolic equation which contains only a first-order time-derivative. In fact, d'Alembert equation is Lorentz-invariant whilst Schrödinger equation is Galileo-invariant.

It is known that this difficulty can be overcome by extending QM to the relativistic framework. This is done, for example, by introducing the Klein-Gordon equation [3], which is a Lorentz-invariant d'Alembert-like equation. To recover the Schrödinger equation from this equation we have to apply the nonrelativistic limit which consists in introducing the *slowly-varying amplitude approximation* [4]. With this method we assume that the solution of the following d'Alembert-like equation

$$\frac{1}{\alpha^2}\frac{\partial^2 \Phi}{\partial t^2} - \nabla^2 \Phi + \beta^2 \Phi = 0 \ , \tag{1.1}$$

with α a real number, has the form

$$\Phi(\vec{r}, t) = \Psi(\vec{r}, t) \exp\left(-i\omega t\right) \ , \tag{1.2}$$

where $\Psi(\vec{r}, t)$ is a very slow function compared to the exponential term (i.e. $|\frac{\partial \Psi}{\partial t}| << |\omega \Psi|$). Once (1.2) is substituted in (1.1) by neglecting the second-order time-derivative (according to the above hypothesis) we obtain the following Schrödinger-like equation

$$i\frac{\omega}{\alpha^2}\frac{\partial \Psi}{\partial t} = -\frac{1}{2}\nabla^2 \Psi + \frac{1}{2}\left(\beta^2 - \frac{\omega^2}{\alpha^2}\right)\Psi \ . \tag{1.3}$$

Consequently, provided that $\alpha = c$, $\omega = mc^2/\hbar$, and $\beta = mc/\hbar$ (m, c, $\hbar$ being the particle rest mass, the speed light and Planck's constant, respectively), (1.1) and (1.3) coincide with Klein-Gordon equation and Schrödinger equation, respectively, for a free particle. This example clearly shows that the slowly-varying amplitude approximation represents the non-relativistic limit of the d'Alembert-like equation (1.1).

However, this approximation can be used also for Helmholtz equation in order to stress the correspondence between e.m. WO and n.r. QM. In fact, starting from Helmholtz equation for the e.m. field in a general medium, the propagation of a monochromatic e.m. beam of frequency ω and wavenumber k, can be described by introducing the slowly-varying amplitude approxi-mation in such a way to assume that the e.m. field amplitude is a slowly-varying function along the propagation direction compared to the scale length $2\pi/k$. Sometimes, this approximation is referred as to *paraxial appro-ximation* [5]. This way a two-dimensional Schrödinger-like equation for the (transverse) e.m. field amplitude is obtained, the so-called Fock-Leontovich equation [6], which describes the evolution of the e.m. beam through a general linear (nonlinear) optical medium, whose linear (nonlinear) refractive index $N(\vec{r}, t)$ plays the role of an *effective potential* (see TABLE I).

TABLE I

Electromagnetic Beam Optics	Charged-Particle Beam Transport	Nonrelativistic Quantum Mechanics
λ	ϵ	$\hbar$
diffraction parameter	emittance	Planck's constant
$\frac{d^2}{dz^2}W + \eta(z)W - \frac{\lambda^2}{W^3} = 0$	$\frac{d^2}{dz^2}R + K(z)R - \frac{\epsilon^2}{R^3} = 0$	$\frac{d^2}{dt^2}\sigma(t) + k(t)\sigma(t) - \frac{\hbar^2}{\sigma(t)^3} = 0$
Envelope equation	Envelope equation	Ehrenfest's theorem
$W P_W \geq \lambda$	$R P_R \geq \epsilon$	$\sigma P_\sigma \geq \hbar$
$i\,\lambda\frac{\partial}{\partial z}\Phi = -\frac{\lambda^2}{2}\nabla_\perp^2\Phi + N\,\Phi$	?	$i\hbar\frac{\partial}{\partial t}\Psi = -\frac{\hbar^2}{2}\nabla_\perp^2\Psi + U\,\Psi$ $(m = 1)$
Fock-Leontovich equation		2-D Schrödinger equation
$\lambda \to 0$	$\epsilon \to 0$	$\hbar \to 0$
Light-rays equation	Electronic-rays equation	Classical trajectory

The correspondence contained in this description provides to replace Planck's constant with the inverse of the wavenumber $\lambda \equiv \lambda/2\pi \equiv 1/k$, and the time with the coordinate, say z, of the propagation direction (see

TABLE I), respectively.

Since the evolution of an e.m. beam is governed by a Schrödinger-like equation for a complex e.m. field amplitude Φ , it is easy to prove the following well-known uncertainty relation:

$$W \, P_W \geq \lambdabar \quad , \tag{1.4}$$

where

$$W \equiv \left[\frac{\int_{-\infty}^{\infty} r^2 |\Phi|^2 \, d^2 r}{\int_{-\infty}^{\infty} |\Phi|^2 \, d^2 r}\right]^{1/2} \equiv < r^2 > \quad , \tag{1.5}$$

is the effective beam radius and

$$P_W \equiv \left[\lambdabar^2 \frac{\int_{-\infty}^{\infty} |\nabla_\perp \Phi|^2 \, d^2 r}{\int_{-\infty}^{\infty} |\Phi|^2 \, d^2 r}\right]^{1/2} \quad , \tag{1.6}$$

is the total e.m. averaged transverse momentum associated with the e.m. beam (r being the radial-cylindrical coordinate), which in the case of Gaussian beams becomes

$$P_W \equiv \left[\left(\frac{dW}{dz}\right)^2 + \frac{\lambdabar^2}{W^2}\right]^{1/2} \quad . \tag{1.7}$$

Note that (1.4) corresponds to the Heisenberg uncertainty principle if $\lambdabar$ corresponds to $\hbar$. (1.4) plays an important role in optical fibers [7],[8] and in Quantum Optics [9], especially in considering the so-called *coherent states* which minimize this inequality [10]-[12].

Furthermore, in the envelope equation for the spot-size at the location z of a quasi-monochromatic e.m. beam travelling in a medium, an extra-term due to the beam-medium interaction appears. In general this interaction is nonlinear [13], but in the small-amplitude approximation and for Gaussian beams, by following [13] and assuming a perturbative form of the refractive index, like $N(r, z) = N_0 + \Delta N(r, z) \equiv N_0 + \frac{1}{2}\eta(z)r^2$ (linear lens), we easily have

$$\frac{d^2 W}{dz^2} + \eta(z)W - \frac{\lambdabar^2}{W^3} = 0 \quad . \tag{1.8}$$

In particular for the propagation in vacuo ($\eta = 0$) :

$$W(z) \; = \; W_0 \left(1 + \frac{z^2}{z_R^2}\right) \quad , \tag{1.9}$$

where W_0 is the minimum spot size (waist) and z_R is the so-called *Rayleigh length* ($W_0^2 - \lambdabar \, z_R$).

Finally, in QM the behaviour of the averaged quantities is given by Ehrenfest's theorem. To this end, let us consider a single quantum particle moving on a plane under the action of a linear force ($\vec{F} = -k(t)\vec{r}$, where the k is a function of t). If we assume for simplicity the mass $m = 1$ and a Gaussian initial state in polar coordinates, we have in exact analogy to the previous case the equation

$$\frac{d^2\sigma}{dt^2} + k(t)\sigma - \frac{\hbar^2}{\sigma^3} = 0 \quad . \tag{1.10}$$

In this case also, setting $k(t) = 0$ we have

$$\sigma^2(t) = \sigma_0^2 \left(1 + \frac{t^2}{t^{*2}}\right) \quad , \tag{1.11}$$

where $\sigma_0^2 = \hbar\, t^*$. In Eqs. (1.10) and (1.11) $\sigma^2(t)$ stands for the quantum expectation value of r^2 at the time t; in particular $\sigma_0^2 \equiv\, < r^2 >_{t=0}$. In this case, to show the complete analogy also for the previous uncertainty relation it is worth to formulate the Heisemberg uncertainty principle in polar coordinates (see TABLE I).

It is worth to recall that the limits $\lambda \to 0$ and $\hbar \to 0$ (see TABLE I) recover geometrical optics (light-rays equation) and classical mechanics (classical motion equation), respectively. This means that the physical meaning of λ is thus given in terms of diffraction parameter. In fact, the condition $\lambda \neq 0$ in paraxial approximation is connected to a weak displacement of light rays from the beam propagation direction in such a way to produce a mixing between them. In the exact geometrical optics limit ($\lambda = 0$) the ray would be straight lines parallel to the propagation direction when the beam is travelling in vacuo, whilst for λ finite the rays mixing (diffraction effect) produces (in vacuo) a hyperbolic hyperboloid around the z-axis which corresponds to a typical caustic shape described by (1.9) [5].
Correspondingly, by means of the analogy between this kind of wave description and QM, we give to $\hbar$ a similar role of diffraction parameter for WM, according to the spreading effect described by Schrödinger equation for a wavepacket associated to a *quantum* particle.

On the basis of the discussion developed above the correspondence between e.m. WO in paraxial approximation and n.r. QM is thus fully constructed, but it could be implemented by considering an additional subject: the relativistic charged particle beam transport (optics and dynamics).

2. Analogy with Charge Particle Beam Transport (Optics and Dynamics)

Within the framework of the conventional theory, the collective transverse dynamics of a particle beam travelling through an optical magnetic device can be described in terms of an equation for its effective radius R (r.m.s. of the transverse particle distribution) called envelope equation [14]. In particular, for a cylindrically-symmetric relativistic beam propagating along z-axis through a quadrupole-like device, which produces on the particles a linear focusing (defocusing) force, the envelope equation is given by [14]

$$\frac{d^2 R}{dz^2} + K(z)R - \frac{\epsilon^2}{R^3} = 0 \ , \tag{2.12}$$

where $K(z)$ is the quadrupole strength, $z = ct$, and ϵ is a constant called *transverse emittance*, connected to the measure of the ensemble area associated with the transverse beam dynamics in the phase space [14]. Physically, ϵ is connected to the thermal spreading of the particle trajectories within the beam due to the finite temperature of the system [14]. Typically, the thermal transverse motion is nonrelativistic compared to the longitudinal (relativistic) motion. In the case of free motion (motion in vacuo) we have $K = 0$ (no external force is acting on the system), and the solution for the beam spot size $R = R(z)$ is similar to (1.9) and reads

$$R^2(z) \ = \ R_0^2 \ \left(1 + \frac{z^2}{\beta^{*2}}\right) \qquad R_0^2 = \epsilon\, \beta^* \ . \tag{2.13}$$

The physical origin of this solution is exactly due to the above thermal spreading of the trajectories around the beam propagation direction. In fact, the thermal spreading (emittance spreading) causes a small mixing of the trajectories in such a way to produce a picture fully similar to the one obtained with the paraxial diffraction in the e.m. beams provided to replace the particle trajectories (*electron rays*) with the light rays. In this sense, the (transverse) emittance plays the role analogous to a diffraction parameter whilst the small deviation of the electron rays from the propagation direction represents the *paraxial approximation* behind the particle beam optics. To this end, note that in the limit $\epsilon \to 0$ (2.12) recovers the *electron ray equation*. In this limit, which occurs for example in an oscilloscope, the particle trajectories within the beam are all parallel to the propagation direction and the beam can be assumed cold.

In addition, if now the quadrupole-like device is switched on, we have to add the driven motion due to the external linear force to the free motion above discussed. This composition produces the well known betatron motion for which the associated particle distribution is Gaussian [14].

3. Uncertainty relation for particle beams

Remarkably, the envelope equations (1.8) and (1.10) presented above can be easily derived from Fock-Lentovich equation and from Schrödinger equation, respectively. The corresponding uncertainty inequalities can be also easily derived from these wave equations, respectively. Nevertheless, still remaining in the framework of the conventional description of charged particle beam transport, it is very easy to prove that the following uncertainty inequality holds

$$RP_R \geq \epsilon \ , \tag{3.14}$$

where

$$P_R = \left[\left(\frac{dR}{dz} \right)^2 + \frac{\epsilon^2}{R^2} \right]^{1/2} . \tag{3.15}$$

The differences between (3.14), on a side, and (1.4) and the Heisenberg's uncertainty relation, on the other side, is that the first one has been consistently obtained within the framework of the conventional theory, without any wave description, whilst the second ones are supported by a suitable wave description. To recognize this important point, let us perform the following argument.

First of all, observe that (2.12) can be obtained by starting from the following hamiltonian:

$$\mathcal{H}(R, P_r) = \frac{P_r^2}{2} + \frac{\epsilon^2}{2R^2} + V(R) \ , \tag{3.16}$$

for $V(R) = \frac{1}{2}KR^2$. The other envelope equation (1.8) and (1.10) can be obtained with similar hamiltonian provided the substitution of R with W and σ, respectively. (3.16) formally corresponds to the hamiltonian of a classical particle with unit mass which moves under the action of the central potential $V(R, z)$ in polar coordinates R and ϕ (z being the curvilinear coordinate). By denoting with L_ϕ the momentum conjugate to ϕ (note that P_r is the momentum conjugate to R, namely the radial component of the total linear momentum), Hamilton's equations give

$$L_\phi = \epsilon = constant \ , \tag{3.17}$$

(angular momentum conservation) and

$$\frac{d^2 R}{dz^2} + \frac{\partial V}{\partial R} - \frac{L_\phi^2}{R^3} = 0 \ . \tag{3.18}$$

By using (3.17) we obtain the expression for the *azimuthal* component P_ϕ of the linear momentum

$$P_\phi = \frac{L_\phi}{R} . \tag{3.19}$$

Thus, the total linear momentum P_R is given by:

$$P_R = \left(P_r^2 + P_\phi^2\right)^{1/2} = \left[\left(\frac{dR}{dz}\right)^2 + \frac{L_\phi^2}{R^2}\right]^{1/2} . \tag{3.20}$$

Consequently, by substituting (3.17) in (3.20), relations (3.14) and (3.15) are immediately obtained.

4. The Thermal-Wave Model

The impressive formal analogy discussed above clearly shows that what is missing in the correspondence among the columns of TABLE I is just a wave description for charged particle beam transport. It would be natural to conjecture such a kind of description in order to complete the above formal correspondence. On the other hand, a wave description would give a deeper physical meaning behind the envelope description of charged particle beam optics and dynamics. Following this idea, it seems natural to construct a *quantum-like* framework, only based on the above analogy, whose *Ehrenfest's theorem* would correspond to the above envelope description of charged particle beam optics and dynamics. This way, between conventional theory of particle beams and this possible wave description would exist a relationship fully similar to the one existing between QM and CM, or between WO and GO. In fact, by following this natural conjecture, recently a wave model, of thermal nature, the so-called Thermal Wave Model (TWM), has been developed for charged particle beam transport [15].

In order to remove the question mark present in TABLE I (fifth row and second column), TWM assumes that the stationary configuration of a relativistic-charged-particle beam of emittance ϵ, travelling along the z-axis with velocity βc ($\beta \approx 1$), under the action of a potential $u(r, z, \phi)$ (r, z, ϕ being the usual cylindrical coordinates), is described by the following two-dimensional Schrödinger-like equation for a wave function $\Psi(r, z, \phi)$, the so-called beam wave function (BWF) :

$$i\,\epsilon\,\frac{\partial}{\partial z}\Psi = -\frac{\epsilon^2}{2}\nabla_\perp^2\Psi + U(r, z, \phi)\Psi , \tag{4.21}$$

where

$$U(r, z, \phi) = \frac{u(r, z, \phi)}{m_0\gamma\beta^2 c^2} . \tag{4.22}$$

In the quantum analogy z and ϵ play the role of *time* and *Planck's constant*, respectively.

Bearing in mind the well-known *norm conservation law* for a Schrödinger-like equation, i.e.

$$\frac{d}{dz}\mathcal{N} = 0 , \tag{4.23}$$

with

$$\mathcal{N} \equiv \|\Psi\| = \left[\int_0^{2\pi} d\phi \int_0^{\infty} r \, dr \, |\Psi|^2 \right]^{1/2} , \qquad (4.24)$$

we get the physical meaning of BWF which can be related to the density of the beam. In fact, if N is the total number of particles, the quantity

$$\Sigma(r, z, \phi) = \frac{N}{\mathcal{N}^2} |\Psi(r, z, \phi)|^2 \qquad (4.25)$$

represents the two-dimensional particle-number density of the beam (number of particles per unitary-beam transverse section). It yields the transverse profile of the beam density. For simplicity let us fix $\|\Psi\| = 1$ (normalized BWF).

Equation (4.21) has to be coupled to an equation for the total field force $\vec{F}_\perp$ acting on the system:

$$\vec{F}_\perp = -(m_0 \gamma \beta^2 c^2) \vec{\nabla}_\perp U . \qquad (4.26)$$

In general $U = U(r, z, \phi)$ is not known. We distinguish two kind of problems related to the system of equations (4.21) and (4.26). One includes the problems of *nonlinearity* corresponding to some dependance of U on $|\Psi|^2$. The other one includes the problems of *linear kind* corresponding to consider U as a given function (external force).

This model has been successfully applied to a number of linear and nonlinear problems of particle beam transport [15]-[22]. In particular, it has been used in the transverse dynamics for describing the optics of a charged particle beam in a thin quadrupole-like lens with small sextupole and octupole deviations [20], and for estimating the luminosity in linear colliders [17]. TWM has also been applied for describing the self-consistent interaction of a relativistic electron (positron) beams with a collisionless, overdense plasma [16], reproducing the main results for the beam filamentation threshold and the self-bunching equilibrium [16].
TWM seems also useful for describing longitudinal particle bunch dynamics in both conventional accelerators and plasma-based accelerators for which coherent instability, in terms of modulational instability, and soliton formation has been investigated [18], [19], [21].
Recently, TWM has been used for describing the longitudinal dynamics of a relativistic charged particle bunch in a circular accelerating machine, when the *radio-frequency* potential well and the self-interaction (wake fields) are present [19] as well as synchrotron radiation damping and quantum excitation (photon noise) are taken into account [22].

Some preliminary useful definitions, in complete analogy with QM, can be introduced in TWM.

-*The Hamiltonian operator and its averaged value:*

$$\hat{H} = -\frac{\epsilon^2}{2}\nabla_\perp^2 + U \quad , \tag{4.27}$$

$$\mathcal{E}(z) = \int_0^{2\pi} d\phi \int_0^\infty r\, dr\ \Psi^* \hat{H} \Psi = i\,\epsilon \int_0^{2\pi} d\phi \int_0^\infty r\, dr\ \Psi^* \frac{\partial}{\partial z}\Psi. \tag{4.28}$$

-*The total, averaged, linear transverse momentum of the system:*

$$P(z) = \epsilon \left[\int_0^{2\pi} d\phi \int_0^\infty r\, dr\ |\nabla_\perp \Psi|^2 \right]^{1/2}. \tag{4.29}$$

-*The averaged kinetic and potential energy of the system:*

$$\mathcal{E}_k(z) = \frac{P^2(z)}{2} \quad , \tag{4.30}$$

$$\mathcal{V}(z) = \int_0^{2\pi} d\phi \int_0^\infty r\, dr\ U\, |\Psi|^2 \quad , \tag{4.31}$$

so that

$$\mathcal{E}(z) = \mathcal{E}_k(z) + \mathcal{V}(z) \quad . \tag{4.32}$$

-*The effective beam transverse dimension (beam spot-size):*

$$R(z) = \left[\int_0^{2\pi} d\phi \int_0^\infty r\, dr\ r^2 |\Psi|^2 \right]^{1/2}. \tag{4.33}$$

Taking (4.21), its complex conjugate and differentiating twice the square of (4.33), after some simple manipulations we obtain the equation

$$\frac{d^2}{dz^2} R^2(z) = 4\mathcal{E} - 4\mathcal{V} - 4\pi \int_0^\infty \left(r\frac{\partial}{\partial r}U \right) |\Psi|^2\, r\, dr \quad , \tag{4.34}$$

where cylindrical symmetry has been assumed for simplicity and definitions (4.27)-(4.33) have been used.

Note that $\mathcal{E}$ varies with respect to z only if also U does it. In fact, differentiating (4.28) with respect to z, one easily finds

$$\frac{d\mathcal{E}}{dz} = \left< \frac{\partial}{\partial z}U \right> = 2\pi \int_0^\infty \left(\frac{\partial}{\partial z}U \right) |\Psi|^2\, r\, dr \quad . \tag{4.35}$$

The case of a very slow variation of U with respect to z frequently occurs in accelerator devices, such as quadrupoles. Thus the contribution of $\partial U/\partial z$ can be neglected and $\mathcal{E}$ becomes an adiabatic invariant.

Eq. (4.34) is the general spot-size particle-beam equation when an external potential $U(r, z)$ is acting on the system.

In the case of free beam motion $(U = 0)$, and in the case in which the beam passes through a quadrupole-like device $(U = \frac{1}{2}Kr^2)$ (4.34) and (4.35) recover (2.13) and (2.12), respectively.

By using (4.26) and (4.32), Eq. (4.34) can be cast in the form

$$\frac{d^2}{dz^2} R^2(z) = 2\mathcal{E}_k + <\vec{r} \cdot \vec{F}_\perp> \quad , \qquad (4.36)$$

which is formally identical to the *virial equation* for a classical particle of unit mass [23].

The next section concerns with one of the several relevant successful applications of TWM to the transverse dynamics of particle beams, recently proposed in literature. We present it in order to give just an idea of both the intrinsic validity and capability of the model in solving concrete physical problems without any physical contradiction (intrinsic coherence of the model). This procedure would lead to naturally conclude that the formal correspondence constructed above, including charged particle beam transport, makes physical sense.

5. A relevant application of TWM

In Ref. [20] TWM has been applied in order to study the propagation of a particle beam throughout an optical device, of total length L, made of a quadrupole with sextupole and octupole deviations, which take a length l, plus a drift space of length $L - l$, and compare the results with those of conventional tracking simulations. To this end, the standard perturbation techniques were used in solving the Schrödinger-like equation for BWF to determine the transverse space-distribution of a purely Gaussian incoming beam at the entrance of the optical device. The theoretical prediction, performed up to first order of the perturbation theory, were then compared with the results of a particle tracking simulations making use of a standard *kick* code [24]. We briefly present the procedure used in [20] to make this comparison.

For the sake of simplicity and without loss of generality, we can consider only one transverse dimension, say x. In this case, the beam wave equation (4.21) becomes

$$i\epsilon \frac{\partial \Psi}{\partial z} = -\frac{\epsilon^2}{2} \frac{\partial^2}{\partial x^2} \Psi + U(x, z)\Psi \quad , \qquad (5.37)$$

where

$$U(x, z) = \frac{1}{2!} k_1 x^2 + \frac{1}{3!} k_2 x^3 + \frac{1}{4!} k_3 x^4 \quad . \qquad (5.38)$$

This corresponds, by using the standard notation, to a quadrupole lens of focusing strength k_1, with sextupole aberrations strength k_2, and octupole aberrations strength k_3. If the density profile at the beginning $(z = 0)$ is purely Gaussian, i.e. $|\Psi(x,0)|^2 = \exp(-x^2/2\sigma_0^2)/\sqrt{2\pi\sigma_0^2}$, corresponding to an incoming particle beam into the lens without any *divergence* of the envelope shape with respect to z-axis, the normalized BWF at the exit of the optical device $(z = L)$ results to be [20] :

$$\Psi(x,z) = \frac{\exp\left[-\frac{x^2}{4\sigma_v^2(z)} + i\frac{x^2}{2\epsilon\rho_v(z)} + i\phi_v(z)\right]}{[2\pi\sigma_v^2(1 + 15\tau^2 + 105\omega^2)^2]^{1/4}} \times$$

$$\times \; \Sigma_{n=0}^4 b_n H_n\left(\frac{x}{\sqrt{2}\sigma_v}\right) \exp\left(i2n\phi_v(z)\right) \quad, \tag{5.39}$$

where

$$\sigma_v(L) = \left[\left(\frac{\epsilon^2}{4\sigma_0^2} + K_1^2\sigma_0^2\right)(L - l)^2 - 2K_1\sigma_0^2(L - l) + \sigma_0^2\right]^{1/2} \quad,$$

$$\frac{1}{\rho_v(L)} = \frac{1}{\sigma_v}\frac{d\sigma_v}{dz}\bigg|_{z=L} \quad,$$

$$\phi_v(L) = -\frac{1}{2}\left\{\arctan\left[\left(\frac{\epsilon}{2\sigma_0^2} + \frac{2K_1^2\sigma_0^2}{\epsilon}\right)(L - l) - \frac{2K_1\sigma_0^2}{\epsilon}\right]\right.$$

$$\left. + \; \arctan\left[\frac{2K_1\sigma_0^2}{\epsilon}\right]\right\} \quad, \tag{5.40}$$

with $\tau \equiv \sigma_0^3 K_2/6\epsilon$, $\omega \equiv \sigma_0^4 K_3/24\epsilon$; $K_j \equiv k_j l$ $(j = 1, 2, 3)$ are the *integrated multipole strength*; $b_0 = 1 - i3\omega$, $b_1 = -i3\tau\sqrt{2}$, $b_2 = -i3\omega$ $b_3 = -i\frac{\tau}{2\sqrt{2}}$, $b_4 = -i\frac{\omega}{4}$; H_n are Hermite polynomials. The theoretical probability distribution $|\Psi(x, L)|^2$ (transverse density profile) available from (5.39) has been compared with the results of a standard tracking technique [24]. To this end a *numerical experiment* has been carried in which the optical device considered, can be thought as made of three thin multipoles: a quadrupole of integrated strength K_1, a sextupole of integrated strength K_2, and an octupole of integrated strength K_3. 30000 particles with starting coordinates x and x' randomly distributed on a 2-dimensional Gaussian have been used to simulate the beam. A tracking simulation of these particles through the device has been done by means of a standard *kick* code [24]: this is generally more than adequate when thin lens approximation can be applied; the coordinates of all particles have been recorded at the exit of each lens.

In Fig.1a the starting distributions of x are displayed: the histogram, properly normalized to take into account the total number of particles and the bin width, represents the *experimental* data (its best fit is the continuous curve), and the dotted line stands for the theoretical starting distribution according to $|\Psi(x,0)|^2$. The agreement between the two makes us feel confident that the description of the beam by means of its $\sigma_0 = 0.046$ m, $\epsilon = 5. \; 10^{-5}$ m and the two normalizations, are done consistently.

The simulated beam distribution and the theoretical one after passing through a pure quadrupole (squared modulus of (5.39) for $\tau = 0$ and $\omega = 0$) of integrated strength $K_1 = 0.025$ m^{-1} and a drift space of length $L = 25$ m , respectively, are shown in Fig.1b. Also here, a complete agreement is noticeable: the distributions are much sharper, but they both keep purely Gaussian with same σ's and heights. In Figs.2 both the *experimental* and the theoretical distributions are shown at the exit of the full device for two different strengths of sextupole and octupole and compared with the corresponding aberrationless distribution (dashed lines). Here the histogram, as usual, represents the experimental distribution, and the solid line is its best fit which results to be a Gaussian function times a polynomial of order 8. The distribution predicted by the *thermal wave model* and given by squared modulus of (5.39) is, instead, represented by a dotted line. In particular, in Fig.2a for integrated aberration strengths $K_2 = 0.05$ m^{-2} $(\tau = .016)$ and $K_3 = 1. \; m^{-3}$ $(\omega = 3.73 \; 10^{-3})$, the agreement between the predictions of the model and the tracking results is quite impressive. In Fig.2b, where the inequalities $\sigma_0 K_2/(3K_1) \ll 1$ and $\sigma_0^2 K_3/(12K_1) \ll 1$ are not strictly satisfied any more, $K_2 = 0.1$ m^{-2} $(\tau = .032)$ and $K_3 = 2. \; m^{-3}$ $(\omega = 7.46 \; 10^{-3})$, the dotted line is barely visible, whilst in Fig.2a it superimposes perfectly to the best fit curve and the only sign of its presence is the slight thickening of the line. It should be also said that, in order to produce a detectable effect after a single pass, both sextupole and octupole strengths as well as starting value σ_0 used exceed by far the values typical of circular accelerators, where usually the beam makes millions of turns.

In conclusion, this simple but very significant numerical experiment presented in [20] has proved the capability of the recently proposed *thermal wave model* [15] to describe the charged-particle beam transport quite accurately.

6. Remarks and Conclusions

In this paper a careful analysis of the analogy among Quantum Mechanics, e.m. Wave Optics, and Charged-Particle Beam Transport has been presented.

It has been pointed out that, usually, the extension to a wave description of the correspondence between Classical Mechanics and Geometrical Optics (starting from Hamilton-Jacobi equation) allows us to write correctly only the stationary Schrödinger equation, to be put in correspondence with Helmholtz equation of Wave Optics. In fact, the non-stationary quantum

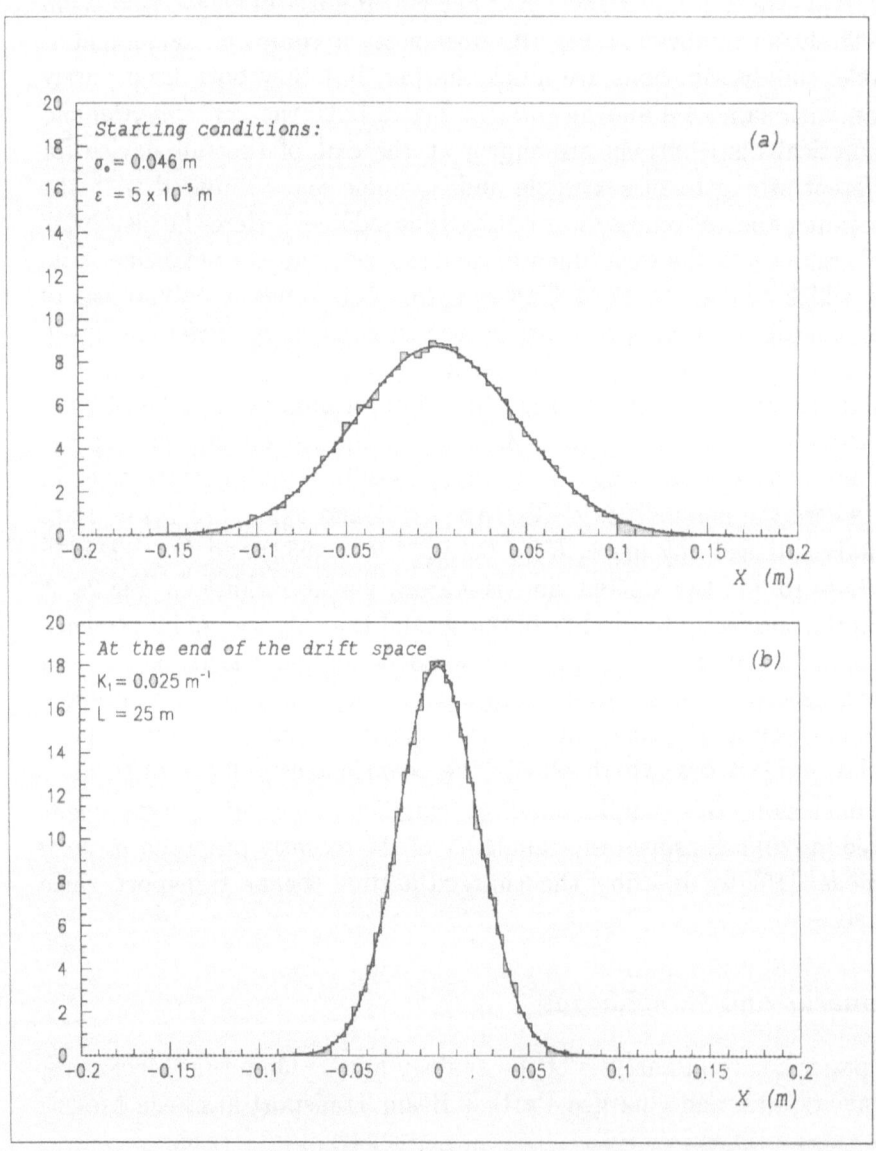

Figure 1.

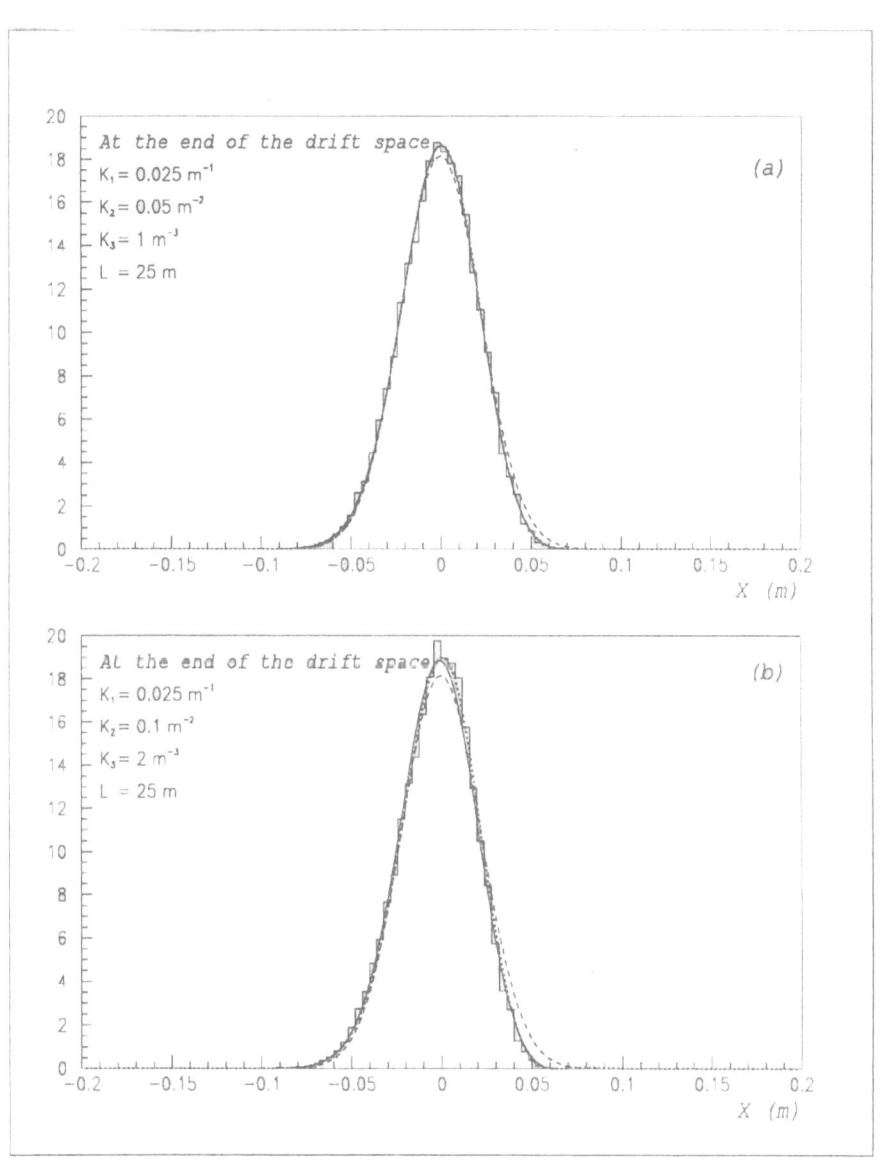

Figure 2.

phenomena in a non relativistic framework are described by the non-stationary Schrödinger equation which does not correspond to d'Alembert
equation describing non-stationary e.m. wave phenomena. In order to recognize the correct correspondence between n.r. QM and WO, the non
relativistic limit of Klein-Gordon equation has been considered. To this
end, it has been pointed out that this non relativistic limit, which allows
us to recover the quantum non-stationary Schrödinger equation, is a sort of
slowly-varying amplitude approximation capable of giving a Schrödinger-
like equation from a d'Alembert-like equation. By following this idea, it
has been recognized that the slowly-varying amplitude approximation can
be also applied to the Helmholtz equation. The result of this procedure
allows us to recover Fock-Leontovich equation [6] which is a Schrödinger-
like equation for the e.m. field amplitude, and the above slowly-varying
amplitude approximation coincides here with the paraxial approximation.
This way, a deeper comparison between WO and QM can be made. According to what is synthetically presented in TABLE I, we conclude that, for
non relativistic limit, the correspondence between these two theories is given
in terms of a paraxial description. In particular, we have observed that: (a)
quantum diffraction (introduced by $\hbar$) can be put in correspondence with
the paraxial diffraction of an e.m. monochromatic beam (introduce by χ),
whose physical meaning is a small mixing of the light rays; (b) uncertainty
relation and envelope equation valid for each theory fully correspond each
other.

In addition this paraxial correspondence has been implemented with the
charged-particle beam transport. On the basis of the conventional *paraxial*
description of particle beams given in literature without any wave picture
[14], a further analogy between this subject, on a side, and n.r. QM and
e.m. Beam Optics, on the other side, has been presented in such a way
to recognize that the recently proposed Thermal Wave Model (TWM) for
charged-particle beam transport [15] completes the above analogy giving a
paraxial wave interpretation of the optics (electron rays) and the dynamics
of charged particle beams. In addition, it has been pointed out that this
implemented analogy makes sense because TWM has been successfully
applied to a number of problems [15]-[22] and, in particular, to this end,
one of its most relevant applications in particle accelerators [20] has been
presented.

Consequently, the impressive correspondence, synthetically presented
in TABLE I and implemented with TWM, suggests, at least in paraxial
approximation, to develop a formal unified framework which allows us:
(1) to describe together different *optical* and *dynamical* phenomena; (2)
to understand better, from a physical point of view, one subject by using
the same language and similar concepts developed in the other ones; (3)

to transfer the most powerful quantum computational techniques into the other quantum-like theories in order to solve concrete physical problems. This is, for example, what has been done by transferring the know-how of QM to the e.m. optics to solve important problems of quantum optics [9]-[12] and optical fibers [7]-[8], and to TWM to solve concrete problems of charged-particle beam transport in particle accelerators and in plasma physics [15]-[22].

References

1. de Broglie, L. (1925). Ann. de Physique, **10** (3), (22).
2. Schrödinger, E. (1926). Ann. der Physik, **79**, (361), (484), (734).
3. Messiah, A. (1976). *Quantum Mechanics*, vol.I, North-Holland Publ., Amsterdam.
4. Whitham, G.B. (1974). *Linear and Nonlinear Waves*, John Wiley and Sons, New York.
5. Tarasov, L.V. (1983). *Laser Physics*, MIR Publ., Moscow, pp.100–142.
6. Leontovich, M. and Fock, V. (1946). *Sov. J. Phys.* **10**, (13).
7. Marcuse, D. (1972). *Light Trasmission Optics*, Van Nostrand, New York.
8. Arnaud, J.A. (1976). *Beam and Fiber Optics*, Academic Press, New York.
9. Sudarshan, E.C.G. (1993). *Uncertainty relations, zero point energy and the linear canonical group*, in Proceedings of Second International Workshop on Squeezed States and Uncertainty Relations,Moscow, 25-29 May, 1992, D.Han,Y.S.Kim and V.I.Man'ko (eds.), NASA Conference Publication 3219, (241).
10. Glauber, R. (1963). *Phys. Rev. Lett.*, **10** (84).
11. Sudarshan, E.C.G. (1963). *Phys. Rev. Lett.*, **10** (177).
12. Klauder, J.R. (1964). *J. Math. Phys.*, **5** (177).
13. Shen, Y.R. (1984). *The principles of nonlinear optics*, Wiley-Interscience Publication, New York, pp.307–309
14. Lawson, J. (1988). *The physics of charged particle beams* Clarendon Press, Oxford, (2nd edition).
15. Fedele, R. and Miele, G. (1991). *Nuovo Cimento D*, **13** (1527).
16. Fedele, R. and Shukla, P.K. (1992). *Phys. Rev. A*, **44** (4045).
17. Fedele, R. and Miele, G. (1992). *Phys. Rev. A*, **46** (6634).
18. Fedele, R., Palumbo, L. and Vaccaro, V.G. (1992). "Proc. EPAC 92", (Berlin, 24-28 March 1992) H. Henke, H. Homeyer and Ch.Petit-Jean-Genaz (eds.), Editions Frontieres, (762).
19. Fedele, R., Miele, G., Palumbo, L. and Vaccaro, V.G. (1993). *Phys. Lett. A*, **179** (407).
20. Fedele, R., Galluccio, F. and Miele, G. (1994). *Phys. Lett. A*, **185** (93).
21. Fedele, R. and Vaccaro, V.G. (1994). *Physica Scripta*, **T52** (36).
22. Fedele, R., Miele, G. and Palumbo, L. (1994). *Phys. Lett. A*, **194** (113).
23. Alonso, M. and Finn, E.J. (1974). *Fundamentals of University Physics*, Inter European Sciences, Amsterdam.
24. Mais, H., Ripken, G., Wrulich, A., Schmidt, F. (1985). "Proc. CERN Accelerator School", Oxford, Sept. 16-27, 1985, Turner Ed., CERN 8703 Vol.2 (690).

QUESTIONING NONLOCALITY:

AN OPERATIONAL CRITIQUE TO BELL'S THEOREM

C. GAROLA

Dipartimento di fisica - Università di Lecce

Via per Arnesano - 73100 - Lecce, Italy

Abstract. Bell's theorem is proved to rest on a metatheoretical assumption (MCP) regarding the validity of physical laws that is compatible with the worldview of Classical Physics, not with the worldview of Quantum Physics (QP). A new general principle (MGP) is stated here that is consistent with the basic operational philosophy of QP. By using MGP, which does not modify the observative content of QP, some sample proofs of Bell's theorem are invalidated. We conclude that the adoption of a more rigorous quantum attitude leads to give up with some features, as nonlocality and noncompatibility with any form of realism, that are usually retained to be unavoidable (and somewhat paradoxical) consequences of QP.

PACS numbers: 03.65.Bz, 02.10.By, 02.10.Gd.

1. Introduction

The existence of paradoxes is a crucial problem in Quantum Physics (QP). Indeed, the interpretation of the mathematical apparatus of QP is often counterintuitive or problematical, and paradoxes suggest that this theory contradicts some largely accepted epistemological requirements regarding physical theories (it is important to note that a paradox must not be confused with an antinomy, which is an internal contradiction of the theory that invalidates the theory itself).

The most important paradoxes in QP, both from a conceptual and an historical viewpoint, are those connected with the thought experiment proposed by Einstein, Podolski and Rosen (EPR) in 1935, with the aim of proving the incompleteness of QP.

At the best of my knowledge, the earliest argument aiming to show that a paradox follows from the quantum treatment of the EPR experiment

C. Garola and A. Rossi (eds.), The Foundations of Quantum Mechanics, 273-285.
© *1995 Kluwer Academic Publishers.*

was proposed by Furry (1936a, 1936b). Later on, many authors proposed new versions of the Furry argument, or invented new arguments (see in particular Bohm, 1951; Bohm and Aharonov, 1957). More recently Bell (1966) obtained some results that should prove that QP is necessarily a *contextual* theory, in the sense that the value of each observable belonging to a set ϕ of observables that are measured on a physical system in a given state depends on the whole set ϕ, and cannot be thought of as prefixed (Bell-Kochen-Specker, or Bell-KS theorem). Two years before Bell (1964) had also proved, by producing a famous inequality which is violated in QP, that contextuality occurs even in the case of compound quantum systems whose elements are far apart, which implies that QP conflicts with an intuitive *locality* assumption (Bell's theorem). These results can be considered either paradoxical or not, depending on the epistemological attitude of the researcher. But nonlocality at least leaves in many physicists a feeling of uneasiness (Sakurai, 1985), while contextuality seems to imply some kind of mysterious "conspiracy of nature" (Mermin, 1993).

Apart from their puzzling consequences, the Bell and the Bell-KS theorems have deep philosophical implications. To understand better this point, let us consider an individual sample x of a given physical system in the state S. Then, the contextuality of QP implies that there are properties of x that cannot be thought of as true or false a *priori*, independently of the choices of the observer, since their truth values depend on the set of the measurements that one decides to perform. Hence, the Bell and the Bell-KS theorems seem to imply that QP requires the adoption of a verificationist truth theory for the language of QP, so that one of the basic philosophical choices in the standard interpretation of QP can be presented as imposed by the mathematical structure of the theory.

It is well known, however, that verificationism, which identifies *truth* with *epistemic accessibility* (here, briefly, *testability*) can be severely criticized from an epistemological viewpoint (e.g. Popper, 1969). In particular, the identification between truth and testability can be legitimately suspected to be the deep root of many quantum paradoxes. Thus, notwithstanding the above results, one may try to construct a language for QP endowed with a correspondence truth theory, so that truth and testability can be distinguished; this would actually realize a new approach to QP (according to which, in particular, every interpreted sentence in the language of QP has a truth value, so that, opposite to the standard interpretation, no "meaningless" statement exists), and one must obviously require that the new approach, besides solving canonical problems and avoiding paradoxes, also explains how the Bell and Bell-KS theorems can be circumvented, so that it is possible to adopt a correspondence truth theory.

I have attempted to carry out the above program in a number of papers

(Garola, 1991; 1992a; 1992b; 1992c; 1993). I have called my epistemological position *Semantical Realism* (SR) in order to put in evidence the adoption of a correspondence truth theory, which is compatible with some forms of realism, from one side, and the absence of any ontological assumption on physical reality, from the other side. In addition, I have recently integrated SR with a new general principle (MGP) regarding the truth mode of empirical physical laws (Garola, 1994; 1995a). An improvement of the system of axioms and definitions is also needed in order to make SR more suitable for dealing with the challenge represented by the Bell and Bell-KS theorems, and it will be presented in a forthcoming paper (Garola, 1995b).

Even in the former version SR provides a suitable background for solving a number of open problems in QP. For instance, it leads to classify Quantum Logic as a theory of testability in QP rather than a theory of truth in competition with Classical Logic. Furthermore, it entails that QP is an incomplete theory in a well defined technical sense. Finally, it allows one to avoid the classical EPR paradoxes by Furry and Bohm-Aharonov. But I would rather discuss here MGP and show that it invalidates Bell's theorem, hence one of the results that could be used in order to criticize SR (I retain that analogous arguments can be adopted in order to invalidate the Bell-KS theorem, and I am presently working on this subject). More specifically, I will introduce and justify MGP in Section 2 on a purely intuitive ground, avoiding the complications following from the formalized treatment of SR that I have supplied in other papers. I will then show in Section 3 how a typical sample proof of Bell's theorem [that is, the proof provided by Sakurai (1985), which can be considered a version of the canonical Wigner's proof (1970)] can be invalidated in a SR context by using MGP. Finally, I will briefly discuss in Section 4 how MGP can be used in order to invalidate different proofs, be they based on inequalities or not (e.g. Greenberger *et al.*, 1990).

2. The Metatheoretical Generalized Principle

Let us assume that the following sentence states a physical law:

for every individual sample x of a given physical system in the state S, x has the testable physical property E_1 iff it has the testable physical property E_2, and the testable physical property F_1 iff it has the testable physical property F_2.

This sentence can be expressed in formal logic by the formula:

$$V =_{df} (\forall x)(S(x) \rightarrow ((E_1(x) \leftrightarrow E_2(x)) \wedge (F_1(x) \leftrightarrow F_2(x))))$$

Let us discuss critically the canonical use of a law of this kind in physics. To this end, let x be interpreted in a laboratory i (space-time domain in the actual world) on an individual sample of the given physical system (briefly, a *physical object*) such that $S(x)$ is true, and let us consider three physical situations in which different *premises* are introduced.

- i) $E_1(x)$ is assumed to be true (or known to be true, for instance because a physicist in i has performed a suitable test of E_1 on x).
- ii) $E_1(x)$ and $F_2(x)$ are assumed to be true (or known to be true, for instance because a physicist in i has performed suitable tests of E_1 and F_2 on x).
- iii) $E_1(x)$ and $F_1(x)$ are assumed to be true (or known to be true, for instance because a physicist in i has performed suitable tests of E_1 and F_1 on x).

In Classical Physics (CP) one can easily use law V in order to predict that $E_2(x)$ must be true in case i), $E_2(x)$ and $F_1(x)$ must be true in case ii), $E_2(x)$ and $F_2(x)$ must be true in case iii). But let us consider the above cases from a quantum viewpoint and let us assume that the following commutation rules hold:

$$[E_1, E_2] = [E_1, F_2] = [F_1, E_2] = [F_1, F_2] = 0, \quad [E_1, F_1] \neq 0 \neq [E_2, F_2]$$

(here the physical properties and the projections that represent them in QP are denoted by the same symbols). Then, cases i) and ii) can occur in QP, and one can make the same predictions made in CP. On the contrary, one can never know that $E_1(x)$ and $F_1(x)$ are simultaneously true, as assumed in case iii), since E_1 and F_1 are not compatible. Thus, one has two possible choices: [a] one can accept the standard interpretation of QP and retain that it would be meaningless to say that $E_1(x)$ and $F_1(x)$ are conjointly true, so that no prediction can be made in this case; [b] one can admit that it is legitimate to consider a situation in which $E_1(x)$ and $F_1(x)$ are conjointly true, though this assumption can never be checked. Choice a) implies accepting *a priori* a verificationist truth theory, which has been criticized in the Introduction, together with contextuality and a number of other puzzling consequences. Choice b), which is unavoidable if a SR viewpoint is adopted, seems to imply at once that $E_2(x)$ and $F_2(x)$ must be true, as in CP. But if one looks deeper into the matter, one sees that this prediction depends on assuming implicitly a classical epistemological viewpoint according to which, roughly speaking, a physical law and all its consequences must be true in every laboratory (*Metatheoretical Classical Principle*, or, briefly, MCP; note that MCP is not consistent with choice a)). This conception is adequate to the worldview of CP, not to the operational philosophy underlying QP (which can be accepted even if the standard interpretation of QP is questioned), as case iii) itself clearly illustrates: in fact, one cannot guarantee the validity of V as a physical law in case iii), since it is impossible to know whether it actually occurs, hence to test V

(even indirectly) in this case. Thus, deducing that $E_2(x)$ and $F_2(x)$ are both true whenever $E_1(x)$ and $F_1(x)$ are simultaneously true sounds completely arbitrary if one accept an operational viewpoint.

The above discussion is important for our aims in this section. In fact, it suggest that, if a SR viewpoint is adopted, a new, more general conception of physical laws must be introduced. In order to work out and express this new conception properly, one should distinguish between *theoretical* and *empirical* physical laws and discuss the respective role of these laws in physics. I will not deal with this subject here (some syntetic hints on this topic are provided in Garola, 1995b). I limit myself to observe that the law V considered above must be classified empirical iff a testable physical property E exists such that $E(x)$ is logically equivalent to $((E_1(x) \leftrightarrow E_2(x)) \wedge (F_1(x) \leftrightarrow F_2(x))$, theoretical otherwise, and to state the new general principle (*Metatheoretical Generalized Principle*, or, briefly, MGP) in a nonformalized and simplified form, this statement being intuitively supported by the above criticism.

MGP. *Let V be a theoretical physical law and let V_A be an empirical physical law deduced from V. Then, V_A can be asserted to be true in every laboratory where, for every physical object x, compatible (equivalently, conjointly testable) premises are stated.*

MGP essentially states that the validity of an empirical physical law can be asserted only in observable physical contexts, while it can be questioned whenever nonobservable contexts are considered. This seems consistent with the basic philosophy of QP (unlike an implicit adoption of MCP), and does not imply any change in the observative content of QP. More formally, one can say that MGP supplies a new characterization of the truth mode of empirical physical laws, which is weaker than the classical mode formalized by MCP (see also Garola, 1991); as a consequence of MGP, the validity of these laws is limited in QP, where there are testable properties that cannot be tested conjointly, while it has no limit in CP, where no restriction exists on the compatibility of physical properties.

3. The critique to a sample proof of Bell's Theorem

In order to realize the program mentioned at the end of the Introduction, I will firstly generalize in this section the Bell inequality provided by Sakurai (1985) and yield a proof of Bell's theorem that clearly exemplifies the standard reasoning in this kind of proofs.

Let us consider a compound physical system consisting of two subsystems, say 1 and 2. Let A, B, C be observables of the whole system, A_1, B_1, C_1 observables of subsystem 1, A_2, B_2, C_2 observables of subsystem 2, such that $A = A_1 + A_2$, $B = B_1 + B_2$, $C = C_1 + C_2$. Let A_1, B_1, C_1, A_2,

B_2, C_2 have discrete non-degenerate spectrum, and let us assume that:

- i) $[A_1, B_1] \neq 0 \neq [A_2, B_2]$; $[A_1, C_1] \neq 0 \neq [A_2, C_2]$; $[B_1, C_1] \neq 0 \neq [B_2, C_2]$;
- ii) three eigenvalues a, b, c exist of A, B, C, respectively, that share a common eigenstate, say S, which is a second type (or nonfactorizable, or entangled) pure state.

Let a_{1j}, b_{1m}, c_{1q} be eigenvalues of A_1, B_1, C_1, respectively, and let a_{2k}, b_{2n}, c_{2r} be eigenvalues of A_2, B_2, C_2, respectively, such that $a_{1j} + a_{2k} = a$, $b_{1m} + b_{2n} = b$, $c_{1q} + c_{2r} = c$. Then, consider an ensemble of samples of the physical system that is described by the pure state S. Whenever an ideal measurement is made of A_1 on 1 in a given sample x, which yields the result a_{1j}, one can deduce that A_2 has value $a_{2k} = a - a_{1j}$ and predict the state S_{jk} of x after the measurement by using the projection postulate. Analogous conclusions hold whenever an ideal measurement is made of B_1 or C_1 on 1.

Now, let us introduce the following properties,

- A_{1j}: "the observable A_1 of subsystem 1 has value a_{ij}",
- A_{2k}: "the observable A_2 of subsystem 2 has value $a_{2k} = a - a_{1j}$",
- $A_{1j}^{\perp}$: "the observable A_1 of subsystem 1 has not the value a_{1j}",
- $A_{2k}^{\perp}$: "the observable A_2 of subsystem 2 has not the value $a_{2k} = a - a_{1j}$",

and let B_{1m}, B_{2n}, $B_{1m}^{\perp}$, $B_{2n}^{\perp}$, C_{1q}, C_{2r}, $C_{1q}^{\perp}$, $C_{2r}^{\perp}$, be defined analogously (by particularizing these properties, together with the physical system and the state, one can recover the cases that are usually found in the literature; for instance, the object x could be a sample of a system of two spin 1/2 particles in the singlet state, A, B, C could be the total spin in different directions $\hat{a}$, $\hat{b}$, $\hat{c}$, respectively, A_{1j} and $A_{1j}^{\perp}$ could be the properties "particle 1 has spin up in the $\hat{a}$ direction" and "particle 1 has spin down in the $\hat{a}$ direction", respectively, etc.).

Let us explicitly accept now the following assumptions.

R. *The results of all conceivable measurements are simultaneously prefixed (even in the case of incompatible observables).*

LOC. *Whenever a physical system consists of two subsystems 1 and 2 that are sufficiently far apart, a measurement made on subsystem 1 (or 2) does not modify the pre-fixed values of the observables of subsystem 2 (or 1).*

Because of R, each property above either holds or does not hold for every object x, independently of any act of observation. Because of LOC, each property of 1 can be tested without modifying the properties of 2, whenever 1 and 2 are sufficiently far apart.

Standing on R and LOC, let us introduce some further symbols, the form of which is justified by technical reasons that will not be discussed here. In every laboratory i, let us denote by $n_i(S(x))$ the number of physical objects in i that are in the state S, by $n_i(E_1(x) \wedge ... \wedge E_h(x)/S(x))$ the number of physical objects in i that are in the state S and have the physical properties $E_1, ..., E_h$, and put:

$$n_i(S(x)) \cdot f_i(E_1(x) \wedge ... \wedge E_h(x)/S(x)) = n_i(E_1(x) \wedge ... \wedge E_h(x)/S(x))$$

(hence, $f_i(E_1(x) \wedge ... \wedge E_h(x)/S(x))$ denotes a relative frequency, the value of which is uniquely defined whenever $n_i(S(x)) \neq 0$.

By using these symbols, one can write:

$$f_j((A_{1j}(x) \wedge B_{2n}^{\perp}(x))/S(x)) = f_i((A_{1j}(x) \wedge B_{2n}^{\perp}(x) \wedge C_{1q}(x))/S(x)) + f_i((A_{1j}(x) \wedge B_{2n}^{\perp}(x) \wedge C_{1q}^{\perp}(x))/S(x))$$

Indeed, the number of physical objects in the state S that have the properties A_{1j} and $B_{2n}^{\perp}$, is equal to the number of objects in S that have the properties A_{1j}, $B_{2n}^{\perp}$ and C_{1q} plus the number of objects in S that have the properties A_{1j}, $B_{2n}^{\perp}$, and "not C_{1q}" (i.e., $C_{1q}^{\perp}$). Analogously, one obtains:

$$f_i((A_{1j}(x) \wedge C_{1q}^{\perp}(x))/S(x)) = f_i((A_{1j}(x) \wedge B_{2n}(x) \wedge C_{1q}^{\perp}(x))/S(x)) + f_i((A_{1j}(x) \wedge B_{2n}^{\perp}(x) \wedge C_{1q}^{\perp}(x))/S(x))$$

and

$$f_i((B_{2n}^{\perp}(x) \wedge C_{1q}(x))/S(x)) = f_i((A_{1j}(x) \wedge B_{2n}^{\perp}(x) \wedge C_{1q}(x))/S(x)) + f_i((A_{1j}^{\perp}(x) \wedge B_{2n}^{\perp}(x) \wedge C_{1q}(x))/S(x))$$

By comparing all these equalities, one gets:

$$f_i((A_{1j}(x) \wedge B_{2n}^{\perp}(x))/S(x)) \leq f_i((A_{1j}(x) \wedge C_{1q}^{\perp}(x))/S(x)) + f_i((B_{2n}^{\perp}(x) \wedge C_{1q}(x))/S(x)) \tag{1}$$

Properties A_{1j} and $B_{2n}^{\perp}$ are compatible in QP, since they refer to different subsystems of a compound system. Therefore, they can be tested conjointly, and QP yields the probability that a physical object in the state S has both the properties A_{1j} and $B_{2n}^{\perp}$. Let us denote by $p((A_{1j} \cap B_{2n}^{\perp})(x)/S(x))$ this probability. By assuming from now on that the laboratory i is such that many samples of the given physical system are produced in it, one gets:

$$f_i((A_{1j}(x) \wedge B_{2n}^{\perp}(x))/S(x)) \cong p((A_{1j} \cap B_{2n}^{\perp})(x)/S(x))$$

Analogously

$$f_i((B_{2n}^{\perp}(x) \wedge C_{1q}(x))/S(x)) \cong p((B_{2n}^{\perp} \cap C_{1q})(x)/S(x))$$

Let us consider now $f_i((A_{1j}(x) \wedge C_{1q}^{\perp}(x))/S(x))$. Since S is an eigenstate of C corresponding to the eigenvalue $c = c_{1q} + c_{2r}$, it seems obvious to predict that 1 has the property C_{1q} iff 2 has the property C_{2r}. It follows:

$$f_i((A_{1j}(x) \wedge C_{1q}^{\perp}(x))/S(x)) = f_i((A_{1j}(x) \wedge C_{2r}^{\perp}(x))/S(x)) \qquad (2)$$

Therefore, being A_{1j} and $C_{2r}^{\perp}$, compatible, one gets:

$$f_i((A_{1j}(x) \wedge C_{1q}^{\perp}(x))/S(x)) \cong p((A_{1j} \cap C_{2r}^{\perp})(x)/S(x))$$

Thus, finally, one obtains the *Generalized Bell Inequality* by substituting probabilities to frequencies in the inequality (1):

GBI. $p((A_{1j} \cap B_{2n}^{\perp})(x)/S(x)) \leq p((A_{1j} \cap C_{2r}^{\perp})(x)/S(x)) +$
       $p((B_{2n}^{\perp} \cap C_{1q})(x)/S(x))$

GBI reduces to a standard Bell inequality whenever suitable specific systems and observables are considered (e.g., whenever one considers a system of two spin 1/2 particles in the singlet state, GBI reduces to the Bell inequality supplied by Sakurai).

In order to prove Bell's theorem it is now sufficient to observe that the probability predicted by QP may violate GBI in the particular cases mentioned above, so that GBI conflicts with QP. Thus, one is induced to retain that at least one of the explicit assumptions on which our proof of GBI is based, that is R and LOC, is not consistent with QP (*Bell's theorem*).

Our first goal in this section is thus achieved. Let us come now to our second goal, that is, to showing that the above proof is not correct whenever the arguments in Section 2 are accepted. To this end, let us analyze the reasoning that leads to eq.(2) in detail.

First, one considers all samples of the system such that 1 and 2 have the (compatible) properties A_{1j} and $C_{2r}^{\perp}$, respectively, and concludes that in each of them 1 also has the property $C_{1q}^{\perp}$, since S is an eigenstate of C corresponding to the eigenvalue $c = c_{1q} + c_{2r}$. Hence he gets the inequality:

$$f_i((A_{1j}(x) \wedge C_{2r}^{\perp}(x))/S(x)) \leq f_i((A_{1j}(x) \wedge C_{1q}^{\perp}(x))/S(x)) \qquad (3)$$

Second, one considers all samples such that 1 has the properties A_{1j} and $C_{1q}^{\perp}$, (which can occur in our framework but it cannot be checked, since A_{1j} and $C_{1q}^{\perp}$ are not compatible) and concludes that in each of them 2 has the property $C_{2r}^{\perp}$. Hence he gets the inequality:

$$f_i((A_{1j}(x) \wedge C_{1q}^{\perp}(x))/S(x)) \leq f_i((A_{1j}(x) \wedge C_{2r}^{\perp}(x))/S(x)) \qquad (4)$$

Putting together inequalities (3) and (4) one deduces eq.(2).

The above analysis shows that GBI not only depends on R and LOC, but also on the unrestricted validity of the following empirical physical law (*Perfect Correlation*, or PC law).

PC. *For every sample of the physical system in the state S, if subsystem 1 has the property $C_{1q}^{\perp}$, then subsystem 2 has the property $C_{2r}^{\perp}$, and conversely.*

This law can be formalized as follows,

$$(\forall x)(S(x) \rightarrow (C_{1q}^{\perp}(x) \leftrightarrow C_{2r}^{\perp}(x)))$$

and can be considered an empirical physical law deduced from the general theoretical law:

$$(\forall x)(S(x) \rightarrow (((A_{1j}(x) \leftrightarrow A_{2k}(x)) \wedge (A_{1j}^{\perp}(x) \leftrightarrow A_{2k}^{\perp}(x))) \wedge ((B_{1m}(x) \leftrightarrow B_{2n}(x)) \wedge (B_{1m}^{\perp}(x) \leftrightarrow B_{2n}^{\perp}(x))) \wedge ((C_{1q}(x) \leftrightarrow C_{2r}(x)) \wedge (C_{1q}^{\perp}(x) \leftrightarrow C_{2r}^{\perp}(x)))))$$

(the proof that the above law is theoretical requires full use of the axioms introduced in the SR approach to QP, and depends, in particular, on the assumption that the state S is entangled; I do not insist on this point here for the sake of brevity).

The unrestricted validity of the PC law is unquestioned in the proof of GBI provided above because of the implicit adoption of the principle called MCP in Section 2. More precisely, assumptions R and LOC from which GBI is deduced imply accepting as physically meaningful situations in which a given physical object is supposed to have some noncompatible properties (even if one cannot recognize by means of empirical tests whether a situation of this kind occurs), which generalizes the choice labelled with b) in Section 2; then, one proves GBI assuming implicitly MCP, which guarantees that the PC law holds in every physical situation. But it has been shown in Section 2 that MCP is not consistent with the operational philosophy of QP, so that, whenever choice b) is made, one must substitute MCP with the more general principle MGP. Therefore, one must renounce to MCP here and adopt MGP, which obliges to reconsider the above reasoning, since MGP limits the validity of empirical physical laws in QP, hence, in particular, of the PC law.

Let us consider the deduction of (3). Here, the set of physical objects is taken in account that have the properties A_{1j} and $C_{2r}^{\perp}$, which are compatible, hence constitute conjointly testable premises for each element of the set. MGP then states that the PC law is true in a laboratory i where only physical objects of this kind are considered, so that each object of the set must also have the property $C_{1q}^{\perp}$. Inequality (3) holds unaltered.

Let us consider the deduction of (4). Here the set of physical objects is taken in account that have the properties A_{1j} and $C_{1q}^{\perp}$, which are not compatible, hence constitute premises that are not conjointly testable for each element of the set; MGP then does not assure that the PC law is true in a laboratory i where these physical objects are considered, so that one cannot assert that each object of the set must also have the property $C_{2r}^\perp$. Inequality (4) could be false.

A breakdown in the deductive process that leads to GBI has thus been found. Indeed, eq.(2) could now be false, hence GBI, which is obtained by replacing $f_i((A_{1j}(x) \wedge C_{1q}^\perp(x))/S(x))$ with $p((A_{1j} \cap C_{2r}^\perp)(x)/S(x))$ in the inequality (1) (the correctness of which is not challenged by MGP) could also be false. Therefore, GBI does not follow from R and LOC.

The above result implies that one cannot assert that the conflict between QP and GBI implies a conflict of QP with R and LOC, hence it invalidates the proof of Bell's theorem supplied in the first part of this section, so that our second goal is achieved (of course, the invalidation extends to all proofs of Bell's theorem of which our proof is a generalization, e.g. Wigner, 1970, and Sakurai, 1985). By adopting the terminology introduced in Sections 1 and 2, one can summarize our reasoning by saying that the above proof of Bell's theorem rests on postulating an SR context (assumption R and LOC) and then adopting a classical viewpoint regarding physical laws (MCP), so that it is invalidated when a more consistent quantum viewpoint (MGP) replaces MCP.

I would like to conclude this section by noticing that a weaker inequality holds in our framework in place of GBI. Indeed, since MGP questions the validity of the PC law only in non observable contexts, one can still replace, $f_i((A_{1j}(x) \wedge B_{2n}^\perp(x))/S(x))$ and $f_i((B_{2n}^\perp(x) \wedge C_{1q}(x))/S(x))$ with $p((A_{1j} \cap B_{2n}^\perp)(x)/S(x))$ and $p((B_{2n}^\perp \cap C_{1q})(x)/S(x))$, respectively, in inequality (1), thus getting the following *Weakened Bell Inequality*.

WBI. $p((A_{1j} \cap B_{2n}^\perp)(x)/S(x)) \leq f_i((A_{1j}(x) \wedge C_{1q}^\perp(x))/S(x)) + p((B_{2n}^\perp \cap C_{1q})(x)/S(x))$

WBI is not an empirical law in the usual sense of the word, it cannot conflict with the predictions of QP (that regard probability values), and cannot be contradicted by any experimental result, since A_{1j} and $C_{1q}^\perp$ are not compatible, and the frequency $f_i((A_{1j}(x) \wedge C_{1q}^\perp(x))/S(x))$ (that might change with the laboratory, unlike probability) cannot be measured.

4. The general critique to the Bell theorem

As seen at end of Section 3, MGP invalidates the (generalized) Sakurai proof of Bell's theorem by questioning the unrestricted validity of the PC law that is (implicitly) adopted in the proof. Therefore, one expects that MGP also invalidates some recent proofs that do not use inequalities, but still need to assume (more or less explicitly), the unrestricted validity of the PC law. Let us consider for instance the proof supplied by Greenberger Horne and Zeilinger (GHZ proof), as reported in the paper by Greenberger et al. (GHSZ paper) quoted in the bibliography. Limiting ourselves to a nonformalized and intuitive approach, let us note that in this proof the following physical law (which is a general theoretical law) is stated:

$$E^\psi(\hat{n}_1, \hat{n}_2, \hat{n}_3, \hat{n}_4) = -cos(\phi_1 + \phi_2 - \phi_3 - \phi_4) \tag{5}$$

(see eq. 9 in the GHSZ paper). Then, a set of derived physical laws (that can be considered empirical laws whenever the angles ϕ and θ are specified) is deduced from eq.(5) by assuming locality, reality and completeness (but the last assumption does not seem strictly needed), as follows.

$$\begin{aligned}
A_\lambda(0)B_\lambda(0)C_\lambda(0)D_\lambda(0) &= -1 \\
A_\lambda(\phi)B_\lambda(0)C_\lambda(\phi)D_\lambda(0) &= -1 \\
A_\lambda(\phi)B_\lambda(0)C_\lambda(0)D_\lambda(\phi) &= -1 \\
A_\lambda(2\phi)B_\lambda(0)C_\lambda(\phi)D_\lambda(\phi) &= -1 \\
A_\lambda(\theta + \pi)B_\lambda(0)C_\lambda(\theta)D_\lambda(0) &= +1
\end{aligned} \tag{6}$$

(see eqs. 12a, 12b, 12c, 12d, and 17, respectively, in the GHSZ paper). Finally, a contradiction is proven to occur when simultaneously applying eqs.(6):

$$\begin{aligned}
A_\lambda(\pi) &- +A_\lambda(0) \\
A_\lambda(\pi) &= -A_\lambda(0)
\end{aligned}$$

According to GHSZ, this contradiction should show that the quantum law (5) is inconsistent with locality, reality and completeness. To be precise, it proves the inconsistency of these assumptions with the perfect correlation between spin directions that occur because of eq.(5) in special cases. But GHZ assume, when simultaneously applying eqs.(6), that perfect correlation simultaneously holds in different directions, that is, they implicitly assume the simultaneous unrestricted validity of all empirical laws deduced from the theoretical law (5). This is not correct if the analysis in Section 2 is accepted. Indeed, realism and locality imply that the choice labelled with b) in Section 2 is made, so that MGP must be adopted. Now, whenever one of the empirical physical laws (6) is assumed to hold in a laboratory i, it constitutes some premises in i that are easily seen to be noncompatible with the premises that one introduces further in order to deduce predictions from another law of the set. Hence, because of MGP, one cannot use all the aforesaid empirical laws conjointly in the laboratory i, which is sufficient to invalidate the GHZ argument. Thus, no inconsistency occurs between QP, here represented by the theoretical law (5), and the assumptions of locality and reality.

Let us compare the above intuitive invalidation of the GHZ proof by means of MGP with the more formal invalidation of the Sakurai proof in Section 3. It is then apparent that the reality and locality assumptions here correspond to assumptions R and LOC, respectively, in Section 3, while the perfect correlation here corresponds to the PC law in Section 3. In both cases, the standard viewpoint obtains a contradiction by introducing some assumptions (R and LOC, or reality and locality) and using a quantum physical law (PC law, or perfect correlation), and interpretes it as a conflict between the assumptions and QP, since it implicitly assumes MCP and does not recognize that this principle can be questioned. Whenever MGP is accepted, the contradiction is removed and no conflict arises between the assumptions and QP.

Let us now abandon the GHZ proof and add some hints on the invalidation of other proofs that can be found in the literature. More specifically, consider the proofs in which a Bell inequality appears of the form $\Delta \leq 2$, where Δ depends on a suitably defined *correlation function* (e.g. Selleri, 1988). In this case, MGP leads to invalidate the identification between the correlation function that appears in Δ and the correlation function introduced in QP, so that, if one substitutes the quantum correlation function in the expression of Δ, one obtains a new function Δ_Q, and no inconsistency occurs if $2 < \Delta_Q$ in some particular cases.

I would like to close this paper with a final remark on MGP. The introduction of this new principle has been motivated in Section 2 by the need of adopting a viewpoint more consistent with the basic philosophy of QP. But, surprisingly enough, MGP leads then to invalidate the claim that QP contradicts R and/or LOC. Should the Bell-KS theorem be also invalidated by MGP, as I intend to prove in a forthcoming paper, even the claim that QP necessarily implies contextuality, at least for nonseparated quantum systems, would be invalidated. This justifies the SR attempt of constructing a language for QP endowed with a correspondence truth theory and provides a context that is compatible with some kind of realism, though it does not imply it. Thus, one can conclude that a stricter adherence to the fundamental attitudes of QP opens the way to a more intuitive and "classical" interpretation of this theory.

References

Bell, J.S. (1964). On the Einstein Podolsky Rosen Paradox, *Physics*, 1 (195).
Bell, J.S. (1966). On the Problem of Hidden Variables in Quantum Mechanics, *Rev. Mod. Phys.*, **38** (437).
Bohm, D. (1951). *Quantum Theory*, Prentice-Hall, Englewood Cliffs (N.J.).

Bohm, D. and Aharonov, Y. (1957). Discussion of Experimental Proofs for the Paradox of Einstein, Rosen, and Podolsky, *Phys. Rev.*, **108** (1070).

Einstein, A., Podolsky, B., Rosen, N. (1935). Can Quantum Mechanical Description of Reality be Considered Complete?, *Phys. Rev.*, **47** (777).

Furry, W.H. (1936a). Note on the Quantum-Mechanical Theory of Measurement, *Phys. Rev.*, **49** (393).

Furry, W.H. (1936b). Remarks on Measurements in Quantum Theory, *Phys. Rev.*, **49** (476).

Garola, C. (1991). Classical Foundations of Quantum Logic, *Int. Journ. of Theor. Phys.*, **30** (1).

Garola, C. (1992a). Semantic Incompleteness of Quantum Physics, *Int. Journ. of Theor. Phys.*, **31** (809).

Garola, C. (1992b). Quantum Logics Seen as Quantum Testability Theories, *Int. Journ. of Theor. Phys.*, **31** (1639).

Garola, C. (1992c). Truth versus Testability in Quantum Logic, *Erkenntnis*, **37** (197).

Garola, C. (1993). Semantic Incompleteness of Quantum Physics and EPR-like Paradoxes, *Int. Journ. of Theor. Phys.*, **32** (1863).

Garola, C. (1994). Reconciling Local Realism and Quantum Physics: a Critique to Bell, *Teoreticheskaya i Matematicheskaya Fizika*, **99** (285).

Garola, C. (1995a). Criticizing Bell: Local Realism and Quantum Physics Reconciled, to appear in *Int. Journ. of Theor. Phys.*, **34**.

Garola, C. (1995b). Pragmatical versus Semantical Contextuality in Quantum Physics, to appear in *Int. Journ. of Theor. Phys.*.

Greenberger, D.M., Horne, M.A., Shimony, A., Zeilinger, A. (1990). Bell's Theorem without Inequalities, *Am. Journ. of Phys.*, **58** (1131).

Mermin, N.D. (1993). Hidden Variables and the Two Theorems of John Bell, *Rev. of Mod. Phys.*, **65** (803).

Popper, K.R. (1969). *Conjectures and Refutations*, Routledge and Kegan Paul, London.

Sakurai, J.J. (1985). *Modern Quantum Mechanics*, W.A. Benjamin, Reading (Mass.).

Selleri, F. (1988). History of the Einstein-Podolsky-Rosen Paradox, in *Quantum Mechanics Versus Local Realism* (F. Selleri ed.), Plenum Press, New York.

Wigner, E.P. (1970). On Hidden Variables and Quantum Mechanical Probabilities, *Am. Journ. of Phys.*, **38** (1005).

BELL'S INEQUALITY FOR TRICHOTOMIC OBSERVABLES

A. GARUCCIO
Dip. di Fisica dell'Università and Sez. INFN, Bari
Via Amendola 173 - 70126 - Bari, Italy
E-mail: GARUCCIO@bari.infn.it

and

L. DE CARO
Centro Nazionale Ricerca e Sviluppo Materiali
SS 7 Appia km. 712 - 72100 - Brindisi, Italy
E-mail: DECARO@cnrsm.csata.it

Abstract. In order to give a more complete description of an actual experiment on EPR paradox, the upper limits for Bell and Clauser-Horn-Shimony-Holt inequalities are deduced in the case of three-valued observables. This limit results a function of the supplementary assumptions. The different cases of "random no-detection processes" and "parameter-independent selection" are discussed. A new stronger limit $\eta = 0.811$ is deduced for the quantum detection efficiency in order to perform a "loophole-free" experiment.

1. Introduction

Let A and B be two trichotomic observables assuming the values $\{\pm 1, 0\}$. For example, in a typical EPR experiment using a source of correlated photon pairs, the values $A = \pm 1$ ($B = \pm 1$) can be associated to the transmission through the ordinary and extra-ordinary channels of the polarizer 1 (2) with axis orientation a (b), respectively, and subsequent detection; whereas $A = 0$ ($B = 0$) can be associated to the case of no count in the detector 1 (2), which means either absorption in the polarizer or not detection.

C. Garola and A. Rossi (eds.), The Foundations of Quantum Mechanics, 287-304.

Let the element of reality λ be variable over the set Λ, with density $\rho(\lambda)$ such that

$$\int_\Lambda d\lambda \rho(\lambda) = 1. \tag{1}$$

Let the result of measurements of observables A and B be dependent on the experimental parameters a and b, respectively, and on the element of reality. In general, the experimental parameters a and b are fixed in the structure of the apparatus in any given run, although they can be varied over different runs.

The expectation value of the observable A is by definition (e.g. De Grott, 1986)

$$E[A] = \int_\Lambda d\lambda \rho(\lambda) A(a,\lambda) \equiv m(a), \tag{2}$$

and, analogously, its variance is:

$$\sigma^2(a) = E[(A-m(a))^2] = E[A^2]-m^2(a) =$$

$$\int_\Lambda d\lambda \rho(\lambda) A^2(a,\lambda) - \left(\int_\Lambda d\lambda \rho(\lambda) A(a,\lambda)\right)^2. \tag{3}$$

By definition, the covariance of A and B is given by

$$R(a,b) = E[(A-m(a))(B-m(b))], \tag{4}$$

whereas the correlation of A and B is defined as

$$C(a,b) = \frac{R(a,b)}{\sigma(a)\sigma(b)}. \tag{5}$$

It is very important to stress that the correlation functions $C(a,b)$ coincide with the expectation values $E[AB]$ if and only if are satisfied the following conditions: $\sigma^2(a) = \sigma^2(b) = 1$ and $m(a) = m(b) = 0$, does not matter if the observable is dichotomic or trichotomic. Moreover, let us observe that the second condition is valid if and only if the observer 1 (2) has the same experimental frequencies for the cases A = 1 and A = -1 (B = 1 and B = -1). In the trichotomic case, the above condition is satisfied if the correlated pairs emitted by the source can be described by a quantum state rotationally invariant with respect to the propagation direction, and if the ordinary and the extraordinary channels are symmetric (the same number of photons is loss). In the dichotomic case, if A = 1 or B = 1 means transmission through the single-channel polarizer 1 or 2, and A = -1 or B = -1 means absorption in the polarizer 1 or 2, respectively, in order to have m(a) = m(b) = 0, the experimental apparatus must satisfy two conditions. The first is to use a source which generates correlated pairs described by a quantum state rotationally invariant with respect to the propagation direction. The second is to have the sum of the transmittances for photon polarization parallel and perpendicular to the polarizer axis equal to one. In all the experimental tests of the local inequalities obtained for a dichotomic observable, this second condition has been only approximately satisfied. In this case from Eq. (3) it follows that we cannot also have $\sigma^2(a)$ = $\sigma^2(b)$ = 1 and, of course, the correlation function C(a,b) coincident with the expectation value E[AB]. Bell's observable for trichotomic observables could be defined as a linear combination of the correlation functions: C(a,a';b,b') = |C(a,b) - C(a,b')| + |C(a',b) + C(a',b')|, as in the dichotomic case (Bell, 1965). Nevertheless, the Bell's inequality has been always defined as a linear combination of the functions E[AB]. Therefore, in order to obtain results directly comparable with those of the dichotomic case, also for the trichotomic case we will define the Bell's observable as a function of the expectation values E[AB].

2. Bell inequality for three-valued observables

Let us define the Bell's observable as follows:

$$\Delta_E(a,a';b,b') \equiv E[A(a)B(b)]-E[A(a)B(b')]| +|E[A(a')B(b)]+E[A(a')B(b')]|. \qquad (6)$$

From the properties of the integrals and from the definitions of $\Delta_E(a,a';b,b')$ and E[AB], it immediately follows:

$$\Delta_E(a,a';b,b') \leq$$

$$\int_\Lambda d\lambda\rho(\lambda)\Big[|A(a,\lambda)||B(b,\lambda)-B(b',\lambda)|+|A(a',\lambda)||B(b,\lambda)-B(b',\lambda)|\Big]. \qquad (7)$$

Now, let us define $\Lambda_0(a)$ and $\Lambda_0(a')$ as the two subset of Λ in which $A(a,\lambda) = 0$ and $A(a',\lambda) = 0$, respectively. Thus, we can subdivide the set Λ in two subsets: the subset $\Lambda_0(b) \cap \Lambda_0(b')$ of λ's such that both $A(a,\lambda) = 0$ and $A(a',\lambda)$ are zero; and the subset $\Lambda' = \Lambda - \Lambda_0(b) \cap \Lambda_0(b')$ of λ's such that either $A(a,\lambda)$ or $A(a',\lambda)$ or both are different from zero. When $\lambda \in \Lambda_0(a) \cap \Lambda_0(a')$ the contribution to the integral vanishes, therefore we can substitute the r.h.s. of (7) for the larger integral

$$\int_{\Lambda'} d\lambda \rho(\lambda) \big[\big| B(b,\lambda) - B(b',\lambda) \big| + \big| B(b,\lambda) - B(b',\lambda) \big| \big]$$

If $\Lambda_0(b)$ and $\Lambda_0(b')$ are the two subset of Λ in which $B(b,\lambda) = 0$ and $B(b',\lambda) = 0$, respectively, when $\lambda \in \Lambda_0(b) \cap \Lambda_0(b')$ the function $|B(b,\lambda)-B(b',\lambda)|+|B(b,\lambda)+B(b',\lambda)|$ is always equal to zero and we can calculate the previous integral only over the subset $\Lambda'' = \Lambda' \cap (\Lambda - \Lambda_0(b) \cap \Lambda_0(b'))$. It is a simple matter to show that, in this subset, $|B(b,\lambda)-B(b',\lambda)|+ |B(b,\lambda)+ B(b',\lambda)| = 2$. Thus, we obtain:

$$\Delta_E(a,a';b,b') \leq 2 \int_{\Lambda''} d\lambda \rho(\lambda) \equiv 2\mu(\Lambda''), \tag{8}$$

where $\mu(\Lambda'')$ denotes the measure of Λ''. From Eq. (1) and from the definition of Λ'' it follows that:

$$\Delta_E(a,a';b,b') \leq 2(1-\mu_0(a,a') -\mu_0(b,b') + \mu_0(a,a',b,b')). \tag{9}$$

where

$$\mu_0(a,a') = \mu(\Lambda_0(a) \cap \Lambda_0(a')) = \int_{\Lambda_0(a) \cap \Lambda_0(a')} d\lambda \rho(\lambda) \tag{10a}$$

$$\mu_0(b,b') = \mu(\Lambda_0(b) \cap \Lambda_0(b')) = \int_{\Lambda_0(b) \cap \Lambda_0(b')} d\lambda \rho(\lambda) \tag{10b}$$

$$\mu_0(a,a',b,b') = \mu(\Lambda_0(a) \cap \Lambda_0(a') \cap \Lambda_0(b) \cap \Lambda_0(b')) = \int_{\Lambda_0(a) \cap \Lambda_0(a') \Lambda_0(b) \cap \Lambda_0(b')} d\lambda \rho(\lambda) \tag{10c}$$

We would stress that the local upper limit given by Eq. (9) is stronger than the analogous limit valid in the dichotomic case. Only if all the $\mu_0(...)$ are equal to zero the upper limit becomes 2. But, in the most general case the $\mu_0(...)$ are different from zero and it will be necessary to evaluate their value in the different cases.

3. CHSH-type probabilities for three-valued observables

As the original Bell's inequality cannot be actually applied to a concrete experiment, Clauser, Horne, Shimony and Holt (CHSH) in 1969 introduced a deterministic approach in order to obtain a stronger inequality. The CHSH inequality, given by the following equation, follows from the hypothesis of local realism and the famous CHSH no-enhancement assumption:

$$-D(\infty,\infty) \leq \Delta_{CHSH}(a,a';b,b') \leq 0, \qquad (11)$$

where

$$\Delta_{CHSH}(a,a';b,b') \equiv D(a;b)-D(a;b')+D(a';b)+D(a';b')-D(a',\infty)-D(\infty,b),$$

and $D(a,b)$ denotes the joint transmission and detection probability with the lincar single-channel polarizers setted to the angles a and b, and ∞ the physical situation when the correspondent polarizer is removed.

In this section we will generalize the CHSH approach to the trichotomic case in order to obtain an inequality for the joint detection probabilities with no further additional assumption. In the next section, we will discuss how this inequality changes as a function of the additional hypotheses.

Let us define:

- $p(a_+,b_+)$ as the probability that both the photons of a pair will cross their respective polarizers with axes a and b through the ordinary channels, and that will be subsequently detected;
- $p(a_+,b_-)$ as the probability that both the photons of a pair will cross their respective polarizers with axes a and b through the ordinary and the extra-ordinary channels, respectively, and that will be subsequently detected;
- $p(a_+,b_0)$ as the probability that the first photon of a pair will cross its polarizer with axis a through the ordinary channel and will be subsequently detected, while the other photon will be absorbed;
- analogous definitions hold for $p(a_-;b_+)$, $p(a_-;b_-)$, $p(a_-;b_0)$, $p(a_0;b_+)$, $p(a_0;b_-)$ and $p(a_0;b_0)$.

When a trichotomic choice holds between: transmission through the ordinary channel <u>and</u> detection; transmission through the extra-ordinary channel <u>and</u> detection; no transmission <u>or</u> detection; we can write the following relations

$$\Sigma_{ij}\, p(a_i;b_j) = 1, \quad \Sigma_j\, p(a_i;b_j) = p_1(a_i), \quad \Sigma_i\, p(a_i;b_j) = p_2(b_j),$$

$$p(a_i;b_+)+p(a_i;b_-) = p(a_i;-), \quad p(a_+;b_j)+p(a_-;b_j) = p(-;b_j), \tag{12}$$

$$p(a_+;b_+)+p(a_+;b_-)+p(a_-;b_+)+p(a_-;b_-) = p(-;-),$$

where: the indexes i, j∈ I≡{+,-,0}; Σ_j and Σ_{ij} stand for the sum over I and I⊗I, respectively; and the symbol –, differently from that one of Eq. (11), means the sum of detected counts in the ordinary and extra-ordinary channels of the correspondent polarizer.

It is important to stress that in Eqs. (12) we have assumed that the integral sum of the counts in the ordinary and the extra-ordinary channel does not depend on the polarizer orientation a or b. The above simplifying hypothesis is directly related to the properties of rotational symmetry of the detecting apparatus. This hypothesis is analogous to that one made by CHSH in their work where they assumed that the sum of probabilities of transmission and absorption in a polarizer are independent of the polarizer orientation when the other polarizer has been removed, This kind of hypothesis permits only to simplify the mathematics but does not have a fundamental role in the EPR paradoxical result. On the other side, Quantum Mechanics gives the same predictions when one polarizer is missing and, therefore, we require that local hidden-variable theories satisfy them too. Moreover, the validity of the above assumption can be easily verified with an experimental test, in which the coincidences $p(a_+,-)$, $p(-,b_+)$ and so on, should be measured at different values of a and b.

Using Eqs. (12) it is a simple matter to show that the expectation value E(a;b) and p(-;-) can be expressed, respectively, by the following equations:

$$E(a;b) \equiv p(a_+;b_+)-p(a_+;b_-)-p(a_-;b_+)+p(a_-;b_-) =$$
$$4p(a_+;b_+)-2p(a_+;-)-2p(-;b_+)+p(-;-), \tag{13}$$

$$p(-;-) = 1-p(a_0; -)-p(-; b_0)-p(a_0; b_0) = 1-p_1(a_0)-p_2(b_0)+p(a_0; b_0). \tag{14}$$

Let us observe that in Eq. (13) we can simplify the notation neglecting the + symbols. It should be noted that in the left member of Eq. (14) there is no explicit dependence on the experimental parameters a and b. This is a consequence of the simplifying hypothesis that the sum of counts in the ordinary and extra-ordinary channels does not depend from a and b. Therefore, in all the functions on the right member of Eq. (14), we must have the same independence from a and b.

Finally, inserting Eq. (13) in the definition of Bell's observable for the trichotomic case, given by the Eq. (9), we obtain:

$$L(a,a';b,b') \le \Delta_{TRIC}(a,a';b,b') \le U(a,a';b,b'), \tag{15}$$

where

$$\Delta_{TRIC}(a,a';b,b') \equiv p(a;b)-p(a;b')+p(a';b)+p(a';b')-p(a',-)-p(-,b),$$

$$L(a,a';b,b') = -[1-\mu_0(a,a') -\mu_0(b,b') + \mu_0(a,a',b,b') + p(-,-)]/2, \tag{16}$$

$$U(a,a';b,b') = [1-\mu_0(a,a') -\mu_0(b,b') + \mu_0(a,a',b,b') - p(-,-)]/2. \tag{17}$$

The above inequality is quite different from the original CHSH one. This is due to the fact that it has not been obtained via the additional CHSH assumption.

Another important difference is related to the fact that the probability $p(-;-)$, being the sum of the detected counts in the ordinary and extra-ordinary channels, is different from $D(\infty;\infty)$ which appears in the original CHSH inequality (Eq. (11)). This latter is a double detection probability with the single-channel polarizers of the dichotomy case removed. However, if all the detectors have the same quantum efficiency η, both these two probabilities for real polarizers near to an ideal behavior should be equal to η^2, because almost all the photons absorbed in a single-channel polarizer would be those which would have travelled in the extra-ordinary channel of a two-channel polarizer. It is a simple matter to verify that analogous results hold for $p(a_+,-)$ and $D(a_+,\infty)$, $p(a_+,b_+)$ and $D(a_+,b_+)$, and so on. Thus the function $\Delta_{TRIC}(a,a';b,b')$ assumes the same values of $\Delta_{CHSH}(a,a';b,b')$ for near-ideal polarizers. In the following we will make this assumption.

4. Evaluation of the boundary limits

In this section, we will calculate $\mu_0(a,a')+\mu_0(b,b')-\mu_0(a,a',b,b')$ in order to evaluate the functions $L(a,a';b,b')$ and $U(a,a';b,b')$. We will consider two important physical situations. In correspondence with these two important cases we will obtain the maximum and the minimum values that the function $\mu_0(a,a')+\mu_0(b,b')-\mu_0(a,a',b,b')$ can assume. In all the remaining physical situations $\mu_0(a,a')+\mu_0(b,b')-\mu_0(a,a',b,b')$ will assume values between the calculated boundary ones.

Moreover, the two considered physical situations differ only in the additional hypoteses. Therefore, we will prove that lower and upper limits of the inequality (15) are functions of the additional hypoteses. Thus, our approach is able to include any additional hypoteses in the deduction of the inequality for locality.

4.1. RANDOM SELECTION OF THE SUBSET OF COINCIDENCES

The first important physical case concerns the physical situation for which $\Lambda_0(a)$, $\Lambda_0(a')$, $\Lambda_0(b)$ and $\Lambda_0(b')$ are completely randomly determined.

Let us denote with $\alpha(a)$, $\alpha(a')$, $\beta(b)$ and $\beta(b')$ the measure of the sets $\Lambda_0(a)$, $\Lambda_0(a')$, $\Lambda_0(b)$ and $\Lambda_0(b')$, respectively, i.e., the function $\alpha(a)$ is the probability for a photon to have the reality element λ belonging to the subset $\Lambda_0(a)$ of Λ. Moreover, the function $\mu_0(a,a')$ is the joint probability that λ will belong to $\Lambda_0(a)$ and $\Lambda_0(a')$. Therefore, the conditional probability $p(a'|a)$ that a λ, belonging to $\Lambda_0(a)$, will belong also to $\Lambda_0(a')$, is given by (if $\alpha(a) \neq 0$) $p(a'|a) = \mu_0(a,a')/\alpha(a)$. In the case of a $\Lambda_0(a)$ and a $\Lambda_0(a')$ completely randomly determined, we will have that the conditional probabilities are $p(a'|a) = \alpha(a)$, and

$$\mu_0(a,a') = \alpha(a)\alpha(a'). \qquad (18a)$$

Repeating the same reasoning for $\mu_0(b,b')$ and $\mu_0(a,a',b,b')$, we obtain

$$\mu_0(b,b') = \beta(b)\beta(b') \qquad (18b)$$

$$\mu_0(a,a',b,b') = \alpha(a)\alpha(a')\beta(b)\beta(b') \qquad (18c)$$

From now, we will refer to this physical situation as: "random selection of the subset of coincidences from all photon pairs"; or much briefly with: "random selection".

Moreover, from the assumed rotational symmetry property of our apparatus, the sum of counts in the ordinary and in the extra-ordinary channels of each detector does not depend on the orientation. Thus, for example, if we denote with $\Lambda_+(b)$ and $\Lambda_-(b)$ the subsets of in which $B(b,\lambda) = 1$ and $B(b,\lambda) = -1$, respectively, the above condition is equivalent to say that $\mu(\Lambda_+(b))+\mu(\Lambda_-(b))$

does not depend on b. Therefore, from the obvious condition $\Lambda_+(b)+\Lambda_-(b)+\Lambda_0(b)$ = Λ, we can also conclude that, even if $\Lambda_0(b)$ depends on b, its measure $\beta(b)$ does not depend on b. In other words, there are different reality elements which contribute to $\Lambda_0(b)$ and $\Lambda_0(b')$, but the probability of no detection is always the same β, does not matter the values of b and b'. The above conclusion permits us to write:

$$\mu_0(a,a')+\mu_0(b,b')-\mu_0(a,a',b,b') = \alpha^2 + \beta^2 - \alpha^2\beta^2. \tag{19}$$

In order to evaluate the functions U and L in the inequality (15) we have to calculate the values of $p(-;-)$ in the considered physical situation, or equivalently, from (14) the probabilities $p_1(a_0)$, $p_2(b_0)$ and $p(a_0,b_0)$. But the probability $p_1(a_0)$ of having no single count at the polarizer 1 is given by the measure of the subset $\Lambda_0(a)$, denoted with α. Analogously, $p_2(b_0)$ is given by β, and $p(a_0,b_0)$ is given by $\mu_0(a,b) = \mu(\Lambda_0(a)\cap\Lambda_0(b)) = \alpha\beta$. Therefore:

$$p(-;-) = 1-\alpha-\beta+\alpha\beta = (1-\alpha)(1-\beta). \tag{20}$$

Inserting Eqs. (19) and (20) into Eqs. (16) and (17), we have:

$$L(a.a',b,b') = -\frac{1}{2} [(1-\alpha)(1-\beta)((1+\alpha)(1+\beta) +1)] \tag{21a}$$

$$U(a,a',b,b') = \frac{1}{2} [(1-\alpha)(1-\beta)((1+\alpha)(1+\beta)-1)]. \tag{21b}$$

In Eqs. (21) we have the lower and the upper limit of the local inequality for trichotomic observables as a function of two variables: α and β. At this point, in order to make a quantitative evaluation of the functions L and U, we will make the simplifying hypothesis that $\alpha = \beta$ (symmetrical measuring apparata). In this case $p(-,-) = (1-\alpha)^2$, and the inequality (15) becomes:

$$-\frac{1}{2} p(-,-)[(1+\alpha)^2 +1)] \leq \Delta_{TRIC}(a,a';b,b') \leq \frac{1}{2} p(-,-)[(1+\alpha)^2 -1] \tag{22}$$

Thus, when we make the additional assumption that there is a "random selection of the subset of coincidences from all photon pairs", not only the lower but also the upper limit for the observable $\Delta_{TRIC}(a,a';b,b')$ depends on $p(-,-)$.

If we divide all the terms of inequality (22) by $p(-,-)$, we obtain the following inequality expressed in terms of the experimentally measurable detection rates $R(i,j) = p(i,j)/p(-,-)$:

$$-\frac{1}{2}\left[(1+\alpha)^2 +1)\right] \leq R_{TRIC}(a,a';b,b') \leq \frac{1}{2}\left[(1+\alpha)^2 -1\right] \qquad (23)$$

where

$$R_{TRIC}(a,a';b,b') = R(a;b) - R(a;b') + R(a';b) + R(a';b') - R(a',-) - R(-,b).$$

4.2. POLARIZER-ORIENTATION INDEPENDENCE OF THE SUBSET OF DETECTED PHOTONS

The second considered case is when all the elements of reality λ (and, therefore, all the photon pairs) which give no count, are always the same, does not matter the value of the orientation a, a' of the first apparatus, and the value of the orientation b, b' for the second one. This can be expressed by the following mathematical requirements:

$$\Lambda_0(a) \cap \Lambda_0(a') \equiv \Lambda_0(a) \equiv \Lambda_0(a') \qquad \forall a, a',$$

$$\Lambda_0(b) \cap \Lambda_0(b') \equiv \Lambda_0(b) \equiv \Lambda_0(b') \qquad \forall b, b', \qquad (24)$$

$$(\Lambda_0(a) \cap \Lambda_0(a')) \cap (\Lambda_0(b) \cap \Lambda_0(b')) \equiv \Lambda_0(a) \cap \Lambda_0(b).$$

From now, we will indicate this physical situation with the phrase: "polarizer-orientation independence of the selected subset of coincidences"; or much briefly with: "parameter-independent selection". Thus, as in the previous sub-section, the probability for a photon to have the element of reality belonging to the subset $\Lambda_0(i)$ of Λ, does not depend on i. But, contrary to the previous case, the conditional probability $p(i'|i)$ that a λ, belonging to $\Lambda_0(i)$, will belong also to $\Lambda_0(i')$, now is equal to one.

Then, from Eq. (24) we have

$$\mu_0(a,a') = p(a'|a)\alpha(a) = p(a|a')\alpha(a') = \alpha,$$

$$\mu_0(b,b') = p(b'|b)\beta(b) = p(b|b')\beta(b') = \beta, \tag{25}$$

$$\mu_0(a,a',b,b') = \mu(\Lambda_0(a) \cap \Lambda_0(b)) \equiv \mu(\alpha,\beta) \leq \min \{\alpha,\beta\}.$$

From Eqs. (14) and (25), it follows that

$$p(-,-) = 1 - \alpha - \beta + \mu_0(a,b). \tag{26}$$

Inserting Eqs. (25), (26) into Eqs. (16) leads to $L(a,a',b,b') = - p(-,-)$ and $U(a,a',b,b') = 0$. Thus, when we make the additional hypothesis of a "polarizer-parameter independence of the selected subset of coincidences", we have:

$$-p(-,-) \leq \Delta_{TRI}(a,a';b,b') \leq 0. \tag{27}$$

If we divide all the terms of the inequality (28) by $p(-,-)$, we obtain the following inequality expressed in terms of the   experimentally measurable detection rates $R(i,j) = p(i,j)/p(-,-)$:

$$-1 \leq R(a;b)-R(a;b')+R(a';b)+R(a';b') -R(a',-)-R(-,b) \leq 0 \tag{28}$$

In other words, we obtain an inequality for the experimental rates very similar to the CHSH inequality. The boundary limits in the inequality (28) are much more stronger than the corresponding ones in the inequality (23). But, these two inequalities have been obtained starting from the local inequality given by the Eqs. (15-16), and making different "ad-hoc" additional assumptions. This result simply shows the great influence of the additional hypoteses on the boundary limits of the CHSH-type inequalities.

In particular, the value zero for the upper limit of the CHSH inequality is obtained when:

1) the joint detection probability $p(-;-)$ is equal to 1, i.e. in the dichotomic case;
2) we assume that Λ is the sum of the two disjoint subsets, Λ' and Λ_0, such that for all $\lambda \in \Lambda'$ both photons of a correlated pair are detected, and for all $\lambda \in \Lambda_0$ we have <u>never joint detection</u>, whatever is the orientation of the two analyzers. Only in this case, we can take into account only the photon pairs with $\lambda \in \Lambda'$, and redefine a new density function

$$\rho'(\lambda) = \rho(\lambda)/\mu(\Lambda'), \tag{29}$$

reducing our trichotomic observables A and B to dichotomic ones.

5. What inequality represents correctly the locality?

As the lower and upper limits which bound the function $\Delta_{TRIC}(a,a';b,b')$ depend on the hypothesis of local realism as well as on an additional assumption, the choice of the inequality, which correctly represents the locality, becomes the main problem. In fact, the experimental tests of a local inequality could become meaningful, if this inequality stands on additional assumptions which are not valid for the considered experimental set-up. In the following sections we will give some examples.

In each case, we are dealing with an infinite number of inequalities, depending on the further additional assumption that we make. What of these inequalities we should choose in order to have a meaningful comparison with the quantum-mechanical predictions and/or the experimental tests?

The answer to this question is strictly related to the physical system used as test of the hypothesis of locality. Up to now, in all the approaches to experimental tests of the EPR paradox the procedure was the following: first, starting from the hypothesis of locality, an inequality was deduced for dichotomic observables; second, supplementary hypoteses were introduced in the local theory in order to deduce a new inequality with measurable quantities. At the same time the value of these measurable quantities was deduced in the quantum-mechanical theory using some other additional hypoteses. So in conclusion the experimental tests up to now performed have compared the quantum-mechanical predictions versus those of Einstein's locality plus some additional hypoteses. But, as in the case of the tests of the Einstein's locality, when we compare two different theories, the correct approach should be to avoid any supplementary hypotheses or, if it is possible, to use additional hypotheses either experimentally testable or explicitly in agreement with both theories, in order to play a "neutral" role on the problem of the locality. In order to satisfy the above mentioned requirement for a correct comparison, in the following section we will use in our approach the non-paradoxical quantum mechanical predictions as supplementary assumption, in order to select the inequality which represents correctly the locality.

6. Additional hypothesis selection

In the actual experiments with photon pairs the quantum efficiency η of the photomultipliers plays a fundamental role for determining the no detection events $A(a,\lambda) = 0$ and/or $B(b,\lambda) = 0$. Quantum Mechanics predicts for the joint and the single detection probability with no polarizers: $D(\infty;\infty) = \eta^2$ and $D_1(a,\infty) = D_2(\infty,b) = \eta$, respectively, if the detection apparata are symmetric ones. Therefore, Quantum Mechanics predicts that the conditional probability that a photon of a pair will be detected by the detector 2 when the first has just been detected by detector 1 is given by $p(\infty|\infty) \equiv D(\infty;\infty)/D_1(\infty) = \eta$. Thus, Quantum Mechanics, when there are no polarizers, assumes a "random selection of the subset of coincidences from all photon pairs"!

This assumption is also considered valid when there are polarizers. In fact, in order to obtain the quantum mechanical predictions for the joint detection and transmission probabilities, it is sufficient to multiply the joint transmission probabilities by η^2. In other words, for a photon pair which emerges from the polarizers, $p(-,-) = \eta^2$.

Therefore, it should be correct to compare the quantum-mechanical results for the EPR paradox with the inequalities (22) and (23), which have been obtained under the same additional physical hypothesis, concerning the independence of the two events of photon detection at the respective detectors.

More precisely, if we choose to assume identical additional assumptions in both the theories, we cannot compare the quantum-mechanical results with the inequalities (27) and (28). In fact, this latter has been obtained assuming the "polarizer-orientation independence of the selected subset of coincidences", which leads to a conditional probability $p(\infty|\infty) = 1$. Thus, inequality (28) stands on an additional assumption which is not compatible with the "non-paradoxical" predictions of the Quantum Theory. Obviously an additional hypothesis which leads to non-paradoxical empirical predictions in contrast with the quantum-mechanical results must be discarded.

7. Quantum Mechanics versus local realism

In section 3. we have stressed that the function $\Delta_{TRIC}(a,a';b,b')$ can assume the same values than $\Delta_{CHSH}(a,a';b,b')$, if the polarizers have a nearly ideal behavior, and all the detectors have the same quantum efficiency. This result is related to the fact that, when we are dealing with near-ideal polarizers, almost all the photons absorbed in a one-channel polarizer would be those which would have travelled in the extra-ordinary channel of a two-channel polarizer.

Under the above simplifying hypothesis, in the following sub-sections we will compare the upper limit of inequality (23) with the quantum-mechanical predictions for two important cases.

7.1. THE CASE OF PHOTON PAIRS IN SINGLET STATE

In figure 1 we have reported the quantum-mechanical prediction for the maximum value that the observable $\Delta(a,a';b,b')$ can reach , as a function of $\alpha = (1-\eta)$, for a singlet state with positive parity. In fact, for this state Quantum Mechanics predicts:

$$\Delta_{CHSH}(a,a';b,b')_{MAX} = \frac{\sqrt{2}-1}{2}\eta^2, \qquad (30)$$

or equivalently

$$R_{CHSH}(a,a';b,b')_{MAX} = \frac{\sqrt{2}-1}{2}. \qquad (31)$$

The above value has been compared with the local upper limits of the inequality given by Eq. (23). We obtain that the quantum-mechanical predictions violate the inequality (23) if and only if $\eta > \sqrt[4]{2} - 1 = 0.811$. This limit is the lowest one deduced for a "loophole free" experiment with symmetric entangled states and it is not far to the actual values of the quantum efficiency of some new photodetectors.

Let us observe that all the experimental tests made using atomic cascade photon sources and, therefore, photon pairs in singlet state have been compared with the CHSH inequality or, equivalently, with Eq. (28) (Freedman and Clauser, 1972; Clauser, 1976, Aspect et al., 1981; Aspect et al., 1982; Duncan et al., 1985). The experimental results are almost all in agreement on the violation of the upper limit equal to zero of this inequality. Moreover, also Quantum Mechanics violates inequality (28), and its predictions are in agreement with the experimental results. The above result has been considered as an experimental proof of the violation of locality and as the confirm of the validity of the quantum-mechanical predictions concerning correlated atomic systems. But, in the previous sections we have shown that the inequality (28) stands on an additional hypothesis which is in contradiction with Quantum Mechanics also for the non paradoxical predictions.

Thus, the fact that the experimental results violate the upper limit of the inequality (28), simply means that the additional hypothesis of a "polarizer-orientation independence of the selected subset of coincidences" is wrong. Therefore, nothing can be concluded about the validity of the hypothesis of locality because, in order to have a meaningful test of the locality, we need to have detectors with a quantum efficiencies greater than 0.811.

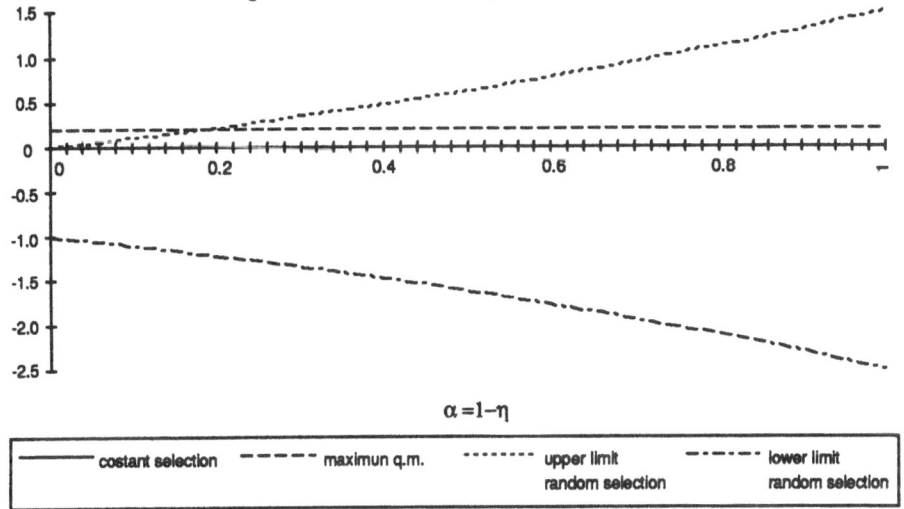

$$\alpha = 1 - \eta$$

| ———— costant selection | ———— maximun q.m. | ·········· upper limit random selection | —·—·—· lower limit random selection |

Fig. 1

The lower (dot and dash line) and the upper limits (dot line) for the observable $R_{CHSH}(a,a';b,b')$ defined with the measurement rates in the case of the additional hypothesis of a "random selection" , as a function of the no-detection probability $\alpha= 1-\eta$. The upper limit for the hypothesis of "polarizer-independent selection" is the continuous line which overlaps the x-axis. The dash line represents the maximum quantum mechanical value for the observable $R_{CHSH}(a,a';b,b')$.

7.2. THE CASE OF DOWN-CONVERTED PHOTON SOURCES

In the recent years, down-converted correlated photon pairs in a non-linear crystal have been used for performing tests of Einstein's locality via Bell's inequality (Ou and Mandel, 1988; Ou *et al.*, 1988; Shish and Alley, 1988; Tan and Walls, 1989). The state of the emerging pairs is given by:

$$|\psi> = \sqrt{T_xT_y} \, |x_1>|y_2> + \sqrt{R_xR_y} \, |y_1>|x_2> -$$
$$i \sqrt{T_xR_y} \, |x_1>|y_1> + i \sqrt{R_xT_y} \, |x_2>|y_2> \qquad (32)$$

where: R_x, R_y and T_x, T_y are the beam-splitter reflectivities and transmissivities, respectively, with $R_x+T_x= R_y+T_y= 1$; $|x_i>$ [$|y_i>$] is the polarization state along the x-direction [y-direction] for the photon in the i-th output channel of the beam-splitter.

When a down-conversion photon source is used in order to obtain a correlated pair of photons, like that one described by the quantum state expressed in the Eq. (32), and we are dealing with ideal detectors, at each polarization apparatus the choice is between three possibilities: 1) transmission (+1); 2) absorption [or detection along the orthogonal direction] (-1); 3) no detection because both the photons travel along the other channel of the beamsplitter reaching the other polarization apparatus (0). In this case it is a simple matter to verify that, for ideal detectors (De Caro and Garuccio, 1994):

$$D(\infty; \infty) = p(-,-) = \frac{1}{2}. \tag{33}$$

If we assume as correct the predictions of quantum mechanics about the distribution of the photon pairs in the two output channels of the beamsplitter, then

$$\mu_0(a,a') = \mu(\Lambda_0(a) \cap \Lambda_0(a')) = \frac{1}{4}.$$

$$\mu_0(b,b') = \mu(\Lambda_0(b) \cap \Lambda_0(b')) = \frac{1}{4}.$$

$$\mu_0(a,a',b,b') = \mu(\Lambda_0(b) \cap \Lambda_0(b') \cap \Lambda_0(a) \cap \Lambda_0(a')) = 0$$

and that the upper limit for the function $\Delta_{TRIC}(a,a';b,b')$ becomes:

$$U(a,a';b,b') \leq \frac{1}{2}\left[1 - \frac{1}{4} - \frac{1}{4} - \frac{1}{2}\right] = 0, \tag{34}$$

while the lower limit becomes

$$L(a,a'.b,b') = -\frac{1}{2}\left[1 - \frac{1}{4} - \frac{1}{4} - \frac{1}{2}\right] = -\frac{1}{2} = -p(-,-), \tag{35}$$

so we reobtain the CHSH inequality.

It is important to stress that:

i) in the deduction of the previous limits, the assumption of perfect photon detection plays a fundamental role; without this assumption we are not able to deduce the used values for $\mu_0(a,a')$, $\mu_0(b,b')$, $\mu_0(a,a',b,b')$;

ii) we have made the implicit assumption that the hidden variable λ determines the fate of each photon of the pair when it impinges on the beam-splitter, or, in other terms, we have assumed a "polarizer-independent selection of the subset of coincidences"

iii) the above upper limit is stronger than that one reported in De Caro and Garuccio, 1994. This is due to the fact that here we have used the inequality (9) which is stroger than the original Bell's inequality used in that paper.

The above result is, for a given point of view, surprising because the state (34) is a factorizable one and it has been proved that factorizable states always satisfy Einstein's locality and, therefore, Bell-type inequalities (Capasso *et al.*, 1973). Moreover, the fact that a factorizable state gives predictions which satisfy the inequality (28) represents an immediate proof of the role of the additional assumptions. The relation between entangled states and supplementary hypotheses will be argument of further studies.

References

Aspect, A., Dalibard, J. and Roger, G. (1982). *Phys. Rev. Lett.*, **49** (1804) .

Aspect, A., Grangier, P., and Roger, G. (1981). *Phys. Rev. Lett.*, **47** (460).

Bell, J.S. (1965). *Physics (N.Y.)*, **1** (195).

Capasso, V., Fortunato, D. and Selleri, F. (1973). *Int. J. Theor. Phys.*, **7** (319).

Clauser, J.F. (1976). *Phys. Rev. Lett.*, **36** (1223).

Clauser, J.F., Horne, M.A., Shimony, A. and Holt, R.A. (1969). *Phys. Rev. Lett.*, **23** (880).

De Caro, L. and Garuccio, A. (1994). *Phys. Rev. A.*

De Groot, M.H. (1986). *Probability and Statistic*, Addison-Wesley publishing Company, pp. 179, 219.

Duncan, W., Perrie, H., Beyer, J. and Kleinpoppen, H. (1985). *Fundamental processes in atomic collision physics*, 555 Plenum, New York.

Freedman, S.J. and Clauser, J.F. (1972). *Phys. Rev. Lett.*, **28** (938).

Ou, Z.Y., Hong, C.K. and Mandel, L. (1988). *Optics Commun.*, **67** (159).

Ou, Z.Y. and Mandel, L. (1988). *Rev. Lett.*, **61** (50).

Selleri, F. (1988). *Quantum Mechanics versus Local Realism: The Einstein, Podolsky, and Rosen Paradox*, F. Selleri (ed.), Plenum , New York.

Shish, Y.H. and Alley, C.O. (1988). *Pyhs. Rev. Lett.*, **61** (2921).

Tan, S.M. and Walls, D.F. (1989). *Optics Commun.*, **71** (235).

QUANTUM MECHANICS AND NONLOCALITY - THE EPR ARGUMENT RECONSIDERED

G. GHIRARDI

Dep. of Theoretical Physics - University of Trieste
International Centre for Theoretical Physics - Trieste
Interdisciplinary Laboratory of the ISAS - Trieste, Italy

and

R. GRASSI

Dep. of Physics - University of Udine
Udine, Italy

Abstract. We analyze the consequences of the nonlocal aspects of quantum mechanics as far as the possibility of attributing objective properties to individual physical systems is concerned. We reconsider the argument of the celebrated EPR paper with reference to a relativistic context and we show that, while the conclusion that quantum strict correlations and locality imply incompleteness is correct, the common further conclusion that requiring completeness implies the acceptance of some spooky action at a distance is inappropriate.

1. Introduction

As is well known, the assumption that the quantum predictions about the outcomes of measurements on entangled composite systems (whose costituents are far apart) are correct implies that nature exhibits nonlocal features. This can easily be proved by the following straightforward argument [1].

Let us consider a standard EPR-Bohm like set up, and let us denote by Q.M. the assumption that the quantum predictions for it hold true and by {100% Corr} the assumption that, if the two particles are subjected to spin measurements along the same direction, the corresponding outcomes are strictly

C. Garola and A. Rossi (eds.), The Foundations of Quantum Mechanics, 305-314.
© *1995 Kluwer Academic Publishers.*

anticorrelated. One can then resort to a combined use of the EPR and Bell's arguments, as follows:

i. $Q.M. \supset \{100\% \ Corr\}$

ii. $\{100\% \ Corr\} \wedge$ Locality $\supset$ Determinism   see, e.g., ref.[1].

iii. Locality $\wedge$ Determinism $\supset \{$Bell's Inequality$\}$ Bell's theorem.

iv. $Q.M. \supset \neg \{$Bell's Inequality$\} \supset \{\neg$Locality$\} \vee \{\neg$Determinism$\}$ by iii.

v. $\neg$Determinism $\supset \neg$Locality   by ii and i.

vi. $Q.M. \supset \neg Locality$ by iv and v.

This fundamental fact that quantum mechanics implies the acceptance of nonlocality leads quite naturally to raise the question of its compatibility with relativity. In this paper we will reconsider the EPR argument by taking into account the nonlocal features of natural phenomena. As we will see, such an analysis leads us to conclude that the cohexistence of quantum theory with relativity is even more peaceful than usually thought.

2. The EPR Argument

We recall some features of the argument which are relevant for our subsequent considerations.

• Measurements are always assumed to give definite outcomes.

• Free will is accepted.

• The analysis is performed within a non relativistic context, i.e. assuming an absolute time ordering between physical events, and no backward causation.

The starting point of the EPR paper is the very definition of individual elements of physical reality [2]: *if, without in any way disturbing a system we can predict with certainty (i.e., with a probability equal to unity) the value of a physical quantity, then there exists an element of physical reality corresponding to this physical quantity.* The sentence quoted above makes reference to the possibility of predicting something about a physical system, a fact that, even if it is perfectly acceptable within a nonrelativistic context in which there is an absolute time order, can give rise to difficulties in connection with relativistic considerations. We then rephrase the EPR criterion in a way which is perfectly adherent to the original formulation but does not make reference to time ordering.

2.1. A SHORT DIGRESSION: COUNTERFACTUALS

In the next Subsection, this reconsideration of the EPR criterion will lead us to deal with counterfactual statements. We will not discuss here the basic features of counterfactuals, we will refer the reader for this purpose to the beautiful book by D. Lewis [3]. We simply mention that counterfactuals are statements of the "if ... then" type in which the antecedent is known (or expected) to be false. In our use of it it will always be false. As a consequence, on the naive formalization using $\supset$ for "if ... then ..." all counterfactual statements would be true, contrary to our intuitive understanding of them. The way out comes from treating them as variably strict conditionals in the usual possible worlds semantic for modal logic. This requires to make precise [4] how the truth-value at a given possible world of a counterfactual depends on the truth values at various possible worlds of its antecedent and consequent.

Let us denote the counterfactual "If ϕ were true, then ψ would be true" as "$\phi \;\square\!\!\rightarrow \psi$" for propositions ϕ and ψ. Then Lewis proposes the following truth condition: $\phi \;\square\!\!\rightarrow \psi$ is true at world w iff either (i) there are no possible worlds in which ϕ is true or (ii) some world where both ϕ and ψ are true is more similar ("closer") to w than any world in which ϕ is true and ψ is false. Obviously one has to specify the possible worlds one is taking into account; this is done by assigning to each world w a set of worlds S_W called [3] the sphere of accessibility around w.

We agree with Lewis that the concept of similarity between worlds is to some extent vague, but in various cases, in particular in those we will consider in what follows, this vagueness can be advantageously resolved by appropriate natural assumptions.

2.2. PROPERTY ATTRIBUTION AND COUNTERFACTUALS

To analyze in full detail the EPR criterion for the attribution of elements of physical reality to individual physical systems and to investigate how one should adapt it to a relativistic context, we discuss a very simple example.

Suppose we are dealing with a spin-1/2 particle of which we know that, at the initial time $t=t_0$, its state vector is not an eigenstate of σ_z. Assuming that the spin is a constant of the motion, let us consider the following situation:

i). At time t_1 a measurement of σ_z is performed and yields the outcome +1;

ii). At time $t_M>t_1$, an observer who knows that σ_z has been measured at t_1 repeats the measurement and, obviously, he finds again the outcome +1;

iii). Finally, at time t_2 another measurement of σ_z is performed (and its outcome is, once more, +1).

The observer performing the measurement at t_M and finding +1, can predict that in the subsequent measurement at time t_2 the outcome will be +1, and thus, according to the EPR criterion, he can attribute the corresponding property to the system just before t_2. According to his knowledge the same observer can retrodict that the outcome of the measurment at t_1 has been +1, but he cannot legitimately state that just before t_1 the system possesses an element of physical reality corresponding to $\sigma_z=+1$. In fact, if one assumes that the quantum description is complete then, one must recognize that, due to the initial conditions, one could have obtained the outcome -1 in the measurement. This last assertion is clearly a counterfactual statement (since it refers to an alternative world, not the actual one). This shows how naturally counterfactuals enter in the problem of attributing elements of physical reality to individual physical systems[1].

Counterfactuals enter in a very natural way also in connection with the interpretation of the physical laws of quantum mechanics. Indeed if one maintains that counterfactuals must be avoided, one must take a position à la Bohr according to which any analysis of a quantum phenomenon must take into account the whole experimental set up. Moreover, one has to say that one can meaningfully speak of probabilities of outcomes of a certain observable, only if an instrument designed to measure it is actually in place (note that in taking such a position one must identify probabilities with limits of relative frequencies). According to the lucid analysis of B. d'Espagnat [5]:

> "as a rule, however the physical probability laws are not of this kind. In classical physics, and also in most of the usual formulations of quantum mechanics, when we say that the probability of such and such a value of an observable A is equal to some number p what we mean is that if an appropriate instrument were placed at an appropriate location and if the experiment were repeated a large number of times, the value in question, let us call it a, would be obtained in a fraction of the cases (approximately) equal to p. And we consider such a statement to be meaningful even in the cases in which no instrument is actually set that way ...".

`To understand better the implications of the above example let us take now into account the case in which only the measurement at t_M is performed and the outcome +1 is obtained. The observer can make the following counterfactual assertion: if a measurement of σ_z is performed at time t_2 then the outcome is +1. In fact, in the considered case, it is possible and appropriate to define the accessibility spheres in terms of matching up to a time shortly before the time of the counterfactual's antecedent - in our case t_2 - and preserving the laws of nature or at least spin being a constant of motion, an assumption which is usually referred to [3] as inevitability at time t. Thus, in all the accessible worlds the outcome of the measurement at t_M ($t_M < t_2$) is the same as in the actual world (i.e. +1) and then, in all accessible worlds in which the measurement at t_2 is performed, the outcome is +1. This proves the truth of the counterfactual assertion. Of course, according to the Lewis criterion for truth conditions, any counterfactual statement specifying an outcome of a spin-measurement at time t_1, is not true. Thus, we are led to link, with d'Espagnat [5], the attribution at time t of the property corresponding to $\sigma_z = +1$ to the truth of the counterfactual assertion: if a measurement of σ_z were performed at time t, then the outcome would be +1. This criterion for the attribution of elements of physical reality to individual physical systems derives quite naturally from the EPR criterion and it is equivalent to it in those cases, like the present one, in which there is a definite absolute time ordering among events. But it allows us to deal with situations in which the EPR criterion, based on the possibility of predicting the outcome of a measurement, cannot be used because a definite time ordering is lacking. Therefore we consider it fully appropriate to base the analysis of the EPR paradox in a relativistic context on this criterion for the attribution of elements of physical reality to individual systems.

It is convenient, in order to prepare the ground for further developments, to give a concise formal expression of the above argument. To this purpose we introduce some notational shortcuts:

$\mathcal{M}_S^{\mathcal{A}}(t)$: A measurement of the observable $\mathcal{A}$ of the physical system S is performed at time t.
$O_S^{\mathcal{A}=a}(t)$: The outcome of $\mathcal{M}_S^{\mathcal{A}}(t)$ is $\mathcal{A}-a$.
$\mathcal{P}_S^{\mathcal{A}=a}(t)$: The system S possesses, at time t, the objective property (or the element of physical reality) $\mathcal{A}=a$.

With these premises we can now formulate the criterion for attributing objective properties to individual quantum systems:
Property Attribution - Definition: We relate the possibility of attributing a specific objective property to an individual physical system to the validity of a counterfactual statement[2]:

$$\textbf{P.A.1}: \quad [\mathcal{M}_S^{\mathcal{A}}(t) \ \Box \rightarrow \ O_S^{\mathcal{A}=a}(t)] \ \supset \mathcal{P}_S^{\mathcal{A}=a}(t) \qquad (2.1)$$

We would like to stress that the above criterion **P.A.1** for property attribution allows inferences from statements involving propositions which constitute the natural universe of discourse for quantum theory (i.e. conditional statements about measurement outcomes), to the enlarged propositional system which allows sentences referring to objective elements of reality.

More generally, we will assert that the system S possesses an element of physical reality referring to the observable $\mathcal{A}$, and we will write $\mathcal{P}_S^{\mathcal{A}}(t)$, in accordance with the following prescription. Let us denote by $\hat{A}$ the self-adjoint operator associated to the observable $\mathcal{A}$. The corresponding eigenvalue equation is

$$\hat{A}|a_k\rangle = a_k|a_k\rangle \qquad (2.2)$$

The spectrum of the operator $\hat{A}$, which, for simplicity, we assume to be purely discrete, will be denoted by $Sp\{\hat{A}\}$. We now put:

$$\textbf{P.A.2}: \quad \mathcal{P}_S^{\mathcal{A}}(t) \equiv df \bigvee_{a_k \in S_p\{\hat{A}\}} \mathcal{P}_S^{\mathcal{A}=a}{}_k(t) . \qquad (2.3)$$

Thus, the truth of $\mathcal{P}_S^{\mathcal{A}}(t)$ is equivalent to the fact that S possesses an element of physical reality pertaining to $\mathcal{A}$. Note that knowing that $[\mathcal{M}_S^{\mathcal{A}}(t) \ \square \rightarrow O_S^{\mathcal{A}=a}(t)]$ holds true is a sufficient but not a necessary condition for making true the statement $\mathcal{P}_S^{\mathcal{A}}(t)$. In fact, within quantum mechanics, one could know that the state vector is an eigenstate of $\mathcal{A}$ without knowing to which eigenvalue it belongs; or one could be dealing with a classical theory in which one can always claim $\mathcal{P}_S^{\mathcal{A}}(t)$, even though one has no specific information about the system under consideration and about the outcomes of hypothetical measurements.

2.3. THE GALILEAN CASE

With reference to the EPR-Bohm singlet state set-up we assume, for simplicity, that the measurements at L and at R both refer to the z-spin components of the two particles (accordingly we will skip the subscript z from the spin observables we will consider). Both in the actual and in the accessible worlds a measurement is performed at R at time t_R, and we are interested in counterfactual statements concerning the outcome of a possible measurement performed at L at time $t_L > t_R$. Thus:

i. $\mathcal{M}_{SR}^{\sigma}(t_R)$: A measurement of the z-spin component is performed on the particle at R at time t_R.

ii. $\mathcal{M}_{SR}^{\sigma}(t_R) \supset O_{SR}^{\sigma=+1}(t_R) \vee O_{SR}^{\sigma=-1}(t_R)$: Such a measurement has as its outcome one of the two possible eigenvalues of σ_z.

The argument makes use of various assumptions which we make explicit and which we express, once more, by resorting to appropriately abbreviated notations:

iii. $\forall\, t_\in (t',t'')[(L,t)\{Iso\}(R,t_R)]$: Particle L is isolated from particle R for an appropriate time interval (t',t'') including both the time t_R at which particle R is subjected to the measurement and the time t_L.

iv. $\{100\%\ Corr\}$: It is a law of nature (to be preserved in the accessible worlds) that the outcomes of spin measurements at L and R are 100% correlated. In other words, the quantum predictions for the singlet state are correct.

v. G-Loc: A system cannot be affected by actions on a system from which it is isolated. In particular, elements of physical reality referring to a system cannot be influenced by actions on systems from which it is isolated.

The argument is then straightforward. Let us express it formally:

1. $O_{SR}^{\sigma=+1}(t_R) \lor O_{SR}^{\sigma=-1}(t_R)$ (by ii)

2. $\{O_{SR}^{\sigma=+1}(t_R) \supset [\mathcal{M}_{SL}^{\sigma}(t_L)\ \Box \rightarrow O_{SL}^{\sigma=-1}(t_L)]\} \supset \mathcal{P}_{SL}^{\sigma=-1}(t_L)$ (by iv and P.A.1)

3. $\{O_{SR}^{\sigma=-1}(t_R) \supset [\mathcal{M}_{SL}^{\sigma}(t_L)\ \Box \rightarrow O_{SL}^{\sigma=+1}(t_L)]\} \supset \mathcal{P}_{SL}^{\sigma=+1}(t_L)$ (by iv and P.A.1)

4. $\mathcal{P}_{SL}^{\sigma}(t_L)$ (by 2,3 and P.A.2)

5. $\mathcal{P}_{SL}^{\sigma}(t^{\#})$ for some appropriate $t' <t^{\#}<t_R$ (by v and spin constant of motion)

6. The state vector at $t^{\#}$ is the singlet state. (hypothesis)

7. Quantum Mechanics is incomplete. (see below)

In steps 2 and 3 one derives, by iv, the truth of a counterfactual statement concerning $O_{SL}(t_L)$ from a premise concerning $O_{SR}(t_R)$. This, as in the case of a single system discussed above, is perfectly legitimate since, within a Galilean context, there is an absolute time ordering.

The obvious conclusion expressed in point 7 of the above analysis derives from the following considerations: the singlet state does not contain any formal element referring to the element of physical reality for S_L which we have been compelled to recognise as possessed by it. In particular, there is nothing in the singlet state characterising the possessed property for the direction z. It follows that the theory is incomplete in the sense that it cannot account for existing objective elements of reality.

2.4. THE RELATIVISTIC CASE WITH THE LOCALITY ASSUMPTION

In this case the argument parallels strictly the one of the previous Subsection. Obviously, some of the previous assumptions must be adapted to the new context, but this raises no problems and the way to do it is quite natural. We have simply to replace assumptions iii and v by the following ones:

iii. $[(L,t_L)\{Space\text{-}Like\}(R,t_R)]$: The space-time regions (L,t_L) and (R,t_R) are space-like separated.

v. L-Loc: An event cannot be influenced by events in space-like separated regions. In particular, the outcome obtained in a measurement cannot be influenced by measurements performed in space-like separated regions; and analogously, possessed elements of physical reality referring to a system cannot be changed by actions taking place in space-like separated regions.

With these premises the argument can be developed. We must, however, remark that, in spite of the similarity with the previous case, there is now an essential difference in the use one makes of the locality assumption. In fact, in the present case, such an assumption must enter into play from the very beginning. In the counterfactual argument leading to the assertion that there is an element of physical reality at (L,t_L), the antecedent is $\mathcal{M}_{SL}^{\sigma}(t_L)$. Accordingly, when defining the sphere of accessible worlds, in our opinion, one can keep as true everything independent of the occurrence of the left measurement. In fact, even though, in accordance with the previous considerations there is a certain vagueness about the definition of the accessibility spheres, due to the locality assumption v, it is quite natural to consider the outcome of $\mathcal{M}_{SR}^{\sigma}(t_R)$ as independent of the occurrence of $\mathcal{M}_{SL}^{\sigma}(t_L)$. This means that in all accessible worlds the considered outcome at right is the same as the one of the actual world.

Once this is clarified one can trivially go on and repeat exactly the steps from 1 to 4 of the previous Section, arriving at the conclusion $\mathcal{P}_{SL}^{\sigma}(t_L)$.

2.5. THE RELATIVISTIC CASE WHEN NONLOCALITY IS TAKEN INTO ACCOUNT

The situation in the case under investigation turns out to be quite different, for a very simple reason. The very starting step of the analysis, i.e. the assumption that for all the accessible worlds the specific outcome at (R,t_R) is the same as the one, e.g. $O_{SR}^{\sigma=+1}(t_R)$, which has occurred in the actual world, when the locality assumption is given up, cannot be maintained. In fact, to accept nonlocality amounts exactly to admit[3] that the outcome at (R,t_R) might depend on the occurrence of $\mathcal{M}_{SL}^{\sigma}(t_L)$. If it were not so, then the quantum correlations would not embody the genuine outcome dependence of the theory but would simply correspond to occasional coincidences of outcomes [6].

The implications of the above argument should be obvious: if the only assumption that is made is that quantum predictions about outcomes are correct, the fact that in the actual world, in which only the measurement at R takes place, $O_{SR}^{\sigma=+1}(t_R)$ holds true, does not justify, by itself, either the counterfactual statement

$$[\mathcal{M}_{SL}^{\sigma}(t_L) \ \Box\!\!\rightarrow O_{SL}^{\sigma=-1}(t_L)] ,$$

or its opposite, i.e., the statement

$$[\mathcal{M}_{SL}{}^{\sigma}(t_L) \ \square \rightarrow O_{SL}{}^{\sigma=+1}(t_L)] \ .$$

Concluding, when one allows for nonlocality, one cannot make property attribution for the particle at L on the basis of one's knowledge about the particle at R. This obviously does not mean that, when nonlocality is accepted, the system at (L,t_L) possesses no objective properties; it simply means that in such a case, an argument of the EPR-type does not lead to its having such properties.

2.6. A SPOOKY ACTION AT-A-DISTANCE?

We recall that Einstein has expressed on various occasions the opinion that the EPR argument implies that, would one keep the completeness assumption about the theory, one must accept that in nature there are "spooky action at - a - distance". As an example among many others we may quote from ref. [7]: *one has to assume that the physically real in B suffers a sudden change as a result of a measurement in A.*

The argument we have just developed should have made clear that, if one assumes, as EPR did, the physically real to denote properties objectively possessed by an individual physical system, the above conclusion is unappropriate. In particular it is obviously illegitimate to claim that in an EPR-Bohm like situation there is instantaneous creation of properties at - a - distance.

This fact is, in our opinion, of some relevance. In particular it shows that, in a sense, the *peaceful coexistence* of standard quantum mechanics with relativity holds to a higher level than the one implied by the well known fact that nonlocality, being of the uncontrollable type, does not allow faster than light signalling.

Notes

[1] You might object that one could avoid counterfactuals by assuming as the criterion for property attribution to an individual physical system its being in a pure state which is an eigenstate of the considered observable. Then, in our example, there are no elements of physical reality referring to σ_Z prior to t_1, and the property $\sigma_Z=+1$ holds after this time. However, we reply, first of all, that we consider it important to be as general as possible, and, in particular, to make use only of the predictions of the theory, without dealing with state vectors. Secondly, we stress that the only reason to resort to the state vector for attributing

objective properties is the fact that only in the case of an eigenstate one can predict with certainty the outcome. So, one in effect goes back to Einstein's criterion.

2 We note that the converse requirement that, when a property is actually possessed, measuring it simply leads to knowledge about its value (expressible as $P_S^{A=a}(t) \supset [M_S^{A}(t) \supset O_S^{A=a}(t)]$) would amount to faithful measurement. Note that this formula involves only standard implications.

3 Obviously, due to the symmetry between R and L, one has also to recognize that outcomes at L might depend on measurements at R.

References

[1]. Redhead, M. (1987). *Incompleteness, Nonlocality and Realism*, Oxford University Press.

[2]. Einstein, A., Podolsky, B. and Rosen, N. (1935). 'Can Quantum Mechanical Description of Physical Reality be Considered Complete?', *Physical Review*, **47** (777).

[3]. Lewis, D. (1986). *Counterfactuals*, Basil Blackwell Ltd., Oxford.

[4]. Clifton, R.K., Butterfield, J.N. and Redhead, M.L.G. (1990). 'Nonlocal Influences and Possible Worlds - a Stapp in the Wrong Direction', *British Journal of Philosophy of Science*, **41** (5).

[5]. d'Espagnat, B. (1984). 'Nonseparability and the Tentative Descriptions of Reality', *Physics Reports*, **110** (201).

[6]. Bell, J.S. (1981). 'Bertlmann's Socks and the Nature of Reality', *Journal de Physique*, *Colloque C2*, **42** (41).

[7]. Born, M. (ed.), (1971). *The Born Einstein Letters*, Walker and Company, New York, p. 158.

SOME REMARKS ON NON-SEPARABILITY

E. GIANNETTO
Dip. di Fisica "A. Volta", Università di Pavia
Via A. Bassi 6 - 27100 - Pavia, Italy
GNSF/CNR, unità di Pavia

Abstract. Starting from Rothstein's analysis of the EPR paradox, the logical incompleteness of quantum mechanics and non-separability which implies "the death of atomism" (Primas) are discussed from both physical and logico-epistemological points of view. Indeed, the indeterminacy of the microphysical, simple-elementary entities as fundamental ground of *physis* involves the breakdown of any kind of reductionism. Reductionism presupposes the possibility of such a determinacy, and so it is nothing else than a particular form of "determinism". The indeterminate universe of physical processes is non-separable and a physics of the universe as a whole seems to be needed. Furthermore, from a historical and philosophical point of view the principle of separability is shown to be related to the meta-physical project of the technical dominion over nature and the emergence of non-separability in physical theories is analysed.

1. Introduction

Since the beginning of the XXth century the appearance of quantum physics has marked an *epochal* change in physics or at least in our insight of it. An epochal change, however, which for many reasons is not yet completely recognized. On one side, already at the level of of its formal (axiomatic) systematization a razionalization process has been operated hiding the general as well as the particular problems. The general problems are related to the breakdown of every conceptual (waves or particles, etc.) schema and of every representation of the world (quantum "reality" is undetermined). Among the "particular" problems we can recall the questions of casuality and discontinuity,

315

C. Garola and A. Rossi (eds.), The Foundations of Quantum Mechanics, 315-324.
© 1995 *Kluwer Academic Publishers.*

the needs of a new mathematics and a new logic, the "violations" of general principles like the ones regarding the uniformity of nature or the conservation of energy, etc[1].

On the other side, physicists' community has been reducing quantum physics just to a new, mere calculus and experimentation technics: technics able to secure us with a wider, technical and practical, dominion over nature even inside its fine structure. The microphysical domain becomes a new source and fund for new instrumental uses of nature at disposal of the human *will to power* . This was pointed out in Heidegger's analysis, but in my opinion it holds only at the level of a *sociological* description of the practices of physics[2].

Moreover, some physicists do not conceive quantum theory as a "universal theory" but, by cutting out the semantical differences, they consider it as a theory with a very limited domain of validitity in respect of the wider domain of classical physics; or they also try to delegitimate quantum theory just reducing it to a wider classical theoretical framework.

Such a "psychological" inertia to recognize the epochal change involved in quantum physics, in my opinion, is breaking down as long as is showing itself as devoid of any ground. Indeed, quantum physics allows us to have a new insight also on classical and relativistic physics. Furthermore, the present "revolution" related to the so-called *chaos physics* is showing us that the specific features of classical and relativistic physics as long as believed to be in opposition to the ones of quantum physics are only *illusory*. Predictability, completeness, and indeed determinism too, break down also for classical and relativistic physics: they reveal themselves as theoretically and experimentally non-effective. The link of chaos physics with quantum physics does not lay only on the ground of the conception of quantum physics as a historical example of an indeterministic physics, but is also historically realized within the works of Max Born since fifties. Indeed, starting from the dialogue with Einstein on quantum physics, Born has shown the unpredictability, indeterminism and incompleteness within classical and relativistic physics: he showed this just for the most simple systems, *one-degree-of-freedom* systems, and already for a *linear* error propagation from initial conditions[3]. The existence of this link of chaos physics with quantum physics is the reason I have said that quantum physics leads to an epochal change in physics or *at least in our insight of it.*

This epochal change, in my opinion, involves the relations between *physis* and *logos* themselves and leads to the breakdown of the logic, of the epistemology and of the ontology of (classical) physics. And we have to remember that classical physics has to be conceived as the actual realization of a process of razionalization of the world, which is proper own to the history of western metaphysics, and of the correlated process of the technical dominion over nature by mankind[4].

From this point of view, I would like to deal with the so-called problem of *non-separability* as it rises within quantum physics. I will briefly discuss the Einstein-Podolski-Rosen paradox from a theoretical and epistemological perspective. Indeed, in my opinion, a *principle of separability* of the world has been the first hidden presupposition at the ground of all the physics since its origins. A principle which is the theoretical counterpart of the political and technical *divide et impera* at the ground of the whole razionalization process of the world and of the technical dominion over nature.

2. EPR and non-separability

The question of non-separability within quantum mechanics, related to the analysis of the Einstein-Podolski-Rosen paradox, has been discussed from several points of view. In my opinion, a technical and conceptual, right solution to EPR was given on one hand - at least partially - on a paper by J. Rothstein and, on the other hand, an independent, deep interpretation of EPR pointing out non-separability was furnished by H. Primas[5].

Here, it is possible to remember very briefly that Rothstein has analysed an EPR version which involves the separate measurements of complementary observables on both partial subsystems derived from the "splitting" of a unique global system: the macroscopically registered results are revealed by a set of macroscopic "observers" as mutually non-interfering and Lorentz-equivalent. These observers are within equivalence classes, that is all the members of the same class agree on the proper quantum descriptions of the results. However, members belonging to different classes give mutually inconsistent, contradictory descriptions, except the case in which wave functions present in their descriptions are replaced by mixtures, that is density matrices. *If one states that quantum descriptions by wave functions were complete, they become mutually contradictory. Viceversa, if the quantum descriptions are given by statistical mixtures and hence they are incomplete, they become non-contradictory.*

The quantum mechanical description of individual subsystems S_1 and S_2 of a previous composed system S_{12} can be given only by terms of mixtures as objective states of $S_1 \times S_2$ (non-pure states) just because the measure process which allows us to obtain informations is an irreversible one. Rothstein has stated that in some way this argument allows to deduce the second law of thermodynamics, through relativity, as a necessary condition for the operational consistency of quantum mechanics.

There is, as noted by Rothstein, the breakdown of the myth of the theoretical completeness and non-contradiction features of quantum mechanics, which has been constituting the rationalization of Bohr's complementarity and Heisenberg's indeterminacy relations. The presumpted completeness of quantum mechanics, which was used as an epistemological obstacle in relation to the trials of formulating other theories (like Einstein's unified field or hidden variables ones) more complete than quantum mechanics can no longer be accepted. However, even if such incompleteness has to be partially understood as Einstein said, that is as a constitutive inability of quantum mechanics to characterize completely the presumpted physical "reality", its deep meaning has to be related to a *Gödel-like incompleteness*[6].

Indeed, since Gödel's theorems were published in 1931 (if not since Löwenheim-Skolem appeared pointing out the fundamental limits of any formal theory with a finite or enumerable number of axioms), it should be clear that theoretical "completeness" is a myth. These theorems, as well known, show the synthactical and semantical incompleteness (otherwise the theory would be contradictory), the presence of undecidable enunciates and the impossibility of proofs of validity, consistency and non-contradiction, the expressive unadequacy, the non-characterizability and non-unicity of models of any formal ("finite") theory[7].

In the EPR case, we have to deal with the undecidability between the two mutually contradictory enunciates which correspond to the different descriptions given by the two observers which could perform two different measurements. Thus, from an epistemological point of view, the quantum mechanics incompleteness does no more seem as an unpleasant feature to be avoided, but a structural, logical characteristic of any theory: here, incompleteness is not mathematically or logically *a priori* given, but it follows from the logic of *experimental*, physical *possible* operations on individual subsystems of a global system.

However, what does imply the need to use density-matrices to describe all the subsystems of the world? Indeed, *the concept of state for any individual subsystem loses meaning, breaks down. That is, non-separability of the world is the physical counterpart, correspondent feature of the formal use of density-matrices* (non-separability is implied also in relation to "spacelike" distances which do not allow any exchange of physical interactions).

Coming now to Primas' analysis, it could be said that the world is a whole which is not constituted by (separable) parts. From such a perspective, quantum physics radically differs from the reductionism and mechanicism involved as the paradigmatic background of classical physics, by which complex phenomena were treated in terms of few fundamental, elementary objects and of their interactions. Indeed, as Primas pointed out, within the used quantum field theories and also grand unification theories physicists go on to represent physical processes in a very superficial way, just ignoring such non-separability feature of quantum physics. Nuclei are conceived as constituted by protons and neutrons, atoms as constituted by nuclei and electrons, and so on. However, we are able only to break from the outside atoms in protons, neutrons and electrons, and these particles as free particles behave in a very elementary way under the action of the proper kinematical group of transformations. *Atomism is definitely dead, and we cannot perform a tensorial-product decomposition of physical "reality".* The fact that we can disintegrate nuclei, atoms, molecules in elementary systems does not imply that nuclei, atoms, molecules are constituted by elementary sub-systems: it is not possible to attribute any kind of individual existence to such sub-systems. For example, a molecule can be described only as an entangled, non-separable system of protons, neutrons and electrons together[8].

3. Non-separability and the quantum-relativistic framework

At the quantum-mechanical level, indeed, non-separability implies the unobservability, non-measurability of physical variables related to individual subsystems of the world[9]. If we consider the quantum-relativistic-field theoretical framework, we have to recognize the *unobservability, non-measurability of fields related to individual particles* which represent the new dynamical variables of the theory.

This implies that *regarding quantum mechanics we can use the wave-function language only for the universe as a whole, whereas, regarding quantum relativistic field theory, we should use field language only for the universe as a whole. Otherwise, we have to use density-matrices for individual particles.* That is, it would become necessary to consider a quantum-relativistic-field theory of the universe as a whole: indeed, such a theory has independently begun to be analysed for other purpose in the framework of a quantum cosmology as a so-called *third quantization theory*[10].

Furthermore, another point has to be stressed regarding quantum-relativistic-field theory. Indeed, in quantum chromodynamics (QCD) and then also in quantum electrodynamics (QED) there appeared correlation phenomena which cannot be eliminated also over spacelike distances: these phenomena have been explained just introducing the concept of a dynamical confinement of fields and of correspondent confined phases of the theory. We say that there is a breakdown of the cluster property (for so-recalled unphysical fields) and of asymptotic *completeness*. Indeed, we could say that these phenomena involve a sort of *dynamical* non-separability. In my opinion, the analogy is very deep and fundamental and we can give a similar treatment: on one side, it suggests us the possibility to conceive interaction phenomena on different scales as quantum non-separability phenomena by operating a sort of *kinematization* of interactions. On the other side, it suggests a new insight on *quantum non-separability as a sort of confinement (separability would involve also an infinite indeterminacy on energy) and quantum incompleteness as a sort of asymptotic incompleteness*[11].

Indeed, a deep analysis, at an axiomatic level, on the dynamical confinement phenomenon has pointed out that individual separate fields are non-local just because they are related to uneliminable specific interaction correlations, and they cannot be considered as observables, measurable quantities: they are not *physical* , because they imply the breakdown of the Lorentz-Poincaré symmetry. That is, the non-physical nature of such individual, separate, (non-local by interaction) fields, is needed from the point of view of the preservation of the Lorentz-Poincaré fundamental symmetry of the theory[12]. Separate fields are non-physical, fictitious, mathematical constructions: they physically correspond to a whole complex, non-elementary, non-separable "reality".

Thus, the relativistic analysis of EPR given by Rothstein must be modified: *the individual subsystems of the world are non-physical , just because their definition imply a breakdown of Lorentz-Poincaré symmetry; they are confined within the universe.* At a quantum-relativistic level, the reality of particles loses meaning: only the universe is physical; *non-separability is the non-particle-like structure of the universe, the non-divisibility by parts of the world.*

Is however possible a quantum-relativistic-field theory (as well as a wave-function theory) of the universe? The answer has to be negative because the universe as a whole is not observable, it is not measurable as long as there exists nothing outside: any observer is always within it, and the world as a whole is completely indeterminate. The only viable description is by density-matrices, related to the various chosen observers, which show the impossibility of a separate description of the observer and of the remaining part of the universe. As David Finkelstein said, a quantum universe does not exist, but only a *universal quantum* , that is, we have to deal with an irreducible multiplicity of non-separable physical processes which can never be described as a uni-verse or a totality[13].

4. Conclusions

First of all, we have to reconsider some questions. Is it true that in classical and relativistic physics atomism, reductionism and mechanicism are actually realized? That is, whithin classical and relativistic physics does it actually hold a principle of separability? No, it does not. Both Mach and Poincaré have respectively shown that classical and relativistic dynamics imply non-separability[14]. Indeed, Mach has shown that the origin of inertia has to be related to the presence of all the masses in the universe; the universe is mechanically non-separable:

$$d^2/dt^2 \left(\Sigma_i \, m_i \, r_i / \Sigma_i \, m_i \right) = 0.$$

Poincaré, following this kind of argumentation has shown that the relativity principle for the universe is a non-separability principle and is like a "gauge-invariance" principle whose transformations reduce to identity when one considers only physical, invariant, relational-"universal" variables.

Moreover, Whitehead has given us a completely relational formulation of special relativity and has further stressed that the relativity principle is nothing else than a principle of universal relatedness, that is a principle of non-separability of the world[15]. Recently, the logician Martin Davis has shown that the indeterminacy principle is nothing else than a relativity principle, and viceversa I have shown that the relativity principle is nothing else than a indeterminacy principle[16]. Indeed, non-separability was already a feature of Leibniz' dynamics, related to the actual meaning of the so-called pre-established harmony[17]. From this point of view, quantum and relativistic physics can be understood in terms of radical processes of de-construction of the meta-physical Newtonian interpretation of the language of mechanics and so also in terms of radical modifications of Newtonian classical physics by introduction of Leibnizian relationist developments. Thus, there is a breakdown of the last illusions of classical and relativistic mechanics, as features of the process of rational and technical dominion over the world, related to the western meta-physics.

Indeed, since its Greek origins, at least after Anaximander, physical theoretization, as related to a philosophical physics or to a scientific physics after Galilei, became meta-physics and nothing else than a research of the *arché*, that is of material or formal, onto-logical or logico-epistemological grounds, from which starting to reconstruct and to explain the variety of physical phenomena.

The indeterminacy of microphysical, elementary and simple entities as material grounds of *physis* does not only imply the breakdown of any kind of reductionism, but also the breakdown of such a project of razionalization and dominion over nature. Reductionism presupposes the possibility of a determination of the microphysical entities and therefore is nothing else than a particular specification of determinism. However, non-separability as a further indeterminacy of the physical processes implies the *non-physicity* itself of elementary, simple, local and individual entities through which it would be possible to (linearly) reconstruct the world. There is not only the impossibility of a material ground for nature, but also of any formal one, that is of any onto-logical or logico-epistemological ground. Indeed, the unobjectivability of any first-quantization physical property regarding individual physical entities as logical subjects, related to unobservability (for non-separability), implies the impossibility of any formal ground like the one exemplified by energy and the breakdown of any subject-predicate logic. The unobjectivability of any second quantization individual entities as unobservable and non-physical (related to non-separability) implies the breakdown of any logical reconstruction of the world in terms of the couple subject-object (for non-separability) and so of any epistemological ground. The unobjectivability of the third quantization world as a totality-universe implies also the breakdown of any holistic ontology of the world being and of any representation of the world: that is the breakdown of any reduction of the world to representation or image. From a logical point of view, hence, there is no possibility to define the objects themselves of a physical universe of discourse[18].

The world shows itself as non-separable and indeterminate, like Anaximander's το απειρον: indeed, the separability of the world by parts or particles was the first hidden presupposition of any numerical-quantitative determination of physical processes, the first presupposition of any measurement, of any λογος as ratio (for the first time introduced by Pythagorean mathematical physics), of the determinacy principle which underlies the non-contradiction principle, as well as of any *principium individuationis* of both objectivity or subjectivity. Thus, by the recognition of a quantum radical non-separability of the world, there is the breakdown of the presuppositions themselves of any physical, philosophical or scientific, representation-theoretization and therefore of the presuppositions of the quantum physics itself.

Hence, at variance with Heidegger's analysis which was embedding quantum physics within a technical-scientific paradigm of dominion over nature which would not be possible to overcome as long as it has its roots in the lifeworld (*Lebenswelt*) of the human will to power; at variance with the seeming sociological features of the "quantum establishment" within the physicists' community; at variance with the appearent new forms of the technical dominion over the world allowed only by a *pragmatical and statistical use* of quantum physics; quantum physics, from indeterminacy to non-separability, has given actually rise to an epochal change, pointing out the lack of any foundation ground of the process of the rational and technical dominion over nature by which the human species has been constituting its peculiarity, and showing the unavoidable limits which nature presents in respect to its reduction to a resource fund (*Bestand*) of exploitation at disposal of the human will to power, and in respect to the constitution of its absolute mastery by men (*Ge-stell*)[19]. We have no more to comprehend the world but just to be com-prehended within the world: non-separability implies a sort of *physical hermeneutics* , related to a physical, non-ontological *Dasein* (indeterminate, non-separable part of the world).

Would it be possible that such a consciousness, required more and more from our world transformed by technics itself as realized by such a science, overcomes and transforms the lifeworld itself which has given rise to it? Or should we wait for the complete breakdown of such historically and genetically determined lifeworld under the heaviness of its proper own contradictions? Could be enough for us in such a waiting and hope the consciousness of the physical irreducibility of nature (of the illusory character of lifeword dominion over nature), whereas nature is even subjected to any seeming - even if non-effective - violent dominion of the lifeworld at the inter-specific, intra-specific, ecological, biological, social and political levels? In any case, physics, beyond any ethical indication, shows us a *physis* of which, either conscious or not, we are parts belonging to that secret love which is the non-separability of the world.

Notes and references

[1] For these issues, see for example: Giannetto, E. *La logica quantistica tra fondamenti della matematica e della fisica*, in *Foundations of Mathematics & Physics*, ed. by U. Bartocci and J. P. Wesley, Wesley, Blumberg 1990, pp.107--127; Giannetto, E. *Toward a Quantum Epistemology*, in *Atti del Convegno S.I.L.F.S. Temi e Prospettive della Logica e della Filosofia della Scienza*, ed. by M.L. Dalla Chiara and M.C. Galavotti, CLUEB, Bologna 1988, pp.121--124; Giannetto, E. (1991), *L'epistemologia quantistica come metafora antifondazionistica*, in *Immagini Linguaggi Concetti*, ed. by S. Petruccioli, Theoria, Roma, pp.301--322; Giannetto, E. *On Truth: A Physical Inquiry*, in *Atti del Congresso 'Nuovi problemi della logica e della filosofia della scienza'*, I, ed. by C. Cellucci and M. Dalla Chiara, CLUEB, Bologna 1991, pp.221--228; Giannetto, E. *Il crollo del concetto di spazio-tempo negli sviluppi della fisica quantistica: l'impossibilità di una ricostruzione razionale nomologica del mondo*, in *Aspetti epistemologici dello spazio e del tempo*, ed. by G. Boniolo, Borla, Roma 1987, pp.169--224.

[2] Heidegger, M. *Die Frage nach der Technik*, in *Vorträge und Aufsätze*, Neske, Pfullingen 1954; Heidegger, M. *Wissenschaft und Besinnung*, in *Vortrage und...*, op. cit.; Giannetto, E. *Heidegger and the Question of Physics*, Proceedings of the "Conference on Science and Hermeneutics" (Veszprém 1993), ed. by O. Kiss and L. Ropolyi, Reidel, Dordrecht (in press).

[3] Giannetto, E. *Note sulla complessità: Max Born e la nascita della nuova fisica del caos*, in *Atti del congresso S.I.L.F.S. Logica e filosofia della scienza: problemi e prospettive*, ed. by C. Cellucci, M.C. Di Maio and G. Roncaglia, Edizioni ETS, Pisa 1994, pp.317--330.

[4] See notes 1, 2, 3.

[5] Rothstein, J. *Physics of Selective Systems: Computation and Biology*, Int. J. Theor. Phys. **21** (1982), pp.327--350; Primas, H. in *Sixty-Two Years of Uncertainty*, edited by A. I. Miller, Plenum Press, New York, 1990, p.233; Primas, H. *Chemistry, Quantum Mechanics and Reductionism*, Springer Verlag, Berlin 1983; Primas, H. *Time-Asymmetric Phenomena in Biology: Complementary Exophysical Descriptions Arising from Deterministic Quantum Endophysics*, preprint 1988, LFC-Zürich; for a general overview, see also: D'Espagnat, B. *Conceptual Foundations of Quantum Mechanics*, Benjamin, Menlo Park, 1971; Bohm D. and Hiley, J.B. *The Undivided Universe*, Routledge, London 1993; R. Kitchener (ed.), *The World View of Contemporary Physics*, State University of New York Press, Albany 1988.

[6] See, for example: Giannetto, E. *The Epistemological and Physical Importance of Gödel's Theorems*, in *First International Symposyum on Gödel's Theorems*, ed. by Z. W. Wolkowski, World Scientific, Singapore 1993, pp.136--147.

[7] See references given in the paper quoted in the previous note 6.

[8] See papers quoted in note 5.

[9] See also the demonstration of the impossibility to attribute a wave function to a single (one-particle) well-defined physical system in quantum mechanics, given in the following paper: Preparata, G. *What is Quantum Physics? Back to the QFT of Planck, Einstein and Nernst*, lecture given at the IX Winter School on Hadron Physics, Folgaria (Italy) 6-13 February 1994, MI-TH 94/3 preprint. However, in my opinion, it is not enough to realize a many-particle theory to deal with the non-separability problem.

[10] See for example: Caderni, N. and Martellini, M. (1984), "Third Quantization Formalism for Hamiltonian Cosmologies", *Int. J. Theor. Phys.*, **23** p.223.

[11] Giannetto, E. (1992), *Teoria quanto-relativistica delle fasi macroscopiche della materia condensata: la transizione fluido-solido*, ph. d. thesis, University of Messina, Messina; Giannetto, E. and Giaquinta, P.V. (1993), "Towards a Quantum-Relativistic Understanding of the Phases of Matter: The Fluid-Solid Transition", *Physics Essays*, **6** pp.98--109.

[12] See note 11 and Kugo, T. and Ojima, I. (1979), *Suppl. Prog. Theor. Phys.*, **66** p.1.

[13] Finkelstein, D. *The Universal Quantum*, in R. Kitchener (ed.), *The World View...*, op. cit., pp.75--89.

[14] Mach, E. (1883), *Die Mechanik in ihrer Entwickelung historisch-kritisch dargestellt*, Brockhaus, Leipzig; Poincaré, H. (1912), "L'éspace et le temps", *Scientia*, **12** pp.159--171; Giannetto, E. (1994), *Henri Poincaré and the rise of special relativity*, lecture delivered in Protvino (Russia), 27 June 1994, at the *International Seminar Devoted to the 140th Birthday of Henri Poincaré*, in press; Giannetto, E. (1993), *Lectures on Relativity*, mimeographed paper, University of Pavia, Pavia.

[15] Giannetto, E. (1993), *Mach's Principle and Whitehead's Relational Formulation of Special Relativity*, conference delivered at the International Congress on 'Mach's Principle', Tübingen, July 1993 (in press); Whitehead, A.N. (1919), *An Enquiry on the Principles of Natural Knowledge*, Cambridge University Press, Cambridge; Whitehead, A.N. (1920), *The Concept of Nature*, Cambridge University Press, Cambridge; Whitehead, A.N. (1922), *The Principle of Relativity with applications to Physical Science*, Cambridge University Press, Cambridge.

[16] Giannetto, E. (1992), *On Relativity Theories and Leibniz*, conference delivered at the International Conference 'Albert Einstein', Ulm, March 1992 (in press); Giannetto, E. *Heidegger and...*, op. cit.; Giannetto, E. *Note sulla...*, op. cit.; Davis, M. (1977), "A Relativity Principle in Quantum Mechanics", *Int. J. Theor. Phys.*, **16** p.867.

[17] Leibniz, G.W. *Mathematische Schriften*, ed. by C. G. Gerhardt, Halle 1850-63; Cassirer, E. *Leibniz' System in seinen wissenschaftlichen Grundlagen*, Elwert, Marburg 1902 and Wissenschaftliche Buchgesellschaft, Darmstadt 1962; Cassirer, E. *Erkenntnisproblem in der Philosophie und Wissenschaft der neuren Zeit*, Berlin 1911-1920; Giannetto, E. (1988), *Lectures on Leibniz*, mimeographed paper, University of Messina, Messina; Giannetto, E. (1991), *Lectures on the History of Energy Concept*, mimeographed paper, University of Messina, Messina.

[18] See references given in note 1.

[19] See references given in note 2.

UNSHARP ORTHOALGEBRAS AND QUANTUM MV ALGEBRAS

R. GIUNTINI

Dipartimento di Filosofia, Università di Firenze
Via Bolognese 52 - 50139 - Firenze, Italy
E-mail: giuntinirisc.idg.fi.cnr.it

Abstract. We prove that any unsharp orthoalgebra gives rise to a quantum MV algebra (QMV algebra) and that any QMV algebra determines in natural way an unsharp orthoalgebra. Some properties of the QMV algebra of all effects of a Hilbert space are also investigated.

1. Introduction

Unsharp orthoalgebras (called also *effect algebras* [4] and *D-posets* [7]) are partial algebraic structures, where the basic operation (⊞) is not always defined. Unsharp orthoalgebras were introduced in [6] in order to develop a formal language for unsharp properties and *effects* of a Hilbert space. Since the operation ⊞ is partial, the question naturally arises whether ⊞ can be extended in a natural way to a total operation. The main aim of this paper is to show that unsharp orthoalgebras can be naturally extended to *quantum MV algebras* ([5]). Quantum MV algebras (QMV algebras) represent a generalization of MV (=multi-valued) algebras, first introduced by Chang [1] in order to prove a completeness theorem for infinite-valued logic (*Łukasiewicz logic*).

As proved in [6], the class $E(\mathcal{H})$ of all effects of any Hilbert space $\mathcal{H}$ determines an unsharp orthoalgebra. As a consequence of the extension theorem (Theorem 4.2), $E(\mathcal{H})$ gives rise to a QMV algebra. However, the QMV algebra of effects is not an MV algebra.

C. Garola and A. Rossi (eds.), The Foundations of Quantum Mechanics, 325-337.
© 1995 *Kluwer Academic Publishers.*

2. Unsharp orthoalgebras

Unsharp orthoalgebras are partial algebraic structures, where the basic operation ($\boxplus$) is not always defined. When $\boxplus$ is defined for two elements a, b we will write $\exists(a \boxplus b)$.

Definition 2.1 An *unsharp orthoalgebra* is a partial algebraic structure $\mathcal{A} = \langle A, \boxplus, 1, 0 \rangle$, where 1 and 0 are two distinct elements of A and $\boxplus$ is a partial binary operation on A which satisfies the following conditions.

(UO1)   *Weak commutativity*
   $\exists(a \boxplus b) \implies \exists(b \boxplus a)$ and $a \boxplus b = b \boxplus a$.

(UO2)   *Weak associativity*
   $[\exists(b \boxplus c)$ and $\exists(a \boxplus (b \boxplus c))] \implies [\exists(a \boxplus b)$ and $\exists((a \boxplus b) \boxplus c)$ and $a \boxplus (b \boxplus c) = (a \boxplus b) \boxplus c]$.

(UO3)   *Strong excluded middle*
   For any a, there exists a unique x s.t. $a \boxplus x = 1$.

(UO4)   *Weak consistency*
   $\exists(a \boxplus 1) \implies a = 0$.

An orthogonality relation, a partial order relation and a generalized complement can be defined in any unsharp orthoalgebra.

Definition 2.2 Let $\mathcal{A} = \langle A, \boxplus, 1, 0 \rangle$ be an unsharp orthoalgebra and let $a, b \in A$.

 (i) a is *orthogonal* to b ($a \perp b$) iff $a \boxplus b$ is defined in A.

 (ii) a *precedes* b ($a \sqsubseteq b$) iff $\exists c \in A$ s.t. $a \perp b$ and $b = a \boxplus c$.

(iii) The *generalized complement* of a is the unique element a' s.t. $a \boxplus a' = 1$ (the definition is justified by the strong excluded middle condition (UO3)).

The notion of unsharp orthoalgebra morphism (UO morphism) is defined in the standard way.

Lemma 2.1 *Let $\mathcal{A} = \langle A, \boxplus, 1, 0 \rangle$ be an unsharp orthoalgebra and let $a, b \in A$. The following properties hold.*

 (i) $a \perp b \implies b \perp a$.

 (ii) $a'' = a$.

 (iii) $1' = 0$ and $0' = 1$.

 (iv) $a \perp 0$ and $a \boxplus 0 = a$.

 (v) $a \perp b,\ a \boxplus b = 0 \implies a = b = 0$.

Lemma 2.2 *Let $\mathcal{A} = \langle A, \boxplus, 1, 0 \rangle$ be an unsharp orthoalgebra and let $a, b \in A$ s.t. $a \perp b$. Then,*

$$a \perp (a \boxplus b)' \quad and \quad b' = a \boxplus (a \boxplus b)'.$$

Lemma 2.3 *Let* $\mathcal{A} = \langle A, \boxplus, 1, 0 \rangle$ *be an unsharp orthoalgebra and let* $a, b \in A$. *Then,*

$$a \perp b \quad \textit{iff} \quad a \sqsubseteq b'.$$

Lemma 2.3 shows that the orthogonality relation has here the usual meaning.

Lemma 2.4 (Orthomodularity)
Let $\mathcal{A} = \langle A, \boxplus, 1, 0 \rangle$ *be an unsharp orthoalgebra and let* $a, b \in A$. *Then,*

$$a \sqsubseteq b \Longrightarrow b = a \boxplus (a \boxplus b')'.$$

Lemma 2.5 (Cancellation law)
Let $\mathcal{A} = \langle A, \boxplus, 1, 0 \rangle$ *be an unsharp orthoalgebra and let* $a, b, \in$ *s.t.* $a \perp c$ *and* $b \perp c$. *Then,*

  (i) $a \boxplus c = b \boxplus c \Longrightarrow a = b$,
  (ii) $a \boxplus c \sqsubseteq b \boxplus c \Longrightarrow a \sqsubseteq b$.

Theorem 2.1 *Let* $\mathcal{A} = \langle A, \boxplus, 1, 0 \rangle$ *be an unsharp orthoalgebra. Then,* $\langle A, \sqsubseteq, ', 1, 0 \rangle$ *is an involutive bounded poset.*

Definition 2.3 A *scale algebra* is an unsharp orthoalgebra which is totally (linearly) ordered by $\sqsubseteq$.

We now present some examples of unsharp orthoalgebras. As standard examples of unsharp orthoalgebras, we will consider the unsharp ortho-algebras determined by the real interval $[0, 1]$ and by the class $E(\mathcal{H})$ of all *effects* of any Hilbert space $\mathcal{H}$.

Example 2.1 (*Standard scale algebra*)
Let $[0, 1] \subset \mathbb{R}$. A partial operation $\boxplus$ can be defined on $[0, 1]$. For any $x, y \in [0, 1]$:

$$\exists (x \boxplus y) \quad \Longleftrightarrow \quad x + y \in [0, 1],$$

where $+$ is the usual sum on the reals.

$$\exists (x \boxplus y) \quad \Longrightarrow \quad x \boxplus y = x + y.$$

The structure $\mathcal{A}_{[0,1]} = \langle [0, 1], \boxplus, 1, 0 \rangle$ is a scale algebra, called *standard scale algebra*. It turns out that $\forall x \in [0, 1]$: $x' = 1 - x$. Further, the partial order relation $\sqsubseteq$, defined according to Definition 2.2(ii), coincides with the usual order of $\mathbb{R}$. The structure $\langle [0,1], \sqsubseteq, ', 1, 0 \rangle$ is an involutive distributive bounded lattice (called also De Morgan lattice), where $\forall x, y \in [0,1]$:

$$\inf(\{x, y\}) = \text{Min}(\{x, y\})$$

and

$$\sup(\{x, y\}) = \text{Max}(\{x, y\}).$$

Example 2.2 (*Standard unsharp orthoalgebra*)
Let $E(\mathcal{H})$ be the class of all effects of any Hilbert space. Let us introduce on $E(\mathcal{H})$ a partial sum $\boxplus$ in the following way. For any $E, F \in E(\mathcal{H})$:

$$\exists (E \boxplus F) \iff E + F \in E(\mathcal{H})$$

(in other words, the orthoalgebraic sum $E \boxplus F$ is defined iff the usual operator-sum $E + F$ is an effect operator),

$$\exists (E \boxplus F) \implies E \boxplus F = E + F.$$

Finally, 0 and 1 are identified with the null and the identity operator ($\mathbb{O}$ and $\mathbb{1}$, respectively). It is easy to check that the structure $\mathcal{E}(\mathcal{H})_{uoa} = \langle E(\mathcal{H}), \boxplus, 1, 0 \rangle$ is an unsharp orthoalgebra, called *standard unsharp orthoalgebra*. It turns out that for any $E, F \in E(\mathcal{H})$: $E \sqsubseteq F$ iff for all density operators D: $\text{Tr}(DE) \leq \text{Tr}(DF)$. Further, for any $E \in E(\mathcal{H})$: $E' = \mathbb{1} - E$. Differently from the standard scale algebra, the involutive bounded poset $\langle E(\mathcal{H}), \sqsubseteq, ', 1, 0 \rangle$ is *not* a lattice.

Example 2.3 (*Interval unsharp orthoalgebra*)
Let $\mathcal{G} = \langle G, +, \leq, \underline{0} \rangle$ be a *partially ordered abelian group* (*po group*), i.e. an abelian group with a translation invariant partial order $\leq$. The *positive cone* of $\mathcal{G}$, denoted by G^+, is the set $\{x \in G \mid \underline{0} \leq x\}$. Let $\underline{0} \neq u \in G^+$ and

$$G^+[\underline{0}, u] := \{x \in G \mid \underline{0} \leq x \leq u\}.$$

For any $x, y \in G^+[\underline{0}, u]$, let us define:

$$\exists (x \boxplus G) \iff x + y \in G^+[\underline{0}, u],$$

$$\exists (x \boxplus y) \implies x \boxplus y = x + y.$$

It is easy to check that the structure $\mathcal{G}^+[\underline{0}, u] := \langle G^+[\underline{0}, u], \boxplus, u, \underline{0} \rangle$, is an unsharp orthoalgebra, called *interval unsharp orthoalgebra*. It turns out that $\forall x \in G^+[\underline{0}, u]$: $x' = u - x$. Further, the partial order $\sqsubseteq$ in $\mathcal{G}^+[\underline{0}, u]$ coincides with the restriction to $G^+[\underline{0}, u]$ of the partial order $\leq$ on G. Therefore, if $\mathcal{G}$ is a linearly ordered abelian group, then $\mathcal{G}^+[\underline{0}, u]$ is a scale algebra. In particular, if $\mathcal{G}$ is the linearly ordered abelian group of the reals with the usual cone $\mathbb{R}^+$, then $\mathbb{R}^+[0,1]$ is the standard scale algebra. Foulis and Bennett ([4]) proved that if an unsharp orthoalgebra $\mathcal{A}$ admits of an order-determining set of probability measures, then $\mathcal{A}$ is isomorphic to an interval unsharp orthoalgebra. We recall that a probability measure on an unsharp orthoalgebra $\mathcal{A}$ is a morphism $p : A \rightarrow \mathbb{R}^+[0,1]$ of A into the standard scale algebra $\mathbb{R}^+[0,1]$. A set Δ of probability measures on $\mathcal{A}$ is said to be *order-determining* iff, for $a, b \in A$, the condition $s(a) \leq s(b)$ for all $s \in \Delta$ implies that $a \sqsubseteq b$.

3. Quantum MV algebras

The operation $\boxplus$ of an unsharp orthoalgebras is partial. However, $\boxplus$ can be naturally extended to a total operation $\oplus$ which coincides with $\boxplus$, whenever it is defined. In this section we will introduce the notion of quantum MV algebra and we will prove a correspondence theorem between unsharp orthoalgebras and quantum MV algebras. Since the class $E(\mathcal{H})$ of all effects determines an unsharp orthoalgebra, it follows that $E(\mathcal{H})$ determines a quantum MV algebra.

Definition 3.1 A *quantum MV algebra* (*QMV algebra*) is a structure $\mathcal{M} = \langle M, \oplus, {}^*, 1, 0 \rangle$, where M is a non-empty set, 0 and 1 are constant elements of M, $\oplus$ is a binary operation and * is a unary operation, satisfying the following axioms (where $a \odot b := (a^* \oplus b^*)^*$, $a \sqcap b := (a \oplus b^*) \odot b$ and $a \sqcup b := (a \odot b^*) \oplus b$).

(QMV1) $(a \oplus b) \oplus c = a \oplus (b \oplus c)$.
(QMV2) $a \oplus 0 = a$.
(QMV3) $a \oplus b = b \oplus a$.
(QMV4) $a \oplus 1 = 1$.
(QMV5) $(a^*)^* = a$.
(QMV6) $0^* = 1$.
(QMV7) $a \oplus a^* = 1$.
(QMV8) $a \sqcup (b \sqcap a) = a$.
(QMV9) $(a \sqcap b) \sqcap c = (a \sqcap b) \sqcap (b \sqcap c)$.
(QMV10) $a \oplus (b \sqcap (a \oplus c)^*) = (a \oplus b) \sqcap (a \oplus (a \oplus c)^*)$.
(QMV11) $a \oplus (a^* \sqcap b) = a \oplus b$.
(QMV12) $(a^* \oplus b) \sqcup (b^* \oplus a) = 1$.

(QMV8) and (QMV9) represent a weak formulation of the absorption and of the associativity laws, respectively. Generally, $\sqcap$ and $\sqcup$ are not lattice-theoretic operations. (QMV10) and (QMV11) represent a kind of conditional distributivity law of $\oplus$ over $\sqcap$.

Definition 3.2 An *MV algebra* is an algebraic structure $\mathcal{M} = \langle M, \oplus, {}^*, 1, 0 \rangle$ that satisfies (QMV1)-(QMV7) and the following condition:

   (LA) $(a \odot b^*) \oplus b = (b \odot a^*) \oplus a.$ (*Łukasiewicz axiom*)

It is easy to check that any MV algebra is a QMV algebra.

The axiomatization for QMV algebras we have presented is not the most economical one. As one can easily check, (QMV6) follows from (QMV2), (QMV3) and (QMV7); further, (QMV7) follows from (QMV2), (QMV4), (QMV5), (QMV6) and (QMV12).

Example 3.1   (*Standard MV algebra*)
Let $[0, 1]$ be the unit real interval. For all $a, b \in [0, 1]$, let

$$a \oplus b := \text{Min} (\{a + b, 1\})   (\textit{truncated sum})$$

and

$$a^* := 1 - a.$$

The structure $\mathcal{M}_{[0,1]} = \langle [0, 1], \oplus, {}^*, 1, 0 \rangle$ is an MV algebra, called *standard MV algebra*. Further, $\mathcal{M}_{[0,1]}$ is *linear*, i.e. $\forall a, b \in [0, 1]$: $a \leq b$ or $b \leq a$. It turns out that $a \odot b = \text{Max} (\{a + b - 1, 0\})$, $a \cap b = \text{Min} (\{a, b\})$ and $a \cup b = \text{Max} (\{a, b\})$.

Example 3.2   (*Standard QMV algebra*)
Let $E(\mathcal{H})$ be the class of all effects of a Hilbert space. $E(\mathcal{H})$ coincides with the class of all bounded linear operators between $\mathbb{O}$ and $\mathbb{I}$, where $\mathbb{O}$ and $\mathbb{I}$ are the null and the identity operator, respectively. Clearly, $E(\mathcal{H})$ contains the class of all $\lambda \mathbb{I}$ (with $\lambda \in [0, 1]$), where $\forall \psi \in \mathcal{H}$: $(\lambda \mathbb{I})\psi := \lambda \psi$. Thus, it seems quite natural to generalize the definition of the operations $\oplus$ and * of the standard MV algebra $\mathcal{M}_{[0,1]}$ to the whole class $E(\mathcal{H})$ in the following way. For any $E, F \in E(\mathcal{H})$:

$$E \oplus F := \begin{cases} E + F, & \text{if } E + F \in E(\mathcal{H}), \\ \mathbb{I}, & \text{otherwise,} \end{cases}$$

where $+$ is the usual operator-sum. Thus, the operation $\oplus$ can be regarded as a kind of *quantum truncated sum*.

$$E^* := \mathbb{I} - E.$$

One can prove ([6]) that the structure $\mathcal{E}(\mathcal{H})_{qmv} = \langle E(\mathcal{H}), \oplus, {}^*, \mathbb{I}, \mathbb{O} \rangle$ is a QMV algebra, called *standard QMV algebra*. The structure $\mathcal{E}(\mathcal{H})_{qmv}$, however, is not an MV algebra, for it violates the crucial axiom (LA), which is the responsible for the lattice-theoretic behavior of the operations $\cap$ and $\cup$ of an MV algebra.

It turns out that $E \preceq F$ iff $E \sqsubseteq F$, where $\sqsubseteq$ is the partial order of $\mathcal{E}(\mathcal{H})$, understood as an unsharp orthoalgebra. Moreover:

$$E \wedge F = \begin{cases} E, & \text{if } E \sqsubseteq F, \\ F, & \text{otherwise,} \end{cases}$$

and

$$E \uplus F = \begin{cases} E, & \text{if } F \sqsubseteq E, \\ F, & \text{otherwise.} \end{cases}$$

Recall that $E \oplus F = E + F$ iff $E + F \in E(\mathcal{H})$ iff $E \sqsubseteq F' = F^*$.

The MV algebra determined by the class of all $\lambda \mathbb{I}$ (where $\lambda \in [0, 1]$) is isomorphic to the standard MV algebra and determines an MV subalgebra of $\mathcal{E}(\mathcal{H})$.

In the following, we will list some arithmetical properties of general QMV algebras ([5]). It should be noticed that the proof of Theorems 2.3)-2.6) makes use of the axioms (QMV1)-(QMV7) only (the "elementary" axioms of QMV algebras).

Definition 3.3 Let $\mathcal{M}$ be a QMV algebra. For all $a, b \in M$:

$$a \preceq b \quad \text{iff} \quad a = a \wedge b.$$

Theorem 3.1 *Let $\mathcal{M}$ be a QMV algebra. The following properties hold.*

(i) $a \odot b = b \odot a$.
(ii) $a \odot (b \odot c) = (a \odot b) \odot c$.
(iii) $a \odot a^* = 0$.
(iv) $a \odot 0 = 0$.
(v) $a \odot 1 = 1$.
(vi) $a \wedge 1 = a = 1 \wedge a$.
(vii) $a \wedge 0 = 0 = 0 \wedge a$.
(viii) $a = a \wedge a$.
(ix) $(a \uplus b)^* = a^* \wedge b^*$.
(x) $(a \wedge b)^* = a^* \uplus b^*$.
(xi) If $a \preceq b$, then $a = b \wedge a$.

It should be noticed that, in general, $a = b \wedge a$ *does not* imply $a = a \wedge b$.

Theorem 3.2 *Let $\mathcal{M}$ be a QMV algebra. The following properties hold.*

(i)   If $a \oplus b = 0$, then $a = b = 0$.
(ii) If $a \odot b = 1$, then $a = b = 1$.
(iii) If $a \uplus b = 0$, then $a = b = 0$.
(iv) If $a \wedge b = 1$, then $a = b = 1$.

Theorem 3.3   (*Cancellation law*)
Let M be a QMV algebra. For any $a, b, c \in M$: if $a \oplus c = b \oplus c$, $a \preceq c^*$ and $b \preceq c^*$, then $a = b$.

Theorem 3.4 *Let M be a QMV algebra. If $a \preceq b$, then $a^* \oplus b = 1$.*

It should be noticed that, in general, $a^* \oplus b = 1$ does not imply $a \preceq b$.

We can now prove some Theorems that require further axioms besides (QMV1)-(QMV7).

Theorem 3.5 *Let M be a QMV algebra. The following properties hold.*

   (i)   If $a \preceq b$, then $b^* \preceq a^*$.
   (ii) $a \preceq b$ iff $b = b \uplus a = a \uplus b$.
   (iii) $a \sqcap (b \uplus a) = a$.

Theorem 3.6 $\langle M, \preceq, ^*, 1, 0 \rangle$ *is an involutive bounded poset.*

It should be noticed that the reflexivity and the antisymmetry of $\preceq$ depend on the axioms (QMV1)-(QMV7), whereas the axiom (QMV9) is essential to prove that $\preceq$ is transitive.

Theorem 3.7 *Let M be a QMV algebra. The following properties hold.*

   (i)   If $a \preceq b$, then $\forall c \in M$: $a \sqcap c \preceq b \sqcap c$. (weak monotony of $\sqcap$)
   (ii)   If $a \preceq b$, then $\forall c \in M$: $a \uplus c \preceq b \uplus c$. (weak monotony of $\uplus$)

It should be noticed that, in general, $a \sqcap b \not\preceq a$, $a \not\preceq a \uplus b$ and $a \sqcap b \not\preceq b \uplus a$.

Theorem 3.8   (*Monotony of $\oplus$ and $\odot$*)
Let M be a QMV algebra. The following properties hold.

   (i)   If $a \preceq b$, then $\forall c \in M$: $a \oplus c \preceq b \oplus c$.
   (ii)   If $a \preceq b$, then $\forall c \in M$: $a \odot c \preceq b \odot c$.
   (iii)   If $a \preceq b$ and $c \preceq d$ then $a \oplus c \preceq b \oplus d$.
   (iv) $a \preceq b$ and $c \preceq d$ then $a \odot c \preceq b \odot d$.

Theorem 3.9 *Let M be a QMV algebra. The following properties hold.*

   (i) $a \odot b \preceq a$.
   (ii) $a \preceq a \oplus b$.
   (iii) $a \odot b \preceq a \sqcap b$, $a \odot b \preceq b \sqcap a$.
   (iv) $a \uplus b \preceq a \oplus b$, $b \uplus a \preceq a \oplus b$.

Theorem 3.10 *Let M be a QMV algebra. Then the following properties hold for any $a, b \in M$.*

   (i) $(a \sqcap b) \oplus (b \sqcap a)^* = 1$.
   (ii) $(a \sqcap b) \sqcap (b \sqcap a) = b \sqcap a$.

Theorem 3.11 *Let $\mathcal{M}$ be a QMV algebra. The following conditions are equivalent.*

(i) $\mathcal{M}$ is an MV algebra.
(ii) $\forall a, b \in M$: if $a^* \oplus b = 1$, then $a \preceq b$.

Corollary 3.1 *If $\mathcal{M}$ is a linear QMV algebra, then $\mathcal{M}$ is an MV algebra.*

Definition 3.4 A QMV algebra $\mathcal{M} = \langle M, \oplus, {}^*, 1, 0 \rangle$ is said to be *quasi-linear* (or *quasi-totally ordered*) iff $\forall a, b \in M$: if $a \npreceq b$, then $a \,\text{\textcircled{m}}\, b = b$.

In other words, a QMV algebra $\mathcal{M}$ is quasi-linear iff the following holds:

$$a \,\text{\textcircled{m}}\, b = \begin{cases} a, & \text{if } a \preceq b, \\ b, & \text{otherwise}, \end{cases}$$

or

$$a \,\text{\textcircled{w}}\, b = \begin{cases} a, & \text{if } b \preceq a, \\ b, & \text{otherwise}. \end{cases}$$

One can easily check that an MV algebra $\mathcal{M}$ is quasi-linear iff $\mathcal{M}$ is linear. Since we know that there are MV algebras that are not linear, we can conclude that not every QMV algebra is quasi-linear. Both the standard MV algebra and the standard QMV algebra are quasi-linear.

Theorem 3.12 *Let $\mathcal{M}$ be a QMV algebra. The following conditions are equivalent.*

(i) $\mathcal{M}$ is quasi-linear.
(ii) $\forall a, b \in M$: if $a \oplus b \neq 1$, then $a \prec b^*$.
(iii) $\forall a, b, c \in M$: if $a \oplus c = b \oplus c \neq 1$, then $a = b$.

4. QMV algebras and unsharp orthoalgebras

In the following theorems, we will show that any unsharp orthoalgebra can be extended to a (quasi-linear) QMV algebra and that any QMV algebra determines an unsharp orthoalgebra.

Theorem 4.1 *Let $\mathcal{M} = \langle M, \oplus, {}^*, 1, 0 \rangle$ be a QMV algebra. Then, the structure $\mathcal{M}_{uoa} = \langle M, \boxplus, 1, 0 \rangle$ is a unsharp orthoalgebra, where*

(i) $\boxplus$ is a partial operation s.t. $a \boxplus b$ is defined iff $a \preceq b^*$;
(ii) if $a \boxplus b$ is defined, then $a \boxplus b := a \oplus b$.

Proof. (UO1) Let us suppose that $a \boxplus b$ is defined; then, $a \preceq b^*$ and $a \boxplus b = a \oplus b$. By Theorem 3.5(i), $b \preceq a^*$ so that $b \boxplus a$ is defined and $b \boxplus a = b \oplus a = a \oplus b = a \boxplus b$.

(UO2)   Let us suppose that $a \boxplus b$ and $(a \boxplus b) \boxplus c$ are defined. Then, $a \preceq b^*$ and $(a \boxplus b) \preceq c^*$ so that $(a \boxplus b) \boxplus c = (a \oplus b) \oplus c$. We want to show that that $b \preceq c^*$ and $a \preceq (b \oplus c)^*$, i.e. $b \boxplus c$ and $a \boxplus (b \boxplus c)$ are defined.
By hypothesis $(a \oplus b) \preceq c^*$. By Theorem 3.9(ii) $b \preceq (a \oplus b)$ so that $b \preceq c^*$. It remains to prove that $a \preceq (b \oplus c)^*$.

$$
\begin{aligned}
a \barwedge (b \oplus c)^* &= a \barwedge [b \oplus (c \barwedge (a \oplus b)^*)]^* && \text{(hp)} \\
&= a \barwedge [(b \oplus c) \barwedge (b \oplus (a \oplus b)^*)]^* && \text{(QMV10)} \\
&= a \barwedge [(b \oplus c) \barwedge (a^* \sqcup b)]^* \\
&= a \barwedge [(b \oplus c)^* \sqcup a] && \text{(hp)} \\
&= a. && \text{Theorem 3.5(ii)}
\end{aligned}
$$

Thus, $(a \boxplus b) \boxplus c = (a \oplus b) \oplus c = a \oplus (b \oplus c) = a \boxplus (b \boxplus c)$.
(UO3)   Let $a \in M$. Now, $a \boxplus a^*$ is defined and $a \boxplus a^* = 1$. Let us suppose that there is an element $b \in M$ s.t. $a \preceq b^*$ and $a \boxplus b = 1$. Then, by the cancellation law (Theorem 3.3): $b = a^*$.
(UO4)   If $a \boxplus 1$ is defined, then $a \preceq 1^* = 0$ so that $a = 0$.   □

We want now to show that any unsharp orthoalgebra determines a quasi linear QMV algebra.

Theorem 4.2 *Every unsharp orthoalgebra determines a quasi-linear QMV algebra.*

Proof. Let $\mathcal{A} = \langle A, \boxplus, 1, 0 \rangle$ be an unsharp orthoalgebra. Let us define the following structure: $\mathcal{A}^{qmv} := \langle A, \oplus, ^*, 1, 0 \rangle$, where * is the generalized complement $'$ of $\mathcal{A}$ (see Definition 2.2 (iii)) and $\forall a, b \in A$:

$$
a \oplus b = \begin{cases} a \boxplus b, & \text{if } a \boxplus b \text{ is defined,} \\ 1 & \text{otherwise.} \end{cases}
$$

Let us put $a \odot b := (a^* \oplus b^*)^*$.
By Lemma 2.3, $a \boxplus b$ is defined iff $a \sqsubseteq b'$, where $\sqsubseteq$ is the partial order of $\mathcal{A}$.
Let us define

$$
a \barwedge b = (a \oplus b^*) \odot b
$$

and

$$
a \sqcup b = (a \odot b^*) \oplus b.
$$

Let us define

$$
a \preceq b \quad \text{iff} \quad a = a \barwedge b.
$$

We want to prove that $\forall a, b \in A$: $a \sqsubseteq b$ iff $a \preceq b$.
Let us suppose $a \sqsubseteq b$. By Lemma 2.3, $a \perp b^*$. We want to show that $b = b \sqcup a$. By Lemma 2.4, $b = a \boxplus (a \boxplus b^*)^*$. Thus, $b = a \oplus (a \oplus b^*)^* = a \oplus (a^* \odot b) = b \sqcup a$.

Conversely, let us suppose that $a = a \cap b = [(a \oplus b^*)^* \oplus b^*]^*$. We want to show that $a \sqsubseteq b$. Let us suppose that $a \not\sqsubseteq b$. Then $a \not\perp b^*$ so that $a \oplus b^* = 1$. Thus, $a = (1^* \oplus b^*)^* = b$, contradiction.

One can easily show that

$$a \cap b = \begin{cases} a, & \text{if } a \sqsubseteq b \text{ or equivalently } a \preceq b, \\ b, & \text{otherwise,} \end{cases}$$

and

$$a \cup b = \begin{cases} a, & \text{if } b \sqsubseteq a, \\ b, & \text{otherwise.} \end{cases}$$

Consequently, if $\mathcal{A}^{qmv}$ is a QMV algebra, then it is quasi-linear. We now prove that $\mathcal{A}^{qmv}$ is a QMV algebra.

Axioms (QMV2)-(QMV7) are trivially satisfied.

(QMV1).   Two cases are possible:

1) $a \oplus b = 1$,

2) $a \oplus b \neq 1$.

Case 1).   The right-hand side of the equation is clearly equal to 1. Let us suppose, by contradiction, that $a \oplus (b \oplus c) \neq 1$. By (UO2), $1 \neq a \oplus (b \oplus c) = a \boxplus (b \boxplus c) = (a \boxplus b) \boxplus c = (a \oplus b) \oplus c$, contradiction.

Case 2).   If $(a \boxplus b) \oplus c = 1$, then $a \oplus (b \oplus c) = 1$ since otherwise, $a \oplus (b \oplus c) = a \boxplus (b \boxplus c) = (a \boxplus b) \boxplus c \neq 1$. If $(a \boxplus b) \oplus c \neq 1$, then $1 \neq (a \oplus b) \oplus c = (a \boxplus b) \boxplus c = a \boxplus (b \boxplus c) = a \oplus (b \oplus c)$.

(QMV8).   If $b \sqsubseteq a$, then $a \cup (b \cap a) = a \cup b = a$. If $b \not\sqsubseteq a$, then $a \cup (b \cap a) = a \cup a = a$.

(QMV9). $(a \cap b) \cap c = (a \cap b) \cap (b \cap c)$.

Two cases are possible:

1) $b \sqsubseteq c$,

2) $b \not\sqsubseteq c$.

Case 1).   If $a \sqsubseteq b$, then $a \sqsubseteq c$ since $\sqsubseteq$ is transitive. Thus, $(a \cap b) \cap c = a \cap c = a = a \cap b = (a \cap b) \cap (b \cap c)$. If $a \not\sqsubseteq b$, then $(a \cap b) \cap (b \cap c) = b \cap b = b = b \cap c = (a \cap b) \cap c$.

Case 2).   Since $b \not\sqsubseteq c$, we have that $b \cap c = c$. Hence: $(a \cap b) \cap (b \cap c) = (a \cap b) \cap c$.

(QMV10) $a \oplus (b \cap (a \oplus c)^*) = (a \oplus b) \cap (a \oplus (a \oplus c)^*)$.

Two cases are possible:

1) $a \oplus c = 1$,

2) $a \oplus c \neq 1$.

Case 1). $a \oplus (b \cap (a \oplus c)^*) = a \oplus (b \cap 0) = a$ and $(a \oplus b) \cap (a \oplus (a \oplus c)^*) = (a \oplus b) \cap (a \oplus (a \oplus c)^*) = (a \oplus b) \cap (a \oplus 0) = [(a \oplus b) \oplus a^*] \odot a = [(b \oplus (a \oplus a^*)] \odot a = (b \oplus 1) \odot a = a$.

Case 2).   We have two subcases:

2a) $b \sqsubseteq (a \oplus c)^* = (a \boxplus c)^*$,

2b) $b \not\sqsubseteq (a \oplus c)^* = (a \boxplus c)^*$.

Subcase 2a). By hypothesis $a \oplus (b \cap (a \oplus c)^*) = a \oplus b$ and $(a \oplus b) \cap [a \oplus (a \oplus c)^*]$ $= (a \oplus b) \cap [a \oplus (a \boxplus c)^*]$. If $a \oplus (a \boxplus c)^* = 1$, we are done. Thus, we can suppose that $a \oplus (a \boxplus c)^* \neq 1$. Thus, $a \oplus (a \boxplus c)^* = a \boxplus (a \boxplus c)^*$. By Lemma 2.2, $a \boxplus (a \boxplus c)^* = c^*$. Therefore, $(a \oplus b) \cap (a \oplus (a \boxplus c)^*) = (a \oplus b) \cap c^*$. By hypothesis, $b \sqsubseteq (a \boxplus c)^*$. Thus, $b \boxplus (a \boxplus c)$ is defined. Then, by (UO3) and (UO1), $(a \boxplus b) \boxplus c$ is defined so that $(a \boxplus b) \sqsubseteq c^*$. Hence: $(a \oplus b) \cap c^* = (a \oplus b)$.
Subcase 2b). By hypothesis, we have that $a \boxplus (b \cap (a \boxplus c)^*) = a \oplus (a \boxplus c)^*$. If $a \oplus b = 1$, then $(a \oplus b) \cap (a \oplus (a \boxplus c)^*) = a \oplus (a \boxplus c)^*$ and we are done. Thus, we can suppose that $a \oplus b \neq 1$ so that $a \oplus b = a \boxplus b$. Let us suppose, by contradiction, that $(a \oplus b) \cap (a \oplus (a \boxplus c)^*) \neq a \oplus (a \boxplus c)^*$. Then, $(a \oplus b) \cap (a \oplus (a \boxplus c)^*) = a \oplus b = a \boxplus b$. Hence, $(a \boxplus b) \sqsubseteq a \oplus (a \boxplus c)^*$. By Lemma 2.2, $a \perp (a \boxplus c)^*$ and $a \boxplus (a \boxplus c)^* = c^*$. By (UO2), we obtain $b \perp (a \boxplus c)$, contradiction. Thus, $(a \oplus b) \cap (a \oplus (a \oplus c)^*) = a \oplus (a \oplus c)^*$. (QMV11) and (QMV12) are easily verified. $\square$

Theorem 4.1 shows that every QMV algebra $\mathcal{M} = \langle M, \oplus, *, 1, 0 \rangle$ can be transformed into an unsharp orthoalgebra $\mathcal{M}_{uoa} = \langle M, \boxplus, 1, 0 \rangle$. Theorem 4.2 guarantees that $\mathcal{M}_{uoa}$ can be transformed into a *quasi-linear* QMV algebra $\mathcal{M}_{uoa}^{qmv} = \langle M, \oplus', *, 1, 0 \rangle$, where $*$ is the generalized complement of $\mathcal{M}_{uoa}$. In general, $\mathcal{M}$ is not isomorphic to $\mathcal{M}_{uoa}^{qmv}$ (in other words, $\oplus \neq \oplus'$) since, otherwise, every QMV algebra would be linear, and this is not the case. One can easily see that $\mathcal{M}$ is isomorphic to $\mathcal{M}_{uoa}^{qmv}$ iff $\mathcal{M}$ is quasi-linear. Thus, in this precise sense, we have that

UNSHARP ORTHOALGEBRAS = QUASI-LINEAR QMV ALGEBRAS

In [2] Chang proved that every linear MV algebra is isomorphic to the MV algebra determined by a positive interval of a totally ordered abelian group. Consequently, any scale algebra is isomorphic to an interval unsharp orthoalgebra. Further, Chang proved that given any equation α, α holds in the class of all MV algebras iff α holds in the standard MV algebra. As to QMV algebras, we do not know whether there exists an equation that holds in the class of all standard QMV algebras but fails in a particular quasi-linear QMV algebra.

However, any QMV algebra determined by the class $E(\mathcal{H})$ of all effects of any Hilbert space satisfies the *regularity property*, where a QMV algebra is said to be *regular* iff $\forall a, b$:

(*regularity*) $a \preceq a^*$ and $b \preceq b^* \implies a \preceq b^*$.

It is an open question whether the class of all regular QMV algebra is equational. Every regular QMV algebra can contain at most one fixed element of the operation $*$. Since there are (quasi-linear) QMV algebras with two fixed points of $*$, we conclude that, differently from MV algebras, not every QMV is regular.

One can define a *regular* unsharp orthoalgebra as an unsharp partial algebra which satisfies the following condition:

(*regularity*) $\exists(a \boxplus a)$ and $\exists(b \boxplus b) \implies \exists(a \boxplus b)$.

Theorem 4.1 and Theorem 4.2 can be easily extended to the case of regular unsharp orthoalgebras and regular quasi-linear QMV algebras. Thus,

REGULAR UNSHARP ORTHOALGEBRAS =
REGULAR QUASI-LINEAR QMV ALGEBRAS

References

1. Chang, C.C. (1957). "Algebraic analysis of many valued logics", *Transactions of the American Mathematical Society*, **88** pp.467–490.
2. Chang, C.C. (1958). "A new proof of the completeness of Lukasiewicz axioms", *Transactions of the American Mathematical Society*, **93** pp.74–80.
3. Dalla Chiara, M.L. and Giuntini, R. "Unsharp quantum logics", to appear in *Foundations of Physics*.
4. Foulis, D.J. and Bennett, M.K. "Effect algebras and unsharp quantum logics", to appear in *Foundations of Physics*.
5. Giuntini, R. "Quantum MV algebras", *(preprint)*.
6. Giuntini, R. and Greuling, H. (1989). "Toward a formal language for unsharp properties", *Foundations of Physics*, **20** pp.931–935.
7. Kôpka, F. and Chovanec, F. (1994). "D-posets", *Mathematica Slovaca*, **44** pp.21–34.

INTRODUCTION TO NELSON STOCHASTIC MECHANICS AS A MODEL FOR QUANTUM MECHANICS

F. GUERRA

Dip. di Fisica Università and INFN - Roma "La Sapienza"
Piazzale Aldo Moro, 2 - 00185 - Roma, Italy.
E-mail: guerra@roma1.infn.it

Abstract. We give a short presentation of Nelson stochastic mechanics, as a generalization of classical mechanics, based on the theory of stochastic processes and stochastic variational principles. Stochastic mechanics can be connected to quantum mechanics through a very simple physical interpretation scheme. From this point of view, stochastic mechanics can be seen as a quantization procedure for mechanical systems, different, but physically equivalent, to the usual operator quantization. Then we deal with the problems related to the possibility of considering stochastic mechanics as a complete physical theory. Through a discrete generalization, we show how the main features of the postulated underlying Brownian motion, at the origin of quantum fluctuations, can be derived also as a consequence of stochastic variational principles. We also discuss the problem of formulating stochastic mechanics in representations different from the configuration representation, and show how the different representations, related by unitary transformations in ordinary quantum mechanics, are connected in stochastic mechanics through stochastic measure preserving transformations. Finally, we show how the basic aspects of the measurement problem, in particular the wave packet reduction, can be interpreted in the frame of generalized stochastic mechanics. Here, the wave function collapse is not instantaneous, but is ruled by a well defined dynamical scheme, with a time asymptotic relaxation behavior. Moreover, the relaxation is slower, if the result of the measurement is more uncertain. Finally, we discuss the possibility of embedding stochastic mechanics into a generalized scheme of Schrödinger stochastic processes. [1]

[1]Research supported in part by MURST (Italian Minister of University and Scientific and Technological Research) and INFN (Italian National Institute for Nuclear Physics).

C. Garola and A. Rossi (eds.), The Foundations of Quantum Mechanics, 339-355.
© 1995 Kluwer Academic Publishers.

1. Introduction

In a remarkable paper, published on Physical Review in 1966 [1], Edward Nelson introduced a general scheme of stochastic mechanics, which could be considered as a quantization procedure for classical dynamical systems, based on the theory of stochastic processes and stochastic dynamics.

While we refer to [1,2,3,4], and references quoted there, for an outline of stochastic mechanics in its historical evolution, we think that the most useful presentation of the subject, in its present status, must depart from a purely historical perspective, and be based on few clear conceptual assumptions.

It is very well known that variational principles of Lagrangian type [5] provide a solid foundation for the whole structure of classical mechanics and classical field theory. In a series of previous papers [6,7,8,9] (see also [4]), we have investigated a generalization of the variational principles, in the case where the smooth deterministic trial trajectories of classical mechanics are replaced by the very irregular trajectories of random diffusions in configuration space. In this approach, the main emphasis is put on the average hydrodynamic field variables, as the density and the forward and backward drifts of the diffusions. Through them one can define the action and its variations.

A very surprising feature of these stochastic variational principles is that the critical processes, making stationary the action under appropriate time boundary conditions, are strictly related to states of the associated quantum dynamical system, according to the general scheme of Nelson stochastic mechanics [1,2,3,4]. Therefore, in the frame of a suitable physical interpretation, as outlined in Section 3 (see also [4,10,11]), we can say that stochastic variational principles provide a kind of stochastic simulation of quantum mechanical behavior.

From this point of view, stochastic mechanics is nothing but a generalization of classical mechanics, completely independent in its foundation from quantum mechanics. However, the programming equations coming from the stochastic variational principle are formally identical to the equations of the Madelung fluid, and these are equivalent to the quantum Schrödinger equation for the wave function. Then it is easy to develop an interpretation code, which transforms the whole structure of stochastic mechanics into a suitable theoretical model for quantum mechanical behavior. Of course, here we consider the phenomenological content of quantum mechanics, and not its theoretical aspects in terms of Hilbert spaces and operators. For example, in the most elementary case of a particle in a potential, the phenomenological content of quantum mechanics will deal, among other things, with the existence and properties of bound states, scattering states, cross sections and so on. All these phenomenological aspects are, in principle, related to suitable experimental verification.

The interpretation code of stochastic mechanics is very simple, and gives results in complete agreement with standard quantum mechanics, as originally noted by Nelson [1], and further developed by other authors [10,11]. However, it is quite displeasing that the most peculiar features of the involved stochastic processes, *i.e.* the transition probability densities, do not enter directly into the interpretation scheme.

This is a very serious problem from the point of view of considering stochastic mechanics as an effective physical theory and not only as a mathematical frame. In fact, stochastic mechanics associates a Markov process to each quantum state. Therefore, transition probabilities play a central role. However, if we try to check that the transition probabilities given by stochastic mechanics, for each quantum state, are experimentally correct, we must make repeated position measurements on the quantum system. Each measurement, according to ordinary quantum mechanics, will produce a change of the quantum state, due to the reduction of the wave packet. Since transition probabilities are associated to a definite quantum state, we see that it is impossible, even in principle, to give an experimental verification of some of the basic aspects of stochastic mechanics, in its present formulation. This very important fact was analyzed in a very lucid way by Onofri in [12] (see also [4,10,11]).

Therefore, we are faced with a dilemma, with two escape possibilities. In the first, we must admit that transition probabilities in stochastic mechanics are not physically observable, but are a kind of generalized gauge variables, necessary to express the dynamical content of the theory in the simple and unifying form of stochastic variational principles. But in this way the stochastic content of the theory is pushed into a background, not directly accessible to phenomenological considerations. The second possibility is to modify the stochastic evolution while the measurement procedure is performed.

Moreover, from the very structure of stochastic mechanics, as related to quantum mechanics, other problems arise in a natural way.

First of all, there is the problem of the underlying postulated Brownian motion, affecting the configuration variables in a scheme completely invariant under time reversal. There is no hope to give a physical interpretation of these Brownian disturbances in terms of the effects of some independent fluctuating external field. In fact, in this case time reversal invariance would be violated and quantum coherence would be lost. Therefore, if the Brownian disturbances are to be considered as the effect of the interaction with a fluctuating external field (the background field hypothesis [4]), then this field can not be dynamically independent from the quantum particles, but there must be a complicated mutual interaction, preserving time reversal and quantum coherence.

There is also the problem of representation. In fact, ordinary quantum mechanics can be formulated in any representation, where a complete set of compatible observables is diagonalized. Different representations are connected through canonical unitary transformations. Nelson stochastic mechanics is formulated in the configuration representation. Therefore, the problem arises about how to give a generalization of stochastic mechanics in each representation. Moreover, it would be important to find the stochastic analog of the quantum unitary transformations, in order to be able to connect different representations of stochastic mechanics.

Therefore, we see that the basic problem in the interpretation of stochastic mechanics is related to the basic problem in the interpretation of quantum mechanics: to evaluate the effects of the measurement and explain the mechanism of the wave packet reduction.

The main objective of this paper is to give a short and synthetic presentation of a general scheme for stochastic mechanics, based on our previous work in [13,15], where all mentioned problems can find a very natural solution.

In fact, starting from quantum mechanics in a generic matrix representation, we will show how to associate a discrete Markov stochastic process to each quantum state in dynamical evolution. These stochastic processes can be also independently defined as critical processes for a stochastic variational principle, with a suitable Lagrangian. Therefore, our generalized stochastic mechanics can be introduced independently from quantum mechanics.

There are two main features in this generalized scheme. The first is that the osmotic part of these processes must not be postulated in advance, as in ordinary stochastic mechanics, but is a consequence of the stochastic variational principle. We can also check that, in the continuous limit, this discrete osmotic part reproduces correctly the assumption of universal Brownian motion in ordinary stochastic mechanics. Therefore, in a sense, the main features of the background field hypothesis can be derived also from a stochastic variational principle. This solves, at least partially, the first previously mentioned problem.

The second main feature is quite surprising. In fact, starting from our discrete stochastic variational principle, we find two classes of critical processes. In the first class, the critical processes are those associated to quantum states in their dynamical evolution, according to Schrödinger equation. Beyond the solutions corresponding to the Schrödinger evolution, the stochastic variational principle also provides other solutions, corresponding to random processes, whose behavior resembles, in this general stochastic framework, the behavior of quantum systems subject to measurement. In particular, the asymptotic time behavior of these processes produces mixtures equivalent to those resulting from the wave function collapse in the ordinary formulation of quantum mechanics. Here, the collapse is not instantaneous, but corresponds to a time asymptotic relaxation behavior. Moreover, the relaxation is slower, if the result of the measurement is more uncertain.

If this behavior, found from the stochastic variational principle, is physically correct, then we can give the following general picture of time evolution for physical quantum systems. The undisturbed system evolves according to the usual Schrödinger equation. If a measurement is performed, then the system relaxes toward the collapsed mixture, according to the von Neumann hypothesis [16]. The collapse is not instantaneous, but it is ruled by a well defined dynamical pattern of time evolution, which in principle should be experimentally observable. The relaxation is faster if the result of the measurement is less uncertain.

The organization of this paper is as follows. In Section 2 we briefly recall the main features of Nelson stochastic mechanics, in the formulation based on stochastic variational principles. In particular, we derive the programming equations, and show how the concept of state arises in stochastic mechanics. Section 3 is devoted to a short presentation of the interpretation code of stochastic mechanics, and to the problem of considering stochastic mechanics as a complete physical theory. In Section 4, by following the presentation given in [13,15], we consider quantum systems in a generic matrix representation and show how to associate a Markov random process to each quantum state. Stochastic variational principles of discrete type are introduced in Section 5. We show how the programming equations give rise to critical processes of two types. Processes of the first type are those associated to Schrödinger solutions. Therefore, they have a definition independent from quantum mechanics, according to the stochastic variational principle. Processes of the second type, with relaxational behavior, are studied in Section 6, and connected to measurement processes in quantum mechanics.

Finally, Section 7 deals with possible further developments and applications of the general stochastic scheme outlined here, in particular with respect to the possibility of embedding Nelson Stochastic Mechanics into the general scheme of Schrödinger processes [17,18].

2. Stochastic variational principles and stochastic states

We show how to generalize the classical variational principles of Lagrangian type to the stochastic case, in a scheme, completely invariant with respect to time reversal. For the sake of simplicity we consider a dynamical system with configuration space R^n and classical Lagrangian given by

$$\mathcal{L}(q, \dot{q}) = \frac{1}{2}m\dot{q}^2 - V(q) \tag{1}$$

Let us introduce trial diffusion processes, with density ρ, satisfying the forward Ito stochastic differential equation

$$dq(t) = v_{(+)}(q(t), t)dt + dw(t) \tag{2}$$

where the Brownian motion $dw(t)$ has a diffusion constant given by ν, and $v_{(+)}(x, t)$ is a forward controlling drift field. It is convenient also to define the backward $v_{(-)}(x, t)$, the current $v(x, t)$ and the osmotic $u(x, t)$ fields through

$$v = \frac{1}{2}(v_{(+)} + v_{(-)}), \quad u = \frac{1}{2}(v_{(+)} - v_{(-)}) = \nu \nabla \log \rho \tag{3}$$

With these definitions the density ρ satisfies the following continuity equation, completely equivalent to the forward Fokker-Planck equation if (3) are taken into account,

$$(\partial_t \rho)(x, t) = -\nabla \cdot (\rho v) \tag{4}$$

Then, the average stochastic action is defined, as in the classical case, by

$$A(t_0, t_1; q) = \int_{t_0}^{t_1} \lim E(\frac{1}{2}m(\Delta q/\Delta t)^2 - V(q(t)))dt \tag{5}$$

where the limit is taken as $\Delta t \to 0^+$, and E denotes an overall average. A simple calculation shows

$$A(t_0, t_1; q) = \int_{t_0}^{t_1} dt \int (\frac{1}{2}mv_{(+)} \cdot v_{(-)} - V(x))\rho(x, t)dx + \ldots \tag{6}$$

where $+\ldots$ are infinite terms irrelevant for the variational principle, since they do not depend on the particular trial diffusion and disappear in the variation of the action [4]. In our paper [6] we took (6) as definition of the stochastic action, then Nelson made the very important discovery that our definition was nothing but the averaged classical definition of the action with an irrelevant infinite constant term subtracted out. This is the key point allowing to consider stochastic mechanics as a kind of simple generalization of classical mechanics.

We assume a fixed initial density $\rho(., t_0)$ and let the process evolve according to (2). Then we have the following stochastic variational principle

(Guerra-Morato [6]). For a given fixed initial density, the action (6) is stationary under arbitrary small variations $\delta v_{(+)}$ of the controlling drift field, subject to the constraint $\delta\rho(.,t_1) = 0$ on the final density, if and only if the following conditions are satisfied. The current velocity field of the process can be expressed in the Hamilton-Jacobi form

$$mv = \nabla S \tag{7}$$

where S is a function satisfying the equation

$$(\partial_t S)(x,t) + \frac{1}{2m}(\nabla S)^2 + V(x) - \frac{\hbar^2}{2m}\frac{\Delta\sqrt{\rho}}{\sqrt{\rho}} = 0 \tag{8}$$

with some given arbitrary final condition $S(x,t_1) = S_1(x)$. In (8) the constant $\hbar$ is defined by

$$\hbar = 2m\nu \tag{9}$$

Let us explicitly remark that here the function S_1 plays the role of a Lagrange multiplier associated to the condition that the density does not change at the final time, and that (8) is a kind of generalization of the Hamilton-Jacobi equation of classical mechanics.

For the proof of this statement, a basic role is played by the following general variation formula, as shown in [6],

$$\delta A = \int S(x_1,t_1)\delta\rho(x_1,t_1)dx_1 - \int S(x_0,t_0)\delta\rho(x_0,t_0)dx_0 \tag{10}$$

$$+ \int_{t_0}^{t_1} E((mv - \nabla S) \cdot \delta v_{(+)}(q(t),t))dt$$

Let us also define the forward and backward transport operators

$$D_{(\pm)} = \partial_t + v_{(\pm)}.\nabla \pm \nu\Delta \tag{11}$$

Then the generalized Hamilton-Jacobi principal function $S(x,t)$, as a consequence of (8), satisfies also the antiparabolic equation

$$D_{(+)}S(x,t) = \frac{1}{2}mv_{(+)}^2 + m\nu\nabla \cdot v_{(+)}, \tag{12}$$

with an arbitrary final specification $S(.,t_1)$.

It is also remarkable that the programming equations of the stochastic variational principle have a very peculiar form. The basic variables are the density ρ and the Hamilton-Jacobi function S, they satisfy the time evolution equations (4,8), supplemented by the *initial* condition for ρ and *final* condition for S. Strictly speaking the general (hydrodynamic) state for this system is associated to a time interval, and is completely specified by the initial density and final function S, as made evident by the antiparabolic nature of (12). This allows a connection with the Scrödinger processes, as explained for example in [18]. A situation of this kind arises also in classical mechanics, if we assume the velocity as a control for the trajectories. Then the final momentum appears as Lagrange multiplier. On the other hand, by the very structure of equations (4,8), nothing prevents from taking both ρ and S as given at the initial time, and let them evolve at later times. In this way, the concept of state acquires a more conventional form.

Finally, let us also explicitely remark that (4,8) are invariant under time reversal.

The original Nelson formulation was based on the following stochastic definition of acceleration [1,2]

$$a(x,t) = \frac{1}{2}(D_{(+)}v_{(-)} + D_{(-)}v_{(+)}) \tag{13}$$

where the forward and backward transport operators $D_{(\pm)}$ are defined in (11). Then the stochastic Newton equation

$$ma = -\nabla V \tag{14}$$

was taken as basic dynamical assumption, together with the irrotationality condition (7).

We consider the general scheme developed in this Section as the very basic foundation of stochastic mechanics. Starting from the striking similarity of (4,8) with the equations for the Madelung fluid in quantum mechanics (see for example [3]), we will develop a kind of reduced physical interpretation scheme, connecting stochastic mechanics with quantum mechanics, in next Section.

3. The physical interpretation of Nelson stochastic mechanics

First of all let us state the formal connection of stochastic mechanics with quantum mechanics. We notice that the programming equations (4,8), previously derived, are in fact the same as the equations for the Madelung fluid of quantum mechanics [4]. Therefore, if we exploit the De Broglie Ansatz in the opposite direction, in order to define the wave function

$$\psi(x,t) = \sqrt{\rho(x,t)}\exp(\frac{i}{\hbar}S(x,t)) \tag{15}$$

we can see that (4,8) are equivalent to the quantum Schrödinger equation

$$i\hbar(\partial_t\psi)(x,t) = -\frac{\hbar^2}{2m}\Delta\psi + V(x)\psi \tag{16}$$

Therefore, we find the basic result that the Born interpretation for the square of the wave function in quantum mechanics must not be postulated, but is a simple consequence of the definition (15), because there is no need for any additional interpretation for the density in stochastic mechanics. On the other hand, the wave function, as defined by (15) in terms of quantities appearing in stochastic mechanics, obeys the correct quantum equation (16), therefore it develops in time as the rigth quantum wave function must do. These simple considerations, originally due to Nelson [1] (see also [2,3,4,10,11]), allow us to introduce a very natural physical interpretation of stochastic mechanics. In fact, if we consider for example the scattering of a particle in a potential, we can prepare the stochastic state, according to the physical incoming conditions, and then calculate the probability that the particle will be found in some scattering cones, when time goes to infinity. In this way we can calculate the cross sections. Analogously for all other observables.

Therefore, the whole physical content of a quantum mechanical system can be found from the stochastic formulation. In [14] we have given a detailed description of the methods for deriving all structures appearing in the standard formulation of quantum mechanics starting from the stochastic scheme, as for example the Hilbert space of states, the operator observable algebra, the quantization rules, the implementation of canonical transformations through unitary transformations in Hilbert space, and so on. For all problems of physical interpretation, we refer to [1,2,3,4,10,11,12,13,14,15].

While the situation concerning the physical interpretation, in the restricted frame considered above, looks satisfactory, nevertheless there are unpleasant features. In fact, only the stochastic aspects connected with the probability density enter directly into the physical interpretation. On the other hand, the stochastic processes, introduced in the frame of stochastic mechanics, have as peculiar aspects the transition probability densities, and their properties. Now, it is easy to show that these can not have a direct physical interpretation [12]. In fact, any attempt to measure the position will change the state, as it has been explained in the introduction. Moreover, the dependence of the transition probability densities can have very strange aspects. For example, Nelson [4] has shown that the transition probabilities for a subsystem can depend on the properties of a decoupled far away different subsystem.

These remarks, togheter with the other presented in the introduction, show that some extension is necessary for the structure of stochastic mechanics, if we are willing to consider it as a complete physical theory.

In this Section, we have considered quantum mechanics and stochastic mechanics in the configuration representation. Next Section shows how to extend this scheme to general representations.

4. Stochastic processes associated to the discrete form of the quantum madelung fluid

Here we follow mainly the treatment given in [15], but more details can be found in [13]. The strategy will be the following. Firstly we show how we can associate stochastic processes to quantum states. In the next Section we will give an independent definition of these processes.

Consider a quantum system on a separable, or even finite dimensional, Hilbert space. If ϕ_i, $i = 1, 2, \ldots$, is an orthonormal basis, let us introduce, for a generic wave function ψ, the components $\psi_i = \langle \phi_i, \psi \rangle$. For the matrix elements of the Hamiltonian H of the system we can write

$$H_{ij} = h_{ij} \exp(\frac{i}{\hbar} \alpha_{ij}) \tag{17}$$

where, by selfadjointness, we can assume

$$h_{ij} = h_{ji}, \quad h_{ij} \geq 0, \quad i \neq j, \quad \alpha_{ij} = -\alpha_{ji}, \quad \alpha_{ii} = 0 \tag{18}$$

Then, the Schrödinger equation can be written for each component in the form

$$i\hbar\dot{\psi}_i = \sum_j h_{ij} \exp(\frac{i}{\hbar} \alpha_{ij})\psi_j \tag{19}$$

It is convenient to introduce real variables (ρ_i, S_i), $\rho_i \geq 0$, such that, in analogy with (15),

$$\psi_i = \sqrt{\rho_i} \exp(\frac{i}{\hbar} S_i) \tag{20}$$

Then, the complex equation (19) splits into two real ones

$$\dot{\rho}_i = \sum_j \frac{2h_{ij}}{\hbar} \sqrt{\rho_i \rho_j} \sin \beta_{ij}, \tag{21}$$

$$\dot{S}_i = -\sum_j h_{ij} \sqrt{\frac{\rho_j}{\rho_i}} \cos \beta_{ij} \tag{22}$$

where

$$\beta_{ij} = \frac{1}{\hbar}(\alpha_{ij} + S_j - S_i), \quad \beta_{ij} = -\beta_{ji}, \quad \beta_{ii} = 0. \tag{23}$$

Notice that (21) is a continuity equation analogous to (4), while (22) is the discrete generalization of the Hamilton-Jacobi-Madelung equation (8).

To each quantum state ψ, a Markov process $q(t)$, taking values on $\{i = 1, 2, \ldots\}$, can be easily associated. We assume $\rho_i(t)$ to be the occupation probability for the site i at time t. Let us also introduce the mean forward, backward, current, and osmotic generalized drifts, a_{ij}^+, a_{ij}^-, a_{ij}, and a_{ij}^0. In terms of the forward transition probability $p(i, t; j, t')$, $t \geq t'$, the generator a^+ is defined by the limit, as $\Delta t \to 0^+$,

$$a_{ij}^+ = \lim \left(p(i, t + \Delta t; j, t) - \delta_{ij} \right) / \Delta t \qquad (24)$$

Analogously, we can define a^- in terms of the backward transition probability. We have (see [13] for all details)

$$\pm a_{ij}^\pm \geq 0, \quad i \neq j, \quad \sum_i a_{ij}^\pm = 0 \qquad (25)$$

$$\dot{\rho}_i(t) = \sum_j a_{ij}^\pm \rho_j(t) \qquad (26)$$

The connection between a^+ and a^- is easily found as a consequence of Bayes theorem, connecting forward conditioning with backward conditioning, in the form

$$a_{ji}^- \rho_i + a_{ij}^+ \rho_j = \delta_{ij} \sum_k a_{ik}^+ \rho_k \qquad (27)$$

Let us also introduce

$$a_{ij} = \frac{1}{2}(a_{ij}^+ + a_{ij}^-), \quad a_{ij}^0 = \frac{1}{2}(a_{ij}^+ - a_{ij}^-) \qquad (28)$$

and note that for $i \neq j$

$$a_{ij}\rho_j = -a_{ji}\rho_i, \quad a_{ij}^0 \rho_j = a_{ji}^0 \rho_i, \quad a_{ij}^0 \geq 0, \quad i \neq j \qquad (29)$$

Moreover

$$\dot{\rho}_i(t) = \sum_j a_{ij}\rho_j(t) \qquad (30)$$

Now we give the interpretation of (30) as a continuity equation fully equivalent to (21). Taking into account the antisymmetry property given by the first relation in (29), we can immediately identify

$$a_{ij}(t) = \frac{h_{ij}}{\hbar} \sqrt{\frac{\rho_i}{\rho_j}} \sin \beta_{ij}, \quad i \neq j \qquad (31)$$

On the other hand we know from the definition (24) that

$$a_{ij}^+ = a_{ij} + a_{ij}^0 \geq 0, \quad i \neq j \qquad (32)$$

In the continuous case, the osmotic velocity u in (3) is a function only of the density and not of the phase. This remark, together with (29), leads us to the natural consideration of processes for which the osmotic part a^0 has the form

$$a_{ij}^0(t) = \frac{h_{ij}}{\hbar}\sqrt{\frac{\rho_i}{\rho_j}}, \quad i \neq j \tag{33}$$

In this way we have completely defined the process $q(t)$ associated to each quantum state ψ, by giving its density and the generators (31) and (33).

In the next Section we will show that these processes can be independently characterized as critical processes for a suitably defined stochastic variational principle.

5. The discrete stochastic variational principle

In order to build a stochastic variational principle in the discrete case, we must find the correct action. In the continuous case, in configuration space, the stochastic action is nothing but the averaged classical action, as (5) shows. The procedure for finding the correct stochastic action in the discrete case rests on the so called verification theorem [19] of mathematical engineering, adapted to the present time reversal invariant situation. This procedure is explained in [13] in full details. Here, we give the general frame and the main results.

Let us introduce trial random processes $q(t)$, taking values on $\{i = 1, 2, \ldots\}$, with distribution $\rho_i(t)$ and a generic controlling drift a^+, so that (26) is satisfied for the evolution of ρ in terms of a^+ as infinitesimal generator.

Introduce now external fields h_{ij} and α_{ij}, with the properties given by (18). We need also an overall scale of action, characterized by a constant $\hbar$, later to be identified with Planck's constant.

Let us consider controlling generators a^+ satisfying the following off diagonal bounds

$$a_{ij}^+ a_{ji}^+ \leq \left(\frac{h_{ij}}{\hbar}\right)^2, \quad i \neq j \tag{34}$$

and introduce, for any specification of a^+, the auxiliary field β_{ij}, such that

$$a_{ij}^+ a_{ji}^+ \leq \left(\frac{h_{ij}}{\hbar}\right)^2 \cos^2 \beta_{ij}, \quad i \neq j, \quad \beta_{ij} = -\beta_{ji}, \quad \beta_{ii} = 0 \tag{35}$$

Define the forward Lagrangian

$$\mathcal{L}_i^+ = \sum_j{}'(\hbar f(\beta_{ij}) + \alpha_{ij})a_{ji}^+ - h_{ii} \tag{36}$$

where the sum excludes the value $j = i$, and

$$f(\beta) = \beta - \cos\beta/(1 - \sin\beta) \tag{37}$$

Then the action associated to the trial process $q(t)$, in a given time interval, is defined by

$$A(t_0, t_1; q) = \int_{t_0}^{t_1} \sum_i \mathcal{L}_i^+ \rho_i(t) dt \qquad (38)$$

We refer to [13] for a full motivation of this choice. Then we can state the following stochastic variational result.

Necessary and sufficient conditions for the action (38) to be stationary under variations of the process, corresponding to variations δa^+, with the conditions that the density stays invariant at the boundary times, are the following general programming equations

$$a_{ji}(a_{ij} - a_{ij}^0 \sin \beta_{ij}), \quad i \neq j \qquad (39)$$

$$\hbar \beta_{ij} + \alpha_{ji} + \frac{\hbar a_{ji}^0 (a_{ij} - a_{ij}^0 \sin \beta_{ij})}{\cos \beta_{ij} a_{ij}^+ a_{ji}^+} = S_j(t) - S_i(t) \qquad (40)$$

where S_i arise from Lagrangian multipliers associated to the time boundary conditions, and satisfy

$$\partial_t S_i + \sum_j S_j a_{ji}^+ = \mathcal{L}_i^+ \qquad (41)$$

Let us now consider the solutions of (39,40). An obvious solution is obtained by putting

$$a_{ij} = a_{ij}^0 \sin \beta_{ij} \qquad (42)$$

$$\hbar \beta_{ij} = \alpha_{ij} + S_j(t) - S_i(t), \qquad (43)$$

From (42,35,28) it is immediate to see that (33) must hold. Therefore, we have recovered the osmotic condition (33) as a consequence of the stochastic variational principle. Moreover, (43) coincides with (23), and a simple calculation shows that (41), with the definition (36), is completely equivalent to (22).

Therefore, we have obtained the following important result. The stochastic processes, each associated to a quantum state, according to the procedure outlined in Section 4, can be also independently characterized as critical processes for a suitably defined stochastic variational problem.

Clearly, there are other solutions for the programming equations (39,40), those associated to the vanishing of some a_{ij} in (39). These will be investigated in the following Section.

If the discrete system arises from a lattice approximation for the continuous system of Section 2, where the Laplacian is replaced by its finite difference approximation, then it can be easily shown (see [13]) that the

expression (33) for the discrete osmotic drift gives rise to the correct continuous expression for the osmotic velocity given by (3).

Let us end this Section with a very important remark. We have shown
how to associate stochastic processes to quantum states evolving according
to the Schrödinger equation (19). Now, time evolutions are a particular case
of unitary transformations, those for which the generator is the Hamiltonian
and time is the associated parameter. Let us now consider a generic self-
adjoint operator H. Then (19) can be interpreted as giving the development
of the generated unitary transformation in terms of the associated parameter
t. But now we can still associate, by following the same procedure as before,
Markov random processes to generic unitary transformation groups.

Therefore, we have the important result that in generalized stochastic
mechanics the unitary transformations for states of quantum mechanics
are replaced by Markov random processes, with well defined properties,
indexed by the parameter conjugated to the selfadjoint generator, and not
by time, as in the Schrödinger case, where the generator is the Hamiltonian.
This solves the problem, raised in the introduction, of finding the stochastic
analog of unitary transformations, relating different representations.

6. Stochastic simulation of the quantum wave packet reduction

Let us now consider solutions of the programming equations (39,40) different from those based on (42). For example, let us assume $a_{ij} = 0$, so that
(39) is satisfied. Now we have $\dot{\rho}_i = 0$, i.e. the occupation probabilities stay
constant in time on the average. On the other hand (35,28) give

$$a_{ij}^0 a_{ji}^0 \leq (\frac{h_{ij}}{\hbar})^2 \cos^2 \beta_{ij}, \quad i \neq j. \tag{44}$$

Therefore, from (29) we have

$$a_{ij}^0(t) = \frac{h_{ij}}{\hbar} \sqrt{\frac{\rho_i}{\rho_j}} |\cos \beta_{ij}|, \quad i \neq j, \tag{45}$$

instead of (33).

In this case, (40) reduces to

$$\hbar(\beta_{ij} - \tan \beta_{ij}) = \alpha_{ij} + S_j(t) - S_i(t), \tag{46}$$

which is deeply different from (43).

It is now easy to calculate S_i from (41). We find

$$\dot{S}_i = -\sum_j{}' \eta_{ij} \sqrt{\frac{\rho_j}{\rho_i}} - h_{ii} = -E_i \tag{47}$$

where η_{ij} is the sign of $\cos\beta_{ij}$ and ρ_i are constant. Let us start from some initial time with some given values for η_{ij}. Because of continuity, they stay constant in time for some time interval. In fact, we will find that they never change sign for any time. In the generic case $E_i \neq E_j$, $i \neq j$. Thus, we see that $S_i - S_j \to \pm\infty$ as $t \to \infty$. Therefore, $\beta_{ij} - \tan\beta_{ij} \to \pm\infty$, and all β's never cross values where $\cos\beta$ can change sign, so that E_i stay constant. Necessarily we have $\cos\beta_{ij} \to 0$. Therefore, we have also from (44) that $a^0 \to 0$ and $a^+ \to 0$, because a is already zero. Therefore, we see that in this case as $t \to \infty$ the diffusion disappears completely in the limit, and the process reduces to a mixture of static distributions at each site.

In conclusion, we have the following important result. Starting from a single stochastic variational principle, we can also simulate the formation of mixtures, in analogy with the quantum measurement collapse.

Compared with the ordinary theory of quantum measurement (see for example [16]), here the wave packet collapse, leading to mixture formation, is not instantaneous, but evolves according to a well defined dynamical scheme given by (45,46,47). Notice that the speed of mixture formation depends on the differences $E_i - E_j$. If they are very large, the progression to collapse is very fast. In the particular case of a two level system, we have found that the collapse is faster for states where there is less uncertainty about the result of the measurement. In order to see whether the phenomenon of relaxational solutions found here has some physical implication for the quantum measurement theory, it would be necessary to investigate the possibility of checking, at the experimental level, that the wave packet collapse is not instantaneous and its speed depends on the initial state.

7. Conclusion and outlook

In this paper we have given a short presentation of stochastic mechanics as a kind of generalization of classical mechanics, based on stochastic variational principles. The similarity of the programming equations of stochastic mechanics with the Madelung equations of quantum systems allows us to introduce a suitable physical interpretation, able to explain all phenomenological features of quantum mechanics. Also the details of the standard operator formulation of quantum mechanics can found a very easy explanation in terms of the general structure of stochastic mechanics. With the purpose of investigating the possibility that stochastic mechanics is a kind of fundamental physical theory, we have presented an extension beyond the configuration space, in a generic representation. We have shown that it is possible to find a stochastic Lagrangian for controlled random processes on discrete configuration space, so that the whole structure of quantum mechanical behavior is correctly simulated. Moreover, the stochastic variational principle also gives the expression for the osmotic part of the transition probabilities per unit time. This expression reduces to the usual one of stochastic mechanics in the continuous limit. A relevant aspect of this scheme is the existence of critical processes simulating the quantum measurement wave packet collapse and producing mixtures in the time asymptotic region.

On the other hand, a physical justification for the assumed form of the stochastic Lagrangian seems to be a very difficult task, and surely it will involve new ideas about the origin of the underlying Brownian motion in stochastic mechanics.

Our general scheme can be applied to any quantum unitary transformation, not necessarily to the time evolution. Therefore, the orbits of unitary flows applied to a quantum state have always associated random processes in a stochastic mechanics setting. In this way one can show that these stochastic implementations of unitary transformations can be exploited in order to restore Lorentz invariance in stochastic models of field theory [20], where the Euclidean invariance is apparently involved.

A very interesting possibility of further developments, is to investigate whether stochastic mechanics can be embedded into the general structure of Schrödinger processes [17], as described in [18]. In the dissipative case, Schrödinger processes typically involve a density function evolving forward in time according to the forward Fokker-Planck equation, and an importance function, the adjoint function, evolving backward in time according to the backward Fokker-Planck equation. The structure is canonical, with an undelying symplectic canonical structure, and time reversal invariant in a suitable sense. For the programming equations coming from the stochastic variational principle, we have seen, in Section 2, that the basic quantities involve a forward propagation for the density, and a backward propagation for the Hamilton-Jacobi function. Therefore, it seems conceivable that the original program of Schrödinger [17] is best pursued in the frame of the stochastic formulation of quantum mechanics, as compared to the usual operator formulation. We plan to report on this problem in a forthcoming paper.

References

1. Nelson, E. (1966). *Phys. Rev.*, **150** (1079).
2. Nelson, E. (1967). *Dynamical Theories of Brownian Motion*, Princeton University Press, Princeton, New Jersey.
3. Guerra, F. (1981). *Phys. Rep.*, **77** (121).
4. Nelson, E. (1985). *Quantum Fluctuations*, Princeton University Press, Princeton, New Jersey.
5. Lanczos, C. (1962). *The Variational Principles of Mechanics*, Toronto.
6. Guerra, F. and Morato, L.M. (1983). *Phys. Rev.*, **D27** (1774).

7.   Aldrovandi, E., Dohrn, D. and Guerra F. (1990). *J. Math. Phys.*, **31** (639).

8.   Aldrovandi, E., Dohrn, D. and Guerra F. (1989). "Stochastic Action of Dynamical Systems on Curved Manifolds. The Isokinetic Developing Map on Trajectories", in *Stochastic Processes, Physics and Geometry*, World Scientific, Singapour.

9.   Aldrovandi, E., Dohrn, D. and Guerra F. (1992). *Acta Applicandae Mathematicae*, **26** (219).

10. Blanchard, Ph., Combe, Ph. and Zheng, W. (1987). *Mathematical and Physical Aspects of Stochastic Mechanics*, Lecture Notes in Physics **281**, Springer-Verlag, Berlin.

11. Blanchard, Ph., Golin, S. and Serva, M. (1986). *Phys. Rev.*, **D34** (3732).

12. Onofri, E. (1979). *Lett. Nuovo Cimento*, **24** (253).

13. Guerra, F. and Marra, R. (1983). *Phys. Rev.*, **D29** (1647).

14. Guerra, F. and Marra, R. (1983). *Phys. Rev.*, **D28** (1916).

15. Guerra, F. (1994). "Nelson Quantum Mechanics and the Interpretation of Quantum Mechanics", in *The interpretation of Quantum Theory: where do we stand?*, Istituto dell'Enciclopedia Italiana, Roma.

16. Von Neumann, J. (1955). *Mathematical Foundations of Quantum Mechanics*, Princeton University Press, Princeton, New Jersey.

17. Schrödinger, E. (1931). *Sitz. der preuss. Akad. der Wiss. Phys. Math. Klasse*, pp. 144–153.

18. D'Autilia, R. and Guerra, F., "Generalized Schrödinger processes", in preparation.

19. Fleming, W.H. and Rishel, R. (1975). *Deterministic and Stochastic Optimal Control*, Springer, Berlin.

20. Guerra, F. and Ruggiero, P. (1973). *Phys. Rev. Lett.*, **31** (1022).

THE FERMI-DIRAC STATISTICS:
A SIMULTANEOUS DISCOVERY

An epistemological perspective of an historical case

N. GUICCIARDINI

Dipartimento di Filosofia
Università di Bologna, Italy

and

G. INTROZZI

Dipartimento di Fisica Nucleare e Teorica
Università di Pavia, Italy

Abstract. The very different approaches used by E. Fermi and P. A. M. Dirac to derive the Fermi-Dirac particle statistics are analyzed in detail. We suggest that the stability of theoretical conclusions through conceptual change, as exemplified in this historical case study, plays a role in justifying our belief in theoretical statements.

1. Introduction

In February 1926 Enrico Fermi published in *Rendiconti dell'Accademia dei Lincei* a paper on a new quantum statistics for a monoatomic perfect gas (Fermi, 1926a; for an historical analysis of Fermi's contribute to quantum statistics: Caldirola, 1975; Belloni, 1978; Amaldi, 1987; Mehra and Rechenberg, 1987, pp. 746-771). Paul Adrien Maurice Dirac derived the same results in a different way, and announced them in August 1926. Immediately after, the Italian scientist sent a letter to Dirac, in which he wrote:

C. Garola and A. Rossi (eds.), The Foundations of Quantum Mechanics, 357-367.

"In your interesting paper 'On the theory of Quantum Mechanics' (Proc. Roy. Soc., 112, 661, 1926) you have put forward a theory of the Ideal Gas based on Pauli's exclusion Principle. Now a theory of the ideal gas that is practically identical to yours was published by me at the beginning of 1926 (Zs. f. Phys. 36, p. 902; Lincei Rend. February 1926). Since I suppose that you have not seen my paper, I beg to attract your attention on it." (Mehra and Rechenberg, 1987, p. 767)

Dirac not only recognized Fermi's priority, but also introduced the term *fermion* to denote a particle obeying to the Fermi-Dirac statistics. Dirac admitted he had read Fermi's paper in *Zeitschrift für Physik* before the summer 1926, but he did not realized its importance. Few years later, Dirac recalled such a circumstance:

"When I looked through Fermi's paper, I remembered that I had seen it previously, but I had completely forgotten it. I am afraid it is a failing of mine that my memory is not very good and something is likely to slip out of my mind completely, if at the time I do not see how it could be important for any of the basic problems of quantum theory; it was so much a detached piece of work. It had completely slipped out of my mind, and when I wrote up my work on the asymmetric wave functions, I had no recollection of it at all." (Kragh, 1990, p. 36)

Dirac's discovery was certainly independent from Fermi's one: their conceptual frameworks and mathematical methods were very different. The only common elements between this two researches are the objects[1] and the conclusions.

2. Fermi

Fermi's research on particle statistics was motivated by O. Sackur's, H. Tetrode's and O. Stern's theory on the absolute constant for the entropy of an ideal gas (Belloni, 1978; Desalvo, 1987). Thus, it is within the frame of classical thermodynamics that Fermi begins its studies on the perfect gas (Fermi, 1923; Fermi, 1924). In such a context, he realizes how the failure of Sommerfeld's quantization rules for a helium atom or a monoatomic gas is linked to the fact that these systems *"contengono alcune parti identiche"* [2].

P. Ehrenfest and A. Sommerfeld generalization of N. Bohr's quantization rules states that the phase space integral $\oint p dy$ is a multiple integer of Plank constant h. According to F. Rasetti:

"nel caso di sistemi con particelle identiche, come l'atomo di elio, il periodo completo al quale si deve estendere l'integrale di Sommerfeld deve corrispondere alla ripetizione di uno stato indistinguibile dallo stato iniziale, anche se i due stati differiscono classicamente poichè si passa dall'uno all'altro con uno scambio di particelle."[3] (Quoted by F. Rasetti in the notes to *Fermi's Collected Papers*, Fermi, 1962-65, vol. 1, p. 124)

In Fermi (1924) a perfect gas of n point-like molecules contained in a volume v is considered. The Italian scientist said that it is possible to formulate different hypothesis on the method to quantize such a gas. Fermi subdivided the volume v into cells:

"potremo a piacere per esempio dividere v in n parallelepipedi, mettendo in ciascuno di essi una sola molecola; oppure dividere v in $n/2$ parallelepipedi, e mettere in ciascuno 2 molecole, oppure infine dare a v la forma di un solo parallelepipedo, e lasciare in esso tutte le n molecole."[4] (Fermi, 1962-65, vol. 1, pp. 125-126)

Fermi showed in the same paper that only the first alternative allows to calculate the value for the entropy constant in agreement with the experimental results.

In 1924 S. N. Bose and A. Einstein treating the light quanta and the atoms of an ideal gas as indistinguishable particles, obtained a new statistics, different from the Maxwell-Boltzmann one, capable to reproduce Plank's formula for the black body radiation spectrum. The following year, W. Pauli introduced the exclusion principle, in order to explain at the same time Bohr's atomic model and the regularities of the periodic table of the elements. These are the theoretical basis for two papers written by Fermi on the quantization of a monoatomic gas (Fermi, 1926a; Fermi, 1926b). The Fermi-Dirac statistics is obtained extending Pauli exclusion principle to the atoms of a perfect gas.

Fermi considered a gas constituted by N *"molecole"* of mass m in an elastic central force field. The potential is described by

$$U(r) = 2\pi^2 \nu^2 m r^2 \tag{1}$$

Each molecule corresponds to an oscillator with frequency ν. The motion of each molecule is described by three quantum numbers s_1, s_2, s_3, correlated to the energy of the molecule:

$$W_s = h\nu (s_1 + s_2 + s_3) = h\nu s \tag{2}$$

Each energy level could be obtained using different combinations of the quantum numbers s_1, s_2, s_3 and the total number of combinations for a given energy W_s is:

$$Q_s = \frac{1}{2}(s+1)(s+2) \tag{3}$$

Fermi applied Pauli exclusion principle to such a gas model:

> "ammetteremo [...] che nel nostro gas ci possa essere al massimo una molecola il cui movimento sia caratterizzato da certi numeri quantici [...] due distribuzioni devono essere considerate identiche, se i posti occupati dalle molecole sono identici, devono essere considerate identiche anche quando differiscono per una permutazione delle molecole."[5] (Fermi, 1962-65, vol. 1, p. 183 and p. 190)

Consequently, denoting with N_s the number of molecules with energy W_s, the following inequality must hold:

$$N_s \leq Q_s \qquad (4)$$

Furthermore, the conservations of matter and energy imply

$$\sum N_s = N \qquad (5)$$

$$\sum W_s N_s = E \qquad (6)$$

with N and E constants.

The next problem consisted in the calculation of the number P of distributions such that N_0 molecules have null energy, N_1 have energy $h\nu$, N_2 have energy $2h\nu$, and so on. It is straightforward to find that P is given by the product

$$P = \prod_s \binom{Q_s}{N_s} \qquad (7)$$

Using a well established procedure, Fermi determined the values for N_s that maximize P, under the conditions indicated in Eqn. (4), (5) and (6):

$$N_s = Q_s \frac{\alpha e^{-\beta s}}{\alpha e^{-\beta s} + 1} \qquad (8)$$

In order to give a physical meaning to the constants α and β, Fermi used a formulation of the correspondence principle:

> "Per trovare la relazione tra queste costanti e la temperatura, osserviamo che, per effetto della attrazione verso 0, la densità del nostro gas sarà una funzione di r, che deve tendere a zero per $r = \infty$. Per conseguenza, per $r = \infty$ debbono cessare i fenomeni di degenerazione, e in particolare la distribuzione delle velocità [...] deve trasformarsi nella legge di Maxwell."[6] (Fermi, 1962-65, vol. 1, p. 184)

Fermi demonstrated that it is necessary to impose

$$\beta = h\nu/kT \tag{9}$$

to obtain Maxwell distribution law for the velocities. In addition, he showed that α is a function of the number of molecules in the gas.

At the end of his paper, Fermi derived the state equation for the gas, calculated the specific heat and the absolute value for the entropy of the gas, that results to be in agreement with the value obtained by H. Tetrode and O. Stern.

3. Dirac

Few months after the publication of the paper by Fermi, the article "On the theory of quantum mechanics" by P. A. M. Dirac appeared in the *Proceedings of the Royal Society* (Dirac, 1926). This was the last of a series of works, published in between 1924 and 1925, on the algebraic approach to quantum mechanics (about Dirac's scientific contribution: Kragh, 1990; Mehra and Rechenberg, 1982). It should be stressed that this approach was a third way to present quantum mechanics, after the matrix algebra proposed by W. Heisenberg, and the more recent wave equation suggested by E. Schrödinger. Dirac, probably influenced by his knowledge of O. Heaviside's operators algebra and perhaps by J. Baker's projective geometry (see Mehra and Rechenberg, 1982, pp. 141-147), formulated in 1925 the q-numbers algebra: it is a non-commutative algebra that represents a generalization of Heisenberg's matrix algebra.

In the paper "On the theory of quantum mechanics", Dirac used Schrödinger's wave mechanics, that Heisenberg described to him in a letter dated April 9th, 1926. In a following letter (May 26th, 1926), Heisenberg also showed to Dirac the equivalence between matrix and wave formulations of quantum mechanics (see Mehra and Rechenberg, 1987, pp. 758-759). The aim of Dirac was to find a quantum mechanics approach to systems containing identical particles. Dirac considered the Schrödinger equation for an atom with two electrons, and denoted as (m, n) the state in which one electron is in the m state, and the other is in the n state. If the Coulomb potential between the two electrons is negligeable, it is possible to write a solution as product $(\psi_m(1) \cdot \psi_n(2))$ of eigenfunctions of a single electron atom (where $\psi_k(i)$ is a short form for $\psi_k(x_i, y_i, z_i, t)$). Indeed, since the two electrons are indistinguishable, the states (m, n) and (n, m) should also result indistinguishable[7]. Thus, the general solution for the two electron atom has to be:

$$\psi_{m,n} = a_{m,n}\,\psi_m(1) \cdot \psi_n(2) + b_{m,n}\,\psi_m(2) \cdot \psi_n(1) \tag{10}$$

The two constants $a_{m,n}$ and $b_{m,n}$ have to preserve the indistinguishability of the states (m, n) e (n, m); consequently

$$a_{m,n} = b_{m,n} \quad or \quad a_{m,n} = -b_{m,n} \tag{11}$$

Since the two electrons are indistinguishable, the self-adjoint operators A for the eigenfunction $\psi_{m,n}$ have to be invariant for a permutation of the two electrons (using Dirac's words: A should be "any symmetrical function of the two electrons"). Then:

$$A\,\psi_{m,n} = \sum_{m',n'} \psi_{m',n'}\, c_{m',n',m,n} \tag{12}$$

where $\psi_{m',n'}$'s represent a complete orthonormal set of solutions for the Schrödinger equation of an atom with two electrons. If $a_{m,n} = b_{m,n}$, then $A\,\psi_{m,n}$ is symmetrical and consequently all the $\psi_{m',n'}$'s in Eqn. (12) also have to be symmetrical. On the opposite, if $a_{m,n} = -b_{m,n}$, then $A\,\psi_{m,n}$ is antisymmetrical, and all the $\psi_{m',n'}$'s as well have to be antisymmetrical. Dirac separated two sets of solutions: symmetrical and antisymmetrical; only the second one is in agreement with Pauli exclusion principle:

> "An antisymmetrical eigenfunction vanishes identically when two of the electrons are in the same orbit. This means that in the solution of the problem with antisymmetrical eigenfunctions there can be no stationary states with two or more electrons in the same orbit, which is just Pauli's exclusion principle." (Dirac, 1926, pp. 669-670)

Finally, Dirac extended this conclusion to atoms with more than two electrons.

In the third paragraph of Dirac (1926), "Theory of the Ideal Gas", he considered a gas of N non-interacting molecules, contained in a volume V. He assumed that:

1. The eigenfunctions of the system have to be either symmetrical or antisymmetrical;
2. All the stationary states, each represented by an eigenfunction, have the same a priori probability.

The eigenfunctions for the gas are solutions of the relativistic equation

$$\left(\Box - \frac{m^2 c^2}{\hbar^2}\right)\psi = 0 \tag{13}$$

and the structure of such eigenfunctions is:

$$\psi_{a_1 a_2 a_3} = e^{\left[\frac{2\pi i}{\hbar}(a_1 x + a_2 y + a_3 z - Et)\right]} \tag{14}$$

where a_i's are the momentum components and E is the energy of the molecule. The usual boundary conditions ($0 < x < 2\pi$ for instance) imply

$$a_1 \frac{h}{2\pi} = n + \epsilon \tag{15}$$

with integer n and real ϵ. Each solution of Eqn. (13) is characterized by three translational quantum numbers (a_1, a_2, a_3), such that:

$$a_1^2 + a_2^2 + a_3^2 - E^2/c^2 = -m^2 c^2 \tag{16}$$

If α indicates a triplet (a_1, a_2, a_3) of quantum numbers, the symmetrical eigenfunctions are:

$$\sum_P \psi_{\alpha_1}(1) \ldots \psi_{\alpha_N}(N) \tag{17}$$

where $\sum_P$ is extended to all the possible permutations of the α_i's indices. The antisymmetrical eigenfunctions are instead:

$$\sum_P \epsilon_P \, \psi_{\alpha_1}(1) \ldots \psi_{\alpha_N}(N) \tag{18}$$

with $\epsilon_P = 1$ for even permutations, and $\epsilon_P = -1$ for odd permutations. Dirac observed that:

> "If now we adopt the solution of the problem that involves symmetrical eigenfunctions, we should find that all values for the number of molecules associated with any wave have the same a priori probability, which gives just the Einstein-Bose statistical mechanics. On the other hand we shall obtain a different statistical mechanics if we adopted the solution with antisymmetrical eigenfunctions, as we should have then either 0 or 1 molecule associated with each wave." (Dirac, 1926, pp. 671-672)

Dirac continued by observing that the Bose-Einstein statistics must be "the correct one when applied to light quanta", while the new one, obtained by selecting only antisymmetrical eigenfunctions, has to be "probably the correct one for gas molecules". The choice of antisymmetrical solutions is equivalent to a generalization of Pauli exclusion principle: two molecules can not have the same quantum numbers (a_1, a_2, a_3). Since Pauli exclusion principle proved to be correct for the electrons in an atom and "one would expect molecules to resemble electrons more closely than light-quanta." (Dirac, 1926, p. 672)

Dirac was able, at this point, to derive the gas state equation, following a standard procedure. He divided the eigenfunctions into sets, each associated to molecules of *"about the same energy E_s"*. The distribution probability for N_s molecules with energy E_s is given by:

$$P = \prod_s \frac{A_s}{N_s!\,(A_s - N_s)!} \tag{19}$$

where A_s is *"the number of waves in the s-th set"*.

The entropy $S = k\,log\,P$ has to be maximized under the usual constraints:

$$E = \sum_s E_s N_s \tag{20}$$

$$N = \sum_s N_s \tag{21}$$

Dirac obtained:

$$N_s = \frac{A_s}{e^{(\alpha + \beta E_s)} + 1} \tag{22}$$

where $A_s = 2\pi V (2m)^{3/2} E_s^{1/2} dE_s/(h)^3$. And using the thermodynamic relation $T = \delta E/\delta S$,

$$\beta = 1/kT \tag{23}$$

Dirac finally wrote the two equations:

$$N = \sum_s N_s = 2\pi V \left(\frac{\sqrt{2m}}{h}\right)^3 \int_0^\infty \frac{\sqrt{E_s}\,dE_s}{e^{(\alpha + E_s/kT)} + 1} \tag{24}$$

$$E = \sum_s N_s E_s = 2\pi V \left(\frac{\sqrt{2m}}{h}\right)^3 \int_0^\infty \frac{E_s^{3/2}\,dE_s}{e^{(\alpha + E_s/kT)} + 1} \tag{25}$$

The state equation can be derived by eliminating α in Eqn. (24) and (25), and using $pV = \frac{2}{3}E$.

Dirac concluded by observing:

> "The saturation phenomenon of the Einstein-Bose theory does not occur in the present theory. The specific heat can easily be shown to tend steadily to zero as $T \to 0$, instead of first increasing until the saturation point is reached and then decreasing, as in the Einstein-Bose theory." (Dirac, 1926, p. 673)

4. Invariance and convergence

As A. Pais puts it:

> "With Fermi's two papers the non-quantum mechanical stage of quantum statistics, the time of mere counting, comes to a conclusion." (Pais, 1986, p. 284)

As a matter of fact, Fermi's and Dirac's works on statistics belong to two different periods of the history of quantum theory. One can point out several relevant differences. While Fermi's papers still belong to what is commonly called "old quantum theory", Dirac formulated the new statistics in the context of the new quantum mechanics. As we have seen, Fermi employed a form of correspondence principle in order to determine the constants α and β. Correspondence principles were important in guiding the researches of physicists of Bohr's and Sommerfeld's generation. Dirac did not accept this methodology. He states:

> "The correspondence between the quantum and classical theories lies not so much in the limiting agreement when $h \to 0$ as in the fact that the mathematical operations on the two theories obey in many cases the same laws." (Dirac, 1925, p. 649)

Dirac is not so much interested in a correspondence between classic and quantum mechanics which should manifest itself through limiting arguments (typically $h \to 0$), but rather in algebraic structure similarity between classic and quantum mechanics. The algebra of quantum operators (q-numbers) is derived from Hamiltonian mechanics. Dirac's guiding principle is the extension of classical algebraic structures to the algebra of quantum operators.

Also from the point of view of reference models, the Italian and the British physicists belong to two different stages of the history of quantum mechanics. Fermi makes recourse to an atomistic model, to which he imposes quantum rules and Pauli exclusion principle. Dirac follows Heisenberg and avoids reference to non observables quantities such as particle's positions or orbital frequencies. He writes:

> "The theory[8] thus enables one to calculate just those quantities that are of physical importance, and gives no information about quantities such as orbital frequencies that one can never hope to measure experimentally." (Dirac, 1926, p. 667)

Notwithstanding these striking differences, the equivalence of Fermi's and Dirac's statistics was immediately recognized. Fermi's discovery was reformulated and accepted within a more abstract and more general conceptual framework within which ultimately the connection between spin and statistic was demonstrated. The stability through conceptual change that is exemplified by this historical case study is far from being an exception. Many aspects of the old quantum theory survived the conceptual reworking of Schrödinger, Heisenberg and Dirac.

We would like to suggest that stability through conceptual change plays a role in justifying our belief in theoretical statements. This is analogous to what happens with experimental statements. The stability of experimental statements in different experimental contexts is one of the strongest reasons for supporting our belief in their referential content (see Hacking, 1983). For instance, we believe that a crystal has a reticular structure since we can verify it in different experimental contexts: macroscopic observations, X-ray's diffraction, electron diffraction, "microscopes" based on tunnel-effects, etc. Similarly we can focus on theoretical statements from a plurality of accepted theoretical points of view. The statements which are invariant through theoretical change are assumed to be more fundamental[9] by the scientific community[10].

A nice feature emerges at this point. The (partial) incommensurability between theories can be utilized in a virtuous way in order to enhance our belief in invariant statements. If the same statement can be achieved within different partial incommensurable theoretical contexts, this statement is thought to be more fundamental. Of course, the reply of a relativist would be that we are not justified in talking about the *same* statement in different theoretical contexts, since the meaning is theory dependent. However, such a semantic approach (whose extremal case is meaning holism) does not play a relevant role in real scientific practice and in history of science. Here scientific theories are always seen as embedded into a broad context of experimental practices, relations with people belonging to other disciplines, technicians, engineers and so on. It is in this broad context that a theoretical statement receives its meaning in a pragmatic sense. Converging on a theoretical statement from a plurality of theoretical contexts is therefore one of the most fruitful ways for understanding its role in scientific thought.

Notes

[1] Nowadays it is well understood that Fermi-Dirac statistics holds for half-odd spin particles.

[2] "contain some identical parts".

[3] "in the case of systems with identical particles, as the helium atom, the complete period to which the Sommerfeld's integral should be extended has to correspond to the replication of a state indistinguishable from the initial state, even if the two states are classically different since the passage from one to another is due to particle exchange."

[4] "we could divide, for instance, v into n parallelepipedi, putting a single molecule in each one; or we could divide v into $n/2$ parallelepipedi, putting 2 molecules in each or else we could consider v as a single parallelepiped, containing all the n molecules."

[5] "we will allow [...] in the gas at most one molecule, the motion of which is characterized by a given set of quantum numbers [...] two distributions should be considered identical, if the places occupied by the molecules are identical, they should be considered identical also if they differ for a permutation of the molecules."

[6] "To find the relation among these constants and the temperature we must note that, as an effect of the attraction to 0, the gas density has to be a function of r, that should vanish for $r = \infty$. As a consequence, for $r = \infty$ the degeneration phenomena should disappear, and specifically the velocity distribution [...] has to turn into Maxwell law."

[7] As Dirac noted, a very important feature of quantum mechanics is that only physically measurable quantities are computable by the theory (Dirac, 1926, p. 667). The two

different transitions $(m, n) \rightarrow (m', n')$ and $(n, m) \rightarrow (m', n')$ are not experimentally distinguishable.

[8] Heisenberg's matrix mechanics.

[9] We employ here the term *fundamental* in order to be neutral on the philosophical interpretations of the nature of invariant statements. For instance, a realist might want to say that they capture fundamental aspects of the world, a formalist that they play a fundamental role in the syntactic structure of theories, and so on.

[10] Several versions of quantum mechanics have been proposed since the early days of quantum theory. Is there a set of statements that are left invariant from one to the other? And how can we define it? These are questions that the historians of physics, when dealing with case studies analogous to the discovery of Fermi-Dirac statistics, must face in order to understand the past. We claim that this exercise could be useful also to practicing physicists, to better focus and understand the core of their theories.

[11] One might notice that the interest that some physicists have for history of science is not merely motivated by an antiquarian approach. The stability through conceptual change emerges very naturally in the study of history of physics and allows therefore a useful understanding of concepts, theories and experiments.

References

Amaldi, E. (1987), "The Fermi-Dirac statistics and the statistic of nuclei", in *Symmetries in physics (1600-1980)*, (1st Int. Meeting on the History of Scientific Ideas, Sant Feliu de Guíxols, Spain, 1983), edited by M. G. Doncel, A. Hermann, L. Michel, A. Pais, Seminari d'Història de les Ciències (Universitat Autònoma de Barcelona, 1987), pp.251–277.

Belloni, L. (1978). "Una nota su come Fermi giunse alla statistica di Fermi-Dirac", *Scientia*, **113**, pp.431–438.

Caldirola, P. (1975). "L'evoluzione storica del principio di esclusione in fisica", *Scientia*, **110**, pp.51–67.

Desalvo, A. (1987). "Il problema del gas perfetto monoatomico e la teoria dei quanti: parte III", *Atti del VII congresso nazionale di fisica*, edited by F. Bevilacqua, Pavia, pp.95–103.

Dirac, P. A. M. (1925). "The fundamental equations of quantum mechanics", *Proc. Roy. Soc. (London)*, **A109**, pp.642–653.

Dirac, P. A. M. (1926). "On the theory of quantum mechanics", *Proc. Roy. Soc. (London)*, **A112**, pp.661–677.

Fermi, E. (1923). "Sopra la teoria di Stern della costante assoluta dell'entropia di un gas perfetto monoatomico", *Rend. Lincei*, **32**, pp.395–398.

Fermi, E. (1924). "Considerazioni sulla quantizzazione dei sistemi che contengono degli elementi identici", *Nuovo Cimento*, **1**, pp.145–152.

Fermi, E. (1926a). "Sulla quantizzazione del gas perfetto monoatomico", *Rend. Lincei*, **3**, pp.145–149.

Fermi, E. (1926b). "Zur Quantelung des idealen einatomigen Gases", *Zeitschrift für Physik*, **36**, pp.902–912.

Fermi, E. (1962-65). *Collected papers*, 2 vol., Rome and Chicago.

Hacking, I. (1983). *Representing and intervening*, Cambridge.

Kragh, H. (1990). *Dirac: a scientific biography*, New York.

Mehra, J. and Rechenberg, H. (1982). *The historical development of quantum theory*, vol. IV, New York.

Mehra, J. and Rechenberg, H. (1987). *The historical development of quantum theory*, vol. V, part 2, New York.

Pais, A. (1986). *Inward bound, of matter and forces in the physical world*, Oxford.

ARE THERE SUB-QUANTUM TRAJECTORIES?
THE BOHMIAN INTERPRETATION OF QUANTUM THEORY

G. PERUZZI

Sezione I.N.F.N. Firenze, Italy

Abstract. In 1952 David Bohm suggested a new interpretation of Quantum Mechanics showing explicitly how the indeterministic description of non relativistic wave mechanics could be trasformed into a deterministic one. The essential idea of this interpretation had been advanced by de Broglie in 1927. We present the basic tenets of Bohm theory at the light of Bohmian mechanics approach of D. Dürr, S. Goldstein and N. Zanghì.

1. Introduction

> "Perhaps a thing is simple if you can describe it fully in several different ways [...] We are struck by very large number of different mathematical formulations that are all equivalent to one another [...] Many different physical ideas can describe the same physical reality [...] Thus classical electrodynamics can be described by field view, or an action at distance view, etc...". [R. Feynman]

In the last years we have assisted to a revival in the discussion on the interpretations of quantum mechanics. New interpretations have been proposed (e.g. Ghirardi-Rimini-Weber theory, or, at a different level of elaboration, Gell-Man-Hartle interpretation), but also old interpretations are revisited. In the latter group we find the interpretation developed by Dürr, Goldstein and Zanghì starting from the work of D. Bohm in the 1952, called "Bohmian mechanics" (BM). Chief aim of this paper is a critical exposition of BM.

The discussion between exponents of different interpretations of quantum mechanics is not always easy. There is an (aggressive) attitude to remark the differences instead of the common points – an attitude that could be explained in several ways which are often connected with some sort

369

C. Garola and A. Rossi (eds.), The Foundations of Quantum Mechanics, 369-379.
© *1995 Kluwer Academic Publishers.*

of "psychology of the minority" with respect to the scientific community. On the contrary, there are several intersections between all these (new and old) approaches to the interpretation of quantum theory. First of all, each approach starts with focusing on the same problems: measurements, classical limit, non-locality, (peaceful or fighting) coexistence of quantum theory with (special or general) relativity, and so on. But there are also some intersections in the used methods and in the "methaphysical" options that are more or less explicitly involved. Moreover, all the interpretations that have been proposed lead, at least for the present, to the same predictions for the experimental results, so that there is no way of deciding experimentally between them. This paper proposes to be a contribute to the diffusion of a tolerant and peaceful confront between all these different interpretations. Nobody has in his hands the "ultimate theory", the "truth theory", the "definitive interpretative schema" (if this way of speaking makes any sense at all). We can understand more and more only in the discussion without prejudices.

To begin with we shall present the basic assumptions of BM by contrasting it with the usual (non-relativistic) quantum mechanics. Then we shall summarize some crucial points for the interpretation of BM. In the last part of the paper we shall give a short account of a fundamental (mathematical) result on the global existence of BM, focusing on its more interesting and, in some sense, unexpected consequences.

2. Bohmian mechanics

According to orthodox interpretation of quantum theory – and most non-orthodox interpretations as well – the physical state of a system is *completely* specified by its wave-function, which determines only the *probabilities* of actual results that can be obtained in a statistical ensemble of similar experiments. The wave-function evolves (deterministically) according to Schrödinger's equation only up to the moment in which we want to collect informations in a macroscopic apparatus; after that we have what the orthodox interpretation calls the "collapse" of the wave-function (in orthodox quantum theory there is no generally accepted account of how and when the "collapse" occurs, except for the fact that the "collapse" is necessary to explain the results at the macroscopic apparatus level). The "collapse" of the wave-function introduces a stochastic element in the dynamics, such that the complete dynamical description of a system is stochastic rather than deterministic.

Therefore the usual interpretation of quantum theory, as Bohm [3] pointed out, requires us to give up the possibility of even conceiving precisely what might determine the behaviour of an individual system at the quantum level, without providing adequate proof that such a renunciation is necessary.

The starting point of Bohm's interpretation is to regard orthodox quantum theory as a phenomenological formalism, roughly analogous to the thermodinamic formalism for the description of certain macroscopic regularies. Just as the thermodynamic formalism can be derived, in principle, from the precisely definable behaviours of the microscopic constituents of the macroscopic systems, the quantum formalism should be derived from a fully microscopic theory. Such a theory goes toward an elimination of the ambiguity and the vagueness from which quantum formalism suffers in the statement of its operational details. Moreover this theory is based on the awareness that, as Einstein claimed, it is quite wrong, on principle, to try founding a theory on observable magnitudes alone, because it is the theory which decides what we can observe.

The very existence of such a theory has been declared (physically and mathematically) impossible by von Neumann and many others after him. In this sense, BM is an explicit, even if often ignored, counterexample to this common belief. The "ontology" to which BM refers is very similar to the "ontology" of classical mechanics. To begin with this theory speaks of "particles", and "particles" means particles. On this ground, BM presumes that every material particle in the universe has a perfectly determinate position and trajectory: the account of the motions of the particles is *completely deterministic*. The universe appears to us to evolve probabilistically because we are *ignorant* of the actual positions of the particles (we have, in other words, an *epistemic* interpretation of the probability just as in the case of classical mechanics). In order to produce the description of the particle evolution BM requires (besides the usual force-fields) the Schrödinger's wave-function, regarded as a real physical object (quite distinct from the particles, roughly analogous to – but different from[1] – a force-field) guiding the particles along their trajectories.

From the BM view point the complete state for a system of N particles is given by (Q, ψ), where $Q = (\mathbf{Q}_1, ..., \mathbf{Q}_N) \in \Omega$, with $\mathbf{Q}_i$ the actual position of the particles, Ω the (3N-dimensional) configuration space, and $\psi = \psi(q) = \psi(\mathbf{q}_1, ..., \mathbf{q}_N)$ is the wave function of the system for the generic configuration space variable q.

The specification of the state of the system at any (initial) time, say (Q_0, ψ_0), must determine the state (Q_t, ψ_t) at later time; consequently the time evolution is defined by *first-order differential equations*. The wave-function evolves (without ever collapsing) according to Schrödinger's equation:

$$i\hbar \frac{\partial \psi_t}{\partial t} = H \psi_t, \tag{1}$$

and the actual configuration Q evolves according to the equation:

$$\frac{dQ_t}{dt} = v^{\psi_t}(Q_t),\tag{2}$$

where $v^\psi = (\mathbf{v}_1^\psi, ..., \mathbf{v}_N^\psi)$ is a vector field on the configuration space Ω. The wave-function therefore has the (dynamical) role to generate the motion of the particles, through the vector field on the configuration space, $\psi \to v^\psi$, to which it is associated.

The form of v^ψ is derived by requiring rotation, time-reversal and Galilean invariance, and homogeneity, i.e. two wave-functions of which one is a nonzero constant multiple of the other should be physically equivalent [5]. The simplest possibility is:

$$\mathbf{v}_i^\psi = \frac{\hbar}{m_i} \operatorname{Im} \frac{\nabla_i \psi}{\psi}\tag{3}$$

for a system of N particles of mass m_i $(i = 1, ..., N)$.

It can be shown (see [1], [2], [3], [4] and [5]) that BM has the same empirical content as non-relativistic quantum mechanics does. Moreover, it resolves all the problems associated with the quantum measurement. It accounts for the "collapse" of the wave-function, for the quantum randomness as expressed by Born's statistical law[2], and familiar (macroscopic) reality. From BM the quantum formalism – operators, commutation relations, and so on – emerges (see [5] and [6]) as a "measurement" formalism.

In order to grasp how, starting with the equations (1), (2) and (3), it is possible to arrive at the above conclusions, we have to understand why the familiar distribution $\rho = |\psi|^2$ plays a special role in BM. A first answer comes from the fact that the dynamical system defined by BM is associated with a natural measure, given by the density $|\psi_0|^2$ on configuration space. It plays the role of the *equilibrium measure* and define the notion of *tipicality* [5]. Given the existence of the dynamics for configurations Q_t, the notion of tipicality is time independent by equivariance [5], i.e.

$$\rho_0 = |\psi_0|^2 \implies \rho_t = |\psi_t|^2, \text{ for all } t \in \mathcal{R}$$

where ρ_t denotes the probability density on configuration space at time t – the image density of ρ_0 under the process Q_t. This follows by comparing the continuity equation for an ensemble density $\rho_t(q)$ (of systems which move along the integral curves of v^ψ)

$$\frac{\partial \rho_t(q)}{\partial t} + \sum_{i=1}^N \nabla_i \cdot [\mathbf{v}_i^{\psi_t}(q)\rho_t(q)] = 0\tag{4}$$

with the quantum "continuity equation"

$$\frac{\partial |\psi_t(q)|^2}{\partial t} + \sum_{i=1}^{N} \nabla_i \cdot [\mathbf{j}_i^{\psi_t}(q)] = 0 \tag{5}$$

and noting that the quantum "probability current" $j^\psi = (\mathbf{j}_1^\psi, ..., \mathbf{j}_N^\psi)$ is given by $\mathbf{j}_i^\psi = \mathbf{v}_i^\psi |\psi|^2$.

The above (mathematical) answer has no conclusive meaning if we don't explain the *physical* significance of $\rho = |\psi|^2$ (see [5] and [6] for the complete account). It turns out that, in BM, $\rho = |\psi|^2$ reflects the (*universal quantum*) *equilibrium* distribution of the configuration relative to ψ: when a system has wave-function ψ, its configuration is random, with distribution $|\psi|^2$ (see [5] for an analogy between the quantum equilibrium hypothesis and the Gibbs postulate of statistical mechanics). Let us suppose that in a quantum measurement (involving an interaction between a system S and an apparatus A) ψ is the wave-function and $q = (q_S, q_A)$ is the configuration of the composite system $S + A$. If the system and apparatus are in quantum equilibrium prior to the measurement, at time t_i, then q is random, with probability distribution given by $\rho_{t_i}(q) = |\psi_{t_i}(q)|^2$. When the measuremet has been completed, at time t_f, the configuration – given the deterministic evolution equations (1) and (2) – will still be random (as will tipically be the outcome of the measurement as given by appropriate apparatus variables), and, by equivariance, $\rho_{t_f}(q) = |\psi_{t_f}(q)|^2$ (in complete agreement[2] with the prediction of the quantum formalism for the distribution of q at t_f).

But a deep and general justification (for which we refer again to [5]) of the above particular result can be found only when we understand that BM is properly a description of the entire universe (i.e., the actual Bohmian system consist of all particles of the universe). At this universal level we have to give meaning to the quantum equilibrium hypothesis[3] without recurring to methaphysical hypotheses of ensemble of universes initially satisfying $\rho = |\psi|^2$. We can arrive at the physical meaning of the quantum equilibrium hypothesis by considering the *empirical distributions* – actual relative frequencies within an ensemble of actual events – arising from repetitions of similar experiments, performed at different places or times, within a single sample of the universe (the one we are in). The first step to demonstrate that BM is compatible with the prediction of quantum formalism is that, for at least some choice of initial universal ψ and q, the Bohmian evolution [equations (1) and (2)] leads to events with empirical distribution given by the quantum formalism. In fact in [5] it is proved even more: for every initial ψ, this agreement with the predictions of the quantum formalism is obtained for *typical* – i.e., for the overwhelming majority of – choices of initial q. And the sense of tipicality here is with respect to the quantum equilibrium, that is mathematically natural because equivariant. So, given the meaning of the wave-function ψ of a subsystem (see [5] for the notion of *effective wave-function*[4]), we can close the circle and establish the fact that observed quantum randomness, as expressed by Born's statistical law, is a manifestation of universal quantum equilibrium in the sense of typicality (see [5]).

3. Comments on the Bohmian universe

1 – Bohm's theory [3] postulates an original uncertainty measured by $|\psi|^2$ about the exact initial positions of the particles in the system. This assumption is not motivated in [3] by any fundamental connection between wave-function and probabilities. Bohm [4] suggests a possible justification of this assumption appealing to some sub-quantum level of (actual) random fluctuations. The Dürr-Goldstein-Zanghì's approach, instead, by using the notion of universal typicality (with respect to universal quantum equilibrium), arrives at a deep justification of $\rho = |\psi|^2$ without appealing to any hypothetically more fundamental stochastic process. Moreover, BM gives a relevant contribute to solving the problem of the interpretation and application of quantum theory in cosmology[5].

2 – BM is a highly *non-Newtonian theory* (despite some similarity in the ontological options). In Newtonian mechanics the state of a system is fixed by giving the positions and the velocities of the constituent particles. If we know the forces (or the potentials) we can derive the particle accelerations from which determining the trajectories. In BM, instead, the state of the system (as we have seen above) is fixed by giving the positions and the wave-function; the potentials determine (via the Schrödinger equation) the evolution of the wave-function, and the wave-function (via equations (2) and (3)) guides the particles on their own trajectories.

In this sense, BM permits to overcome some criticisms against Bohm's theory which has been originally formulated analysing the formal analogies between the equations of quantum mechanics and the Hamilton-Jacobi equations of the Newtonian mechanics. In this formulation the motion of the particles can be interpreted in such a way that each particle is acted on, not only by a "classical" potential, but also by a "quantum" potential (depending on ψ). The introduction of the quantum potential – even if permitting a straightforward analysis of the classical limit of quantum mechanics[6] – risks ambiguities (where the quantum potential come from? what is its actual physical status?). BM picture, by exhibiting the radical non-Newtonian status of the new dynamics, concludes that as a matter of fact the quantum potential is redundant.

Moreover it is perhaps interesting to underline that, by virtue of its non-

Newtonian character, BM gives the opportunity to reconsider the onto-logical status of the velocity. A well known criticism to the meaning of the velocity in Newtonian mechanics can be summarize in the following quotation: "Motion consists *merely* in the occupation of different places at different times [...] There is no transition from place to place, no consecutive moment or consecutive position, no such thing as velocity except in the sense of a real number which is the limit of a certain set of quotients. The rejection of velocity and acceleration as physcal facts (i.e. as properties belonging *at each instant* to a moving point, and not merely real numbers expressing limits of certain ratios) involves, as we shall see, some difficulties in the statement of the [Newtonian] laws of motion" (see [7], §447). The ontology of BM (for which only particle positions and wave-functions exist in every, fixed, instant of time) is coherent with this ontology of the "now", in which only the istant of time exists.

3 – BM is explicitly non-local (an event in one place can have immediate effects at arbitrary distance). But the so called Bell's inequality shows us that any theory which recovers the predictions of quantum theory has to be non-local. So we can conclude with the words of Bell: "That the guiding wave, in the general case, propagates not in ordinary three-space but in a multidimensional-configuration space is the origin of the notorious 'nonlocality' of quantum mechanics. It is a merit of the de Broglie-Bohm version to bring this out so explicitly that it cannot be ignored" (see [2], p. 115).

4 – In orthodox quantum mechanics all particles, elementary or composite, must be either fermions or bosons. However, it has recently been discovered that there can be quantum (quasi) particles which are neither fermions nor bosons. Such particles (anyons) can only occur in two spatial dimensions[7]. The mathematical and physical analysis of anyons gives, as a byproduct, a (topological) justification of the existence in the usual three-dimensional space of only fermion and boson particles. We are used to think that the permutation eigenvalue p controls the statistical behaviour of a system. It turns out, however, that the quantity which is crucial to the physics is not p but rather θ, the phase which arises from the (adiabatic) transport of two particles along a path which gives an actual physical exchange. This phase depends upon the topology of the configuration space. The main results come from the study of actual particle paths in the multiply-connected configuration space, naturally arising when we want to describe a system of identical particles [8]. BM accounts for all this stuff (see [9]) much better than orthodox quantum theory and in very natural way (because of the actual existence of particles trajectories, and the fundamental role of the configuration space). A byproduct of this analysis is that in BM (as a consequence of the very existence of trajectories) parastatistics are ruled out, without appealing to a superselection rule (see [9]).

5 – Up to now, a relativistic generalization of BM is problematic. And also other interpretative models (e.g. the Ghirardi-Rimini-Weber theory) suffers of the same problem. On the other hand, this crucial point is, first of all, a concrete example of what said above: different interpretative models use often the same mathemathical devices[8]. The reasons of these similarities between different interpretative models should be carefully investigated. Moreover, in this attempts toward a relativistic generalization we can try to understand more deeply the nature of the difficulties arising also in the relativistic generalization of orthodox quantum theory. Is it possible that at a fundamental (non observational) level Lorentz invariance is violated?

4. On the global existence of Bohmian mechanics

Solving the problem of the global existence of BM is a mathematical results with many physical and philosophical implications. As we have seen above in the section 2, the notion of typicality is time independent by equivariance *given the global existence of BM*. With regard to this condition, it is important to bear in mind the conceptual difference between equations (4) and (5). The continuity equation (4), even without global existence of differentiable trajectories Q_t, holds "locally" on the set where v^ψ is smooth, with ρ_t suitably interpreted. Equation (5), on the other hand, is an identity for every ψ_t which satisfies Schrödinger's equation classically[9]. But, without having established global existence, it is not a continuity equation in the classical sense – despite its name. By establishing global existence, we simultaneously show that the quantum "probability current" j^ψ is indeed a *classical probability current*, propagating the ensemble density $|\psi|^2$ along the integral curves of v^ψ.

The mathematical problem of existence and uniqueness of the motion in BM, consists of establishing that for given Q_0 and ψ_0 at some "initial" time t_0 ($t_0 = 0$), solutions (Q_t, ψ_t) of (1), (2) with $Q_{t_0} = Q_0$ and $\psi_{t_0} = \psi_0$ exist uniquely and globally in time[10]. Our first motivation for addressing this problem is the fact that the velocity field (3) reveals rather obviously possible catastrophic events for the motion: v^ψ is singular at points where $\psi = 0$, i.e., at the nodes of ψ. Furthermore, the solution may cease to exist at singularities of the wave-function (if it has singularities), from the possibility of reaching the boundary of the configuration space Ω (if it has a boundary), and from the possibility of "explosion," that is the escape to infinity of a particle in a finite amount of time – events which have analogues in the N-body problem (of gravitational interaction) in Newtonian mechanics.

Recall that the problem of the existence of dynamics in Newtonian mechanics is notoriously difficult, and no general results exist[11]. Seen in this light, it is remarkable that the situation in the "corresponding" quantum system is very different. In orthodox quantum theory the time evolution of the state ψ_t is given by a one parameter unitary group U_t on an abstract Hilbert space $\mathcal{H}$. U_t is generated by a self-adjoint operator H, which on smooth wave-functions in $\mathcal{H} = L^2(\Omega)$ is given by

$$H = -\sum_{k=1}^{N} \frac{\hbar^2}{2m_k} \Delta_k + V = H_0 + V, \tag{6}$$

i.e., Schrödinger's equation is regarded as the "generator equation" for U_t. Hence the "problem of the existence of dynamics" is reduced to showing that the relevant Hamiltonian H (given by the particular choice of the potential V) is self-adjoint. This has been done in great generality, independent of the number of particles and for large classes of potentials, including singular potentials like the Coulomb potential, which is of primary physical interest[12].

In BM we have not only Schrödinger's equation (1) to consider but also the differential equation (2), governing the motion of the particles. Thus the question of existence of the dynamics of BM draws again nearer to the situation in Newtonian mechanics, as it depends now on detailed regularity properties of the velocity field v^ψ (3)[13].

The problem we have addressed [11] is the following: Suppose that at some arbitrary "initial time" ($t_0 = 0$) the N-particle configuration lies in the complement of the set of nodes and singularities of ψ_0. Does the trajectory develop in a finite amount of time into a singularity of the velocity field v^ψ or does it reach infinity in finite time? We have shown that the answer is negative for "typical" initial values and a large class of potentials, including the physically most interesting case of N-particle Coulomb interaction with arbitrary charges and masses.

The quantity of central importance for our proofs of these results – as well as for the question of self-adjointness of the hamiltonian – turns out to be the quantum flux $J^\psi = (j^\psi, |\psi|^2)$ with the quantum probability current $j^\psi = v^\psi |\psi|^2$. The absolute value of the flux through any surface controls the probability that a trajectories crosses that surface. Surfaces of interest for us are the ones formed by the boundaries of neighbourhoods around all the singular points for BM. In a loose manner of speaking the importance of the quantum flux is grounded in the insight: "If there is no absolute flux into the singular points, the singular points are not reached."

The proof of the global existence for a fundamental completely determinist theory is, in some sense, surprising. From this result we can proceed toward a deeper understanding of the points of contact and departure of the ontology of BM with respect to the ontology of Newtonian mechanics. On the other side, the perspective of BM gives us a deeper (and more natural) understanding of the meaning and the status of the self-adjointness of the hamiltonian [11].

Notes

1. It is enough, regarding the difference between familiar force fields and wave-function, to note that a force-field is described in the (3+1 dimensional) physical space, whereas the wave-function is described in the configuration space (plus the time coordinate).
2. According to orthodox quantum theory the probability that a particle be found, in a position measurement, at the time t, with position q is $|\psi_t(q)|^2$.
3. The quantum equilibrium hypothesis expressed by $\rho = |\psi|^2$ relates two objects which are referred to rather different ontological levels: ρ, the probability distribution, and ψ, the wave-function, which is a real dynamical object from the perspective of Bohm's theory. So some justification is called for (see [5]).
4. The considerations about the connection between effective wave-function (i.e. the wave-function of a physical sub-system) and universal wave-function – the only appropriate in BM – are analogous to the cosmological considerations arising when we treat the problem of the origin of irreversibility (in the form expressed, for example, by the second law of thermodynamics).
5. At a cosmological level, in particular, there is nothing outside of the system to perform the measurements from which $\rho = |\psi|^2$ derives its meaning in orthodox quantum theory.
6. Bohm [3] expressed the classical limit in the (orthodox) form $\hbar \to 0$. And it is well known that this form is not so conclusive in respect to all the problems of the meaning of the classical limit.
7. Two-dimensional space is the physical space for elementary eccitations in low temperature condensed matter systems, like those in which fractional quantum hall effect is observed.
8. For example, Weiss-Dirac-Schwinger-Tomonaga description of the evolution of a system in terms of *functionals* on the set of space-like hypersurfaces, rather than in terms of the more familiar wave-functions.
9. This is seen by calculating

$$\frac{\partial |\psi_t|^2}{\partial t} = \frac{1}{i\hbar}(\psi_t^* H \psi_t - \psi_t H \psi_t^*).$$

10. Note that Schrödinger's equation (1) is independent of the particle motion, while for solving the equation for the particle motion (2) we need the wave-function ψ_t.
11. Apart from the possibility of collision singularities, the N-body problem entails marvellous scenarios of so-called pseudocollisions, where the particles tend to infinity in finite time and oscillate wildly. It is not even known whether initial configurations (of more than four particles) leading to such catastrophies form a set of Lebesgue measure zero – i.e., are improbable.
12. It may be worthwhile to note, however, that the sufficiency of establishing only the self-adjointness of the hamiltonian for a satisfactory physical interpretation has been questioned by Radin and Simon: "Interestingly enough, while Kato's result "solves" the dynamical existence question in the quantum case, it says nothing about the question of $x(t)^2$ remaining finite in time! From its physical interpretation, proof of such regularity property is clearly desirable" [10].
13. *Local* existence and uniqueness of Bohmian trajectories is guaranteed if the velocity field v^ψ is locally Lipschitz continuous. We therefore certainly need more regularity properties of the wave-function ψ than merely having ψ in $L^2(\Omega)$ [11]. *Global* existence is more delicate: Besides the nodes of ψ there are singularities which compare with those of

Newtonian mechanics. Firstly, even for a globally smooth velocity field the solution of (2) may explode, i.e., it may reach infinity in finite time. Secondly, the boundary points of Ω, which usually are the singular points of the potential, are reflected in singular behavior of the wave-function at such points, giving rise to singularities in the velocity field (3).

References

[1] Albert, D. (1992). *Quantum Mechanics and experience*, Harvard University Press, Cambridge, Massachusetts.

[2] Bell, J.S. (1987). *Speakable and unspeakable in quantum mechanics*, Cambridge University Press, Cambridge.

[3] Bohm, D. (1952). 'A suggested interpretation of the quantum theory in terms of "hidden" variables', *Phys. Rev.*, **85** pp.166–193.

[4] Bohm, D. and Hiley, B.J. (1992). *The undivided universe: An ontological interpretation of quantum theory*, Routledge & Kegan Paul, London.

[5] Dürr, D., Goldstein, S. and Zanghì, N. (1992). 'Quantum equilibrium and the origin of absolute uncertainty', *J. Stat. Phys.*, **67** pp.843–907.

[6] Daumer, M., Dürr, D., Goldstein, S. and Zanghì, N. 'On the role of operators in quantum theory', in preparation.

[7] Russell, B. (1903). *The principles of mathematics*, London.

[8] Leinaas, J.M. and Myrheim, J. (1977), 'On the theory of identical particles', *Nuovo Cimento*, **37B** pp.1–23; for a general review see Peruzzi, G. (1994), 'I modelli nella Fisica. Un esempio: gli *anyoni*', in *Logica e filosofia della Scienza: problemi e prospettive*, ETS, Pisa, pp.299–316

[9] Dürr, D., Goldstein, S. and Zanghì, N. 'Bohmian mechanics, identical particles, parastatistics and anyons', in preparation.

[10] Radin, C. and Simon, B. (1978). 'Invariant domains for the time-dependent Schrödinger equation', *J. Diff. Equations*, **29** pp.289–96.

[11] Berndl, K., Dürr, D., Goldstein, S., Peruzzi, G. and Zanghì, N. (1993). 'Existence of trajectories for Bohmian mechanics', *Int. J. Theor. Phys.*, **32** pp.2245–2251; Berndl, K., Dürr, D., Goldstein, S., Peruzzi, G. and Zanghì, N. 'On the global existence of Bohmian mechanics', in preparation.

ALGEBRAIC STRUCTURES AND OBSERVATIONS: QUANTALES FOR A NONCOMMUTATIVE LOGIC - THEORETIC APPROACH TO QUANTUM MECHANICS

M. PIAZZA

Dip. di Filosofia - Università di Genova
Genova, Italy

What meaning should be given to such results, considering that
logic must be as much above suspicion as Caesar's wife?
(R. Omnès, 1990)[1].

Abstract. This paper is a series of intertwining observations about the connection between the logic-theoretic noncommutativity and a logical foundation of quantum mechanics. We will analyze noncommutativity, both from an algebraic and proof-theoretic point of view, w.r.t. the quantum mechanics notion that the order of observation making is central to their description. To this end, we will present the sequential conjunction $\otimes : A \otimes B$ means "A at time t_1 and *then* B at time t_2".

The thread running through our discourse is given by quantales, i.e. algebraic structures introduced by Mulvey as models for the logic of quantum mechanics, which offer an appropriate algebraic (and topological) tool for describing noncommutativity.

1. Introduction

Quantales are partially ordered algebraic structures which generalize both locales (also known as frames or complete Heyting algebras) and ideal lattices. They were introduced in the eighties by C. Mulvey in the ambitious aim of providing a possible common setting for noncommutative C^*-algebra, for constructive foundations for quantum mechanics (QM), and for noncommutative logics. The term itself was coined as a combination of "quantum

381

C. Garola and A. Rossi (eds.), The Foundations of Quantum Mechanics, 381-393.

logic" and "locale". (In terms of quantales, a *locale* is a commutative, right-sided idempotent quantale)[2].

The idea underlying Mulvey's seminal, albeit programmatic, paper is the assumption of an intimate connection between the Gelfand-Naimark representation of noncommutative C^*-algebra, stated by Giles and Kummer in 1971, and the foundations of QM in terms of propositional theories within a constructive, but noncommutative, logic. This connection, he suggests, appears more fundamental than the formalistic links with C^*-algebras within the conventional foundation of QM by way of Hilbert space (Mulvey, 1986)[3]. Another intriguing issue concerns the complex numbers: the interdependence between the logic-theoretic noncommutativity and the logical foundation for QM is made closer by the fact that in the set theories based on quantum logics investigated by Takeuti, the complex numbers are noncommutative. For the sake of convenience, this framework can be summarized in Diagram 1.

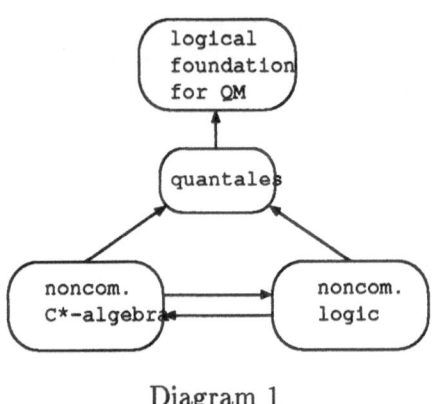

Diagram 1

We will start from the bottom of the above diagram: indeed, the aim of our contribution is to consider a possible link between the noncommutativity of observations in QM and the noncommutative logics in which the formulas of the language are strictly sequential. In particular, we focus on *noncommutative linear logic*, i.e. the logic obtained from a sequent formulation of classical (or intuitionistic) logic by rejecting all the structural rules: *weakening, contraction,* and *exchange* (Abrusci, 1991). Noncommutative linear logic seems to satisfy Mulvey's request of constructiveness, since it enjoys cut-elimination theorem and consequently it is equipped with an appropriate semantics of proofs. Moreover, quantales give an algebraic semantics for linear logic (and in particular, noncommutative quantales do the same for noncommutative linear logic).

This paper will be devoted to uncovering the intimate connections between the treatment of noncommutativity in different settings. Some of these connections can be surprisingly subtle. In this sense, even if this paper should seem somewhat rhapsodical, *Gershwin-like* so to speak, its declared purpose is more to raise questions than answers.

2. The noncommutative paradigm

One of the most natural problems arising out of the investigation of an abstract logic-theoretic framework for QM is that of noncommutativity, that is, roughly speaking, the requirement of a (logic-theoretic) noncommutative operation corresponding more or less to the quantum mechanical notion that the order in which observations are carried out is central to their description. This logical operation "retains a vestige of temporality".

Traditional quantum logic appears quite silent on this point. By a *quantum logic* we mean, as usual, the couple (L, M) where L is an orthomodular σ-lattice and M is a strong set of states on L, i.e. the statement $\{m \in M | m(a) = 1\} \subseteq \{m \in M | m(b) = 1\}$ implies that $a \leq b$, $a, b \in L$. The well-known Jauch-Piron property in the σ-form is also supposed for any state of M. Quantum logic models mathematically the set of all experimentally verifiable propositions about a physical system.

The interdependence between the logic-theoretic noncommutativity and the logical foundation for QM was posed precisely by M.D. Srinivas in the seventies. Previously, in 1957, P. Jordan, still working in a lattice theoretic framework, formulated a theory of skew lattices (i.e. noncommutative lattices) in which the commutativity of the operations is not assumed, in order to provide a quantum propositional calculus (Jordan, 1959)[4].

Also Birkhoff in 1959 suggested that the experimental procedure for the proposition $A \wedge B$ be given by the sequence of the measurements of A and B *in that order* (Birkhoff, 1961). And this is an orthodox noncommutative "reading". But, in practice, Birkhoff has exploited the lattice theoretic commutative conjunction in the propositional calculus, since — he argued — "the set of all states that have unit probability for satisfying the experimental procedure $\{A, B\}$ also have unit probability for satisfying the sequence $\{B, A\}$" (Srinivas 1975, p. 1678).

Nevertheless — Srinivas observed —: "the set of all states that assign unit probability to a proposition do not completely characterize the corresponding experimental procedure... Two experimental propositions can have the same set of (all) the states that assign unit probability to them and still correspond to experimental procedures if the associated measurement transformations are different" (Ibidem).

In his work Srinivas dismissed the lattice-theoretic logics as models for the calculus of quantum experimental propositions: the "logic of experimentally verifiable proposition" leads now to a "logic of operations (or measurement transformations)" on physical systems. This logic of operations arises out of the analysis of the relations among experimental procedures, and it differs radically from the usual lattice-theoretic logics which represent each experimentally verifiable proposition in quantum theory as the set of all states assigning probability one to the proposition. This representation fails to characterize completely the experimental procedures corresponding to a proposition.

In order to correlate the conjunction more closely with empirical procedure, Srinivas defines the proposition $\mathcal{E}_1 \wedge \mathcal{E}_\in$ as follows: "the experimental procedure corresponding to the proposition $\mathcal{E}_1 \wedge \mathcal{E}_\in$ is the procedure in which the system is subjected to the sequence of procedures $\mathcal{E}_1, \mathcal{E}_\in$ in that order" (Srinivas 1975, p. 1679). Schematically:

Let V be the set of self-adjoint trace class operators on a Hilbert space, and let $\circ$ denote the composition of operations. The operation $\mathcal{E}_1 \wedge \mathcal{E}_2$ is thus defined such that:

$$\forall v \in V , \; (\mathcal{E}_1 \wedge \mathcal{E}_2)(v) = (\mathcal{E}_2 \circ \mathcal{E}_1)(v) = \mathcal{E}_2(\mathcal{E}_1(v)) .$$

Obviously, $\wedge$ is associative and, in general, it is not idempotent. Two propositions are compatible, (written $\mathcal{E}_1 \leftrightarrow \mathcal{E}_2$), if $\mathcal{E}_1 \wedge \mathcal{E}_2 = \mathcal{E}_2 \wedge \mathcal{E}_1$.

In the next section we will see how this noncommutative operation $\wedge$ and the noncommutative linear operation $\otimes$ resemble each other.

3. Quantum proof theory?

It is well-known that measurements in a quantum system, carried out in order to find its current state, influence the measured value noticeably, and that if the interaction between the instrument (measuring device) and the system is weak, then this influence is less. We have to increase the period of observation adequately if we want the uncertainty in the estimate of the state of instrument to decrease.

The measurements within a system in a pure state (i.e., a state that can be described by a wave function) are reduced to the identification of the initial conditions, whereas if the interaction between the instrument and the system is strong, the *pure state* of the system becomes *mixed* (i.e., only the probability that the state occurs is known) in the process of measurement. It is interesting to note that if the observable (the measured physical quantity) is defined by an operator commuting with the Hamiltonian of the system, "the originally determined state cannot be perturbed by any subsequent measurements" (Butkovskiy and Samoilenko, 1990 p. 5).

Can we describe logically the participation of observers in the process of measurements? The aim of this section is to present a possible proof-theoretic approach to the quantum theory of measurements: this kind of approach to QM can be completely justified by the question raised by Mulvey: *with what rules of deduction are physical observations naturally manipulated?* From the standpoint of the syntactical manipulability of physical observation, Mulvey's question can be strengthened by another question: *what is, if any, "the physical meaning" of cut-elimination in a sequent calculus for quantum logic?* The latter question boils down to a question of the meaning of the *subformula property* in such a sequent calculus[5].

In QM, both the aspects that we would like to observe (the state of the particle) and the act of observation itself are modelled. Classical logic is not a "logic of observations" since connectives are static operations, manipulating frozen objects oblivious to observation. This is why classical logic cannot formalize precisely any description or reasoning associated with quantum world: in other terms, *there is no definite direction of time.* For example, in classical logic $A \wedge B$ means that the observer sees both A and B *at the same time.*

From a proof-theoretic viewpoint, in a Gentzen-type sequent calculus, the structural rule (i.e. a rule of inference which does not involve any connective), which is harmless in both classical and intuitionistic logic but responsible for the *simultaneity of observations*, is the *exchange rule*[6]:

$$\frac{\Gamma_1, A, B, \Gamma_2 \Rightarrow \Delta}{\Gamma_1, B, A, \Gamma_2 \Rightarrow \Delta} \ (E, L) \qquad \frac{\Gamma \Rightarrow \Delta_1, A, B, \Delta_2}{\Gamma \Rightarrow \Delta_1, B, A, \Delta_2} \ (E, R)$$

where A, B are formulas and Γ and Δ finite sequences of formulas of the language $\mathcal{L}$. This rule says that the order of formulas on the left side (antecedent) and right side (succedent) of the sequent symbol $\Rightarrow$ is immaterial. In terms of physical observation, in a weak reading, the rule (E, L) represents the commutativity of observations: the order of such observations is immaterial to the outcome. In a strong reading this rule says that the observations are simultaneous.

Another structural rule of classical logic that is misleading from the point of view of observation is the *contraction rule*:

$$\frac{A, A, \Gamma \Rightarrow \Delta}{A, \Gamma \Rightarrow \Delta}\ (C, L) \qquad \frac{\Gamma \Rightarrow \Delta, A, A}{\Gamma \Rightarrow \Delta, A}\ (C, R)\ .$$

This rule says that the number of times we carry out an observation has no effect on the outcome.

The first step for a logical foundation of QM is then given by removing the exchange rule and contraction rule from our sequent calculus: this means that the 2-place logical operation $\otimes$ ($\wedge$ and & in resp. Srinivas's and Mulvey's notation)[7], which corresponds to the comma in (E, L), and in (C, L), is no longer commutative and idempotent. The validity of $A \otimes B$, as in Srinivas' idea, is to be regarded as "we have verified A, and then we have verified B".

Thus, we have the noncommutative (and not idempotent) conjunction $\otimes$. The sequential conjunction $A \otimes B$ means "A at time t_1 *and then* B at time t_2".

The sequent calculus rules for $\otimes$ are the following:

$$\frac{\Gamma_1 \Rightarrow \Delta_1, A, \Delta_2 \quad \Gamma_2 \Rightarrow \Delta_3, B, \Delta_4}{\Gamma_1, \Gamma_2 \Rightarrow \Delta_3, \Delta_1, (A \otimes B), \Delta_4, \Delta_2}\ (\otimes, R) \qquad \frac{\Gamma_1, A, B, \Gamma_2 \Rightarrow \Delta}{\Gamma_1, (A \otimes B), \Gamma_2 \Rightarrow \Delta}\ (\otimes, L)$$

if $\Delta_2 = \Delta_3 = \emptyset$, or $\Delta_2 = \Gamma_1 = \emptyset$, or $\Delta_3 = \Gamma_2 = \emptyset$.

Dually, we have the noncommutative (and not idempotent) disjunction $\star$ (*par*) which corresponds to the comma in (E, R) and in (C, R) : $A \star B$ means sequentially "A at time t_1 *or then* B at time t_2". The rules for $\star$ are:

$$\frac{\Gamma_1, A, \Gamma_2 \Rightarrow \Delta_1 \quad \Gamma_3, B, \Gamma_4 \Rightarrow \Delta_2}{\Gamma_3, \Gamma_1, (A \star B), \Gamma_4, \Gamma_2 \Rightarrow \Delta_1, \Delta_2}\ (\star, L) \qquad \frac{\Gamma \Rightarrow \Delta_1, A, B, \Delta_2}{\Gamma \Rightarrow \Delta_1, (A \star B), \Delta_2}\ (\star, R)$$

if $\Gamma_2 = \Gamma_3 = \emptyset$, or $\Gamma_2 = \Delta_1 = \emptyset$, or $\Gamma_3 = \Delta_2 = \emptyset$.

We might also think of Γ, A, B as events (or experimentally verifiable propositions of a statistical physical theory): the happening of the events Γ, A, B in that order causes Δ to happen.

Beyond any particular kind of informal interpretation of the formulas, it should be stressed that, *in absence* of the exchange rule, a sequent may represent *time-ordered* strings of observations on a sub-microscopic scale, as well as events or single-time propositions describing what may occur in a quantum system: we shall read $A \otimes B$ as "A precedes B" (or "B follows A"). If it a pair of formulas A, B exists such that $A \otimes B \simeq B \otimes A$ then we have a sort of *closed timelike loop*[8].

The full removal of the exchange rule from classical logic (and contraction and weakening also) leads to different negations in noncommutative linear logic: the *postnegation* and the *retronegation* $^{\perp}(\text{-})$ (see Abrusci, 1991)[9]:

$(-)^\perp$ -rules:

$$\frac{\Gamma \Rightarrow \Delta, A}{\Gamma, A^\perp \Rightarrow \Delta} \; ((-)^\perp, L) \qquad \frac{A, \Gamma \Rightarrow \Delta}{\Gamma \Rightarrow A^\perp, \Delta} \; ((-)^\perp, R) \; ;$$

$^\perp(-)$ -rules:

$$\frac{\Gamma, A \Rightarrow \Delta}{\Gamma \Rightarrow \Delta, {}^\perp A} \; ({}^\perp(-), R) \qquad \frac{\Gamma \Rightarrow A, \Delta}{{}^\perp A, \Gamma \Rightarrow \Delta} \; ({}^\perp(-), L)$$

De Morgan's laws

- $(b \otimes c)^\perp = c^\perp \star b^\perp$
- $^\perp(b \otimes c) = {}^\perp c \star {}^\perp b$

- $(b \star c)^\perp = c^\perp \otimes b^\perp$
- $^\perp(b \star c) = {}^\perp c \otimes {}^\perp b$

Thus, we obtain also two different implications: the *postimplication* (or right implication): $A \longrightarrow_r B = A^\perp \star B$ and the *retroimplication* (or left implication): $A \longrightarrow_l B = B \star {}^\perp A$.

Intuitively, since noncommutativity has an intrinsic temporal meaning the postimplication $\longrightarrow_r$ may denote forward causality, whereas the retroimplication $\longrightarrow_l$ "retrocausality", i.e. the causal influence back in time, from future to past. The postimplication assumes a situation A *at time t_1* and predicts a situation B at a later *time t_2*; the retroimplication assumes the situation A *at time t_2* and reconstructs the past by concluding with the situation at an earlier *time t_1*.

As Omnès stresses in a different context "this kind of results only holds for a system having a regular dynamics" and "dynamically regular systems behave in an (almost) deterministic way and *one has recovered many cases where determinism holds, although one never left QM*" (Omnès, 1990).

The phenomenological interpretation of postimplication may be summarized in this way: A *post*-implies B when the probability measure of B is equal to 1, once A is granted to happen. Since probability does not have an inherent direction in space or time A *retro*-implies B is interpreted analogously.

It is worth stressing that retrocausality which describes a time-reversed process, is crucial in QM since retrocausality is a way of reacting to a consequence of Bell's theorem: *any interpretation of QM in terms of classical logic must be nonlocal.*

If we remove the exchange rule from our sequent calculus, it would nevertheless be absurd not to recover it in some way since otherwise we would have a logic less expressive than classical logic. To get the strength of classical logic back the *exchange modalities*: ▷ (read *everywhere*) and ◁

(read: *somewhere*), are introduced. These S4-like modal operators mimic and control the exchange. We give here only the basic rules (the reader is referred to Abrusci 1993 for details):

$\triangleright - rules :$

$$\frac{\Gamma_1, \triangleright A, B, \Gamma_2 \Rightarrow \Delta}{\Gamma_1, B, \triangleright A, \Gamma_2 \Rightarrow \Delta} \ (\triangleright, EF) \qquad \frac{\Gamma_1, B, \triangleright A, \Gamma_2 \Rightarrow \Delta}{\Gamma_1, \triangleright A, B, \Gamma_2 \Rightarrow \Delta} \ (\triangleright, EB)$$

$\triangleleft - rules :$

$$\frac{\Gamma \Rightarrow \Delta_1, \triangleleft A, B, \Delta_2}{\Gamma \Rightarrow \Delta_1, B, \triangleleft A, \Delta_2} \ (\triangleleft, EF) \qquad \frac{\Gamma \Rightarrow \Delta_1, B, \triangleleft A, \Delta_2}{\Gamma \Rightarrow \Delta_1, \triangleleft A, B, \Delta_2} \ (\triangleleft, EB)$$

$(-EF)$ and $(-EB)$ are the rules for *forward exchange* and *backward exchange*, respectively.

The successive step is to give an appropriate semantics to these rules: roughly speaking, this is a kind of semantics of observation (see Def.4.7). To this end we exploit the notion of quantale, which can be seen as an algebraic axiomatization of the rules of linear sequent calculus.

4. Quantales

The basic theory of quantales can be found in the book of Rosenthal (Rosenthal, 1990). The notion of *Girard quantale* is due to Yetter (see Yetter, 1990). Nevertheless one of the difficulties in studying Girard quantales is the lack of information about specific examples: proposition 4.5 exhibits an example of Girard quantale not known in the literature, called R-quantale. One would like to have an analogous concrete example of an autonomous quantale, i.e. noncommutative Girard quantales, but at the moment (spring '94) little is known about this structure[10]. It is worth noting in passing that noncommutative algebraic structures are not interesting for their own sake, but rather their study is of a strategic nature. Indeed, for example, a quantum group can be regarded as an endomorphism of a quantum plane whose coordinates form a noncommutative (associative) algebra (see Manin, 1988).

4.1. DEFINITION. A *quantale* is a complete lattice Q together with an associative binary operation $\otimes$ satisfying:

$$a \otimes (\sup_\alpha b_\alpha) = \sup_\alpha (a \otimes b_\alpha) \quad \text{and} \quad (\sup_\alpha b_\alpha) \otimes a = \sup_\alpha (b_\alpha \otimes a),$$

$\forall a \in Q$ and $\{b_\alpha\} \subseteq Q$.
It follows that $- \otimes -$ is order-preserving.
A quantale Q is *commutative* iff $\otimes$ is commutative.

A quantale Q is *unital* iff there is an element $1 \in Q$ s.t. $\forall a \in A$, $1 \otimes a = a = a \otimes 1$. Let P, Q be quantales, $f : P \to Q$ is a quantale homomorphism if and only if it preserves sups and the operation $\otimes$.

Examples: 1) The structure $Q = (\mathcal{P}(M), \subseteq, \circ, \{e\})$ where $(M, \circ, e)$ is a monoid and $\circ$ is the Frobenius's product, is an unital quantale called *phase quantale*; 2) any locale with $\otimes$ as conjunction, is a commutative right-sided idempotent quantale[11]; 3) the lattice of closed left ideals of a C^*-algebra is a quantale w.r.t. multiplication of left ideals.

4.2. DEFINITION. Given a quantale Q, we define $\forall a, b \in A$

i) $a \longrightarrow_r b = \sup\{q \in Q | a \otimes q \leq b\}$

ii) $a \longrightarrow_l b = \sup\{q \in Q | q \otimes a \leq b\}$.

$\forall a \in Q$, the endofunctors $a \longrightarrow_r -$ and $a \longrightarrow_l -$ are respectively the right adjoints and the left adjoints of $a \otimes -$ and $- \otimes a$.

4.3. DEFINITION. By *autonomous quantale* (AQ), we mean every structure $Q = (Q_0, \otimes, \leq, \perp)$ defined as follows:

i) $(Q_0, \otimes, \leq)$ is a quantale

ii) $\perp$ is the *dualizing element* of Q_0, i.e. $\forall a \in Q_0$,

$$(a \longrightarrow_l \perp) \longrightarrow_r \perp = a = (a \longrightarrow_r \perp) \longrightarrow_l \perp .$$

We write $^\perp a$ for $a \longrightarrow_l \perp$ and $a^\perp$ for $a \longrightarrow_r \perp$, and we get resp. the *right* and *left* negation. Let Q be an AQ, $\forall a, b \in Q$ we define the operation $\bigstar$ (*par*) by $a \bigstar b = ^\perp [b^\perp \otimes a^\perp]$. Note that $^\perp [b^\perp \otimes u^\perp] = [^l b \otimes^l u]^l$.

Some properties.

- Every AQ is unital with $1 = \perp^\perp = ^\perp \perp$

- $[\inf_\alpha \{b_\alpha\}]^\perp = \sup_\alpha \{b_\alpha^\perp\}$

- $^\perp [\inf_\alpha \{b_\alpha\}] = \sup_\alpha \{^\perp b_\alpha\}$

- $a \otimes b \leq c \mapsto a \leq c \bigstar^\perp b$

- $a \leq b \bigstar c \mapsto a \otimes c^\perp \leq b$

- $[\sup_\alpha \{b_\alpha\}]^\perp = \inf_\alpha \{b_\alpha^\perp\}$

- $^\perp [\sup_\alpha \{b_\alpha\}] = \inf_\alpha \{^\perp b_\alpha\}$

- $a \otimes b \leq c \mapsto b \leq a^\perp \bigstar c$

- $a \leq b \bigstar c \mapsto ^\perp b \otimes a \leq c$.

4.4. DEFINITION. If, in the previous definition, $\perp$ is also *cyclic* i.e. if $\forall a \in Q_0 : a \longrightarrow_r \perp = a \longrightarrow_l \perp$, then Q is called a *Girard quantale*[12]. $\perp$ is cyclic if $a_1 \otimes \cdots \otimes a_{n-1}, a_n \leq \perp \mapsto a_n, a_1 \ldots a_{n-1} \leq \perp$.

Every Girard quantale is unital with $1 = \perp^\perp$.

If $T = \max Q_0$ and $0 = \min Q_0$ then: $\perp = 1^\perp$; $0 = T^\perp$; $T = 0^\perp$.

4.5. PROPOSITION. $\forall n \geq 1$, the structure $Q^{(n)} = (Q_0, \otimes, \leq, \perp)$ defined as follows:

(i) $Q_0 = \{(h, k) \mid h = 0, \ldots, n - 1; k = 0, \ldots, n\}$

(ii) $\otimes : Q_0 \times Q_0 \to Q_0$ is the binary operation:
   $((h, k), (h'k')) \mapsto (\max\{h + k' - n, 0, h' + k - n\}, \max\{h + h' + 1 - n, 0, k + k' - n\})$

(iii) $(h, k) \le (h', k')$ if $h \le h'$ and $k \le k'$

(iv) $\perp = (n - 1, 0)$,
   is a Girard quantale, called "*R-quantale of order n*", where:

(v) $0 = (0, 0)$; $T = (n - 1, n)$; $1 = (0, n)$

(vi) $(h, k) \longrightarrow\!\!\bullet\ (h', k') = (\min\{n - 1 - h + k', n - 1, h' + n - k\}, \min\{n - h + h', n, n - k + k'\}$
   $(h, k)^{\perp} = (n - 1 - h, n - k)$
   $(h, k) \star (h', k') = (\min\{h + k', n - 1, h' + k\}, \min\{h + h' + 1, n, k + k'\})$.

Proof: left to the reader.

Example:

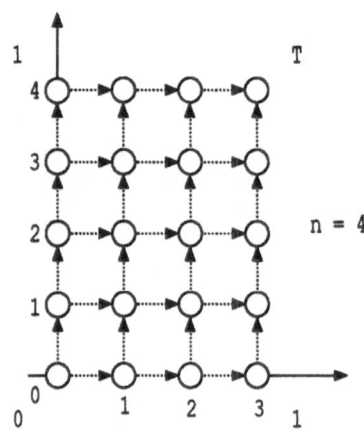

is the *R*-quantale of order 4.

4.6. THEOREM. Given a *complete Boolean algebra* $B = (B_0, \wedge, \vee, (-)', T, F)$, the structure $(Q_0, \otimes, \le, \perp)$ defined as follows:

(i) $Q_0 = B_0 \times B_0$

(ii) $(h, k) \otimes (h', k') = ((h \wedge k') \vee (h' \wedge k), (h \wedge h') \vee (k \wedge k'))$

(iii) $(h, k) \le (h', k')$ if $h \le h'$ and $k \le k'$

(iv) $\perp = (T, F)$

is a Girard quantale.

Proof: see (Piazza, 1994).

   We now introduce topo-autonomous quantales which play a key role in the quantale semantics for noncommutative linear logic plus exchange connectives (E-LLK) since the topo-autonomous quantale is the carrier set in the quantale interpretation of E-LLK.

4.7. DEFINITION. Given an AQ, $Q = (Q_0, \otimes, \leq, \perp)$ we call (*open*) *autonomous unital modality* in Q_0 every $\mu : Q_0 \to Q_0$ s.t.:

i) μ is a coclosure operator (decreasing, idempotent, monotone)
ii) $^\perp[\mu(a^\perp)] = [\mu(^\perp a)]^\perp$, $\forall a \in Q_0$ (*autonomy*)
iii) $\mu(\mu(a) \otimes \mu(b)) = \mu(a) \otimes \mu(b)$, $\forall a \in Q_0$ (*stability*)
iv) $\mu(1) = 1$ (*unitality*).

An (open) autonomous unital modality μ is:

v) *weak* iff $\mu(a) \leq 1$, $\forall a \in Q_0$
vi) *idempotent* iff $\mu(a) \otimes \mu(a) = \mu(a)$, $\forall a \in Q_0$
vii) *central* iff $\mu(a) \otimes b = b \otimes \mu(a)$, $\forall a \in Q_0$

4.8. DEFINITION. By *topo-autonomous quantale*, we mean every structure $(Q_0, \otimes, \leq, \perp, \triangleright)$ defined as follows:

i) $(Q_0, \otimes, \leq, \perp)$ is an autonomous quantale Q
ii) $\triangleright : Q_0 \to Q_0$ is an (open) autonomous unital central modality.

A completeness theorem for E-LLK is stated in (Castellan and Piazza, 1994).

5. Concluding Remarks

It cannot be expected that the theory of quantum mathematical measurements can be clarified in one fell swoop merely by choosing the right kind of logic. However, to employ quantales as a working hypothesis may be useful in order to unify different frameworks: in this sense, Omnès's question of whether we can replace a reasoning based on classical physics by another reasoning based on a consistent quantum representation of logic is actualized by quantales. Theorem 4.6 shows how a Girard quantale can be constructed from a Boolean algebra; it is now a matter of investigating noncommutative Girard quantale in terms of familiar objects.

Notes

[1] R. Omnès 1990, p. 408.

[2] There is a certain family likeness between quantales and the algebraic structures such as residuated lattices (also noncommutative) investigated by Dilworth and Ward in the thirties.

[3] Historically, the formulation of QM by C^*-algebra goes back to the paper of Haag and Kastler in 1964 following a previous suggestion of Segal in the fourties who pointed out that several physically interesting questions (e.g., the determination of spectral value) can be answered without referring to a Hilbert space if one chooses the algebra of observables as a C^*-algebra (Haag and Kastler, 1964).

[4] The works of Jordan on noncommutative lattices go back to the late fourties (see Jordan, 1949).

[5] The proof-theoretical approch to quantum logic is not new; indeed, Nishimura has already presented a quite natural syntactical approach to quantum logic based on a sequential formulation reminiscent of Gentzen's LK and LJ. However, Nishimura's sequential formulation of ortohologic and orthomodular logic fails to satisfy the cut-elimination theorem and so his calculus is not constructive (see Nishimura 1980).

[6] In Gentzen's system for intuitionistic logic LJ the exchange rule is forbidden on the right side of the sequent symbol ⇒; but in the equivalent system LJ* of Maehara this rule is allowed on both sides of ⇒.

[7] We refer also to Stachow's *sequential conjunction* ∩. More generally, in Stachow's sequential quantum logic the *sequential propositions* refer to "the *evolution* of a quantum mechanical system due the Hamiltonian of the system as well as due to a sequence of measuring processes with respect to the system" (Stachow, 1980 p. 252). Stachow shows that the suitable mathematical tool for studying sequential propositions is a Baer- * semiring.

[8] See Isham 1994 for an exposition of a similar idea in another setting.

[9] *Caveat*: this is not a logic aimed at providing a foundation for QM like Nishimura's. This is simply a logic which offers a natural syntactical framework for noncom-mutativity and that is constructive, thus satisfying Mulvey's requirement of a noncommutative and constructive logic. We shall not go into technicalities con-cerning the lack of the weakening and contraction which splits the calculus into a *multiplicative* part and an *additive* part. Note that in Nishimura's calculus the weakening rule is retained.

[10] For an example of a noncommutative quantale the reader is referred to the relational quantales studied in Brown and Gurr (1993).

[11] A quantale Q is *right-sided* if for every element a of Q, $a \otimes T \leq a$, where $T = \max Q_0$. Q is *idempotent* if every element a of Q is idempotent (i.e. $a \otimes a \leq a$).

[12] Girard quantales are precisely the partially ordered examples of the *-autonomous categories in Barr (1979).

References

Abrusci, V.M. (1991). "Phase semantics and sequent calculus for pure noncommutative classical linear propositional logic", *The Journal of Symbolic Logic*, **56** (1403).

Abrusci, V.M. (1993). *Exchange connectives*, preprint, 1993.

Barr, M. (1979). *-Autonomous Categories, Springer, LNM 752.

Birkhoff, G. (1961). "Lattices in applied mathematics", collected in R.P. Dilworth (ed.), *Lattice theory*, American Mathematical Society, Rhode Island.

Brown, C. and Gurr, D. (1993) "A representation theorem for quantales", *Journal of Pure and Applied Algebra*, **85** (27).

Butkovskiy, A.G. and Samoilenko, Yu.I. (1990), *Control of Quantum Mechanical Processes and Systems*, Kluwer Academic Publisher, Dordrecht.

Castellan M. and Piazza, M. "Quantale semantics for noncommutative classical linear logic with exchange connectives and exponentials", *Proceedings of the Workshop "Linear logic and Lambek calculus"*, Rome, June 28-30, 1993, forthcoming.

Haag, R. and Kastler, D. (1964). "An Algebraic Approach to Quantum Field Theory", *Journal of Mathematical Physics*, **5** (848).

Isham, C.J. (1994). "Quantum logic and the histories approach to quantum theory", *Journal of Mathematical Physics*, **35** (2157).

Jordan, P. (1949). "Über nichtkommutative Verbände", *Archiv der Mathematik*, **2** (56).

Jordan, P. (1959). "Quantenlogik und das kommutative Gesetz", collected in Henkin, L, Suppes, P. and Tarski, A. (eds.), *The Axiomatic Method*, North-Holland, Amsterdam.

Manin, Yu.I. (1988). *Quantum groups and noncommutative geometry*, Centre de Recherches Mathematiques, Universite de Montreal.

Mulvey, C.J. (1986). "&", *Supplemento ai Rendiconti del Circolo Matematico di Palermo*, **12** (99).

Nishimura, H. (1980). "Sequential method in quantum logic", *The Journal of Symbolic Logic*, **45** (339).

Omnès, R. (1990). "From Hilbert Space to Common Sense: A Synthesis of Recent Progress in the Interpretation of Quantum Mechanics", *Annals of Physics*, **201** (354).

Piazza, M. (1994). *How to construct a (non trivial) Girard quantale from a Boolean algebra*, manuscript, June.

Rosenthal, K.I. (1990). *Quantales and their application*, Longman Scientific & Technical, Essex.

Srinivas, M.D. (1975). "Foundation of a quantum probability theory", *Journal of Mathematical Physics*, **6** (1672).

Stachov, E.W. (1980). "Logical Foundation of Quantum Mechanics", *International Journal of Theoretical Physics*, **19** (251).

Yetter, D.N. (1990). "Quantales and (noncommutative) linear logic", *The Journal of Symbolic Logic*, **55** (41).

HERMENEUTICS AS A CONCEPTUAL MODEL FOR AN ONTOLOGICAL EVALUATION OF QUANTUM THEORY

A. REBAGLIA
Dip. di Filosofia, Ist. di Filosofia della Scienza, Università degli Studi di Milano
Via Festa del Perdono 7 - 20122 - Milano, Italy

Abstract. The philosophical issues involved in establishing quantum conceptual foundations are first summarised. Then the analogous models historically advanced by quantum physicists to regard those conceptual demands are shortly analysed. The fathers of quantum mechanics had to take into account two different bounding lines, being aware on one side of the needing to refer to some peculiar philosophical framework and on the other of the awful consequence of quantum revolution, which makes all philosophical systems advanced *before* the quantum development unable to satisfy quantum interpretative exigencies. The want of any plausible quantum *model* following from some philosophical features becomes manifest when considering both this goal and the above mentioned limits. Finally, a new model will be presented, more powerful in our opinion, based on the hermeneutical philosophy. It is noteworthy that the authors of quantum 'revolution' never contemplated this kind of model, though hermeneutics took a significant part in the cultural movement which gave rise to quantum theory as well. A careful examination of the hermeneutic model is then required, to evidence its capability in disentangling the complex knot of unsolved conceptual inquiries that quantum perspective involves.

1. Ontological and methodological implications of quantum theory

As a consequence of the destroying role played by quantum theory against the classical conceptual framework, we find that both the constitutive elements of reality and the basic methodological principles 'prescribed' by quantum mechanics are well focused if compared with the most critical, and partly 'heretical', conceptions developed within the traditional philosophical horizon.

C. Garola and A. Rossi (eds.), The Foundations of Quantum Mechanics, 395-405.
© *1995 Kluwer Academic Publishers.*

We refer to the doctrines of the 18th century Irish Bishop George Berkeley and those of his contemporary, the Scottish Sceptical thinker David Hume. Their theses, though in fact belonging to the classical thought, show up a noticeable contrast with the traditionally accepted ontological and methodological principles respectively. So, it is not surprising that they both were carefully considered - more or less explicitly - by the 'fathers' of quantum physics.

1.1 IDEALISM AND LACK OF A PRIORI ONTOLOGICAL CHARACTERS OF THE OBJECT TO BE KNOWN

Einstein's charge against the "Copenhagen interpretation" of quantum mechanics (which is equally extendible to any interpretation that does not involve "hidden variables") concerns the quantum refusal of the "detached" and "objective" role of the experimental observer. His criticism points up a significant comparison with the thought of George Berkeley.   In fact, the reason itself that makes Berkeley's arguments 'awkward' inside the traditional philosophy, is responsible for the 'worrying' image of the quantum theory: the *utter refusal of an object 'in itself', existing and having 'significance'*.

According to Berkeley's Idealism[1], the *existence* of any physical event is due to a direct observation the subject achieves on it. A statement in full agreement with the "orthodox" quantum formalism interpretation, which in fact underlines the need of empirical tests by an observer for granting the *existence* of the physical event. And both these perspectives specify that the *essence* of reality can not be reduced to the concept of matter as an extra-linguistic substance, but it is precisely linguistic and rational. In Berkeley's thought, this is argued by claming that objects - even when unseen by the human observer - are ideas in the mind of God, while within the 'orthodox' quantum formulation the ultimate essence of objects, independently of the sensory impressions, is considered the mathematical equation describing quantum "entanglement" of physical states, an evidently formal and rational essence of nature.

Nevertheless, 'quantum' Idealism cannot be totally reduced to the 'classical' Idealism. The last one, even if we consider its most extreme claims, does not invalidate in any way the thinking of *phenomenon*, considered as what actually reveals to our senses, but already possesses some characters 'in itself', stated in advance and independently of any interaction with the observer. On the contrary, according to the micro physical laws, the essence of things cannot be entirely pre-stated. Quantum laws require that all physical events be free of any property fixed in advance, *a priori*, while observer is allowed not only to grant the existence, but also to characterise the essence of nature[2].

1.2 A-CAUSALITY AND LACK OF *A PRIORI* GNOSEOLOGICAL STRUCTURES OF THE KNOWING SUBJECT

Quantum demand for a conceptual hiatus between empirical observation and causal relation poses a most fundamental challenge to our capability of knowledge, unavoidably leading to a Sceptical view in methodology, quite near to Hume's philosophy. In approaching the problem of knowledge, in fact, quantum theory seems to have inherited the sharp distinction made by Hume between "matters of fact" and "relations of ideas"[3].

According to Hume, there are no necessary causal connections within experience, but our belief in causes is justified on "habit" and custom. Similarly, in quantum research causality turns out to be exclusively significant when considering it as an heuristic concept, completely devoid of any *a priori* character and ontological value. Moreover, in Hume's analysis demonstrable and necessary connections can be established only where experience is not involved. Analogously in quantum view the deterministic framework based on causality maintains its validity solely in the formal description, which is only probabilistically connected with the physical world.

The new broad lines of this quantum based analysis of causality suggest the impossibility for the subject to acquire knowledge by means of "pure", *a priori* categories, whose effectiveness could be justified autonomously, without any link to the empirical world and its interactions. These conclusions can be directly related to the quantum ontological issues we discussed above: while all physical events acquire their "existence" as well as their "essence" in consequence of the *active* role played by the subject, the latter cannot exert an absolutely "creative" activity, because he does not own gnoseological *a priori* categories.

As well as we have shown for the quantum Idealistic ontology, there are also some distinctive features differentiating the quantum Sceptical methodology from the classical one, the most remarkable being its *scientific* way of treating causality. Quantum mechanics, as well as the Humean philosophy, recognises the impossibility of basing causality on the physical space, since no causal relationship links the state of a system before and after a measurement is performed on it. But quantisation locates its representation of the physical world within a special *mathematical* domain, the R-dimension configuration space, R being the degrees of freedom of the considered system. Here the causal principle retains its generalising capability, based on the property of the logical inference "from one to all", which is essential for any scientific perspective and had been lost within the Humean thought[4].

2. Philosophical models historically accomplished in quantum foundations

Two models, that were characterised by a considerable impact, provide the starting-point of our discussion, making clear fundamental epistemological issues about the bases and criteria of a satisfactory quantum picture of the world. These influential models are those introduced by Werner Heisenberg, referring to some original theses of Aristotle, and by Niels Bohr, who was interested in the philosophy of the Danish 19th century thinker Søren Kierkegaard.

An evaluation of those models will show the basic problems that a quantum world-picture has to solve, in order to become really trustworthy. Thus, we will examine their structural foundations and, mainly, their constraints, with the aim of stressing their significance and showing a way for introducing a more suitable modelisation.

2.1 HEISENBERG'S ARISTOTELIAN MODEL

The model Heisenberg assumes, with the task of affording the *ontological* problems arising from quantum formalism, is based on the well-known Aristotelian distinction between *potential* and *actual* being[5].

The situation outlined by the quantum state vector "superposition" - Heisenberg emphasises - can be regarded as correctly described by Aristotle's category of "potential" reality, the intermediate state which is between being and not being. In so doing, Heisenberg finds a reliable guide to assign a specific ontological value to the mathematical formalism of wave function, which represents the true essence of the quantum world, whose features it describes without any direct connection with the empirical experience. One of the two focal points of quantum Idealism is thus maintained and clarified: micro physical events are not made up of a material *essence* "in itself". On the contrary, they consist of a sort of linguistic essence, by which we mean the information that the mathematical function contains.

Heisenberg's Aristotelian model can outline the second main feature of quantum Idealism as well. According to Aristotle, if movement and change are the actualising of potentialities, then substance in its actual state is permanent and not likely to be reversed. In analogy with the Aristotelian statements, all observed quantum physical systems are irreversible and deprived of any possibility of change. In fact, in the experimental observation the quantum state vector - which describes the dynamic of the physical system - is "collapsed". Thus observed systems can be regarded as the being in its "actual" state, and this stresses that the *existence* of micro physical events is not independent of any empirical observation.

Within the analogy so defined, even the main open question coming from the discussion about the 'idealistic' character of quantum foundations - namely the unsolved question concerning the dependence of the *essence* of quantum events upon the empirical experimentation, that we argued referring to the theses of Berkeley - can find a widely positive solution. It is possible taking into account another antithesis which runs through Aristotle's philosophy and is deeply involved with the contrast between potential and actual: the one between matter and form. Pure potentiality belongs only to quality-less matter, a logical abstraction which does not exist in the physical world. In the *actual* individual object, matter and form are inseparably united, organised in a complex. Matter is the undetermined factor which receives its determinations by the structural law, or form. But matter is just *relatively* undetermined: it cannot be supposed to give itself all sorts of specific determinations. There is a *range* of forms - that can be acquired by a certain substratum, and not *any* form can be. The analogy between matter - as an undeveloped potentiality - and superposition in quantum mechanics is quite clear. Moreover, if we consider the micro physical observed event as the individual when finally determined by the form, then we can take into account the item according to which neither existence nor essential characters are *a priori* conferred to quantum systems. Besides, we can lay stress on the fact that the formal quantum essence is originated from a knowledge effort, not from an autonomous creative act of a rational subject: some constraint is present on the range of actual possibilities for a given event, but the physical reality itself imposes this constraint, which is not derived arbitrarily by the rational observer.

The negative side of this analogy concerns the analysis of "substance". According to Aristotle, the matter is the persistent underlying *substratum* in which the development of the form takes place, so maintaining the concept of substance as that which persists through superficial change. This classical concept does not fit quantum requirements, which do not allow the maintenance of continuity in becoming. Some troubles related to the Heisenberg uncertainty principle are well-known: denying the concept of "continuous trajectory", it breaks the evolutive development of an event into a sequence of observational frames, without any guarantee about self-identity and substantial permanence.

2.2 BOHR'S KIERKEGAARDIAN MODEL

If the model proposed by Heisenberg has attempted to describe thematically quantum ontology, then a first step in the direction concerning quantum *methodological* problems has been taken by Niels Bohr, who - as a Danish physicist - referred to the thought of a well-known Danish philosopher of the 19th century, Søren Kierkegaard, in order to make a satisfactory gnoseological model[6]. Employing it, Bohr pointed out that the physical state superposition and the direct empirical observation are rooted in two different gnoseological levels, the former being somehow "deeper" than the latter, while both always show a sort of "complementarity".

The topics by which Kierkegaard investigates our capability of understanding man's essence can be used, according to Bohr, as a model to investigate our capability of understanding the essence of quantum nature.

Kierkegaard distinguishes three "stages" of existence between which there is no continuity[7]. The first one - the "aesthetic stage" - is the one in which man lives for the pleasure of the moment, where human existence is identified with a set of single experiences never organised in a global horizon, or in historical sequences. This sum of discontinuous experiences is analogue to the empirical knowledge in quantum mechanics, because the measured quantum event is always the result of the wave function "collapse". As in the aesthetic stage the true essence of life is not deeply understood, so in measuring micro physical events the subject does not obtain a full knowledge of the profound meaning of reality, and perhaps he risks to misunderstand the essence of the object, using some classical categories of knowledge that do not take into account quantum interaction between physical event and experimental apparatus.

In the two deeper stages described by Kierkegaard - the "ethical" stage and the "religious" one- existence is interpreted in terms of possibility: the stability and *continuity* of life is based on the consciousness that existence is always *individual* and made by a sum of particular decisions and commitments which have to be recognised as *possibilities* realised among others, and that never correspond to logical necessities. In agreement with these terms, in quantum mechanics each event is historically recognisable only as the sum of all its possible physical states, while quantum probability is devised not to any statistical set, but to every *single* micro system.

With respect to the gaps that break any continuity between the stages, Kierkegaard asserts that man can only "jump" from one to the other. Further, each of these stages has a different set of procedures and criteria for judging and behaving - that is to say a different "method". Both these features are sufficient to bring to a possible solution those methodological questions that were partially unsolved when compared with the Humean Scepticism. In quantum theory - as we remarked above - there are two different roles that causality has to play. When considered within the formal knowledge of the wave function, causality retains its validity as a universal and necessary *law*. But the causal link remains a simple heuristic guide, without any ontological implication, with respect to the experimental investigation activity. So each of the quantum levels of knowledge, according to the Kierkegaardian model we described, has a peculiar kind of methodological causal approach, both relevant and useful, but separated by a "gap" that the subject must "jump", suddenly reorganising the conceptual framework.

By insisting on the human existence, Kierkegaard's Existentialism is explicitly foreign to scientific methodology. Therefore, using central themes in Kierkegaard's analysis as a quantum epistemological model is not possible to adequately clarify either level of knowledge effectively describes physical reality. The model seems to show that reality itself does possess two different levels of being, but their properties and reciprocal relations are totally unknown. As for the Aristotelian model proposed by Heisenberg, also in this kind of analogy some influential claims due to the quantum conceptual outcomes still remain unsolved.

3. The hermeneutical model

Heidegger's depth thought, which gave rise to the hermeneutical philosophy, may represent a new, effective quantum conceptual model. A model which can be successful in the attempt to unify ontological and methodological analogies in one and the same framework.

Likewise, as Heidegger's philosophy has its roots in the Aristotelian thought, such as in Kierkegaard's Existentialism, there is a conceptual bridge that links together Bohr's and Heisenberg's cultural frame of reference with hermeneutics; so, developing an hermeneutical quantum model sheds new light on the philosophical issues of the "Copenhagen interpretation", providing deeper foundations for them.

3.1 A RESTATEMENT OF THE ONTOLOGICAL QUANTUM IDEALISM

The basic thought of Heidegger permits us to consider the central points of quantum immaterialism and quantum idealism under a different perspective, and to attain a satisfying solution of the problem about the relationship between the things that are not in themselves and the subject who attempts to understand their nature.

According to Heidegger, the fundamental character of the reality of the thing (*res*) is substantiality. But this idea of being, Heidegger maintains, is incapable of explaining how things can be originally approached, discovered and understood by man: man's factual existence discloses world as a significance-whole[8]. The disclosure of world manifests an articulated complex of references and linguistic relations which expresses the authentic essence of things.

Depending on significance, meanings and ultimately on language (as the total meaning-context which brings the world to light), things have no *a priori* characters: «world» is made possible by man's understanding of it. The analogy with quantum events is clear: they also possess an essence which is linguistic, being related to the mathematical formalism, and depends on the subject (a feature that could not be understood using the Aristotelian model, as we argued above).

Handiness, Heidegger says, is the essential character of the things we meet within our everyday world. The disclosure of the world discovers things not as mere substances, but as the handy presence of useful things: a presence made possible by their linguistic essence, by the totality of their references. Immediately evident is the analogy with quantum objects, when they are experimentally observed. According to Heidegger, each particular thing is connected with the whole «world», which is always understood by man in advance; in the same way every micro physical event in quantum mechanics is connected with the «superposed» formal description, the subject always knows before experience.

Then, the hermeneutic model still contains every positive element belonging to the ontological model exemplified by Heisenberg, and it makes new, fundamental themes to emerge. But there is more to it. This model is essentially connected with the ontological and *equally* with the methodological issues arisen by quantum theory's conceptual implications.

3.2 A RESTATEMENT OF THE METHODOLOGICAL QUANTUM SCEPTICISM

As a philosophical model, Heidegger's thought seeks to overcome also methodological quantum paradoxes by permitting us to analyse even the gap between approaching reality in an observational way and understanding it in the formal way which appeals to causal law.

Man's way to be, according to Heidegger, is to understand being. The subject (*Dasein* or "being there" is how Heidegger defines him) exists in an *actively disclosing* way[9]. It is clear enough that for the subject there is no *a priori* character with the help of which knowing the world. If things and events disclose meanings by virtue of being intrinsically meaningful, then *Dasein* must himself presuppose the irreducibility of language[10]. Furthermore, it is because man's being is disclosed to him as depending on language that it is manifest to him as "being in the world": the world-horizon, which is primarily meaning-giving, precedes experience. These topics provide a very good model to be followed in approaching the quantum problem of knowledge. In fact the essence of micro physical events can be caught solely by using the formal description of the "superposed" entanglement of physical states. However, this formal description is not a free creation of the subject's mind. Instead it is rigorously prescribed by the physical situation, even if it is determined in advance, before every particular experiment.

"Authenticy" and "inauthenticy" - the two basic ways of existing, according to Heidegger - can model both formal and empirical knowledge better than the analogy with Kierkegaard's philosophy, examined above.

The everyday, inauthentic way of existing is characterised by the misunderstanding of the phenomenon, which is seen as related to a thing in itself. This "disowned" world-view misses the coherent whole of the world - constituted by the above - mentioned articulated reference structure. Similarly, experiments and empirical observations constitute our everyday way of attempting to understand the universe, but their data might be misinterpreted, by supposing they belong to an objective, autonomous physical reality, so missing the depth of the "superposed" quantum essence of the world. Moreover, both inauthenticy and empirical knowledge are unavoidable means of relating the world. They represent the way in which "in the first place and for the most part" man understands being.

Man achieves what Heidegger calls «authentic existence» when he can be wholly himself, when he chooses his «ownmost possibility» among the possibilities of his being; the possibility on the basis of which he can «project himself». «*Dasein* is always its own possibility», Heidegger emphasises. What enables man to understand the possibilities of his own being is the source of possibility as such. As Heidegger tells us, it is the impossibility of all of the possibilities of existence, which is death. Correspondingly, in order to have an authentic knowledge of quantum world, subject must ascribe an ontological value to probability (that is to possibility in terms of physics). For attaining it, subject has to take into account the «superposition» of quantum function and the deriving «entanglement» of the physical alternative possibilities in which the event can happen. It is equally required taking account of the «collapse» of the wave function, that states subject's impossibility of knowing all of the statistical possibilities of existence for the physical event.

4. The peculiar, positive role of the hermeneutical model

As Heidegger repeatedly stresses, there is a remarkable «circle structure» which characterises both ontology and methodology. This same circularity can be found at the basis of both quantum ontology and quantum methodology, according to what has been argued above.

The hermeneutical circular path of the problem of being concerns the disclosure of the world, that cannot be conceived without an essential relationship to man, while man is «thrown» into the world. Undoubtedly, this philosophical model allows to get to the root of the concept of quantum physical event, whose essential features are defined within a peculiar relationship to subject, while not entirely depending on him.

Moreover, the hermeneutical circular path of the problem of understanding concerns a process of learning in which subject comes to know in a deeper way what he already knows in a way that is more immediate, not removable, but that can be misunderstood. So he knows as his own possibility what is «projected», choosing it from the totality of involvement that constitute the structure of the world; while his projecting is executed being already engaged in a world already disclosed in disposition. With this methodological model a number of questions about a satisfactory quantum way of knowing become clear. It is possible to separate the formal way of knowing nature from the empirical one without introducing any irreversible break between them (as it was required by the Kierkegaardian model), so giving an adequate conceptual background for discussing categories like probability and certainty, like indeterminacy and truth.

This circularity, Heidegger insists, is totally different from a logical «vicious circle». It is not a weakness to be avoided or suppressed. The task, instead, must be to find the right way of «getting into the circle». This has perhaps to be also the task of quantum philosophy, circularity possibly being responsible of the depth and complexity of the new conceptual framework - depth and complexity which are often mistaken for uncompleteness and paradoxicality.

Notes

[1] See Berkeley [1707] and Berkeley [1710].

[2] According to Feynman, "The observer was sometimes important in prequantum physics, but only in a trivial sense. The problem has been raised: if a tree falls in a forest and there is nobody there to hear it, does it make a noise? A *real* tree falling in a *real* forest makes a sound, of course, even if nobody is there. Even if no one is present to hear it, there are other traces left. The sound will shake some leaves, and if we were careful enough we might find somewhere that some thorn had rubbed against a leaf and made a tiny scratch that could not be explained unless we assumed the leaf were vibrating" (Feynman, Leighton, Sands [1970], vol.III, p.2-13). Obviously, it is exactly this kind of trivial objectivity that is no more present in the subtle quantum world.

[3] See Hume [1739].

[4] In a pregnant statement of Bohr: "The very nature of the quantum theory thus forces us to regard the space-time co-ordination and the claim of causality, the union of which characterises the classical theories, as complementary but exclusive features of the description, symbolising the idealisation of observation and definition respectively" (Bohr [1928], p.580).

[5] For Heisenberg, "In the experiments about atomic events we have to do with things and facts, with phenomena that are just as real as any phenomena in daily life. But the atoms or the elementary particles themselves are not as real, they form a world of potentialities or possibilities rather than one of things or facts" (Heisenberg [1958], p.186).

[6] In several letters from 1909 Bohr reported how his readings of Kierkegaard made a deep impression on him (see Bohr [1972], vol.1).

[7] See Kierkegaard [1843].

[8] According to Heidegger's words: «When beings within-the-world are discovered along with the Being of Dasein (..) we say that they have *meaning (Sinn)*» (Heidegger [1927], p.151).

[9] As Heidegger remarks in one of his later works: "the nature of man is to be conceived and grounded (..) as *the locus* which being requires for its revelation. Man is the in itself open there, into which beings stand" (Heidegger [1958], p.156).

[10] As Heidegger stresses: "we are moving within language, which means moving shifting ground, or, still better, on the billowing waters of ocean" (Heidegger [1954], p.169).

References

Berkeley, G. (1707). *Commonplace Book. 1707-1708*, in A.A. Luce (ed.), Berkeley, *Works*, vol.1, 1948.

Berkeley, G. (1710). *A Treatise Concerning the Principles of Human Knowledge*, Dublin.

Bohr, N. (1928). "The Quantum Postulate and the Recent Development of Atomic Theory", *Nature*, pp. 580--90.

Bohr, N. (1972). *Collected works*, North-Holland, Amsterdam.

Feynman, R. P., Leighton, R. B., Sands, M. (1970). *The Feynman Lectures on Physics*, Addison-Wesley Publishing Company, California Institute of Technology.

Heidegger, M. (1927). *Sein und Zeit*, Max Niemayer Verlag, Tübingen 1963[8].

Heidegger, M. (1954). *Was Heisst Denken?*, Max Niemayer Verlag, Tübingen.

Heidegger, M. (1958). *Einfuehrung in die Metaphysik*, Max Niemayer Verlag, Tübingen.

Heisenberg, W. (1958). *Physics and Philosophy*, Harper Torchbooks, New York 1962.

Hume, D. (1739). *A Treatise of Human Nature*, London, voll.1,2 1739, vol.3 1740.

Kierkegaard, S. (1843). *Enten-Eller*, Copenhagen.

RANDOM PATH QUANTIZATION

M. RONCADELLI

Sez. INFN di Pavia

Via A. Bassi 6 - 27100 - Pavia, Italy

Abstract. We offer a concise account of the *Random Path Quantization* (RPQ), whose motivation comes from the fact that quantum *amplitudes* satisfy (almost) the *same* calculus that *probabilities* obey in the theory of classical stochastic diffusion processes. Indeed – as a consequence of this structural analogy – a new approach to quantum mechanics naturally emerges as the quantum counterpart of the Langevin description of classical stochastic diffusion processes: This is just the RPQ. Starting point is classical mechanics as formulated *a là* Hamilton-Jacobi. Quantum fluctuations enter the game through a certain *white noise* added in the first-order equation that yields the configuration space trajectories as controlled by the solutions of the (classical) Hamilton-Jacobi equation. A *Langevin equation* arises in this way and provides the *quantum random paths*. The quantum mechanical propagator is finally given by a *noise average* involving the quantum random paths (in complete analogy with what happens for the transition probability of a classical stochastic diffusion process within the conventional Langevin treatment). The general structure of the RPQ is discussed, along with a suggested *intuitive picture* of the quantum theory.

1 – A new approach to (nonrelativistic) quantum mechanics – called *Random Path Quantization* (RPQ) – is presented in a rather schematic way. Basically, the RPQ arises quite naturally when a fundamental (formal) analogy between classical and quantum physics is recognized and its implications are worked out (see below). As a consequence, a *novel strategy* to evaluate the quantum mechanical propagator becomes available. Application to concrete physical situations shows that this technique turns out to be (more or less) as effective as the Feynman path integral quantization. Moreover, the RPQ brings out quite sharply a *deeper connection* between classical and quantum dynamics, which leads in turn to a very simple *intuitive picture* of the quantum theory. Some comments about the RPQ will be offered toward the end of the paper.

Perhaps, the most dramatic difference between classical and quantum mechanics – one which was stressed particularly by Feynman [1] – is that

C. Garola and A. Rossi (eds.), The Foundations of Quantum Mechanics, 407-416.
© *1995 Kluwer Academic Publishers.*

classical *probabilities* get replaced by quantum (probability) *amplitudes*. Schematically

$$\begin{matrix} CLASSICAL \\ PROBABILITIES \end{matrix} \quad \longleftrightarrow \quad \begin{matrix} QUANTUM \\ AMPLITUDES. \end{matrix} \tag{1}$$

Yet, a very remarkable *structural similarity* between classical and quantum physics exists, since *amplitudes* satisfy (almost) the *same* calculus [1] that *probabilities* obey in the theory of *classical stochastic diffusion processes* [2] [2]. As regard to this analogy, classical stochastic diffusion processes have to be defined in the *configuration space* $\mathcal{M}$ of the dynamical system under consideration [3]. Symbolically

$$\begin{matrix} CLASSICAL \ PROBABILITY \\ CALCULUS \end{matrix} \quad \simeq \quad \begin{matrix} QUANTUM \ AMPLITUDE \\ CALCULUS. \end{matrix} \tag{2}$$

Before proceeding further it seems worthwhile to clarify the precise meaning of (1) and (2). Everybody knows that the quantum mechanical time evolution is fully specified by an initial *wave function* $\psi(x, t_0)$ and by the *transition amplitude* $\langle x, t | x_0, t_0 \rangle$, since the wave function at a different time is

$$\psi(x, t) = \int_{-\infty}^{\infty} dx_0 \langle x, t | x_0, t_0 \rangle \psi(x_0, t_0). \tag{3}$$

Furthermore, the transition amplitudes enjoy the convolution property

$$\langle x, t | x_0, t_0 \rangle = \int_{-\infty}^{\infty} dx' \langle x, t | x', t' \rangle \langle x', t' | x_0, t_0 \rangle. \tag{4}$$

As $\psi(x, t)$ satisfies the Schrödinger equation, eq. (3) entails that $\langle x, t | x_0, t_0 \rangle$ is just the quantum mechanical *propagator*. Consider now classical stochastic diffusion processes, which are basically Markov processes with continuous sample paths [2]. As is well known, these processes are fully described by an initial *probability density* $P(x, t_0)$ and by the *transition probability* $P(x, t | x_0, t_0)$. In particular, the probability density at a time $t \geq t_0$ is given by

$$P(x, t) = \int_{-\infty}^{\infty} dx_0 P(x, t | x_0, t_0) P(x_0, t_0). \tag{5}$$

Moreover, the transition probabilities obey the *Chapman-Kolmogorov equation*

$$P(x,t|x_0,t_0) = \int\limits_{-\infty}^{\infty} dx' P(x,t|x',t') P(x',t'|x_0,t_0) \tag{6}$$

where it is assumed $t \geq t' \geq t_0$. As a matter of fact, $P(x,t)$ satisfies the *Fokker-Planck equation* [2], and so eq. (5) implies that $P(x,t|x_0,t_0)$ is the *propagator* of that equation. Manifestly, the comparison of eqs. (3) and (4) with eqs. (5) and (6) leads us directly back to (1) and (2), thereby providing their justification. Now, (1) and (2) possess two far-reaching *consequences*

a) quantum mechanics *can* be formulated in (at least) *three* different ways, as it is the case for classical stochastic diffusion processes (see below);
b) these formulations of quantum mechanics should be *structurally very similar* to those of classical stochastic diffusion processes.

What is the *actual import* of these statements? We all know that $\psi(x,t)$ is an arbitrary solution of the Schrödinger equation, while $\langle x,t|x_0,t_0\rangle$ is the associated propagator. Similarly, $P(x,t)$ is an arbitrary solution of the Fokker-Planck equation and again $P(x,t|x_0,t_0)$ is the associated propagator. In addition, both equations are linear and give rise to a time evolution with a semigroup structure. Therefore, the Schrödinger formulation of the quantum theory turns out to be on the *same footing* as the Fokker-Planck treatment of classical stochastic diffusion processes. Schematically

$$\begin{matrix} FOKKER-PLANCK \\ APPROACH \end{matrix} \quad \longleftrightarrow \quad \begin{matrix} SCHRÖDINGER \\ APPROACH. \end{matrix} \tag{7}$$

Alternatively, $\langle x,t|x_0,t_0\rangle$ is given by a path integral [3], [4] and analogously $P(x,t|x_0,t_0)$ can be expressed as a generalized Wiener path integral [5]. In effect, the two treatments are formally *almost indistinguishable*, and so we can write

$$\begin{matrix} WIENER \\ APPROACH \end{matrix} \quad \longleftrightarrow \quad \begin{matrix} PATH\ INTEGRAL \\ APPROACHES. \end{matrix} \tag{8}$$

In spite of the fact that all this is well known, a *crucial point* emerges. As far as classical stochastic diffusion processes are concerned, $P(x,t|x_0,t_0)$ can *also* be expressed as a *noise average* involving the solutions of a *Langevin equation* with a *gaussian white noise* [2]. What the foregoing discussion entails is that *even* in quantum mechanics a *similar representation* of $\langle x, t|$ $x_0, t_0 \rangle$ *should exist*. Surprisingly enough, this strategy has not yet been developed and indeed the RPQ is just an attempt to fill this gap [4]. So we have schematically

$$
\begin{array}{ccc}
LANGEVIN & & RANDOM\ PATH \\
APPROACH & \longleftrightarrow & APPROACH.
\end{array}
\qquad (9)
$$

More explicitly, the RPQ is an attempt to formulate the quantum theory along the same lines of the Langevin description of classical stochastic diffusion processes. Actually – starting from the latter approach – the RPQ can be derived to a large extent by the analogy symbolized by (2) through the correspondence (1), as (9) indeed suggests.

2 – A few preliminaries are now in order. We shall consider throughout a point particle $\mathcal{S}$ (mass m, no spin) with *configuration space* $\mathcal{M} = \mathcal{R}^N$ and described classically by the (nonrelativistic) lagrangian

$$
L(x, \dot{x}, t) = \frac{1}{2} m \dot{x}_i \dot{x}_i + \Omega_i(x, t) \dot{x}_i - \Phi(x, t). \qquad (10)
$$

According to classical mechanics, the Hamilton-Jacobi equation associated with lagrangian (10) is

$$
\frac{\partial}{\partial t} S(x, t) + \frac{1}{2m} \left(\frac{\partial}{\partial x_i} S(x, t) - \Omega_i(x, t) \right)^2 + \Phi(x, t) = 0. \qquad (11)
$$

When a particular (arbitrary) integral $S(x, t)$ of eq. (11) is chosen, a family of trajectories in $\mathcal{M}$ is correspondingly provided by the *first-order* equation

$$
\frac{d}{dt} q_i(t) = \frac{1}{m} \left(\frac{\partial}{\partial x_i} S(x, t) - \Omega_i(x, t) \right) \Big|_{x = q(t)}. \qquad (12)
$$

Let us denote by $q(t; x', t'; [S(\cdot)])$ the solution of eq. (12) with initial condition $q(t') = x'$ and controlled by $S(x, t)$. Then $q(t; x', t'; [S(\cdot)])$ is just the classical dynamical trajectory of $\mathcal{S}$ in $\mathcal{M}$ selected by the initial data $q(t') = x'$, $p(t') = (\nabla S)(x', t')$ [7]. We also remind the reader that the Feynman path integral representation of the quantum mechanical propagator [3] corresponding to lagrangian (10) reads [5]

$$
\langle x'', t'' | x', t' \rangle = \int \mathcal{D}x(t)\ \delta(x'' - x(t'')) \delta(x' - x(t')) \cdot
$$
$$
\exp \left\{ (i/\hbar) \int_{t'}^{t''} dt\ L(x(t), \dot{x}(t), t) \right\}. \qquad (13)
$$

3 – Starting point of the RPQ is classical mechanics as formulated *a là* Hamilton-Jacobi by eqs. (11) and (12) [6]. Central to the RPQ is the idea – naturally suggested by the similarity epitomized in (2) and (9) – that quantum fluctuations can be simulated by a certain *white noise* $\eta(t)$ that perturbs the classical time evolution in $\mathcal{M}$. Actually, this scenario is made much more specific by a *further* assumption: *All* quantum effects are accounted for in this way [7]. As a consequence, the white noise variables $\eta(t)$ should simply be *added* into eq. (12) without altering the already present terms. This means (in particular) that the Hamilton-Jacobi eq. (11) remains *unchanged* within our formulation (more about this, later). So, *quantization* is presently accomplished by turning eq. (12) into the following *Langevin equation*

$$\frac{d}{dt}\xi_i(t) = \frac{1}{m}\left(\frac{\partial}{\partial x_i}S(x,t) - \Omega_i(x,t)\right)\bigg|_{x=\xi(t)} + \left(\frac{\hbar}{m}\right)^{1/2}\eta_i(t) \qquad (14)$$

where $\eta(t) \equiv \{\eta_i(t)\}_{1\leq i\leq N}$ is a *Fresnel white noise* defined by the functional (pseudo) measure

$$\mathcal{D}\mu[\eta(\cdot)] \equiv \mathcal{D}\eta(t)A[\eta(\cdot)] \equiv \mathcal{D}\eta(t)\exp\left\{(i/2)\int_{-\infty}^{\infty}dt\ \eta_i(t)\eta_i(t)\right\}. \qquad (15)$$

Quite analogously to the case of classical stochastic diffusion processes, the Fresnel noise average of a given $\eta(t)$-dependent quantity $(\cdots)$ is defined by

$$\left\langle(\cdots)\right\rangle_\eta \equiv \int \mathcal{D}\mu[\eta(\cdot)](\cdots). \qquad (16)$$

We stress that the Fresnel white noise variables $\eta(t)$ are *real*, while $\mathcal{D}\mu[\eta(\cdot)]$ is manifestly *complex* – this circumstance prevents a standard probabilistic interpretation of eq. (15) [8]. Still, eq. (1) naturally suggests to regard $\mathcal{D}\mu[\eta(\cdot)]$ as an *amplitude* (pseudo) measure, so that $A[\eta(\cdot)]$ is to be understood as the *amplitude distribution* for the Fresnel white noise configurations $\eta(t)$. Obviously, the $\eta(t)$ variables are insensitive to the dynamics of S, but in addition $\mathcal{D}\mu[\eta(\cdot)]$ does not contain any parameter pertaining to S. As a result, the Fresnel white noise has a *universal nature*. Observe that as either $\hbar \to 0$ or $m \to \infty$ the noise decouples from the dynamics of S and the classical behaviour shows up.

Coming back to the Langevin eq. (14), we denote by $\xi(t; x', t'; [S(\cdot), \eta(\cdot)])$ its solution with initial condition $\xi(t') = x'$ and controlled by an (arbitrary) integral $S(x,t)$ of eq. (11). These solutions describe the *quantum random paths* (controlled by $S(x, t)$), which are the basic objects in the present approach. They are *fluctuating curves* (fractals with Hausdorff dimension *two*) [9] quite similar to the erratic trajectories characteristic of macroscopic brownian motion.

According to the RPQ, the quantum mechanical propagator $\langle x'', t'' | x', t' \rangle$ is expressed as a Fresnel noise average involving the quantum random paths, much in the same manner as the transition probability $P(x'', t'' | x', t')$ of a classical stochastic diffusion process arises in the Langevin treatment (namely, as a gaussian noise average involving the solutions of the Langevin equation). Explicitly, we have [10]

$$\langle x'', t'' | x', t' \rangle = \exp\left\{ (i/\hbar)[S(x'', t'') - S(x', t')] \right\} \cdot$$
$$\left\langle \delta(x'' - \xi(t''; x', t'; [S(\cdot), \eta(\cdot)])) \Delta(t''; x', t'; [S(\cdot), \eta(\cdot)])^{1/2} \right\rangle_\eta \tag{17}$$

where the definition

$$\Delta(t''; x', t'; [S(\cdot), \eta(\cdot)]) \equiv \det \left| \frac{\partial}{\partial x'_j} \xi_i(t'', x', t'; [S(\cdot), \eta(\cdot)]) \right| \tag{18}$$

is used [11]. An important feature of the *random path representation* of the propagator (17) is that $S(x, t)$ is an *arbitrary solution* of eq. (11) (indeed, it can be shown that the r. h. s. of eq. (17) does *not* depend on which specific solution $S(x, t)$ is employed). Besides making the random path representation of $\langle x'', t'' | x', t' \rangle$ quite flexible in practical applications, this fact also explains why the dependence on the classical initial momentum $p(t') = (\nabla S)(x', t')$ gets washed out on going over to the quantum theory.

4 – As anticipated at the beginning of this paper, the RPQ offers a very simple *intuitive picture* of quantum mechanics – we would like to discuss this topic very briefly. Even in this respect, the analogy expressed by eqs. (2) and (9) is quite suggestive.

Consider a classical stochastic diffusion process whose deterministic dynamics is unaffected by fluctuations [12]. As is well known, the gaussian white noise describes a very specific kind of fluctuations which are independent of the dynamics of the diffusing particle: For this reason, they will be referred to as *background gaussian fluctuations*. Clearly, all this entails that *no interference* between deterministic and fluctuation effects exists. Now, the Langevin treatment brings out quite sharply the fact that in this case the *intrinsic* dynamics of a diffusing particle is *not* modified when fluctuations are taken into account. Rather, the *same* dynamics – which is deterministic when considered *in vacuo* – is actually supposed to occur in an environment where background gaussian fluctuations are present. They perturb the deterministic trajectory, thereby giving rise to a family of fluctuating random paths (one for each gaussian white noise configuration) which are given by the Langevin equation.

We find it quite tantalizing to imagine that the Fresnel white noise similarly describes *background quantum fluctuations*, which do *not interfere* with any deterministic effect. Observe that they possess a *universal natu-re* [13]. Moreover – even when S is replaced by a many-particle system – it makes sense to think of the background quantum fluctuations as living in *physical space* (this point will be discussed in great detail elsewhere). Now, the drift in the Langevin eq. (14) is – as before – unaffected by these fluctuations, and so the RPQ naturally leads to the conclusion that the *intrinsic* dynamics of S is *always* the *same*, as dictated by eqs. (11) and (12). When this dynamics is considered *in vacuo*, the classical deterministic behaviour emerges. Yet, this turns out to be merely an approximation. In order to get a more detailed description, we have to suppose that *the environment contains background quantum fluctuations*. Accordingly, the classical dynamical trajectory $q(t; x', t'; [S(\cdot)])$ is perturbed by background quantum fluctuations and gets replaced by the family of fluctuating quantum random paths $\xi(t; x', t'; [S(\cdot), \eta(\cdot)])$ (one for each Fresnel white noise configuration). Mathematically, this amounts to consider eq. (14) instead of eq. (12) whereas eq. (11) still holds true. Remarkably enough, all this is in agreement with physical intuition. In fact, the background quantum fluctuations are supposed *not* to interfere with deterministic effects, hence the drift in eq. (14) has to be the *same* as in eq. (12), which implies in turn that eq. (11) should remain *unchanged* (these facts indeed justify our conclusion that the *intrinsic* dynamics of S is always the same). All ordinary quantum fluctuations are thus brought about by the entanglement of classical dynamics with background quantum fluctuations. The following scenario emerges

$$QUANTUM \ MECHANICS = CLASSICAL \ MECHANICS \ + \atop BACKGROUND \ QUANTUM \ FLUCTUATIONS. \tag{19}$$

Although it is not clear whether background quantum fluctuations possess any *physical* meaning, eq. (19) looks as an important achievement to us, since it provides a vivid *intuitive picture* of the quantum theory, which might eventually throw new light on the nature of the quantization.

Before closing this paper, we address a potential source of confusion. According to the RPQ, the function $S(x, t)$ entering eqs. (14) and (17) satisfies the classical Hamilton-Jacobi eq. (11), thereby implying that the latter equation should hold true *even* in the quantum domain. Superficially, this circumstance seems in flat contradiction with the fact that the phase of any wave function obeys a *modified* Hamilton-Jacobi [14]. The best way to clarify this point is to focus the attention on eq. (17). Denoting by $S_Q(x'', t''; x', t')$ the phase of $\langle x'', t''| x', t' \rangle$ [15], eq. (17) gives

$$S_Q(x'', t''; x', t') = S(x'', t'') - S(x', t') + S_{QC}(x'', t''; x', t') \qquad (20)$$

where $S_{QC}(x'', t''; x', t')$ is the phase of $\langle \cdots \rangle_\eta$ [15]. Since $A[\eta(\cdot)]$ in eq. (15) is complex, it follows that $S_{QC}(x'', t''; x', t') \neq 0$, hence eq. (20) entails that $S_Q(x'', t''; x', t')$ should satisfy (in the variables x'', t'') a *modified* Hamilton-Jacobi equation, wherein the quantum correction is ultimately traced back to the *complex* nature of the functional (pseudo) measure $\mathcal{D}\mu[\eta(\cdot)]$ that defines the Fresnel white noise so characteristic of the RPQ [16].

5 – We have outlined the RPQ, stressing its structural similarity with the Langevin approach to classical stochastic diffusion processes. As an immediate consequence, many of the mathematical techniques developed in connection with *stochastic differential equations* can be applied to quantum mechanical problems as well. For instance, the whole WKB expansion of the propagator emerges here as a *small noise expansion* [2] performed on the Langevin eq. (14). Moreover, the RPQ supplies an ideal theoretical setting of the recently proposed δ-*expansion* [10], which is expected to yield a new approximation method whereby quantum mechanical systems can be handled. Several applications and extensions of the RPQ are currently under investigation, including the relativistic generalization.

Acknowledgments

Conversations with S. Bergia, F. Cannata, G. F. De Angelis, A. Defendi, I. Guarneri, F. Guerra, G. Jona-Lasinio, H. Kleinert, G. Parisi, E. Pollack and L. Schulman are gratefully acknowledged.

Notes

(1) Only the normalization condition is different in the two situations.

(2) This statement is true *only* for *unobserved* quantum systems. Questions concerning measurement lie outside the scope of the present paper.

(3) What is most important for the subsequent discussion is that the *deterministic* dynamics underlying diffusion processes of this kind is dictated by a *first-order* equation. Surely, the most famous example of classical diffusion process in configuration space is provided by macroscopic brownian motion in the so-called *Einstein-Smoluchowski approximation* [2].

(4) We point out that the RPQ is *also* hinted at by a very different line of reasoning, concerning the *classical* dynamical origin [6] of the paths which *actually* contribute in a new path integral representation of the quantum mechanical propagator [4].

(5) Contrary to a naïve expectation, only a certain set of *fluctuating paths* with the property $\Delta x(t) \sim (\Delta t)^{1/2}$ *actually* contribute in the Feynman path integral (13) (they are the so-called *Feynman paths*). Observe that amplitudes – but *not* probabilities obeying classical Kolmogorov axioms – can consistently be associated with these paths, which therefore *cannot* be interpreted as trajectories followed by s in the *classical* sense [3].

(6) This is in fact our (somewhat obvious) choice for the quantum counterpart – in the sense of correspondence (9) – of the *deterministic* dynamics underlying a classical stochastic diffusion process. We are employing the Hamilton-Jacobi formulation of classical mechanics because we need the deterministic trajectory of s to be given by a *first-order* equation, in order to further exploit correspondence (9) (recall note 3).

(7) We stress that it is just this assumption that makes RPQ *very different* from Nelson's stochastic formulation of quantum mechanics [8] (the assumption in question does *not* hold true within Nelson's approach).

(8) We would like to emphasize that *quantum fluctuations* come into the game precisely through the *imaginary unit* in eq. (15) (this will be explicitly shown later on). Although eq. (15) can be regarded as a postulate whose justification ultimately comes from the fact that the whole strategy works, it is nevertheless remarkable that eq. (15) can once again be derived by means of the analogy (9). Indeed, the free-particle Schrödinger equation arises from the heat equation with *diffusion constant* D by the replacement $D \to i\hbar/2m$. Correspondingly, the usual probability measure of the gaussian white noise [2] becomes just eq. (15), up to the constant $m/\hbar$ which has been shifted into eq. (14) by a trivial rescaling of the noise (in this fashion, one also understands why the constant $(\hbar/m)^{1/2}$ appears in eq. (14)).

(9) This circumstance follows from the property $\Delta \xi(t; x', t'; [S(\cdot), \eta(\cdot)]) \sim (\Delta t)^{1/2}$, which is in turn implied by eq. (15). Moreover, since for any noise sample $\xi(t; x', t'; [S(\cdot), \eta(\cdot)])$ collapse on $q(t; x', t'; [S(\cdot)])$ as $\hbar \to 0$, we may say that the quantum random paths possess a *classical* dynamical origin. However, they *cannot* be viewed as trajectories followed by s in the *classical* sense (since an *amplitude* distribution is induced on their set by $A[\eta(\cdot)]$ through eq. (14)). Hence, the quantum random paths are quite similar in nature to the Feynman paths (recall note 5). We also remark that the paths which *actually* contribute in the previously-mentioned new path integral [4] (note 4) can be recognized as quantum random paths [6].

(10) We are supposing throughout that $s(x,t)$ is a smooth single-valued function. This is certain the case for $|t'' - t'|$ finite but sufficiently small [7]. So, one can compute $\langle x'', t'' | x', t' \rangle$ first within such a limitation by eq. (17). The convolution property (4) then ensures that the latter result can trivially be extended to arbitrary times.

(11) Perhaps, the simplest way to prove eq. (17) is to show that it indeed leads to eq. (13). This can be done by means of the same formal manipulations whereby the generalized Wiener path integral representation of $P(x'', t'' | x', t')$ is derived within the Langevin description of classical stochastic diffusion processes.

(12) This is e. g. the case of macroscopic brownian motion in configuration space (Einstein-Smoluchowski approximation), but *not* in phase space [2].

(13) This follows from a similar remark made about the Fresnel white noise after eq. (16).

(14) See e. g. ch. 6 of ref. 9.

(15) in units of $\hbar$.

(16) Observe that an ordinary gaussian white noise would *not* produce any *quantum correction* to classical dynamics (this circumstance was anticipated in note 8).

References

1. Feynman, R.P. (1948). *Rev. Mod. Phys.*, **20** (367).
   Feynman, R.P. and Hibbs, A.R. (1965). *Quantum Mechanics and Path Integrals*, McGraw-Hill, New York.
2. Van Kampen, N.G. (1981). *Stochastic Processes in Physics and Chemistry*, North-Holland, Amsterdam.
   Risken, H. (1984). *The Fokker-Planck Equation*, Springer, Berlin.
   Gardiner, C.W. (1985). *Handbook of Stochastic Methods*, Springer, Berlin.
3. Feynman R.P. and Hibbs, A.R. (1965). *Quantum Mechanics and Path Integrals*, McGraw-Hill, New York.
   Schulman, L.S. (1981). *Techniques and Applications of Path Integration*, Wiley, New York.
   Kleinert, H. (1990). *Path Integrals in Quantum Mechanics Statistics and Polymer Physics*, World Scientific, Singapore.
4. Roncadelli, M. (1992). *J. Phys. A*, **25** L997.
5. Gel'fand, I.M. and Yaglom, A.M. (1960). *J. Math. Phys.*, **1** (48).
   Graham, R. (1977). *Z. Phys.*, **B26** (281).
6. Roncadelli, M. (1993). *J. Phys. A*, **26** L949.
7. Arnold, V. (1978). *Mathematical Methods of Classical Mechanics*, Springer, Berlin.
8. Nelson, E. (1967). *Phys. Rev.*, **150** (1066).
   Nelson, E. (1985). *Quantum Fluctuations*, Princeton University Press, Princeton.
9. Messiah, A. (1961). *Quantum Mechanics*, North-Holland, Amsterdam.
10. Bender, C. *et al.* (1991). *J. Stat. Phys.*, **64** (395).

THE LOSS OF INDIVIDUALITY FROM CLASSICAL TO QUANTUM PHYSICS

A. ROSSI

Dipartimento di Fisica, Università di Lecce
Via per Arnesano - 73100 - Lecce, Italy

Abstract. Quantum Mechanics doesn't confine itself, as Relativity Theory, to making some properties of bodies variable and dependent on their state of motion, but it also questions the concept itself of individual object. At variance with Classical Statistical Mechanics, which admits an indistinguishability in fact but not in principle among individual objects, Quantum Mechanics does not distinguish at all among apparently identical atomic particles. However, though the different quantum statistics introduce different correlations among atomic objects, they do not automatically verify specific theoretical models of those objects, such as their supposed loss of identity. This conclusion contradicts the opinion of E. Fermi, the author, with P. A. M. Dirac, of the second quantum statistics, who did not want to question this point seriously in the philosophically disengaged Italian scientific milieu of his time.

1. Introduction: the classical framework

Quantum Mechanics develops to its utmost limits the criticism to the concept of physical object meant as an individual, separate and distinguished entity, which began in the last century with the rise and early growth of theoretical physics. Indeed, nineteenth century theoretical physics had already shown the presence of indistinguishability, in fact even though not in principle, among gas molecules, even suggesting, with E. Mach (1886-1916) in contrast with L. Boltzmann (1844-1904), the merely auxiliary and fictitious character of the concepts themselves of individual atom and molecule[1].

Let me briefly discuss this point. According to Boltzmann, the concepts of atom and molecule are the firm stronghold of physical knowledge[2], even if the constructive and revisible character of these concepts is acknowledged by this author, who looks at them as "mental pictures". Thus there was

417

C. Garola and A. Rossi (eds.), The Foundations of Quantum Mechanics, 417-426.
© *1995 Kluwer Academic Publishers.*

a conflict between the physicists like Boltzmann, who went on attributing some function of approximation to reality to conceptual constructions going beyond experience and empirically unobservable, such as individual, indistinguishable and interchangeable atoms and molecules in kinetic theory of gases, which supplied experience with a certain coherence and even revealed an unsuspected fecundity[3], and others like Mach, who instead thought that such constructions had a merely economical, quite provisional function, doomed as they were to be replaced, soon or later, by mathematical correlations among changing sense data, such as the differential equations of mathematical physics[4]. For these, in their phenomenalist view, the concept of physical object lost its consistency and autonomy in front of changing sense data and their merely formal correlations and invariances.

Truly, Mach's view claimed, contrary to the tendency to critical deepening of conceptual models of the new theoretical physics, from J. C. Maxwell (1831-1879) to H. v. Helmholtz (1831-1894), H. Hertz (1857-1894), J. H. Poincaré (1854-1912) and Boltzmann himself[5], too strict a reducibility of physical concepts to experience, since it did not accept the explicatory and not only euristic use made of them in this new theoretical physics. Nevertheless, it also contributed to undermine the traditional concept of physical object and thus gave way to new developments and critical revisions even beyond the critical viewpoint of theoretical classical physics, which indeed maintained, in particular with Boltzmann, that there is a correspondence between our models and physical reality, though it may be partial and approximate. Boltzmann, in fact, based his reconstruction of thermodynamic phenomena in statistical terms on a rigorous assumption of individuality of atomic objects, even if he retained that they were unobservable as such and considered his particular mental pictures of them as a fertile but largely arbitrary approximation to reality[6]. Thus, he assumed that they were totally independent individuals, whose history was at least in principle autonomously recoverable, so that one could and should distinguish in statistical reckoning the case of two identical atomic particles in two different states from that of the same two particles exchanged of state, even if the difference apparently only consisted of the pure names of the particles and not of the physical situations, as the particles were identical and identically described by the assumed model[7]. Indeed, the realistic assumption of an independent objectivity of the atomic individuals, their intrinsic difference in the sense of Leibniz's principle of the identity of indiscernibles - according to which, if two objects are two and not one, they must have an intrinsic mutual difference, not only an extrinsic and numerical one [8] - made distinguishing them possible even beyond the limits of our ignorance. More precisely, the so to say "transcendental", that is not immediately empirical identity of the particles, allowed one to distinguish between the two cases mentioned above, and then, in the classical context, gave validity to the Maxwell-Boltzmann statistics that assumed such a distinction at the basis of its reckonings. Summing up, Boltzmann's point of view expressed a critical realism which considered the atomic conception as the affirmation of the validity of the classical concept of individual physical object, this concept being a scientifically fertile mental construction, capable to extend our intellectual mastery of reality and revealing, through empirical checks, a not merely fictitious character, even though it was not decidible and certain in its specific traits, or at least not crucially decidible[9].

2. Bose-Einstein

A first partial blow to the conception of the stability, independence and separability of physical objects, which is deeply embedded in the classical model of atoms and molecules - apparently identical, but really distinguished by state of motion and previous history - was inflicted, in my opinion, by the Theory of Relativity of A. Einstein (1879-1955), who was, in his own words, influenced by Ernst Mach's "uncorruptible skepticism"[10]. Indeed, Einstein put in evidence in Special Relativity that some fundamental properties of physical objects as mass and dimensions may change with the state of motion of the objects themselves, and this variability even concerns the particles and elementary components of bodies. Thereafter, a characteristic aspect of the traditional view of physical objects, the permanence of properties, seemed to fail, being substituted by a variability or covariance of properties depending on variations of the state of motion.

But Einstein deepened the question of the representation of apparently identical physical objects such as elementary particles only when he elaborated a new statistics for elementary particles, developing the Indian physicist Bose's work. In fact he then also questioned the identity of physical objects, by admitting, from one side, contrary to Mach and phenomenalists in general, their reality distinguished from mere phenomenic appearances[11], but even analysing critically, from the other side, the concept itself of physical object. In Einstein's view, at variance with Maxwell-Boltzmann statistics, we cannot distinguish in statistical reckoning the case in which two identical particles occupy two different states from the case in which the same particles simply exchange their states. As a consequence, the *a priori* probability of the presence of one particle in each state is less than in Maxwell-Boltzmann statistics, while that of the presence of the two particles together in one single state is greater, that is one third for either state, instead of one fourth as in the older statistics.

As well known, this move was motivated by Bose with the need of making the statistical reckoning on a photon gas fit Plank's law, beyond the limits of validity of Wien's empirical law[12]. Apparently, quantum physics forces us to eliminate the identity of elementary particles, when interpreting Bose's statistical reckoning in its specific terms (which in fact Bose himself didn't dare to use[13]), that is when introducing an absolute indistinguishability of bosons. For equal configurations, you cannot distinguish the material identity (or "genidentity" referring to its previous history[14]) of each particle, as if it were absorbed in a wave. Indeed a wave is considered, even in classical physics, as something that is permanent but continuously changes its specific material composition, that is the fluid particles which compose it [15].

Einstein's view was different and subtler, though it still started from Bose's calculus, since he thought that the anomalous behaviour of bosons was due to the presence of reciprocal correlations, "attractive" in character, which increased the probability of the simultaneous occupation of the same state by the two particles, and thus correlatively reduced the probability of the two particles occupying two different states. The "ondulatory" behaviour of the particles is then due, in Einstein's view, to the presence of positive correlations among the particles rather than to their "loss of identity", there being at work hidden parameters which maintain the identity of particles but hinder its empirical revelation in quantum physics, even more definitely than in classical physics. In short, in Einstein's view, reciprocal dependence of quantum particles, rather than their mere loss of individual identity, allows us to consider them as waves or, better, as driven by "pilot-waves" in their coherent path, according to L. de Broglie's original idea, even though it seems as if a loss of identity had place, if we renounce the classical concept of physical object. The principle of the identity of indiscernibles is thus preserved by admitting that the particles, being correlated among themselves, do not lose in the correlation their material identity based on their previous history or actual specific properties, but maintain it, provided that they are in fact correlated among themselves as in an ondulatory process. In this sense, in Einstein's view, the material identity of particles is taken for granted, contrary to their supposed loss of identity, and is covered by correlations that do not exist in classical physics but only in quantum physics: this is shown by an empirical evidence that quantum theory describes, but it does not succeed in providing with a plausible explanation[16].

3. Fermi-Dirac

As is well known, not all elementary particles have the same statistical behaviour. Therefore the problem of the identity of indiscernibles comes back in a different way when other statistics are considered, forcing one to a further critical examination of the concept of physical object applied to microphysics. This is indeed the case when the Fermi-Dirac statistics is taken into account. At first view, to be true, it seems that the identity of indiscernibles is here rigorously re-established, even at the empirical level, for it is excluded that a half-integral spin particle, different from integral spin bosons, be in the same state as another one, and be then identical to it. A new statistical reckoning corresponds to this new situation, which attributes zero probability to the simultaneous presence of two particles in the same quantum state. For this reason H. Weyl rebaptized Leibniz's principle of the identity of indiscernibles as "the Pauli-Leibniz principle", substantially identifying the former principle with Pauli's exclusion principle that rules the statistical behaviour of half-integral spin particles[17].

However, the new statistics, independently worked out by E. Fermi and P. A. M. Dirac[18], does not make any difference, as the Bose-Einstein one, between the possible configurations of two particles in two different states, that is, it does not consider whether one or the other occupies each state. In fact, in any case, it attributes probability one to such a configuration. The particles are two of number, but nothing changes if they reciprocally exchange their places[19]. This suggested a conception of Fermi particles as "nomological" particles, deprived of proper material individual identity and only defined by their state of occupation, in the sense that they assume the identity of the specific states they happen to occupy, independently of their previous history and further identity, because all particles occupying a certain state are absolutely identical[20]. Thus, since the particles are no longer identifiable with individuals, they cannot be considered as examples of violation of the principle of the identity of indiscernibles.

The above loss of identity of fermions appears, just because of the exclusion principle, much deeper than that of bosons; for, the exclusion principle, which hinders to hypothize the simultaneous presence of identical particles in the same state, avoids from the start the introduction of the hidden parameters that have been introduced in the case of bosons for establishing a difference among particles being simultaneously in the same state and being, at the same time, quantistically identical. It is indeed even empirically evident that there is in this new case a negative correlation ("repulsion") which prevents two fermions from being simultaneously in the same state, independently of a deeper individual identity of the particles, which are in fact interchangeable and identical, either the one or the other being able to be in a certain state, no matter what its previous history has been[21].

As is known, this strict interchangeability is linked in Quantum Mechanics to the loss of space-time localization and to the ensuing impossibility of following the history of each single particle, since there seems to be in this theory an absolute limit to the possibility of exact simultaneous measurement of complementary observables as position and momentum, or time and energy. It is a limit which in some cases even hinders to distinguish not only elementary particles among themselves, even if they are spatially distant, provided they have interacted in the past, but also these quantum objects and the measuring apparatuses, and then, more generally, macroscopic objects among themselves, as the so called EPR and Schrödinger cat paradoxes clearly show if we assume quantum description as a complete one. This limit is clearly implied in the description of quantum states by means of wawefunctions, interpreted probabilistically, whenever this description is considered complete and incompatible with any, more or less deterministic, attempt to complete it. A consequence of this approach is indeed the impossibility of individuating an elementary particle with respect to all the particles which are identical to it, so that we cannot absolutely say which particle occupies, at the moment of the empirical measurement, a certain defined state, with the exclusion of any other particle if they are fermions[22].

4. Statistics and theoretical models

If one does not accept Einstein's faith in the possibility of completing deterministically Quantum Mechanics, interpreted by Einstein as a purely statistical and incomplete theory, or Schrödinger's opposite proposal of abandoning the concept of individual particle in favour of a purely ondulatory conception of quantum theory, totally eliminating the individualization of quantum systems[23], the only escape from the conceptual embarassment engendered by the paradoxical implications of Quantum Mechanics consists in my opinion in refusing to identify our theoretical representations and models, though precise and sophisticated, with reality as such; this attitude should be applied not only to quantum theory but even to classical theory, as some of the greatest theoretical physicists of the later 19th century had already begun to do.

As a consequence, we will recognize the irreducibility of the physical object itself, with its complex behaviours and properties, to its particular theoretical images, even if we do not intend to disregard in any way the euristic and explicative fecundity of different theories and models and, in particular, of different classical and quantum conceptions of the physical object[24]. Thus, we think that the reach of the existing theoretical interpretations and models must be reconsidered, and that one cannot demand any longer that they exhaust physical reality; we will rather recognize the existence of definite bonds which are compatible with different theoretical interpretations and models, and underline the bearing and limits of the bonds themselves in defining and choosing the theoretical possibilities. In the specific case of different statistics and their history, the bonds are, first of all, provided by the correlations among the objects under scrutiny, which are null in Maxwell-Boltzmann, negative in Fermi-Dirac and positive in Bose-Einstein statistics[25]. Certainly they will favour either one or the other of the theoretical solutions or models, but cannot prove or refute them, and no crucial experiment leading to this result can be devised. This, of course, does not mean denying or underestimating the conceptual and theoretical fecundity of theories and models, the role of which does not reduce to having euristic or experimental usefulness.

From an historical viewpoint it is now important to note that the above epistemological position was not shared by E. Fermi, who thought instead that an immediate correspondence exists among experimental facts, theories and models, so that the mathematical formalism together with its most direct physical interpretation can be validated by proper crucial experiments, thus transforming theoretical entities and their interpretations into direct expressions of reality. Among the facts supporting an ortodox quantum mechanical interpretation of his statistics, Fermi particularly insisted on the homeopolar chemical bond between electrically neutral atoms forming molecules as, for instance, the hydrogen molecule. Fermi noted that there are no forces, except quantum mechanical interactions, that can explain this bond, since it simply appears because it is impossible to distinguish the orbiting electrons of the two atoms combined together, according to his statistics, so that there is a continuous exchange between them, which engenders an attraction (exchange force) giving stability to the molecule[26]. Here, as elsewhere[27], Fermi's approach tries to overcome the methodological and theoretical break introduced by the new theory in the link among experience, models, theories and interpretations, which was certainly deeper than that introduced by classical mechanics, stressing its operational fecundity and richness rather than its conceptual aspects (and difficulties).

We observe that Fermi's attitude is similar to that of T. Levi-Civita[28], who greatly contributed to the mathematical and technical development of General Relativity Theory with his Tensorial Absolute Calculus, and it consists in minimizinging the conceptual bearing of the introduced changements. Yet, as in the Levi-Civita case, also in the Fermi case the introduced changements were so decidedly revolutionary that they could question well consolidated conceptual certanties and force deep methodological revisions, if they had received enough attention. For example, Fermi himself originally developed the methods of the second quantization of fields by brilliantly applying them to his theory of beta-decay[29]. Such methods implied the introduction of creation and destruction operators that modify the particle number, which is then no longer considered fixed as in the previous first

quantization approaches. This entails deep changes in the traditional concepts of particle and wave[30], but Fermi avoided discussing the epistemological implications of his work. We conclude that the strong conservativism
and philosophical disengagement that characterized Italian culture in the
first decades of this century lead great scientists, as Fermi and Levi-Civita,
to avoid deepening the break between the old and the new physics. They
privileged technical competence and common sense realism that immediately justified innovations through experimental proofs, but did not evidentiate
the most problematic aspects in the new physics, first of which the failing
of the correspondence among experimental events, theories and models.
This indeed made the cruciality of experiments and a naive realism almost
unbearable and induced a different deepening of the foundations of physics
elsewhere. But the Italian cultural climate, marked by the egemony of neo-
idealistic philosophy, together with the philosophically disengaged operational attitude of most scientists (with very few exceptions as Enriques,
Persico, Majorana), seriously hindered Italian reflections on the foundations
of physics[31].

Notes

[1] Mach (1960), pp.599 ff. and Mach (1943), *passim.*
[2] Boltzmann (1974), pp. 41-53.
[3] Boltzmann (1964).
[4] Mach (1960),p.599 and Mach (1943), *passim.*
[5] D' Agostino (1993) and Rossi (1993).
[6] Boltzmann (1974), pp. 201 ff.
[7] Boltzmann (1964), pp. 297-309.
[8] Leibniz and Clarke (1717), p. 97.
[9] D' Agostino (1990).
[10] Holton (1973), p.223.
[11] Holton (1973).
[12] Bose (1924).
[13] Bergia (1987).
[14] The concept, or still better the term itself of "genidentity" in this sense has been
introduced by K. Lewin with reference to Einstein's relativity (cp. Lewin (1923)) before
Reichenbach (1928) whom van Fraassen (1984) credits with it. I owe D. Howard this piece
of information through private communications.
[15] van Fraassen (1984), pp. 159-62.
[16] van Fraassen (1984), pp. 159-62.
[17] van Fraassen (1984), p, 156.
[18] Guicciardini and Introzzi (1995).
[19] van Fraassen (1984), pp. 157 ff.
[20] Toraldo di Francia (1985), pp. 207 ff.
[21] van Fraassen (1984), pp. 161 f.
[22] Caldirola (1975).
[23] D' Agostino (1993), pp.157 f.
[24] Chevalley (1992).
[25] Van Fraassen (1984), pp. 166 ff., and van Fraassen (1991).
[26] Fermi (1930a).

[27] Fermi (1930b).
[28] Levi-Civita (1918-19).
[29] Fermi (1933).
[30] Butterfield (1993).
[31] Maiocchi (1985).

References

Bergia, S. (1987). Who discovered the Bose-Einstein statistics?, in *Symmetries in Physics (1600-1980)* (1st Int. Meeting on the History of Scientific Ideas, Sant Feliu de Guíxols, Spain, 1983), M. G. Doncel, A. Hermann, L. Michel and A. Pais eds., Seminari d'Història de les Ciències (Universitat Autònoma de Barcelona, 1987), pp. 221-250.

Boltzmann, L. (1964). *Lectures on Gas Theory*, University of California Press, Berkeley and Los Angeles.

Boltzmann, L. (1974). *Theoretical Physics and Philosophical Problems* , Reidel, Dordrecht and Boston.

Bose, S. N. (1924). Planck's Gesetz und Lichtquantumhypothese, *Zeitschr. f. Phys.*, **26**, pp. 178-181.

Butterfield, J. (1993). Interpretation and Identity in Quantum Theory, *Stud. Hist. Phil. Sci.*, **24**, pp. 443-476.

Caldirola, P. (1975). Historical Evolution of the Exclusion Principle in Physics, *Scientia*, **110**, pp. 69-81.

Chevalley, C. (1992). Le conflit de 1926 entre Bohr et Schrödinger: un example de sous-détermination des théories, in *Erwin Schrödinger, Philosophy and the Birth of Quantum Mechanics*, M. Bitbol and O. Darrigol eds., Editions Frontières, Gif-sur-Yvette, pp. 81-94.

D'Agostino, S. (1990). Boltzmann and Hertz on the *Bild*-Conception of Physical Theory, *Hist. Sci.*, **18**, pp. 380-398.

D'Agostino, S. (1992). Schrödinger on Continuity and Completeness in Physical Theory, *Physis*, **29**, pp. 539-561.

D'Agostino, S. (1993). A consideration of the rise of theoretical physics in Europe and of its interaction with the philosophical tradition, in *History of Physics in Europe in the 19th and 20th Centuries*, (1st EPS Conference, Como, Italy, 1992), F. Bevilacqua ed., Società Italiana di Fisica, Bologna, pp. 5-28.

Einstein, A. (1925). Quantentheorie des einatomigen idealen Gases. Zweite Abhandlung, *Berl. Ber.*, **23**, pp. 3-14.

Fermi, E. (1930a). I fondamenti sperimentali della nuova meccanica atomica,*Periodico di Matematiche*, **10**, pp. 71-84.

Fermi, E. (1930b). L'interpretazione del principio di causalità nella meccanica quantistica, *Rend. Lincei*, **11**, pp. 980-985.

Fermi, E. (1933). Tentativo di una teoria dell' emissione dei raggi "beta",*Ric. Scientifica*, **4**, pp. 491-495.

Guicciardini, N. and Introzzi, G. (1995). The Fermi-Dirac statistics: a simultaneous discovery, in this volume.

Holton, G. (1973). *Thematic Origins of Scientific Thought*, Harvard University Press, Cambridge, Mass.

Leibniz, G.W. and Clarke, S. (1717). *A collection of papers, which passed between the late learned Mr. Leibniz and Dr. Clarke. In the years 1715 and 1716 Relating to the Principles of Natural Philosophy and Religion*, London.

Levi-Civita, T. (1918-19). Come potrebbe un conservatore giungere alla soglia della nuova meccanica, *Rend. Se. Mat. Fac. Scien. Univ. Roma*, **5**, pp. 10-28.

Lewin, K. (1923). Die Zeitliche Geneseordnung, *Zeitschr. f. Phys.*, **13**, pp. 62-81.

Mach, E. (1943). *Popular Scientifc Lectures*, The Open Court Publishing Company, La Salle, Ill.

Mach, E. (1960). *The Science of Mechanics*, The Open Court Publishing Company, La Salle, Ill.

Maiocchi, R. (1985). *Einstein in Italia. La scienza e la filosofia italiane difronte alla teoria della relatività*, Franco Angeli, Milano.

Reichenbach, H. (1928). *Philosophie der Raum-Zeit-Lehre*, Veit & Co., Berlin and Leipzig.

Rossi, A. (1993). Kantism, phenomenalism, reductionism and the emergence of theoretical physics in the 19th century, in *History of Physics in Europe in the 19th and 20th Centuries*, (1st EPS Conference, Como, Italy, 1992), F. Bevilacqua ed., Società Italiana di Fisica, Bologna, pp. 279-285.

Toraldo di Francia, G. (1985). Connotation and Denotation in Microphysics, in *Recent Developments in Quantum Logic*, (Proceedings of the International Symposium on Quantum Logic, Cologne, Germany, 1984), P. Mittelstaedt and E. W. Stachow eds., B. I.- Wissenschaftsverlag, Mannheim, Wien and Zürich, pp. 203-214.

van Fraassen, B. C. (1984). The Problem of Indistinguishable Particles, in *Science and Reality: Recent Work in the Philosophy of Science*, J. T. Cushing, C. F. Delaney and G. M. Gutting eds., University of Notre Dame Press, Notre Dame, pp. 153-72.

van Fraassen, B. C.(1991). *Quantum Mechanics: An Empiricist View*, Clarendon Press, Oxford.

THE MYSTERIES OF QUANTUM THEORY:
AN INTRODUCTORY TALK

G. TAGLIAFERRI

Ist. di Fisica Generale Applicata, Università di Milano
Via Celoria 16 - 20133 - Milano, Italy

Abstract. Ever since its birth following the introduction of Planck's constant h in 1900, quantum theory has been legitimated only by the amazing capacity it has exhibited in accounting quantitatively for a huge variety of physical phenomena. The so-called Copenhagen interpretation has allowed to accomodate in a coherent frame the tenets of quantum mechanics, but at the cost of bringing in the demand of abandoning, at the microscopic level, causality, determinism, and physical reality. The results of Aspect and coworkers' ingenious experiments, however, can give further support to the belief in the unlimited validity of the quantum theory. But, for all its successes, the theory remains basically incomprehensible.

I am quite aware that the title of my talk can make me appear presumptuous: and this I would be if it were read as a claim from me of being able to say something original on a subject that has been debated for several tens of years by important scientists and philosophers. On the contrary, my intention is very limited. Having been asked to be the first speaker at this meeting, I thought an admissible opening to give a brief historical survey of the coming into existence of the main enigmas of the quantum theory that still defy our comprehension. These enigmas are well known, and there would be no need to spend time to comment on them, were it not for the fact that many physicists active in research tend to disregard such unresolved difficulties. While it is undeniable that this attitude has brought about a wealth of new knowledge, I hope you will bear with me as I, an old experimental physicist, recall summarily the intriguing questions that have accompanied the progress of the quantum theory.

I don't want to dwell here upon the birth and early developments of quantum physics. Suffice it to mention that since the beginning its concepts appeared

C. Garola and A. Rossi (eds.), The Foundations of Quantum Mechanics, 427-434.

inexplicable: from the introduction of the energy elements to that of the radiation quanta, from Bohr's atomic model to de Broglie's waves, etc. Anyway, after twentyfive years of initial confusion, thanks to the independent constructions of matrix mechanics by Heisenberg in 1925, and undulatory mechanics by Schrödinger soon afterwards, a form of quantum theory arose that succeeded in assembling into a coherent scheme the ad hoc assumptions introduced step by step in the body of nineteenth century physics. This new quantum mechanics, developed and refined unceasingly throughout this century, provided - to use the crisp prose of David Mermin -

"... the unambiguous calculational method that has underlain all the explosive growth and flowering that physical science has enjoyed from 1925 right up to the present moment. But the quantum theory remains deeply mysterious. It is no harder to use than any other branch of physics, and thousands - indeed hundreds of thousands - of people have mastered its computational intricacies since it was first put forth. It is capable, in principle, of predicting the outcome of any experiment one can describe precisely enough to apply the mathematical apparatus of the theory. What makes it mysterious is that in general the quantum theory refuses to offer any picture of what is actually going on out there. If you ask it a question of the form 'If I do this then what will I find if I measure that?' it will give you the answer. But if you ask 'Explain why I get that answer?' or ... 'How can that possibly be the answer?' it is silent." (Mermin, 1988).

The founders of the quantum theory were aware of course of the gnoseological problems raised by their creation. The history of physics of the years following immediately that eventful 1925 records a remarkable abundance of debates, propositions, contentions involving the very founders of that theory; a discussion, theirs, afterwards carried on and enriched to this day by many philosophers and physicists (although it should be noted that the majority of the latter seem to have accepted, at least implicitly, the so-called Copenhagen interpretation). Till now however a definitive answer to the essential question "What does the quantum theory really say on the nature of the physical world?" - an answer, I mean, on which there might be general agreement - is not available.

Chronologically the epistemological problems implied by the formulation of quantum mechanics became common knowledge following the formulation by Heisenberg in 1927 of the uncertainty principle. This is exposed, if I am allowed to recall it, in the famous paper "On the Observeable Content of Quantum Theoretical Kinematics and Mechanics" (Heisenberg, 1927), in whose abstract Heisenberg stated plainly the origin of the irreconcilable difference between quantum and classical mechanics. In fact one reads there:

"In the present work are enunciated first exact definitions of the words: place, velocity, energy etc. (e.g. of the electron), which retain validity also in quantum mechanics, and it is shown that canonically conjugated quantities can be determined simultaneously only with a characteristic uncertainty. This uncertainty is the real ground for the appearance of statistical connections in quantum mechanics...".

And, if somebody had any doubts about what Heisenberg meant by "statistical connections", he can be helped to dispel them by the closing sentence of Heisenberg's paper:

"The true content [of this work] can be characterized as follows: since all experiments are subject to quantum mechanics, and hence to [the formula of the characteristic uncertainty], the invalidity of the causality law will be therefore definitely established by quantum mechanics."

As to this dismissal of causality, Heisenberg makes clear that what is inapplicable is not a weak form of it, like the one that is met when (his words) "from exact data statistical conclusions only can be inferred"; what is invalidated instead is "a strong formulation of causality...[of the type]: if we know exactly the present, we can predict the future".

Prompted by the introduction of the uncertainty relations, Bohr hastened to organize the thoughts he had been entertaining for some time about the wave-particle duality, and gave his ideas coherent formulation with the new notion of complementarity. This amounted to an affirmation that it is in general impossible to represent exactly microscopic processes by using simultaneously mutually exclusive descriptions.

Now, it is obvious that positions like those of Heisenberg and Bohr (accepted also by Born, Jordan, Pauli and Dirac) implying fundamental questions of philosophy, could not leave indifferent other important physicists; and in fact e.g. Planck, Einstein, Schrödinger, de Broglie soon expressed dissent. Also professional philosophers, even though with some delay, did worry at the basic problem raised by the uncertainty principle, namely the physical reality of the microscopic world. Here I have not the time to stray from the main theme of my talk and dwell upon the history of the reactions for or against that principle (although such diversion would attract me on account of its intrinsic interest). At least, however, I can refer you (need I say) to the valuable book written by Franco Selleri (Selleri, 1990). As for me, I will limit myself, as I said above, to some elementary remarks, without venturing in a field requiring specialist knowledge.

The extraordinary practical successes of quantum mechanics with Born's probabilistic interpretation did convince in a short time the physicists' community of its validity as a satisfactory and irreplaceable instrument to deal with phenomena pertaining to the microscopic realm.

Then a situation developed that might be called of pragmatic acceptance, that is one where the conceptual difficulties have been either ignored or relegated to a state of limbo from which it did not seem impelling to extricate them. Even opponents of de Broglie and Schrödinger's calibre accepted for a few years the quantum theory as interpreted by the Copenhagen-Göttingen school (but later on they resumed an adverse attitude).

The question of the physical reality, in fact, can be shelved, but not dismissed altogether. I, a former experimenter not quite familiar with theoretical and philosophical subtleties, would side with those who think decidedly reductive Heisenberg's statement (in the above mentioned paper) that "physics has to confine itself to the formal description of relations among perceptions". Taken to the letter, these words express an impossibility or a renunciation of the true knowledge of the microscopic world. This is a big limitation of the objective and motivation of the study of nature.

"The positivist approach [of the Copenhagen school] - wrote e.g. J.C. Polkinghorne in a very readable little book (Polkinghorne, 1984) - emphasizes perceptions that can elicit intersubjective agreement; its tests of meaningfulness and verity are based on the specification of observable procedures. The contents that such an approach attributes to physical reality must be interpreted in terms of the observer's experience, his research concerns the harmonization of that experience. The world that this presents is populated by pointers moving along dials of measuring instruments and by tracks on photographic plates. The operator puts right the apparatus and records its indications. This is his central role."

If things are so, the situation in which the modern physicist finds himself is oddly fictitious: his world is inhabited only by the scientists engaged in fundamental physics research. I do not believe, however, that there are many physicists who imagine they are living in a world with no sub-stratum of reality. In my younger years I have spent most of my time in research laboratories, and I never met anyone who asserted that he did not consider external reality the crucial factor activating the human capacity to know; and knowledge, so the philosophers tell us, is much more than a matter of mere perceptions. After all, this attitude of many research workers (even if unspoken) is the traditional realist position in epistemology, and also one that best justifies the commitment and devotion of those who choose the challenge of investigating nature. More prosaically, I think that physicists would not be very successful when asking for substantial public funds to build and run their big machines, were they to present as a primary objective that of rendering their own perceptions mutually coherent.

Recapitulating then, the uncertainty principle, as interpreted by the Copenhagen-Göttingen school, requires abandoning, at the microscopic level, causality, determinism, reality. This course appeared so unacceptable to a stubborn realist like Einstein that he was induced after several years' meditation to propose, together with B. Podolski and N. Rosen, the famous "Gedankenexperiment", referred to (improperly) as EPR paradox, meant to limit the validity of quantum mechanics by exposing its incompleteness (Einstein et al., 1935). The discussion started by the EPR argument is well known.

Einstein and coll. had concluded their attack with this blunt affirmation:

"While we have thus shown that the wave function does not provide a complete description of the physical reality, we left open the question of whether or not such a description exists. We believe, however, that such a theory is possible."

(Unfortunately, this belief has not yet passed the stage of wishful thinking).

Bohr was quick to reply, relying also on the support of his complementarity viewpoint, on whose comprehensibility however reservations are not uncommon. He argued that quantum mechanical systems ought not to be considered separately from all the classical instruments and apparatuses envisaged to carry out the observations: if in the middle of a planned measurement one were to change the intended observations, there would arise a new situation even if the system to be observed remained the same. In short: quantum mechanics did not allow separation of the observer from the observed system. Bohr's argumentation, coherent with his somewhat positivistic attitude (but it is fair to remark that he always avoided taking up position in ontological matters), stressed the ineliminable intervention of the classical measuring apparatus. His answer was, so to say, within the frame of the Copenhagen interpretation, and as such seemed to most physicists an apt refutation of EPR rasoning; but in fact it avoided facing the basic problem that had been raised, which was not, after all, so much the completeness of quantum mechanics or the lack of it, as the validity of the intuitively reasonable idea that if two systems have been dynamically separated from each other for a period of time, then a measurement performed on one of these systems cannot affect in any real way the other.

Anyway, for several years the controversy between Einstein and coworkers and the metaphysical superstructure of the quantum theory was considered almost irrelevant by the majority of physicists. Einstein, for instance, complained (perhaps jocularly) that at the Institute for Advanced Studies in Princeton he was regarded as "an old fool". And the point about Einstein's pretended foolishness is specifically made in the following often quoted passage of a letter from Pauli to Born (Pauli, 1954):

"One should no more rack one's brain about the problem of whether something one cannot know anything about exists all the same, than about the ancient question of how many angels are able to sit on the point of a needle. But it seems to me that Einstein's questions are ultimately always of this kind."

Pauli's stinging comment, however, was not to be the final word on the matter. Briefly now, since I am speaking in front of professionals, I will recall the next developments. In the early 50s David Bohm proposed a new version of the EPR argument (Bohm, 1952). It had the merit of being able, at least potentially, to lend itself to perform measurements in laboratories materially existent, so as to solve the controversy between a locally realist theory, as that advocated by EPR thought-experiment, and conventional quantum mechanics. Bohm's work did not cause much of a stir at the time. But some twelve years afterwards a decisive thrust towards undertaking experimental measurements came from an analysis by John Bell, currently known as "Bell's theorem" (Bell, 1964). Quoting Mermin again:

"Bell thereby demonstrated that, Pauli to the contrary notwithstanding, there were circumstances under which one could settle the question of whether "something one cannot know anything about exists all the same", and that if quantum mechanics was quantitatively correct in its predictions, the answer was, contrary to Einstein's conviction, that it does not."

Bell's contribution to solve the controversy brought up by the paper of Eistein-Podolski-Rosen has been fundamental, as all physicists and many philosophers know. In fact prior to his analysis the controversy had been between opposing beliefs as to the completeness of the quantum theory, and therefore which side one chose was in the end a matter of taste; but after his analysis the situation has changed, because from it resulted that one could carry out some experiments conceptually simple but suitable to establish whether the previsions of quantum mechanics are correct, by comparing them with those of any model locally realist within the spirit of Einstein's ideas. Now it became possible for the experimenters to enter the lists, or rather the laboratories; and indeed in the following years several researches were instituted, climaxing in the early 80s with the very cogent experiments of Alain Aspect and coworkers at the Institut d'Optique d'Orsay of the Université de Paris Sud (Aspect et al., 1981, 1982a, 1982b).

Aspect and coworkers' results had a very pronounced impact, since they appeared to confirm that the previsions of quantum mechanics about photon correlations were correct, while those derived from local hidden-variables models were not. If this is the case, and many physicists agree that indeed it is, then quantum mechanics contrives to be more incomprehensible than ever. How can one system influence another separated system without exerting on it any recognizable action?

Bell's own reaction to those disturbing results can be read for instance in his book "Speakable and Unspeakable in Quantum Mechanics" (Bell, 1987), from which I quote:

"For me it is so reasonable to assume that the photons in those experiments carry with them programs, that have been correlated in advance, telling them how to behave. This is so rational that I think that when Einstein saw that, and the others refused to see it, *he* was the rational man. The other people, although history has justified them, were burying their heads in the sand...Einstein's intellectual superiority over Bohr, in this instance, was enormous; a vast gulf between the man who saw clearly what was needed, and the obscurantist. So for me it is a pity that Einstein's idea doesn't work. The reasonable thing just doesn't work."

Of course, one is not bound to assume that the analyses of the experiments carried out to test Bell's inequality are beyond the possibility of criticism, and in fact such analyses have been found wanting by some theoreticians (Selleri, 1990). But it seems right, for people who are not specialists in quantum theory (and I am one of them), to take account of Aspect and coworkers' results in the sense perceived by the majority of physicists, namely that quantum mechanics holds up. Then which attitude can be adopted towards its persistent mysteriousness ?

An easy option might be to refuse being bothered, and this is, I daresay, a widespread attitude that would be difficult to condemn, since in over half a century no new physics has sprouted in the wake of EPR's argument. The mystery of the extraordinary outcome of the measurements of Aspect et al. is (and I quote once more Mermin)

"... that it presents us with a set of correlations [between photons] for which there simply is no explanation. The majority [of physicists] would probably deny even this, maintaining that the quantum theory does offer an explanation. That explanation, however, is nothing more than a recipe for how to compute what the correlations are. This computational algorithm is so beautiful and so powerful that it can, in itself, acquire the persuasive character of a complete explanation...

This should not necessarily be dismissed as a cowardly refusal to face hard questions. When Newton explained the motion of the planets along elliptical orbits as a consequence of the inverse square law of gravitational attraction by the sun, many of his contemporaries took the view that he had explained nothing, but merely provided a powerful computational algorithm for calculating those orbits... Subsequent generations learned that this was not a fruitful question to ask, and the law of gravity acquired the character of one of the few fundamental irreducible facts on which all the rest of our knowledge is based."

Could a similar evolution take place in the future also for the quantum theory ? That is, could the present theory be accepted at its face value, stopping the unrewarding practice of putting questions to which perhaps there are no answers ? Actually, a lot of physicists find reasonable a "wait and see" position. However, there are also some who do not resign themselves to idleness.

In the example of Newton's theory the sun and the planets are at least real entities, whereas in quantum theory one is weighed down by the additional burden of the exclusion of the objective reality from the microscopic world. Thus the problems with quantum theory may not disappear just by waiting. For the time being, in order to attenuate the shock of the Parisian findings on photon correlations, one might resort to some convenient hypothesis, for instance like the following one fantasized by Bell:

"There are influences going faster than light, even if we cannot control them for practical telegraphy. Einstein local causality fails, and we must live with this".

But there is also reason to suspect that just living with the failure of local causality was not Bell's true aspiration. It would be rash to suppose that the whole of his mind can be contained in a short quotation; the one I am going to report, however, might serve as an indicator of his radical thinking:

"...the quantum mechanical description will be superseded. In this it is like all theories made by man. But to an unusual extent its ultimate fate is apparent in its internal structure. It carries in itself the seeds of its own destruction."

Without doubt the everlasting mysteries of quantum mechanics would disappear with its destruction!

References

Aspect, A. *et al.* (1981). *Phys. Rev. Lett.*, **47** (460); (1982a), *ibid.*, **49** (91); (1982b), *ibid.*, **49** (180).

Bell, J.S. (1964). *Physics,* **1** (195); (1987). *Speakable and Unspeakable in Quantum Mechanics*, Cambridge University Press, Cambridge.

Bohm, D. (1952). *Phys. Rev.*, **85** (166) and (180).

Einstein, A. *et al.* (1935). *Phys. Rev.*, **47** (777).

Heisenberg, W. (1927). *Z. Phys.*, **43** (172).

Mermin, N.D. (1988). "Spooky Actions at a Distance", in *The Great Ideas Today*, Britannica Great Books, Ghicago.

Pauli, W. (1954), in *The Born-Einstein Letters* (223). The original expression of this comment is attributed to Otto Stern.

Polkinghorne J.C. (1984). *The Quantum World*, Longman, London. The reported excerpt is a translation back into English from the Italian edition of Polkinghorne's book *Il mondo dei quanti*, Garzanti, Milano, 1986.

Selleri, F. (1990). *Quantum Paradoxes and Physical Reality*, Kluwer Academic Publishers, Dordrecht.

ON THE DIFFERENT FORMS OF QUANTUM ACAUSALITY

G. TAROZZI
Ist. di Filosofia, Università di Urbino
Via Saffi 9 - 61029 - Urbino, Italy

Abstract. Four empirical formulations of the causality principle, due respectively to Laplace, Kant, John S. Mill and Hume, are discussed in relation to the philosophy of quantum mechanics, showing how they are all violated by the basic principles of the Copenhagen interpretation, whereas they are perfectly compatible, with the only exception of Laplacean determinism, with a realistic reinterpretation of the present theory.

1. Introduction

Several authors[1] have critically analysed the role of acausal conceptions on the birth and development of quantum mechanics by stressing the strong influence, with respect to the genesis of this theory, of the openly irrationalistic philosophical background dominating European culture, and in particular the Weimar republic, between the two World Wars.

In the present paper we propose to extend the results of the previous analyses from the context of discovery to the one of justification, by showing how the theses of these historical studies can be confirmed by an epistemological discussion on the conceptual foundations of quantum mechanics.

We shall show, in fact, the existence of four different formulations of *empirical causality* which are contradicted by the basic principles of the Copenhagen interpretation of quantum theory. More precisely we intend to demonstrate that:

(a) *deterministic causality* is refuted by the Heisenberg indeterminacy principle, which seems to exorcise once for all the Laplace demon, by establishing the impossibility to define the initial state, in terms of positions and momenta (or velocities), even in the case of the "smallest atom", but is

435

C. Garola and A. Rossi (eds.), The Foundations of Quantum Mechanics, 435-447.
© 1995 *Kluwer Academic Publishers.*

reintroduced, even if in a non mechanistic form, for the evolution of unobserved microsystems;

(b) causality interpreted as *legality*, on the grounds of the second Kantian analogy of experience according to which "everything that happens ... presupposes something upon which it follows *according to a rule*"[2], appears, in the light of Bohr's principle of complementarity, incompatible with a realistic description based on space and time co-ordination, but can be preserved if we replace the complementary interpretation with a particular micro realistic interpretation of the quantum mechanical wave function;

(c) causality identified with *uniformity between cause and effect*, in the sense that the same causes will produce the same effects, according to Mill's principle of the uniformity of nature, is in conflict with the well known thesis of the completeness of quantum formalism maintaining the identity, through the attribution of the same identical wave function, of physical systems characterised by completely different individual behaviours. It has been, however, recently shown, on the grounds of the assumption of the validity of a very general probabilistic principle of physical reality, that the postulate of the completeness implies the inconsistency of quantum mechanics, due to a conflict between Heisenberg's indeterminacy relations and the superposition principle;

(d) causality in its weak and famous anti metaphysical Humean formulation of *ordinate connection*, which is merely based on the denial of the possibility of an inversion of temporal order, which would allow the occurrence of an effect before the cause generating it, is contradicted by some solutions of quantum paradoxes based on the abolishment of the time arrow.

We propose moreover to show that the violation of causality in the formulation given at point (d), which seems to reintroduce into physical theories a new form of super determinism much stronger that the Laplacean one, appears as a consequence of the refutation of the causal principle in the formulation (b), and that, therefore, it can be avoided if one assumes a realistic interpretation of the quantum mechanical wave function.

A similar proof of incompatibility between some well defined formulations of a fundamental philosophical principle and the basic tenets of the dominant interpretation of the principal theory of modern physics would seem to confirm the thesis, already maintained by the present author in the case of empirical realism and of realistic interpretations[3], of the possibility to provide reformulations endowed with meaning, in a factual sense, of philosophical principles refuted as metaphysical and, therefore, meaningless by the main currents of contemporary analytic philosophy. As a matter of fact, if the previously considered formulations of the causality principle would correspond to statements devoid of meaning, any interpretation based on their assumption could be legitimately maintained without in any way conflicting with orthodox quantum mechanics.

We believe, moreover, that our conclusion does not only show the profound impact of physical theories on philosophical thinking but also stresses how an epistemological analysis on the foundations could bring to the discovery of paradoxes and inconsistencies within these theories, providing in this way an essential contribution to a conceptual clarification of the basic physical concepts and principles.

2. Laplacean causality and the uncertainty principle

According to a radical, and also widespread, point of view, the validity of Heisenberg's uncertainty principle would involve, as an unavoidable consequence, a definitive refutation of the causality principle. The first to express this view was Heisenberg himself, but his position was shared by famous philosophers of science, like Friedrich Waismann, who in a conference on "The Decline and Fall of Causality", regarded 1927 as the year that "saw the obsequies of causality" in contemporary science, and Moritz Schlick, who in his *Philosophy of Nature*, maintained:

"...It is impossible to say that the state of a system can be accurately determined by a measurement. But since a determination of this kind is a prerequisite for the strict application of the principle of causality, it follows that modern science must renounce the exact truth of this principle..."[4]

Such a radical view derives from the reduction of causality to its strong, and restricted, formulation provided by Laplacean determinism, which is exemplified by the famous demon:

"An intelligence which at a given moment knew all the forces that animate nature, and the respective position of the beings that compose it, and further possessing the scope to analyse these data, could condense into a single formula the movements of the greatest bodies of the universe and that of the least atom: for such an intelligence *nothing would be uncertain*, and the past and the future alike would be before its eyes."

In the previous section we have already mentioned, however, three completely different, and weaker, expressions of the causality principle which, as we shall see, are not contradicted by the indeterminacy relations.

It must be added, moreover, that if it is true that Heisenberg's principle exorcises once for all the Laplacean demon, it is equally true that from a strictly logical point of view this principle does not represent a refutation of deterministic causality. As a matter of fact, Laplacean causality is expressed through a conditional statement of the following form: if one can determine the initial conditions of a physical system, then its time evolution will be rigorously predicted. And what Heisenberg's principle establishes is only the invalidity of the premise of the preceding statement, due to the impossibility of a precise determination of the initial conditions of any microsystems in the terms of the two canonical variables position and momentum. Such a refutation of its premise does not imply, in turn, a falsification of Laplacean determinism, but rather its trivial truth, and consequent inapplicability, as a methodological principle, to the theories of microphysics.

As a consequence of this denial of Laplacean causality, due to the assumption of an unconditioned validity of Heisenberg's principle, several authors view the basic feature of quantum physics in its transition from classical determinism to quantum mechanical indeterminism.

At variance with this widespread point of view, I have tried to put in evidence that one of the main peculiarities of quantum mechanics is rather represented by an ambiguous formal coexistence between a deterministic and a probabilistic description, by showing how such an unsolved dualism can be viewed as an essential interpretative key to the main conceptual problems of this theory.[5]

As a matter of fact, and as von Neumann was the first to emphasise in his *Grundlagen der Quantenmechanik*, the measurement problem in quantum physics has its origin, from a logical point of view, in the presence of two different and incompatible kinds of evolution of the quantum mechanical wave function: the former deterministic -- even if not in Laplace's sense, requiring the simultaneous attribution of well defined positions and momenta, or velocities, for the definition of the initial conditions, possibility which is precluded, as we have seen, by Heisenberg's principle -- regulated by the Schrödinger equation for unobserved, or at least unmeasured, systems; the latter probabilistic whenever a measuring operation takes place, without being given an unambiguous specification of precise and mutually exclusive conditions for these, diametrically opposite, kinds of description.

Such a dualistic aspect of quantum theory seems to impose the necessity to choose between a *deterministic non objectivistic description*, because what evolves according to the Schrödinger equation is not in general a well defined state, but a superposition of several different states, and an *indeterministic objectivistic description*, as is the one of the results of our measurements or observations. One of the first who emphasised these epistemological implications of quantum mechanics was the great mathematician and philosopher Federigo Enriques. In his book on causality and determinism in science, he considered determinism an indispensable constituent of scientific theories, and preferred to admit the possibility of renouncing to define objective properties at the microscopic level.[6]

Other scientists who are not satisfied by this alternative between objectivity and deterministic causality, have tried to elaborate objective and causal theories of measurement based either on the hypothesis that the Schrödinger equation is always valid, even in the description of measuring interactions, or on a causal completion of quantum formalism through the introduction of hidden variables. The difficulties of these two class of theories are well known, being the former class explicitly antirealistic, as a consequence of its assumption of a branching universe, the latter class non local, involving the possibility of actions at a distance propagating with superluminal velocity.

There have been, moreover, some significant attempts to disprove the validity of Heisenberg's principle, like the ideal experiments discussed by Einstein, culminating in his famous photon box experiment, and by Popper. A recent proposal to test Feynman's version of the indeterminacy principle, corresponding to the impossibility of building an interferometric device in which the path followed by a micro object can be established without destroying the interference pattern, has been advanced by the present author.[7]

We may conclude this section with the remark that the indeterminacy principle by superseding classical Laplacean determinism, has neither refuted the causality principle, nor denied the possibility of a deterministic non mechanistic evolution of microsystems, like the one regulated by the Schrödinger equation, that appears, nevertheless, incompatible with an objectivistic description, requiring that the microsystem is not represented by a superposition, but by a well defined state.

3. Kant's empirical causality vs. space-time co-ordination

Another non metaphysical, and also more general, formulation of causality is contained in the *Critique of Pure Reason,* where this principle is regarded as a synthetic statement, whose a priori validity can be proved by showing that it represents a condition for the possibility that objects or processes can occur in our experience. In his proof Kant concentrated the attention on the difference between the order in which we see the appearance of a house, when we view it from right to left, or from top to bottom, and the order in which we perceive the positions of a ship as it moves downstream, and stressed the fact that whereas there is no determined order in the series of our perceptions of the house, such an order is instead present in our perceptions of the movement of the ship. The existence of such an order in the latter case can be guaranteed if we are able to "derive the subjective succession of apprehension from the objective succession of appearance"[8], and, this is possible, according to Kant, only if there is some kind of rule determining the sequence of our perceptions in such a way that "I cannot arrange the apprehension otherwise that in this very succession"[9]. The causal principle appears therefore the necessary condition for the possibility of this experience, allowing the distinction between subjective perceptions and objective phenomena:

"... the relation of appearances (as possible perceptions) according to which the consequent event, that which happens, is, as to its existence, necessarily determined in time by something preceding *in conformity with a rule* – in other words, the relation of cause to effect – is the condition of the objective validity of our empirical judgements, in respect of the series of perceptions, and so of their empirical truth; that is to say, it is the condition of experience (under the conditions of succession), as being itself the ground of the possibility of such experience."[10]

Causal knowledge is, in this perspective, conceived as the only guarantee for the existence of natural science, and causality plays therefore a fundamental role in all phases of Kantian philosophy. In his work the presence of other two different, but not empirical, forms of causality has been stressed: the former "transcendental", endorsed in the pre critical period under the influence of Leibniz's and Wolff's rationalism, the latter "metaphysical", introduced in the *Metaphysical Foundations of Natural Science* and also maintained in his *Opus postumum*.

Since the Como Congress of 1927, Bohr formulated his complementarity principle as a consequence of a necessity of a radical revision of the relation between the concepts of observation and definition of the state of a physical system, implied by the fundamental quantum postulate, according to which every physical process is characterised by an essential discontinuity. In this way the definition of the state, that requires the elimination of all external disturbances, involves, on one hand, according to the fundamental quantum postulate, the impossibility of making any observation and, above all, that the concepts of space and time loose their immediate sense. On the other hand, "if, in order to make observation possible we permit certain interactions with suitable agencies of measurement, not belonging to the system, an unambiguous definition of the state of the system is naturally no longer possible, there can be no question of causality in the ordinary sense of the word."[11]

On the grounds of the previous considerations, Bohr arrives thus to his famous formulations of the complementarity principle:

"The very nature of quantum theory thus forces us to regard the space-time co-ordination and the claim of causality, the union of which characterises the classical theories, as complementary but exclusive features of the description, symbolising the idealisation of observation and definition respectively."[12]

It must moreover be added that it is just this identification of causality with a conformity to precise physical rules, which allows Bohr to give a direct operative foundation to his complementarity principle through the well known ideal experiment of the diaphragm with the slit, which, when rigidly connected with the reference system allows a space time but an acausal description, as a consequence of the impossibility of calculating the exchange of energy and momentum between the microsystem and the diaphragm, and when suspended with a very sensible spring allows a causal but not space time description, because of the indeterminate localisation of the slit itself, due to the motion of the diaphragm.

Whereas, as we have mentioned before, there have been several attempts to disprove Heisenberg's relations, this has not been the case of complementarity, that has been usually considered as an additive interpretative (metaphysical) assumption, often not shared even by some of the main exponents of the orthodox interpretation. It has been, however, recently been shown, that the principle of complementarity can be experimentally refuted, and that, moreover, such a refutation would not imply the necessity of a modification of quantum formalism, but only a replacement of the philosophical interpretation of the Copenhagen school with a realistic one[13].

This means of course that since we are not dealing with an ambiguous metaphysical statement, also what complementarity denies is endowed with meaning: we have provided in this way a clear proof of the non metaphysical nature of Kant's empirical causality.

As we shall see in the last section the least acceptable aspect of the complementarity principle is that it does not limit itself to establish an incompatibility between Kantian causality and space-time co-ordination, but implies, if brought to its extreme consequences, even a violation of Humean causality as ordered connection.

4. Uniformity of nature vs. completeness

A further weakened version of empirical causality can be found in John Stuart Mill's *System of Logic,* a work almost entirely devoted to the analysis of inductive procedures, and to the possibility to extend inductive method from physics to other sciences. Induction is defined by Mill as "the operation of discovering and proving general propositions"[14], and considered therefore as the basis of "all proof and all discovery of truth not self evident"[15]. A justification of inductive procedures requires, however, the appeal to another, more general principle:

"if we throw the whole course of any inductive argument into a series of syllogisms, we shall arrive by more or fewer steps at an *ultimate* syllogism, which will have for its major premise *the principle of the uniformity of the course of nature*"[16].

The uniformity of nature, i. e. the fact that the course of Nature is uniform represents for Mill "the fundamental principle, or general axiom, of induction"[17]. In this way by identifying causality with uniformity Mill viewed in the notion of cause "the root of the whole theory of induction"[18].

The principle of the uniformity of nature has been already introduced by Hume in the following stronger ontological form, which appears at least unusual, in his sceptical perspective:
"like objects placed in like circumstances, will always produce like effects"[19].

Mill provided a weakened and explicitly non metaphysical version of the principle:
"...when in the course of this enquiry I speak of the cause of any phenomenon I do not mean a cause which is not itself a phenomenon; I make no research into the *ultimate or ontological cause* of anything.

The only notion of a cause which the theory of induction requires is such a notion as can be gained from experience. The law of causation, the recognition of which is the *main pillar of inductive science*, is but the familiar truth that invariability of succession is found by observation to obtain between every fact of nature and some other fact which has preceded it, independently of all considerations respecting the ultimate mode of production of phenomena and of every other question regarding the nature of things in themselves"[20].

It appears immediately clear that also this very general and reasonable form of causality is in conflict with another fundamental requirement of the Copenhagen interpretation, the well known thesis of the completeness of the present theory. According to this point of view the probabilistic character of quantum mechanical predictions would not be a consequence of an insufficient knowledge of the state of microsystems, but an intrinsic and definitive limitation in the description of nature.

The most classical case is the one of the quantum laws of radioactive decay, defining an average life time for a given class of atomic particle, but not explaining the different individual behaviour for each particle of this class. The most natural attitude would be the one to consider this different individual behaviour as a consequence of some cause presently unknown, regarding in this way quantum mechanics as a statistical and therefore incomplete description of atomic phenomena. But this is precisely what is explicitly denied by the orthodox thesis of the completeness, assuming the necessity that the same wave function provides a complete description of physical objects characterised by different individual behaviour. As stressed by Franco Selleri:
"Today's physics does not provide an understanding of these causes and accepts in fact an acausal philosophy: every decay is a spontaneous process and does not admit a causal explanation. The question about the different individual life of similar unstable systems, like neutrons, will according to this line of thought

remain forever without answer and should indeed be considered as a 'non-scientific' question"[21].

In this way one is lead to the conclusion that identical objects, in identical conditions, i. e. *identical causes, may produce completely different effects*, in open violation of the principle of the uniformity of nature.

The problem of the incompleteness of quantum mechanics was raised for the first time by Einstein, since the Solvay Conference of 1927, finds its most famous discussion in the Einstein-Podolsky-Rosen (EPR) paradox and in Bell's theorem, and represents one of the most serious open questions in microphysics. I would limit myself therefore, in this context, to mention only a recent result, according to which the thesis of the completeness brings to a proof of inconsistency of quantum formalism[22], due to a conflict between the superposition principle and the indeterminacy relations between time and energy.

5. Humean causality vs. retroaction in time

The Humean doctrine of causality, which is generally considered as his main contribution to the history of philosophy, is contained in his *Enquiry Concerning Human Understanding* and consists of two parts: a negative one, criticising the aprioristic and deductive character attributed to this principle, and a positive one, clarifying and founding its experimental nature.

The former is characterised by an essential contrast with all the preceding philosophical tradition which, since Aristotle, interpreted causal relation as a logical and metaphysical inference: Hume dissociates the order of causes with the order of reasons, arguing that no reason *a priori* may infer that from a given thing must necessarily follow the existence of another one, maintaining that experience alone may say to us what really will follow. But if causality is not that productive power that dogmatic metaphysics pretends, and since, according to Hume, we cannot go beyond the realm of our subjective impressions for having access to things in themselves, the only valid conception of causality will be the one which has its starting point in the experimental data in our possession.

Hume rejected in this way the identification of causality with the idea of a necessary connection, maintaining that we are unable "to comprehend any force or power by which the cause operates, or any connection between it and its supposed effect" and proposed therefore to reduce this principle to the perception of constant conjunction:
"We never can, by our utmost scrutiny, discover anything but one event following another."

Several authors have strongly criticised this reduction of causality to constant conjunction, and there is still an open debate on this question. What I shall now try to show briefly is the impossibility to maintain even Hume's anti metaphysical causality if Kantian empirical causality is refuted.

The argument is based on the analysis of Wheeler's delayed choice experiments, which, as I shall show, are a consequence of an integral acceptance of the complementarity principle. It must be stressed, however, that a violation of Humean causality is also implied in the time-symmetric interpretations of the EPR paradox, as a consequence of their proposal of an abolishment of the time arrow, but since this question is well known, it will not discussed here now[23].

Wheeler's conceptual experiment is a variant of the double slit one, where the screen with the two slits has been replaced by a semireflecting mirror $1/2S$ dividing the incoming beam in a transmitted one, propagating towards the mirror B, which reflects it in the direction of the counter P_B, and in a reflected one propagating towards the mirror A, which reflects it again, in this case in the direction of the counter P_A. In this physical situation, which Wheeler defines as "which route?", each photon emitted entering the experimental device will be detected *either by P_A or by P_B*, that is to say that the photon has followed alternatively one of the two route $1/2SAP_A$, when revealed by P_A, or $1/2SBP_B$ when revealed by P_B.

Let us now suppose to perform a second experiment by inserting, before the two counters, a second semireflecting mirror whose thickness has been calculated as a function of the wavelength of the light of the source which has been used, in such a way that the superposition of the transmitted beam coming from B and the reflecting beam coming from A, generates a wave of zero intensity: in this new case the counter P_B, placed in the region of destructive interference will not reveal the arrival of any photon, which will be all detected by P_A, placed in the region of constructive interference. In this second physical situation, defined by Wheeeler as "both routes", each photon interferes with itself rendering impossible its deviation towards P_B and absolutely certain its deviation towards P_A: this result can be explained by admitting that the photon has followed both the "transmitted" route $1/2SBP_B$ and the "reflected" route $1/2SAP_A$. This absence of a localisation of the photon is peculiar of the ondulatory behaviour.

The explanation of these two experiments in the light of the complementarity principle is that the photon, like any other micro object, has either an ondulatory or a corpuscular behaviour, but never the two at the same time. As well known, Bohr extended his complementary interpretation from space-time and causal description to the two conceptions of the nature of light which, according to him, should have to be considered as different attempts to give an interpretation of experimental evidence in which the limitation of classical concepts is expressed: as a matter of fact, on the one hand classical electrodynamics provides a space-time description of light propagation, and on the other hand the conservation laws of energy and momentum, which represent the causal aspect of optical phenomena, find their adequate formulation in Einstein's corpuscular theory of the light quanta.

Wheeler propose to introduce, at this point, a mechanism which allows one to insert or release the second semireflecting mirror with the greatest quickness and at the last moment, when the photon has already interacted with $1/2S$ and is propagating along the transmitted and/or reflected route of our device. Now, if we insert the second mirror in the configuration "which route?" of the first experiment, in which the photon has a corpuscular behaviour and has therefore been already either transmitted or reflected by $1/2S$, this configuration immediately turns into the one "both routes" of the second experiment, in which the photon has an ondulatory behaviour and is, therefore both transmitted and reflected. In the same way, by releasing the second mirror in the configuration "both routes", the ondulatory behaviour is turned into the corpuscular one of the configuration "which route?".

Nevertheless, since this insertion or release is made *after* the interaction of the photon with $1/2\,M$, we are faced with the absolutely paradoxical conclusion that the *future choice* of the position of the second mirror *modifies the past interaction* with the photon and the first one. This possibility of an inversion of the relationship between cause and effect and of the consequent elimination of a temporal order in physical phenomena would imply the breakdown even of Humean causality.

It can be easily seen, however, that the paradoxical conclusions of Wheeler's conceptual experiment arise from the limitations imposed by the complementarity principle, and, as has been stressed by Franco Selleri[24], by the exclusion *a priori* and without discussion of the possibility that the photon could have at the same time both ondulatory and corpuscular properties. The contradiction can be easily avoided once we adopt a different realistic interpretation allowing a positive solution of the wave-particle duality. In the light of such an interpretation, which considers micro objects as localised particles immersed in extended ondulatory phenomena, the delayed choice to insert or release the second mirror does not produce any modification in the previous process of interaction of the photon with the first mirror, whose corpuscular part chooses "which route?" whereas its ondulatory part propagates along "both routes".

With the intention to investigate in other papers some more general aspects in the relations between non metaphysical formulations of the causal principle and empirical realism, I would like to conclude by stressing that we have seen how a realistic interpretation of the fundamental theoretical concept of quantum formalism, i.e. the wave function or state vector ψ, allows us both to maintain the compatibility of space-time co-ordination with Kantian empirical causality and to preserve the arrow of time required by the anti metaphysical formulation of this principle given by Hume.

Notes and references

[1] Jammer, M. (1966), *The Conceptual Development of Quantum Mechanics*, Mc Graw Hill, New York, pp.166--180; Forman, P. (1971), "Weimar Culture, Causality and Quantum Theory (1918-1927)", in *Historical Studies in the Physical Science*, University of Pennsylvania Press, Pittsburgh, vol.III, p.1; Selleri, F. (1989), *La causalità impossibile*, Jaca Book, Milan; Filippini, G. and Tarozzi, G. (1991-92), "Il rifiuto della causalità nella filosofia della fisica prequantistica", *Mem. Accad. Naz. Sc. Lett. Arti di Modena, VII, IX*, p.97.

[2] Kant, I. (1781), *Kritik der reinen Vernunft*; English edition of N. K. Smith, *Immanuel Kant's Critique of Pure Reason*, Macmillan, London, 1929, p.218.

[3] Tarozzi, G. (1981), "Réalisme d'Einstein et mécanique quantique: un cas de contradiction.entre une théorie physique et une hypothèse philosophique clairement définie", *Revue de Synthèse, 101-102*, pp. 125--158; (1980), "Realism as A Meaningful Philosophical Hypothesis", *Mem. Acc. Sc.Ist. Bologna, XIII, VII*, p.1; (1988), "Science, Metaphysics, and Meaningful Philosophical Principles", *Epistemologia, XI*, p.97.

[4] Schlick, M. (1949). *Philosophy of Nature*, Philosophical Library, New York, pp. 69--70.

[5] Tarozzi, G. (1988), *Probability and Determinism in Quantum Theory*, in E. Agazzi (ed.), *Probability in the Sciences*, Kluwer, Dordrecht, p.237.

[6] Enriques, F. (1945). *Causalità e determinismo nella filosofia e nella storia della scienza*, Atlantica, Rome.

[7] Tarozzi, G. (1985), "A Unified Experiment for Testing Both the Interpretation and the Reduction Postulate of the Quantum Mechanical Wave Function", in *Open Questions in Quantum Physics*, ed. by G. Tarozzi and A. van der Merwe, Reidel, Dordrecht.

[8] Kant, I. see 2, p.221.

[9] ibid., p.222.

[10] ibid., p.227.

[11] Bohr, N. (1928), "The Quantum Postulate and the Recent Development of Atomic Theory", *Nature 121*, pp. 580--590, reprinted in *Quantum Theory and Measurement*, ed. by J.A. Wheeler and W.H. Zurek, University Press, Princeton, 1983, p.89.

[12] ibid.

[13] Tarozzi, G. (1982). "Two Proposals for Testing Physical Properties of Quantum waves", *Lett. Nuovo Cimento, 35*, p.53.

[14] Mill, J.S. (1872). *A System of Logic*, Longmans, London, bk 3, p.186.

[15] ibid., p.185.

[16] ibid., p.203.

[17] ibid., p.201.

[18] ibid., p.213.

[19] Hume, D. *Treatise*, p.107.

[20] Mill, J.S. see 11, p.213.

[21] Selleri, F. (1990). *Quantum Paradoxes and Physical Reality,* Kluwer, Dordrecht, pp.33--34.

[22] Tarozzi, G. (1992). "On the Implications of Generalised EPR States for the Completeness and Consistency of Quantum Formalism", Proceedings of the Conference *The Interpretations of Quantum Mechanics. Where Do We Stand?,* New York, April.

[23] Selleri, F. and Tarozzi, G. (1981). "Quantum Mechanics, Reality and Separability", *Rivista Nuovo Cimento*, vol. 4(2), pp.1--54

[24] Selleri, F. (1990). *Quantum Paradoxes and Physical Reality*, Kluwer, Dordrecht, p.116.

Fundamental Theories of Physics

Series Editor: Alwyn van der Merwe, *University of Denver, USA*

1. M. Sachs: *General Relativity and Matter.* A Spinor Field Theory from Fermis to Light-Years. With a Foreword by C. Kilmister. 1982   ISBN 90-277-1381-2
2. G.H. Duffey: *A Development of Quantum Mechanics.* Based on Symmetry Considerations. 1985   ISBN 90-277-1587-4
3. S. Diner, D. Fargue, G. Lochak and F. Selleri (eds.): *The Wave-Particle Dualism.* A Tribute to Louis de Broglie on his 90th Birthday. 1984   ISBN 90-277-1664-1
4. E. Prugovečki: *Stochastic Quantum Mechanics and Quantum Spacetime.* A Consistent Unification of Relativity and Quantum Theory based on Stochastic Spaces. 1984; 2nd printing 1986   ISBN 90-277-1617-X
5. D. Hestenes and G. Sobczyk: *Clifford Algebra to Geometric Calculus.* A Unified Language for Mathematics and Physics. 1984
          ISBN 90-277-1673-0; Pb (1987) 90-277-2561-6
6. P. Exner: *Open Quantum Systems and Feynman Integrals.* 1985   ISBN 90-277-1678-1
7. L. Mayants: *The Enigma of Probability and Physics.* 1984   ISBN 90-277-1674-9
8. E. Tocaci: *Relativistic Mechanics, Time and Inertia.* Translated from Romanian. Edited and with a Foreword by C.W. Kilmister. 1985   ISBN 90-277-1769-9
9. B. Bertotti, F. de Felice and A. Pascolini (eds.): *General Relativity and Gravitation.* Proceedings of the 10th International Conference (Padova, Italy, 1983). 1984
          ISBN 90-277-1819-9
10. G. Tarozzi and A. van der Merwe (eds.): *Open Questions in Quantum Physics.* 1985
          ISBN 90-277-1853-9
11. J.V. Narlikar and T. Padmanabhan: *Gravity, Gauge Theories and Quantum Cosmology.* 1986   ISBN 90-277-1948-9
12. G.S. Asanov: *Finsler Geometry, Relativity and Gauge Theories.* 1985
          ISBN 90-277-1960-8
13. K. Namsrai: *Nonlocal Quantum Field Theory and Stochastic Quantum Mechanics.* 1986   ISBN 90-277-2001-0
14. C. Ray Smith and W.T. Grandy, Jr. (eds.): *Maximum-Entropy and Bayesian Methods in Inverse Problems.* Proceedings of the 1st and 2nd International Workshop (Laramie, Wyoming, USA). 1985   ISBN 90-277-2074-6
15. D. Hestenes: *New Foundations for Classical Mechanics.* 1986
          ISBN 90-277-2090-8; Pb (1987) 90-277-2526-8
16. S.J. Prokhovnik: *Light in Einstein's Universe.* The Role of Energy in Cosmology and Relativity. 1985   ISBN 90-277-2093-2
17. Y.S. Kim and M.E. Noz: *Theory and Applications of the Poincaré Group.* 1986
          ISBN 90-277-2141-6
18. M. Sachs: *Quantum Mechanics from General Relativity.* An Approximation for a Theory of Inertia. 1986   ISBN 90-277-2247-1
19. W.T. Grandy, Jr.: *Foundations of Statistical Mechanics.*
    Vol. I: *Equilibrium Theory.* 1987   ISBN 90-277-2489-X
20. H.-H von Borzeszkowski and H.-J. Treder: *The Meaning of Quantum Gravity.* 1988
          ISBN 90-277-2518-7
21. C. Ray Smith and G.J. Erickson (eds.): *Maximum-Entropy and Bayesian Spectral Analysis and Estimation Problems.* Proceedings of the 3rd International Workshop (Laramie, Wyoming, USA, 1983). 1987   ISBN 90-277-2579-9

Fundamental Theories of Physics

Fundamental Theories of Physics

43. W.T. Grandy, Jr. and L.H. Schick (eds.): *Maximum-Entropy and Bayesian Methods.* Proceedings of the 10th International Workshop (Laramie, Wyoming, USA, 1990). 1991   ISBN 0-7923-1140-X
44. P.Pták and S. Pulmannová: *Orthomodular Structures as Quantum Logics.* Intrinsic Properties, State Space and Probabilistic Topics. 1991   ISBN 0-7923-1207-4
45. D. Hestenes and A. Weingartshofer (eds.): *The Electron.* New Theory and Experiment. 1991   ISBN 0-7923-1356-9
46. P.P.J.M. Schram: *Kinetic Theory of Gases and Plasmas.* 1991   ISBN 0-7923-1392-5
47. A. Micali, R. Boudet and J. Helmstetter (eds.): *Clifford Algebras and their Applications in Mathematical Physics.* 1992   ISBN 0-7923-1623-1
48. E. Prugovečki: *Quantum Geometry.* A Framework for Quantum General Relativity. 1992   ISBN 0-7923-1640-1
49. M.H. Mac Gregor: *The Enigmatic Electron.* 1992   ISBN 0-7923-1982-6
50. C.R. Smith, G.J. Erickson and P.O. Neudorfer (eds.): *Maximum Entropy and Bayesian Methods.* Proceedings of the 11th International Workshop (Seattle, 1991). 1993
ISBN 0-7923-2031-X
51. D.J. Hoekzema: *The Quantum Labyrinth.* 1993   ISBN 0-7923-2066-2
52. Z. Oziewicz, B. Jancewicz and A. Borowiec (eds.): *Spinors, Twistors, Clifford Algebras and Quantum Deformations.* Proceedings of the Second Max Born Symposium (Wrocław, Poland, 1992). 1993   ISBN 0-7923-2251-7
53. A. Mohammad-Djafari and G. Demoment (eds.): *Maximum Entropy and Bayesian Methods.* Proceedings of the 12th International Workshop (Paris, France, 1992). 1993
ISBN 0-7923-2280-0
54. M. Riesz: *Clifford Numbers and Spinors* with Riesz' Private Lectures to E. Folke Bolinder and a Historical Review by Pertti Lounesto. E.F. Bolinder and P. Lounesto (eds.). 1993   ISBN 0-7923-2299-1
55. F. Brackx, R. Delanghe and H. Serras (eds.): *Clifford Algebras and their Applications in Mathematical Physics.* Proceedings of the Third Conference (Deinze, 1993) 1993
ISBN 0-7923-2347-5
56. J.R. Fanchi: *Parametrized Relativistic Quantum Theory.* 1993   ISBN 0-7923-2376-9
57. A. Peres: *Quantum Theory: Concepts and Methods.* 1993   ISBN 0-7923-2549-4
58. P.L. Antonelli, R.S. Ingarden and M. Matsumoto: *The Theory of Sprays and Finsler Spaces with Applications in Physics and Biology.* 1993   ISBN 0-7923-2577-X
59. R. Miron and M. Anastasiei: *The Geometry of Lagrange Spaces: Theory and Applications.* 1994   ISBN 0-7923-2591-5
60. G. Adomian: *Solving Frontier Problems of Physics: The Decomposition Method.* 1994
ISBN 0-7923-2644-X
61 B.S. Kerner and V.V. Osipov: *Autosolitons.* A New Approach to Problems of Self-Organization and Turbulence. 1994   ISBN 0-7923-2816-7
62. A. Heidbreder (ed.): *Maximum Entropy and Bayesian Methods.* Proceedings of the 13th International Workshop (Santa Barbara, USA, 1993) 1995   ISBN 0-7923-2851-5
63. J. Peřina, Z. Hradil and B. Jurčo: *Quantum Optics and Fundamentals of Physics.* 1994
ISBN 0-7923-3000-5

Fundamental Theories of Physics

KLUWER ACADEMIC PUBLISHERS – DORDRECHT / BOSTON / LONDON